Micro/Nano Technology Systems for Biomedical Applications

Micro/Nano Technology Systems for Biomedical Applications

Microfluidics, Optics, and Surface Chemistry

Edited by

Chih-Ming Ho

Henry Samueli School of Engineering and Applied Science, UCLA

OXFORD
UNIVERSITY PRESS

Great Clarendon Street, Oxford OX2 6DP

Oxford University Press is a department of the University of Oxford.
It furthers the University's objective of excellence in research, scholarship,
and education by publishing worldwide in

Oxford New York

Auckland Cape Town Dar es Salaam Hong Kong Karachi
Kuala Lumpur Madrid Melbourne Mexico City Nairobi
New Delhi Shanghai Taipei Toronto

With offices in

Argentina Austria Brazil Chile Czech Republic France Greece
Guatemala Hungary Italy Japan Poland Portugal Singapore
South Korea Switzerland Thailand Turkey Ukraine Vietnam

Oxford is a registered trade mark of Oxford University Press
in the UK and in certain other countries

Published in the United States
by Oxford University Press Inc., New York

First published 2010

British Library Cataloguing in Publication Data
Data available

Library of Congress Cataloging in Publication Data
Data available

Typeset by SPI Publisher Services, Pondicherry, India
Printed in Great Britain
on acid-free paper by
CPI Antony Rowe

ISBN 978–0–19–921969–8

1 3 5 7 9 10 8 6 4 2

CONTENTS

ACKNOWLEDGMENT

The authors are indebted to their students/post-docs and colleagues who have made this book possible. We also thank the following groups for their continuous support: NIH (US), NSF (US), DARPA (US), NASA (US), EPSRC (UK), CNRS (France), ESPCI (France) and Hong Kong-France PROCORE Program (China and France).

Royalties from this book have been donated to Save the Children (http://www.SavetheChildren.org).

LIST OF CONTRIBUTORS

Narayan R. Aluru
Department of Mechanical Science and Engineering
3265 Beckman Institute, MC-251
405 North Mathews Avenue
University of Illinois at Urbana-Champaign
Urbana, IL 61801

Eric P. Y. Chiou
Mechanical and Aerospace Engineering
University of California, Los Angeles
420 Westwood Plaza
37–138 Engineering IV Building
Los Angeles, CA 90095–1597

Robin L. Garrell
UCLA Department of Chemistry and Biochemistry
4077 Young Hall
Los Angeles, CA 90095–1569

Jian Gong
UCLA Mechanical and Engineering Department
420 Westwood Plaza
Los Angeles, CA 90095–1597

Nicolas G. Green
School of Electronics and Computer Science
University of Southampton
Southampton, SO17 1BJ
United Kingdom

Chih-Ming Ho
Mechanical and Aerospace Department
University of California, Los Angeles
420 Westwood Plaza
38–137J Engineering IV Building
Los Angeles, CA 90095–1597

George Em Karniadakis
Division of Applied Mathematics
Box F
Brown University
Providence, RI 02912

Chang-Jin 'CJ' Kim
Mechanical and Aerospace Engineering
University of California, Los Angeles
420 Westwood Plaza
37–134 Engineering IV Building
Los Angeles, CA 90095–1597

Yi-Kuen Lee
Department of Mechanical Engineering
Hong Kong University of Science and Technology
Clear Water Bay
Kowloon, Hong Kong
China

Kelvin Liu
Department of Mechanical Engineering and Biomedical Engineering
Johns Hopkins University
108 Latrobe, 3400 North Charles Street

Heather D. Maynard
UCLA Department of Chemistry and Biochemistry
4505B Molecular Sciences Building
Los Angeles, CA 90095-1569

Carl D. Meinhart
Department of Mechanical Engineering
University of California
Santa Barbara, CA 93106

Scott Miserendino
Department of Mechanical Engineering
California Institute of Technology
Pasadena, CA 91125

Hywel Morgan
School of Electronics and Computer Science
University of Southampton
Southampton, SO17 1BJ
United Kingdom

Christopher M. Puleo
Department of Mechanical Engineering and Biomedical Engineering
Johns Hopkins University
108 Latrobe, 3400 North Charles Street
Baltimore, MD 21218

Tao Sun
School of Electronics and Computer Science
University of Southampton
Southampton, SO17 1BJ
United Kingdom

Patrick Tabeling
MMN-LPS, ESPCI
10 rue Vauquelin
75231 Paris, France

Yu-Chong Tai
Department of Electrical Engineering
California Institute of Technology
Mail Code 116–81
Pasadena, CA 91125

Yuan Wang
Mechanical Engineering Department
University of California
5130 Etcheverry Hall
Berkeley, CA 94720–1740

Steven T. Wereley
School of Mechanical Engineering
Purdue University
585 Purdue Mall
West Lafayette, IN 47907–2088

Tza-Huei Wang
Department of Mechanical Engineering and Biomedical Engineering
Johns Hopkins University
108 Latrobe, 3400 North Charles Street
Baltimore, MD 21218

Ming C. Wu
EECS Department
261M Cory Hall
Berkeley, CA 94720-1740

Hsin-Chih Yeh
Department of Mechanical Engineering and Biomedical Engineering
Johns Hopkins University
108 Latrobe, 3400 North Charles Street
Baltimore, MD 21218

Xiang Zhang
Mechanical Engineering Department
University of California
5130 Etcheverry Hall
Berkeley, CA 94720–1740

1

INTERSECTIONS BETWEEN MICRO/NANO TECHNOLOGIES AND BIOLOGICAL SCIENCES

Chih-Ming Ho

The human body is measured on a length scale of meters, which is six to nine orders of magnitude larger than its micro/nano scale building blocks, e.g. nucleic acids, proteins, and cells. Through the orchestrated management of vast arrays of nano-scale functional molecules that are regulated by genetic information, cells emerge as a fundamental system for the organization of life. The nano-sized cellular components self-assemble into cell systems of incredible complexity, characterized by multiple layers of increasingly more sophisticated functionality. Then, perhaps even more amazing, is the fact that through local communication between individual cells, the information passed initiates their self-organization into ever larger macro-scale systems with functions that transcend its smaller scale building blocks. Cells form tissues, which assemble into organs and then arrange into physiological systems, resulting in a higher level organism. At each stage, the length scale increases and thereby the governing mechanisms of the emerged system become more complex.

The ultimate goal of developing technologies, such as micro/nano electronics, micromachine, or biotechnology devices, is to enrich human lives. The micro/nano technologies provide transducers with dimensions comparable to that of cells or bio-molecules. This capability offers unprecedented opportunities to explore the world in small scales. Yet linking the experimental findings from the micro/nano scale to the macro scale has many challenges.

1.1 Length scale

1.1.1 *Matching of length scales—a fundamental requirement for efficient transduction*

The wavelength of visible light is about half a micron, which is similar in size to the majority of biological components. The invention of the optical microscope enabled the viewing of cells and intra cellular components with a spatial resolution larger than 0.2 microns. With the ability to see living cells, biological research has improved dramatically. In addition to viewing the subject, changing the relative position between particle/cell and its embedded fluid is a commonly required procedure in biological research. Further, efficiently exchanging momentum or size matching between the particles and the transducer are, again, necessary

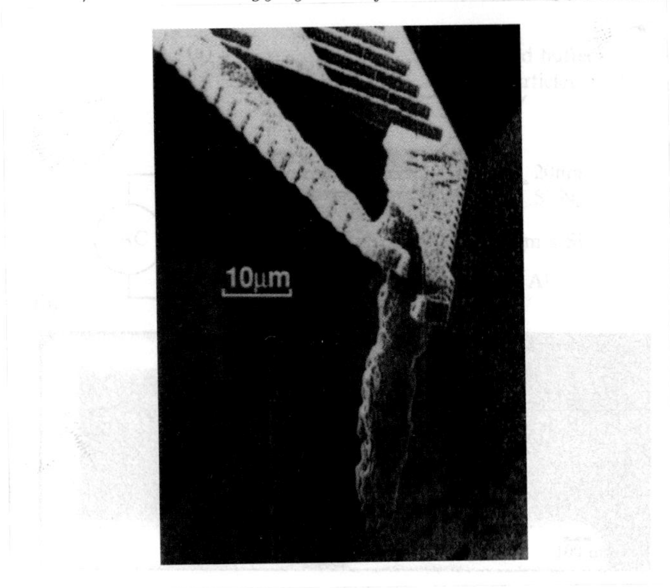

Fig. 1.1: A micro tweezers holding an euglena. (Reprinted with permission from Kim *et al.* 1992, Copyright 1992, IEEE).

actions undertaken in biological experiments. In most transduction processes, matching of length scales is a fundamental requirement.

In the late 1980s, micro-electro-mechanical systems (MEMS) technologies emerged and it became possible to manufacture micron-scale mechanical devices. The picture represented above is of a pair of micro tweezers holding an euglena (Fig. 1.1), which signifies the capability of picking and holding a single cell with a micron device. For biological studies, cells need to be cultured in fluid and thereby fluid needs to be handled efficiently in all areas of biotechnology. With the MEMS technology, we can fabricate fluidic processors, channels, and reactors, as well as transducers in micron sizes (Ho and Tai 1998; Ho 2001; Stone *et al.* 2004).

Traditional biological studies are based on population. The data collected from experiments are often normalized/averaged over a cell population of thousands to millions. With microfluidic bio-reactors and the fluidic transport circuitry, we can place a small number of cells or simply a single cell on a micro machined island in a large array under a precisely controlled micro environment (El-Ali *et al.* 2006; Sims *et al.* 2007; Li and Ho 2008). Information regarding the cell on each island can be obtained by high throughput micro array parallel processing (Huang *et al.* 2007). This capability provides very unique opportunities in the field of small-scale, high-throughput, biological research. Cells in the same sampling batch are not necessarily the same and have been shown to possess a vast diversity based upon the stochasticity of the cells as seen with cell division (Elowitz *et al.* 2002). The averaged values of all the single cell readouts can provide information for traditional population studies. The life cycle or the responses to external stimulations can then be traced and followed on a specific, single cell level for studying these vast differences. This is an example of how the push–pull process of interdisciplinary research can move both fields into a new direction.

With the advancements of silicon based MEMS techniques (see Chapter 4) and the introduction of soft lithography (Xia and Whitesides 1998; Whitesides *et al.* 2001), we have possessed the capability to fabricate intricate microfluidic circuitry for biological applications since around the year 2000. However, a microsystem integrated with channels, reactors, and transducers may not be necessary to provide a properly functional system. In biological flows, the biological macromolecules embedded in fluid medium usually interact with the surface along the flow paths, resulting in surface fouling. On the other hand, we can take advantage of the large surface-to-volume ratio in microsystems to modify the surface property for accomplishing an assay (Wong and Ho 2009). For example, specific cell types can be captured in a region on the surface through functionalizing that surface with antibodies against a pinpointed cell membrane protein (Jung *et al.* 2001, Falconnet *et al.* 2006). After the year 2000, research focused mainly on engineering the nanoscale surface properties for ensuring efficient operations of the microfluidic systems.

The length scale involved in surface molecular property modifications or biomarker detections is in nanometers. Within the nanometer region, using visible light to identify particles, bio-molecules, or cellular organelles presents great experimental challenges. This roadblock is a consequence from the fact that these particles are usually smaller than the diffraction limit of visible light. Electronic microscopy and atomic force microscopy, in fact nanoscopy, have a resolution of a nanometer and can map out nano subjects in order to 'visualize' the morphology. However, these instruments have their own limitations, which restrict their applicability to view live cells. The recent development of near field optics based nanoscopy (see Chapter 8) provides an opportunity to see living objects in the tens of nanometers range, but the particle needs to be in very close proximity to the objective lens. In addition, a set of new modalities (Ho *et al.* 2006) to physically handle nano scale objects are in development for particle placement, as well as generating relative motion between nano particles the surrounding fluid.

1.1.2 *Transcending through length scales*

On the one hand, we are furthering the understanding and the control of the micro/nano world. On the other hand, it is necessary to transcend the length scale limitations to ensure that these research efforts will be beneficial to improving human wellbeing on the macro scale.

Transducers of different scales are made by various manufacturing processes. The majority of traditional engineering techniques are top-down fabrication. Pressure and shear stress are used in making macro-scale mechanical parts. While the separation between microns and nanometers is only three orders of magnitude, this difference in length scales dictate the threshold demarcating the top-down or bottom-up fabrication processes, which use very different force fields. The top-down MEMS lithographic based fabrication enables us to produce micron scale subjects. Nature uses molecular forces and self-assembly to produce very intricate nano structures through bottom-up fabrication. The technologies of manufacturing nano-scale engineering structure are still in very primitive stages. The pathway and the limitations of integrating these fabrication technologies are not obvious and non-trivial, because the force fields are so different while the length scales span nine orders of magnitude (Wong *et al.* 2009a). Developing effective processing technologies across these length scales proves to be the present and future challenge.

1.2 The nanometer range forces dictating the interactions of molecules

1.2.1 *Intermolecular forces*

In the case of molecular interactions, molecular bonding can be classified into different categories based upon their respective physical mechanisms. Bonding is identified as purely electrostatic, quantum mechanical in nature, or arisen from dipole–dipole interactions (Israelachvili 1992). Thermal energy is used to gauge the range and strength of these interactions. Above the absolute zero degree, molecules move in random thermal motions. The thermal energy is measured in the units of kT, where k is the Boltzmann constant ($k = 1.381 \times 10^{-23}$ J/K) and T is the absolute temperature. In order for the molecules to initiate any kind of interaction, such as bonding onto the surface, the interaction energy between the molecules has to overcome this intrinsic thermal energy barrier. According to Israelachvili's formulation based on Trouton's rule (Israelachvili 1992; Atkins *et al.* 2002), when the interaction energy between two molecules in close proximity exceeds $3kT/2$ at standard temperature and pressure, the two molecules will condense into a solid or liquid phase.

Among the intermolecular interactions, covalent interaction is the strongest. This interaction is quantum mechanical in nature and arises from electron cloud sharing between atoms. The typical covalent bond strength is in the order of 100 kT with the effective range in the order of an atomic radius (i.e. ~ 0.1–0.2 nm). An example of a man-made material benefiting from a covalent interaction is the carbon nanotube (Iijima 1991) with a measured Young's Modulus in the order of 1 TPa (Treacy 1996; Wong *et al.* 1997).

In addition to covalent interactions, a Coulomb interaction is another strong intermolecular force, which is electrostatic in nature. For isolated K^+ and Cl^- ions in a vacuum, the interaction energy between this ion pair is about 170 kT (Lide 1994). For the pair of ions, the potential function decays slowly based upon the inverse of the distance between the two ions. It is, therefore, a long range intermolecular force in that it takes about 60 nm to decrease to 1 kT, while the sum of the two ion radii is only 0.33 nm. Recently, the actuation energy of an electrostatically actuated synthetic linear motor molecule, rotaxane, was measured and had the value of 110 kT (Brough *et al.* 2006).

van der Waals force originates from a dipole–dipole interaction and, therefore, can be found between any type of atoms or molecules. Unlike the previously mentioned strong covalent and Coulomb interactions, van der Waals interactions are much weaker in terms of strength, averaging 0.4 kT. The potential function decreases to the inverse of the distance to the sixth power. In general, the van der Waals interactions are effective within the atomic spacing 0.2 nm to $\sim$ 10 nm.

Hydrogen bonding is another type of intermolecular interaction which is significant in determining the three-dimensional structures of macromolecules. The double helical DNA structures are built upon the hydrogen bonding interactions between intramolecular bases on the sugar backbone components (Watson and Crick 1953). This type of interaction arises when a hydrogen atom electrostatically interacts with two electronegative atoms (i.e., atoms that tend to attract electrons to itself, such as oxygen, fluorine, or nitrogen). The strength of hydrogen bonding interactions is between 4 kT–16 kT (Joesten and Schaad 1974). The effective range of this particular type of interaction is between that of covalent and van der Waals interactions. For example, the hydrogen bond length of O-H is about 0.176 nm. Comparatively, the sum of the van der Waals radii of O and H is approximately 0.26 nm, whereas the covalent bond length of O-H is 0.10 nm.

A hydrophilic molecule is most likely a polar or locally polar molecule in that it can produce a transient hydrogen bond with water, which is a highly polarized molecule. A hydrophobic molecule lacks this type of molecular force interaction.

In addition to the above mentioned intermolecular forces, there exist other types of interactions, such as hydration, steric, and hydrophobic interactions, etc. (Israelachvili 1992). In general, the effective range of all intermolecular interactions rarely exceeds 100 nm. Surface properties are determined by the presence of the atoms/molecules within this distance, and the competition of the affinity forces of these molecules.

1.2.2 *Functional macromolecules self-assembled through molecular forces*

Nucleic acids and proteins are macromolecules, which carry out key cellular functions. These functional molecules self-assemble through intermolecular forces into complex three-dimensional structures. The shape complementarity of the three-dimensional structure provides additional constraints, similar to that of the lock and key mechanism, which makes the hybridization between strands of nucleic acid and the antibody-antigen interaction very specific. This specific recognition among the bio-molecules serves as the foundation of the ordered life functions and also as the mechanism for specific recognition used in molecular diagnostic techniques.

1.2.2.1 *Nucleic Acid* A nucleic acid is a polymer made from nucleotides, with a sugar-phosphate backbone joined by phosphodiester bonds. Each nucleotide consists of three components: a nitrogenous heterocyclic base, a pentose sugar, and a phosphate group. The most common nucleic acids in cells are deoxyribonucleic acid (DNA) and ribonucleic acid (RNA). They differ in both the structure of the sugar backbone and the types of bases in their nucleotides. DNA contains 2-deoxyriboses, while RNA contains a ribose. Adenine (A), cytosine (C), and guanine (G) are found in both RNA and DNA, while thymine (T) only occurs in DNA and uracil (U) is only found in RNA.

In the canonical Watson–Crick base pairing (Watson and Crick 1953), A only pairs with T, while G only pairs with C in DNA. In RNA, T is replaced by U. The GC pair has three hydrogen bonds, whereas the AT or AU pair has only two; therefore, the GC pair is more stable. The base pairs can form a linear strand with a string of code consisting of A, C, G.... Another strand can hybridize with the first strand, but only if it has a complimentary string of code, hence the lock and key mechanism. About 1 nN force is needed to pull complementary 20-bp strands apart (Lee *et al.* 1994). Nucleic acids are usually either single-stranded or double-stranded, though structures with three or more strands can form. Not only base pairing but also base stacking contribute to the formation and stability of these structures. Base stacking is the interaction between the aromatic rings of the bases in adjacent nucleotides in the nucleic acid. Although it is a non-covalent interaction, base stacking plays an important role in the stabilization of various nucleic acid structures, from double helixes in DNA to stem-loops in RNA.

1.2.2.2 *Protein* Twenty different types of amino acid linked by peptide bonds are classically known as the building blocks of proteins. A protein is a linear polymer, which folds into three-dimensional structures. These structures consist of four levels of organization; 1) primary structure, 2) secondary structure, 3) tertiary structure, and 4) quaternary structure (Branden *et al.* 1999). The unique geometry of the protein structure and the intermolecular forces dictate the specific types of molecules with the characteristics needed to interact with the protein. This constraint determines the signaling/regulatory pathway functions in a regulated manner. If the protein folding process is disturbed, structural protein changes lead toward un-programmed molecular interactions, which may drive the cell to deviate from the homeostatic state.

1.2.2.3 *Aptamer-Ligand* Aptamers are oligonucleotides that can bind, with a high affinity and specificity, to a great variety of target biomolecules ranging in size from small molecules to complex multimeric structures, such as proteins. While aptamers exist in nature, these components are usually isolated from combinatorial oligonucleotide libraries through the SELEX (Systematic Evolution of Ligands by Exponential Enrichment) process, which was developed by Craig Tuerk and Larry Gold in 1989 (Gold and Tuerk 1989).

The high resolution three-dimensional structures of aptamer-ligand complexes (Hermann *et al.* 2000) have revealed that the enclosure of large parts of a ligand by the aptamer creates numerous discriminatory intermolecular contacts, including stacking, steric hindrance, shape complementarity, electrostatic interactions, and hydrogen bonding. It is these intermolecular contacts that confer aptamers' affinity and specificity comparable to that of antibodies. Since the aptamer is an oligonucleotide, its shape complementarity to the ligand is achieved through base pairing and stacking, as described in the previous section. The affinity constant of an aptamer to a ligand, which ranges from μM to sub-nM, is usually determined empirically. A resource for obtaining the affinity of existing aptamers is the Aptamer Database, developed by the Ellington group (Lee *et al.* 2004a).

1.2.2.4 *Protein-Ligand* The ability of a protein to bind to a ligand with a high specificity and affinity is due to the formation of a set of non-covalent bonds and the shape complementarity between the protein and the ligand. These non-covalent bonds are often hydrogen bonds, ionic bonds, van der Waals attractions, and hydrophobic interactions. Due to the diversity of the 20 amino acids found in proteins, a more complex shape complementarity is provided in proteins as compared to aptamers, which consist of only four structurally uniform nucleotides. The binding affinity of a protein to a ligand usually ranges from μM to pM, corresponding to $13.8\sim 27.6\,\mathrm{kT}$. While the binding affinity is usually obtained via experimental methods, computational methods can also be used to predict this value, if the high resolution structural information is available (Ajay and Murcko 1995).

1.3 Microfluidic forces—enabling the transcendence from molecular scale to micro/macro scales

Fluid is the life supporting medium for all organisms, most importantly the means of transporting molecules or particles, e.g. cells, to accomplish physiological functions. The molecular interactions used to accomplish these goals are forces acting on the nanometer scale. The action range is extended to larger distances by fluid transportation. For example, within the human circulatory system, serum is the fluid used to move molecules and blood cells through a network of vessels and capillaries to achieve nano-scale molecular reactions over various regions of the human body, over meters in transport.

In MEMS based engineering systems, such as medical diagnostic platforms, fluid moves embedded functional macromolecules, e.g. proteins, RNA/DNA, around the circuitry in order to concentrate target molecules, mix with reagents (see Chapter 2), and produce signals on a sensor surface (see Chapter 10). Producing fluid motion on a micro scale is far different from that on the macroscopic side (Ho 2001). For example, the hydrodynamic pressure drop (see Chapter 4) needed to push liquid through a micron wide and a few mm long channel can reach the 1000 psi range. Designing and fabricating mechanical pumps/valves for satisfying these requirements is non-trivial. The diagnostic tool for characterizing flow field in microscale has been well developed (see Chapter 9). Electrokinetic forces (see Chapter 5) and surface tension (see Chapter 6) become effective on small length scales and can serve as alternatives to the hydrodynamic pressure. On the other hand, several of these forces depend upon the gradient of electrical potentials. For

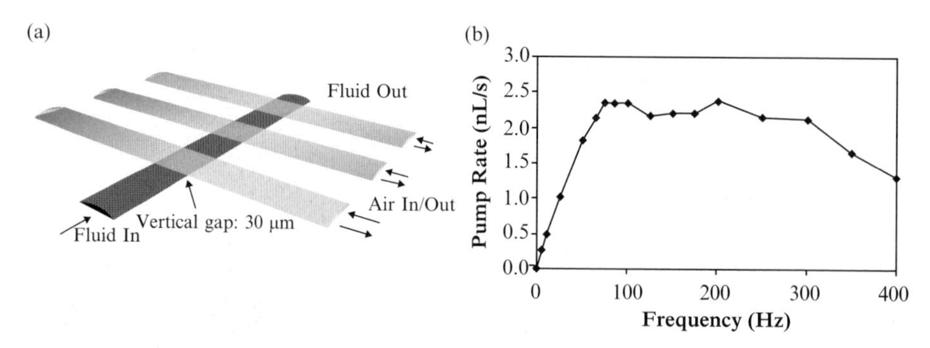

Fig.1.2: (a) A 3D diagram of an elastomeric peristaltic pump. (b) Pumping rate of a peristaltic micropump versus various driving frequencies. (Reprinted with permission from Unger *et al.* 2000 Copyright 2000, AAAS).

fluids with high ionic concentration, like many of the bioreagents, the efficiency drops significantly due to the difficulties of maintaining this potential gradient.

1.3.1 *Hydrodynamic pressure*

Moving/stopping fluids, mixing, concentration, and separation are necessary fluidic processes in micro bioengineering systems. Pressure is the most common driving force in fluidic engineering. In microfluidic circuitry, inertial energy is much smaller than the viscous term, i.e. low Reynolds number flow (see Chapter 3). The pressure is used to compensate the viscous shear stress, which is usually very high due to the large transverse velocity gradient in a micron wide channel.

In incompressible flows, the applied pressure is instantaneously felt everywhere within the entire system. The bulk fluid and the embedded particles can all be moved together under this pressure gradient. The common on-chip driving device is a peristaltic pump (Yang *et al.* 1998, Unger *et al.* 2000, Figure 1.2), because positive displacement and rotary pumps are not easily manufactured or integrated in microfluidic systems.

1.3.2 *Surface tension*

Fluid is traditionally transported in a continuous stream, while pressure is used as the driving force. Kim (Lee *et al.* 2002) suggested moving fluid and the embedded particles in a discrete droplet form to improve the efficiency of particle movement. Surface tension dominates when droplet size reduces and, thus, becomes an effective driving force.

In addition, the contact angle between the liquid droplet and the solid can be modified by the surface electrical potential, which is known as the electro-wetting phenomenon (see Chapter 5). A surface electropotential gradient will produce different contact angles across a droplet resulting in an imbalanced surface tension.

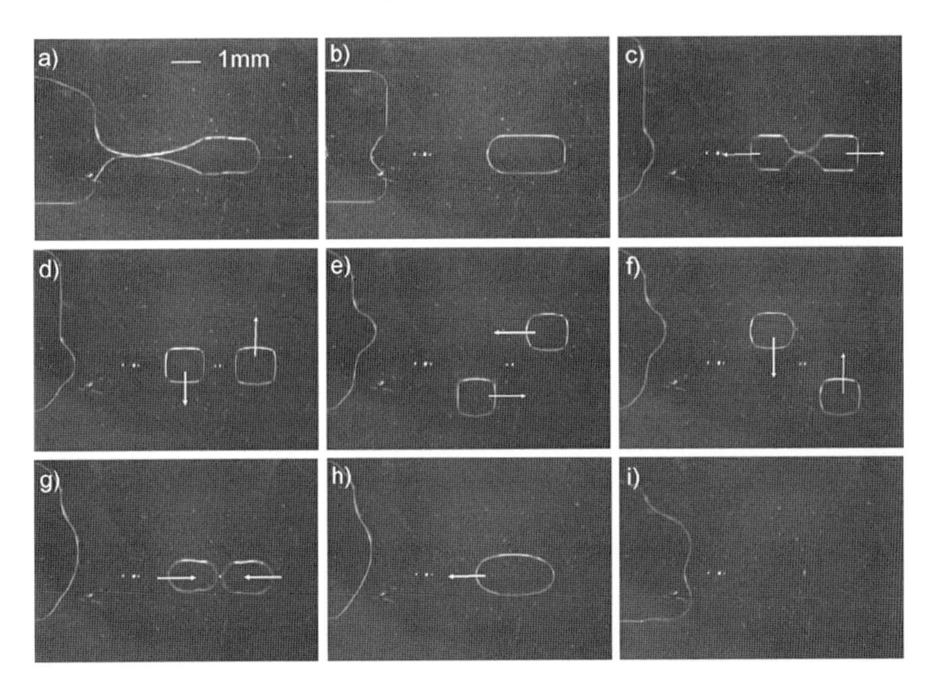

FIG. 1.3: Droplet operations in digital fluidic circuit. (a) and (b) Droplet creation from reservoir; (c) droplet cutting; (d)–(f) simultaneous and independent transport of two droplets; (g)–(i) merging of the droplets and return to reservoir. (Reprinted with permission from Fan *et al.* 2003 Copyright 2003, IEEE).

This force moves the droplet along the electro-gradient prescribed by the surface electrode pattern. Cho *et al.* (2003) and Fan *et al.* (2003, Fig. 1.3) demonstrated creation, transportation, cutting, and merging of fluid droplets such that mixing, separation, and concentration can all proceed in a digital fluidics form.

1.3.3 *Electrokinetic forces*

Some of the fluidic processes, e.g. concentration and mixing, require relative motions between particles and fluid. Actuators, other than pumps or valves, are needed because pressure moves all constituents together within a fluid medium. Various electrokinetic effects (see Chapter 5) can directly apply forces to move charged or polarized particles embedded in a fluid and are therefore very useful in micro/nanofluidics (Morgan and Green 2003; Li 2004; Wong *et al.* 2004 a b).

1.3.3.1 *Electrophoresis* When a DC electric field is applied to fluid containing charged particles, the particles experience an electrostatic force. This phenomenon of moving charged particles within a fluid by force is termed electrophoresis. At the same time, the particle experiences a frictional drag, due to the relative velocity between particle and surrounding fluid (Manz *et al.* 1992; Probstein 1994). Owing to the existence of electrostatic charges in a variety of biomolecular entities, electrophoresis presents a very powerful way to manipulate molecules within a fluid. Molecules such as DNA, proteins, and peptides (Bousse 2000) have been directed in fluidic devices to accomplish a specific process. In recent studies, the electrophoretic force has been used to transport proteins and molecules inside nano fluidic channels (Fan *et al.* 2005).

1.3.3.2 *Dielectrophoresis* For a non-charged dielectric particle placed in an inhomogeneous electric field, the particle can be polarized and it then experiences a dielectrophoretic (DEP) force (Pohl 1978). The higher the spatial gradient of the electrical field, the larger the DEP force felt by the particle. On the other hand, ionic fluid smears out the electrical gradient, resulting in a small force. The DEP force is a volume force, which is proportional to the size of the particle. Studies suggest that the DEP force is not effective in moving particles less than 100 nanometers. The DEP force depends on the difference in the complex permittivities of the particle and medium, in that the force acting on the particle can be either repulsive or attractive. This feature provides great flexibility in manipulating particle motion in a fluid. Carbon nanotubes (Vijayaraghavan 2007), proteins (Asokan 2003), DNA (Washizu and Kurosawa 1990), and viruses (Hughes 1998) have been successfully concentrated by the DEP force manipulation technique for single molecule studies or in accomplishing manufacturing processes. Based upon the dielectric properties of the embedded entities, the DEP force can be utilized for particle separation (Huang *et al.* 2002).

1.3.3.3 *DC Electroosmosis* The electrical charges on a solid surface attract counterions and repel co-ions. A thick, electrical double layer (EDL) is formed near the surface and shields the potential field. The original charge-free fluid carries excess co-ions. When a DC electrical potential gradient is applied, counterions in the EDL will be moved in the direction of the gradient. Fluid molecules will be dragged by counterions forming the electroosmotic flow, termed electroosmosis. Electroosmosis is a non-mechanical fluidic pump.

Electroosmotic flow is commonly used for sample injection in electrophoresis-based separation (Harrison *et al.* 1993). More complex manipulation of fluids through electroosmotic pumping can be achieved by strategically applying electrical potentials in different locations across a microfluidic channel.

1.3.3.4 *AC electroosmosis* If an AC electric field is applied in a microfluidic device with surface charge (Ramos 1998), as opposed to a DC field, the sign of the counterions in the electrical double layer changes with the applied AC field. This interaction generates a force in the electrical double layer and produces a net fluid motion. At high frequency, AC potential field ions, especially the ones with a large mass, cannot efficiently follow the driving force, due to low acceleration and high viscous drag. Generally, the effective operating frequency for AC electroosmosis is below 1 MHz.

1.4 Fluidic processes

Sample collection, sample preparation, and sensing are the three stages used to achieve biomolecular detection and disease diagnoses. At each stage, the mismatch in the length scales between fluid volumes handled and the target particles size results in great challenges for device design. In particular, the design principles of the devices, choice of forces, and the integration need to accommodate the orders of magnitude changes in length scales and force levels.

1.4.1 *Sample collection*

Sample collection is the task of concentrating target particles from a diluted medium. The sample collection procedure typically occurs at the macro and micro ranges. For example, concentrating the micron scale spores containing bio-warfare agents from thousands or millions of cubic feet of air/water at a very low concentration level is tedious and painstaking. Similarly, sorting cells from tens of milliliters of body fluid results in challenges. Traditionally, using a centrifuge to collect target particles and then transporting the material into the next stage fluidic processor requires human intervention and is, therefore, not easy to be automated. Filtering is an alternative approach. However, passing a large volume of fluid through holes to collect small solid particles will result in a high energy dissipation and clogging. A non-traditional filter design (Fig. 1.4) with micron sized openings has solved some of these challenges (Yang *et al.* 1999). The silicon nitride filter is only one micron thick and coated with parylene. This robust structure is able to withstand a 4 psi pressure drop and needs very low power to push the fluid through the small holes due to the thin passages. The DEP force generated by the electrodes (Desai *et al.* 1999) can move the collected particles easily on the parylene coated surface without clogging the filter openings.

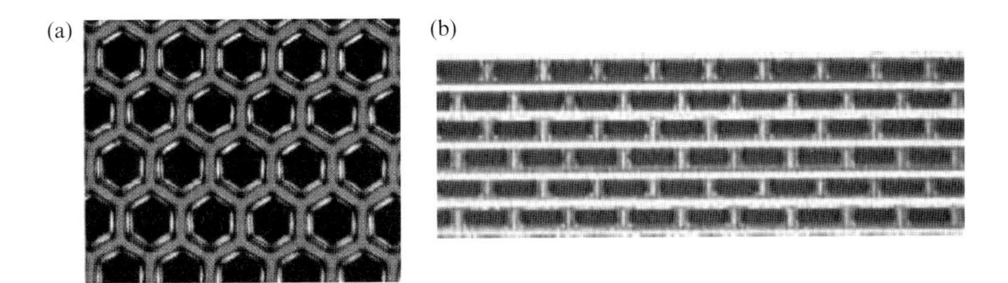

Fig. 1.4: (a) filter with parylene coating with 10 micron openings, (b) filter with micro electrodes for particle transport (Reprinted with permission from Yang *et al.* 1999; Desai *et al.* 1999 Copyright 1999, Elsevier).

1.4.2 *Sample preparation*

Sample preparation is the step in the process whereby the collected sample is readied for sensor detection. The collected and concentrated target particles, nucleic acids, proteins, or cells need to be conjugated with catch probes if surface immobilization detection is used. The target particles also need to be marked by signaling probes for detection purposes. Introducing these reagents containing molecular probes into the fluidic system requires the development of several fluidic processes for facilitating the diagnostics.

1.4.2.1 *Mixing* During the processing of biological samples, introducing one type of fluid into another type of fluid for homogenizing constituents is termed mixing. The time needed to finish the process is commonly referred to as incubation time. Target particles must be within a close proximity to the probes in order for the conjugation to succeed. The homogenization can be accomplished by diffusion, which is driven by intrinsic Brownian motion. Brownian motion is caused by collisions between a particle and a thermally agitated fluid molecule surrounding the particle. When a particle concentration gradient exists, particles will diffuse along this gradient. The gradient diminishes with time, as does the diffusion speed. Diffusion is typically a slow process, which is even more pronounced for larger particles. Generally, incubation times are extended to ensure complete diffusion and conjugation. Incubation time is, therefore, the limiting factor with regard to time in the mixing process.

Recently, many microfluidic mixing methods have been developed to shorten the detection time. This approach involves stretching the fluid for thinning the separation between streams containing different species. Small separation shortens the time needed to diffuse across the stream, resulting in molecular level mixing. Turbulence cannot take place in low Reynolds number microfluidic flows. Chaotic flow (Lee *et al.* 2007) is the alternative to achieve efficient mixing (Fig. 1.5). Chaos can happen in 2-D unsteady or 3-D steady and unsteady flows. In 3-D steady flows, chaotic fluid stretching can be produced by flowing a fluid through a channel with a corrugated surface (Stroock *et al.* 2002). Another approach requires the production of an unsteady velocity shear to stretch fluid streams at different rates to increase the interface.

1.4.2.2 *Separation* Instead of bringing particles together to facilitate a biochemical reaction, increasing the distance between particles is also commonly used in the sample preparation procedures. For example, the initial ratio of signaling probes to target particles is usually large to ensure that each target particle is effectively tagged. However, removal or separation of the excess signaling probes is necessary to avoid a false positive or high background. One way to achieve this is through immobilization of the target particles onto a surface. A flow is generated to wash away the excess signaling probes and non-specific bindings, which produce undesirable noise. The wash is a delicate process, which

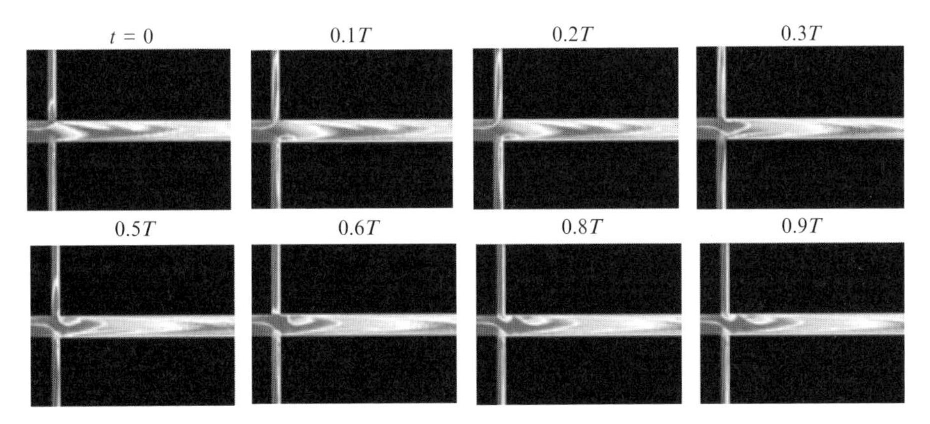

FIG. 1.5: Generation of stretching and folding of the fluids (glycerine solution with and without Rhodamine 6G dye) in a micromixer at different stages during one period (T =1s) at Re $= 2.4 \times 10^{-4}$ and St $= 0.08$. (Reprinted with permission from Lee *et al.* 2007 Copyright 2007, Cambridge University Press).

needs to produce enough force to remove the non-specific binding molecules but leave the immobilized target particles unscathed.

Separating cells of various types (Tsutsui and Ho 2009) is also a common practice. For example, fetal cells can be found in the maternal blood for prenatal diagnosis, which can offer useful medical information for the parents and health-care professionals. However, only one fetal cell out of a million or more red blood cells can be found in the maternal blood. Here, it is necessary to first recognize the particles of interest, fetal cells, and then to separate them from a large pool of unwanted components. With the help of immunocytochemical labeling to specifically recognize the fetal cells, these marked cells go through micro cytometry or imaging analysis to efficiently diagnose the rare fetal cells circulating in the maternal blood (Merchant and Castleman 2002).

1.5 The surface—the molecular effects at the interface between fluid and solid

The surface plays a major functional role in micro/nano devices whose surface to volume ratio is inversely proportional to the transverse dimension. The large surface to volume ratio indicates that the fluid molecule in a micro/nanofluidic device will be more likely to interact with the surface than inside the macro dimensional flow (Ulmanella and Ho, 2008; Wong *et al.* 2009c). This property presents opportunities, e.g. sensitive and specific biomolecular recognition, and also challenges, e.g. biofouling. Both challenges and opportunities are the result of competitive molecular interactions on the surface. A deep understanding of the surface forces and knowledge regarding proper control of the surface chemistry presents vast opportunities to improve the performances of diagnostics and therapeutics.

1.5.1 *Surface modification*

The purpose of surface modification is to render desired physical and chemical interactions between biological substances and engineered devices, thus achieving better performance. Applications of such surface modifications range from reducing bio-fouling, patterning biological macromolecules/cells, and improving biosensors performance, to surface force-driven fluidic manipulations (see Chapter 11). In the case of surface passivation, the surface will be treated to alleviate adsorption of unwanted biosubstances, nucleic acids, proteins, and cells. In the case of immobilized diagnosis, we will treat the surface only to capture certain target bioparticles for signal generation, but reject all other non-specific molecules/cells in order to reduce noise. Implanted devices may require an appropriate surface modification to prevent host immune rejection, cytotoxic effects, and isolation from the targeted tissue. Understanding the molecular forces and the properties of macromolecules can greatly facilitate the progress of surface property control. Due to the needs of handling numerous types of cells and molecules, this area remains the most challenging task in developing bio-medical transducers.

Bio-fouling occurs on the surface of most substrate materials, such as silicon, glass, PDMS, and SU-8, found on microfluidic devices, which readily adsorbs biomolecules. Once a surface is coated with biomolecules, cells readily adhere causing bio-fouling, which will interfere with the device measurements. Fouling can reduce analyte concentration in the bulk fluid due to surface adsorption, clogging of fluidic paths, and deterioration of biosensor surfaces.

Surface patterning places DNA, proteins, and cells at selected island areas in an array arrangement for high throughput bio-marker sensing, drug screening, and cell–matrix interactions. Fabrication of such arrays involves defining non-fouling areas and adhesion-promoting areas on a substrate by photolithographic or soft lithographic patterning (Whitesides 2001).

1.5.2 *Techniques of modifying surface properties*

Surface modification techniques can be categorized into two major approaches: dynamic coating and permanent coating (Wong and Ho 2009a). Dynamic coating involves the physical adsorption of a polymer or surfactant applied prior to or simultaneously with the analysis, and is thus a simple process to implement. Chemical modification of surfaces are highly dependent upon the substrate materials such as glass, polymethylmethacrylate (PMMA), and poly(dimethylsiloxane) (PDMS) (Belder and Ludwig 2003). For example, successful dynamic coatings were demonstrated with cellulose derivatives such as hydroxyethylcellulose (HEC) and hydroxypropylmethylcellulose (HPMC) for a glass substrate. Cellulose derivatives and poly(ethylene glycol) and its derivatives can be dynamically coated on PMMA devices. Coating of 2-morpholinoethanesulfonic acid

(MES) and polybrene/dextrane sulfate layers in PDMS device has also been demonstrated. On the other hand, permanent coating is a covalent modification process, which can render stable and homogeneous coatings, although such chemical processing could be more time consuming. Typically, permanent coating on glass and PDMS, which have surface silanol groups, takes advantage of the well-established silanization chemistry, followed by subsequent grafting of polymers such as polyacrylamide (PAA), poly(vinyl alcohol) (PVA), or PEG derivatives. The PMMA surface can first be derivatized with an amine- or carboxylic- group through chemical reaction or a pulsed UV excimer laser, respectively. Then various coating compounds can be covalently attached to such functional groups on PMMA. Physical adsorption and covalent modification can be either a thin monolayer coating (i.e. self assembled monolayer) or a thick polymer grafted layer.

1.5.2.1 *Self assembled monolayer (SAM)* Self assembled monolayers (SAM) form on reactive surfaces through spontaneous organization of molecules with active chemical moieties (Ulman 1996). Due to its ease of preparation, low cost, reproducibility, and versatility, SAMs have attracted enormous research interest dedicated to many disciplines over the past two decades (Senaratne *et al.* 2005). The most common examples of SAM systems are organosilane species, including trichlorosilanes, triethoxysilanes, and trimethoxysilanes derivatives on oxidized glass or PDMS surface, organosulfur species (thiol-derivatives) on noble metals, and carboxylic acid derivatives on metal oxides, grown in either a liquid or vapor phase. Both thiol and silane chemistries introduce a moiety at one end of the molecule, which has an affinity to the surface and a terminal functional group, presented on the other end of the molecule, which determines the final interfacial characteristics, such as wetting properties, chemo/bio adhesiveness, surface potential, and reactivity for further modification. These surface terminal functional groups are vitally important in enabling various technologies based on specific interactions between an analyte and surface, such as biosensors, cell/protein adhesion and patterning, and microfluidic biological applications.

1.5.2.2 *Polymer coatings* Compared to SAM and physical absorption technologies, covalent polymer coatings exhibit superior mechanical and chemical robustness. These properties are also coupled with a high degree of synthetic flexibility toward the introduction of a variety of functional groups (Pallandre *et al.* 2006). Polymer coatings can be prepared either by directly tethering end-functionalized polymer chains onto reactive surfaces (termed 'grafting to') or by inducing *in situ* polymerization from surface initiation sites (termed 'grafting from') (Edmondson *et al.* 2004). Both methods require pre-functionalizing the bare surface through SAM, which renders it reactive before polymer grafting.

Examples of 'grafting to' approaches include direct tethering end-functionalized polymers or block copolymers of poly(ethylene glycol) (PEG) or dextran covalently onto a reactive surface to ensure non-bio-fouling. The major advantage of the 'grafting to' method is its experimental simplicity and the large selection of commercially available functionalized polymer molecules. However, this approach usually suffers from low grafting density, due to the crowding of initial grafted chains, which prevent further insertion of polymer chains onto the surface, and film thickness limitation (less than a hundred nanometers) by the molecular weight of the grafted polymer chains. An alternative for preparing thick polymer coatings for biotechnological applications is the 'grafting from' technique (also known as surface initiated polymerization (SIP)), in which polymerization of monomers originates from the surface-anchored initiation sites, mostly immobilized through self-assembly chemistry (Brough *et al.* 2007). Since polymer chains grow in close proximity to each other, repulsive interactions between neighboring chains keep the chains stretching away from the surface, avoiding overlap, and thereby a thick and dense layer of polymer layer can be formed.

1.5.3 *Surface patterning*

Surface modification and fabrication techniques enable us to spatially define surface features and functionalities in micro/nano scale arrays. These arrays allow us to detect or to process biomolecules/cells in a parallel fashion. Biomarker sensor arrays can sensitively and specifically diagnose diseases for clinical applications. DNA/protein microarrays become standard screening tools in genomic and proteomic research. Surface physical and chemical property modifications have been demonstrated to affect cellular developments, such as proliferation, differentiation, and migration. In addition, living cell arrays have found promising applications in both cell-based biosensors and cell-based bioassays for drug screening.

Surface patterning for biotechnology applications is usually implemented by producing ordered micro/nano features of biospecific (or bioadhesive) and non-biofouling islands, where biomolecules and cells can be selectively localized (Christman *et al.* 2006). Physical adsorption of biological materials onto a solid substrate without specific recognition or a receptor of interest is regarded as non-specific binding, which produces undesirable background noise, or a false positive, and thus hampers the performance of biosensors or cell array analysis. Developing non-biofouling materials, as well as corresponding chemistries of immobilization of these materials onto patterned surfaces, is the main task of array technology.

Fabrication techniques for surface patterning include standard photolithography, electron beam lithography, soft lithography, micro-contact printing (μCP), scanning probe based lithography, dip pen lithography (DPL, Piner *et al.* 1999), and nano-imprint. While each of these techniques presents their own advantages, potential limitations of each method also exist and must be addressed for various applications.

1.5.3.1 *Photo/electron lithography* Early demonstration of surface patterning of a SAM is through photolithography, where a patterned photoresist serves as a sacrificial stencil layer for generating organosilane patterns on silicon based surfaces. The photoresist is eventually dissolved in an organic solvent and the exposed region can be backfilled by silanization with another type of silane reagent. Similar results can be achieved by deep ultra-violet (DUV) lithography without the use of a photoresist. Homogeneous coating of amino-silane SAM is exposed to DUV through a mask to induce photochemical changes of the exposed portion of the amino-silane monolayer, which is used for modification with other silane reagents. Higher resolution patterning in nanometer scale can be achieved by electron beam lithography. Analogous to UV and DUV photolithography, electron beam surface patterning with or without a photoresist has been effectively demonstrated. However, this method suffers from low fabrication efficiency due to its serial processing nature.

1.5.3.2 *Soft lithography* Soft lithography (Xia and Whitesides 1998) represents a group of techniques relying on the utilization of 'soft' elastomeric poly (dimethylsiloxane) (PDMS) stamps for patterning. Among soft lithographic techniques, micro contact printing (μCP) is the most popular, due to its simplicity, cost-effectiveness, parallel processing, and versatility. In μCP, a PDMS stamp molded with microstructures is inked with self assembly molecules (such as alkanethiol or organosilanes) and brought into conformational contact with a solid substrate for pattern transfer. Further modifications, such as backfilling the unstamped area with non-biofouling molecules (such as PEG-thiol) or further attachment with functional proteins or biospecific molecules, render the surface applicable for cell patterning, protein microarrays, and microscopic biochemical assays. Certain limitations of μCP encompass low feature resolution (typically sub-microns), low reproducibility in a large area (couple square centimeters), the need for designing special alignment systems for multi-ink printing, and potential contamination of substrates with low molecular weight PDMS residues.

1.5.3.3 *Scanning probe lithography and nano-imprint* For higher resolution, down to tens of nanometers, with a high throughput and low cost, nano-imprint is a powerful alternative to the aforementioned methods (Chou *et al.* 1995). In nano-imprinting, a thin photoresist cast on the substrate is pressed under pressure against a hard master, fabricated with nanostructures, to transfer the pattern from the master to photoresist. This high pressure pressing is then followed by anisotropic etching on the resist to further transfer the pattern to the substrate for surface modification. On the other hand, resolution down to molecular scale can be achieved through scanning probe based lithography, such as dip pen lithography (Piner *et al.* 1999) and nano-shaving (Liu *et al.* 2000). Taking advantage of the high spatial accuracy of tip movement in scanning probe microscope (SPM), self assembling molecules can be directly delivered from the tip to the surface, with a resolution down to 30 nm. An opposite implementation of this technique is nano-shaving, where local interactions between molecules and the tip are enhanced significantly to selectively displace the molecules from the surface. Despite the ultimate advantage of spatial resolution of these two methods, their serial processing nature and the need of specially designed SPM devices limit their potential in biological applications.

1.6 Nano/micro technology based engineering system—synergistically organizing building blocks to provide high level functionalities

With matched length scales, nano/micro transducers can greatly facilitate the studies of interactions of biomolecules in the cellular network, as well as the understanding of cellular system functions. In order to achieve these tasks, sensors, actuators, and fluidic reactors must be integrated into an engineering system to efficiently interact with the cell or its components. Examples of these functionalities include manipulating cells, controlling their extracellular environment, and performing analytical functions such as sorting, separation, mixing, and concentration of biological analytes. The fast advancements in this area, such as the lab-on-a-chip or micro-total-analysis-system, have led to revolutionary developments of engineering systems for detecting biomolecules as a diagnostic tool or for cellular culture to understand their behavior.

1.6.1 *System integration*

Manipulating and sensing the interactions of nucleic acids, proteins, and other molecules are the key processes for detecting biomarkers or for studying cellular functions. In order to achieve sensitive and specific molecular recognition at the nano scale, samples must be collected, e.g. micro-scale cells, from body fluid in milliliter volumes, and target molecules must be separated from the background particles. These target molecules should be transported and well mixed with probes in microfluidic channels and reactors with sizes from hundreds of microns to the millimeter range. Modification and patterning of the surface properties becomes important for fouling prevention of the fluidic devices and decreased background noise. The processors and sensors span the length scales from nanometers to millimeters. Molecular forces determine the molecular recognitions. Fluid transport, mixing, and various fluidic processes are carried out by electrokinetic forces or hydrodynamic pressure. Design and development of individual devices have been very successful in the past. A greater challenge here is the multi-scale integration from molecules and devices to form a system used in performance optimization. In fact, the issues to be overcome always occur at the interface. Examples are: 1) How to efficiently transduce molecular recognition events to an electronic or optical signal? 2) How to select and concentrate a small number of miniature particles from a large amount of fluid? 3) How to couple different types of force fields? 4) How to design the electrical and fluidic interconnects between the devices?

1.6.2 *Biomolecular detection system*

In medical diagnoses or bio-threat sensing, either the presence or the increase in biomarker concentration indicates a positive detection. In order to achieve detection of these small molecules, biomarkers must be very specific to a particular disease. Generally, multiple markers are needed to reduce the false positive/negative rate. Large scale micro arrays or mass spectrometer screenings are the typical schemes to sort out a small portion of RNA or proteins specifically marking the disease from the vast population of biomolecules. Tremendous amounts of research have been focused on this important topic. However, commonly accepted biomarkers are still extremely scarce.

After the biomarkers have been decided upon, the detection of these marker molecules is extremely tedious. The disease or bio-threat must be detected in its early stage when the marker is in very low concentrations and surrounded by millions of other noise-generating nano particles. The signal-to-noise ratio is the key parameter for achieving the desired limit of detection (LOD). The fluidic process of separation is commonly used to concentrate the targets, i.e. separating the targets from other particles. Mixing is employed to ensure that target molecules have the opportunity to reach probes in order to increase the signal level. The surface modification techniques have proved very effective at preventing non-specific binding for noise reduction.

The lab-on-a-chip system offers advantages such as reduction of reagent consumption, increased assay speed, a potential for mass production with low cost, and the ability to achieve high level automation. The successful realization of such a fully integrated system would greatly facilitate clinical molecular diagnostics (Mao *et al.* 2009). Xie *et al.* (2005) demonstrated an integrated parylene based gradient liquid chromatography (LC) system, which offers the advantages of integrating sample handling, preparation, mixing, and separation into single platform. This system can achieve the resolution of a commercial high pressure liquid chromatography (HPLC) (Agilent 1100 series) resolution, but with significant reduction in processing time and sample consumption. Three electrolysis-based electrochemical pumps can load the sample and deliver the solvent gradient (Fig. 1.6). Platinum electrodes were integrated to drive these pumps and the electrospray nozzle. A diffusion-based static mixer with as little as 1.5 nL volume can efficiently achieve uniform mixing of the solvent gradient. Full integration of all fluidic components with column and electrospray nozzle greatly reduced the sample volume and thus increased the functional efficiency. Automating this process led to the successful operation of pumps and the electrospray nozzle to sequentially perform the column wash and re-equilibrium, sample loading, sample wash, and gradient elution of the separated components.

The microfluidic based system can be expanded to include thousands of integrated micromechanical valves and control components to form microfluidic large-scale integration (mLSI) (Melin *et al.* 2007). The components and operating processes can be designed to stitch together in a seamless and transparent fashion, analogous to the electronic LSI, which facilitates biological assay automation. mLSI enables hundreds of assays to be performed in parallel with multiple reagents in an automatic manner. Quake's group demonstrated such an mLSI system in which cell isolation, cell lysis, mRNA purification, cDNA synthesis, and cDNA

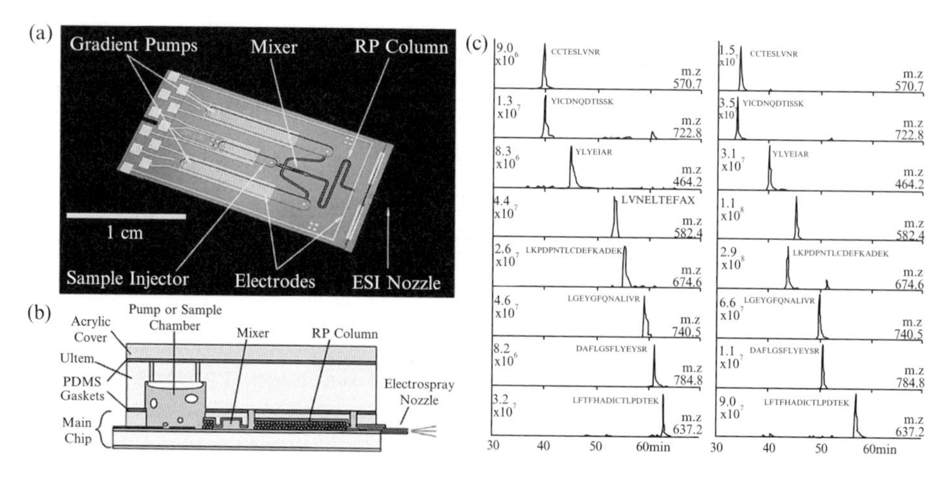

Fig. 1.6: Microfluidic liquid chromatography-Tandem Mass Spectrometry (LC-MS/MS) platform. (a) Photograph of fabricated LC chip. (b) Schematic of LC chip assembly with integrated components. (c) Performance comparison of microfluidic LC (left) with an Agilent 1100 series HPLC system (right) for the extracted ion chromatograms (EIC) of eight tryptic peptides from BSA. Amino acid code is labeled to indicate the peptide sequence based on analysis of the corresponding MS/MS spectrum. (Reprinted with permission from Xie *et al.* 2005 Copyright 2005, American Chemical Society).

purification functions were integrated in a monolithic chip with parallel processing capability (Fig. 1.7). Solid-phase cDNA synthesis was implemented on-chip with purified NIH/3T3 mRNA by utilizing the beads as both primers and a support. This device has very high sensitivity to detect low (5–10 copies/cell) and medium (~1800 copies/cell) transcripts from single NIH/3T3 cells.

1.6.3 *Cell assay system*

In a biomolecular detection system, we are interested in either collecting the molecules in the fluid medium or lysing the cells to extract the interior molecules for detection purposes. In cell assay systems, we need to culture the cells to sustain them in a controlled environment and to study their functionalities. Handling live cells with an engineered system is a much more challenging task than that in a biomarker detection system.

PDMS becomes an extensively used structural material, due to its biocompatibility, low toxicity, high thermal stability, permeability to gas, and optical transparency. The material has gained much popularity since it can be processed by simple soft-lithographic fabrication techniques. Whitesides' group has studied a wide variation in PDMS processing conditions, i.e. duration and temperature of

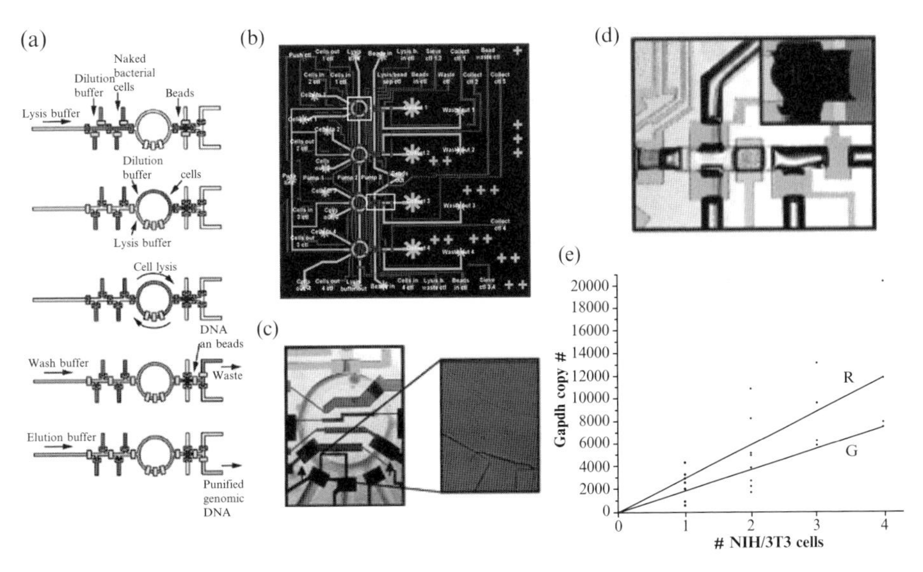

FIG. 1.7: Integrated microfluidic platform for mRNA isolation from single mammalian cells. (a) Step-by-step schematic of a single bioprocessing unit for cell loading, lysis, DNA isolation, and purification. (b) AutoCAD drawing of a 4plex mRNA isolation/cDNA synthesis device with key component highlighted in white boxes: the cell lysis chamber with an NIH/3T3 cell captured (c) and the affinity column area (d). (e) The level of GAPDH gene expression in individual NIH/3T3 cells. Line G indicates the average copy number per cell obtained from a bulk experiment. Line R indicates the average copy number for microchip experiment adjusted for efficiency of the process of each sample group. (Reprinted with permission from Marcus *et al.* 2006 Copyright 2006, American Chemical Society).

curing, the ratio of base to curing reagent, surface oxidation on the attachment, and growth of several types of cells. These studies revealed the cell culture performance for different fabricating procedures (Lee *et al.* 2004b).

Cell-based assays, such as microflow cytometry and whole cell biosensing, usually require transportation, manipulation, and sometimes immobilization of cells onto a solid surface in parallel. With devices of matching in physical dimensions, cell manipulation technologies can be used to control the specific position of a cell in solution. Magnetic, optical, mechanical, and electrical forces have been successfully demonstrated (Yi *et al.* 2006, Huh *et al.* 2005). Among these technologies, dielectrophoretic (DEP) force and optoelectronic tweezers (see Chapter 9, Chiou *et al.* 2005) have garnered the most attention, due to the label-free and noncontact scheme used to precisely position large amounts of cells in a parallel fashion with high throughput and high spatial resolution.

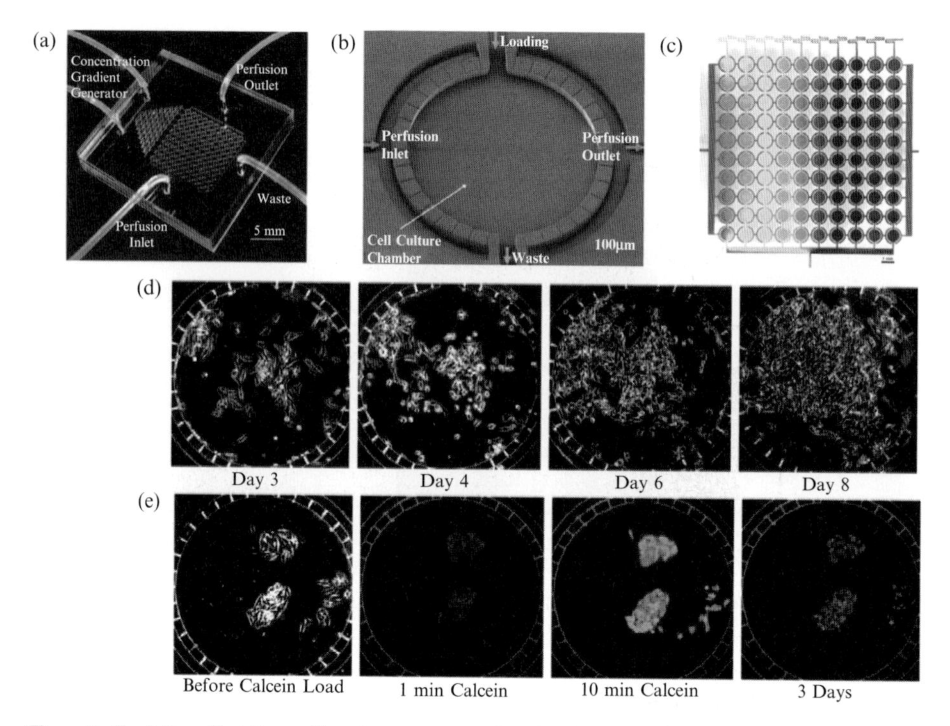

Fig. 1.8: Microfluidic cell culture array for high-throughput cell-based assays. (a) Photograph of a 2cm × 2cm PDMS device with microchamber for a 10 × 10 cell culture array. (b) SEM picture of a single culture unit with multiple perfusion channels surround the main culture chamber. (c) Concentration gradient generated across 10 columns with red dye initially perfused from left to right and blue/yellow dyes loaded from the two separate ports at the top. (d) Continuous growth of human Hela cells under microculture chamber at 3, 4, 6, and 8 days after loading. (e) Time series of cell taking calcein AM introduced by fluidic system showing the cell-based assay capability of integrated cell culture array. (Reprinted with permission from Hung *et al.* 2005 Copyright 2005, John Wiley & Sons, Inc.).

Cells respond to intracellular and/or extracellular stimuli with time, from a matter of seconds to a few days or longer. An engineering system with integrated transducers can control the cellular microenvironment and perform analytical experimental measurements with high accuracy. Conventional biological experimental approach studies a large number of cells and pools the time or population average of the desired information. However, it may ignore the substantially characteristic differences among each cell, such as their status in the cell cycle, lineage, history, or epigenetic modification etc. (Chen *et al.* 2005). Therefore, single-cell assays can uncover individual phenotypic variations and dynamically

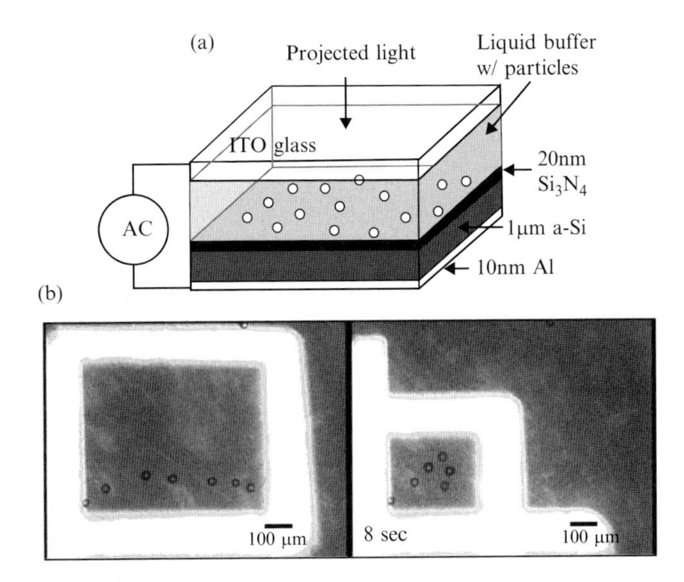

Fig. 1.9: (a) Schematic overview of the optoelectronic tweezers. The optical images shown on the digital micromirror display (DMD) are focused onto the photosensitive surface and create a non-uniform electric field for DEP manipulation. (b) Collection of 25 μm diameter polystyrene beads by OET. (Reprinted with permission from Ohta *et al.* 2007 Copyright 2007, IEEE).

monitor the living cells' responses to external stimuli over various time courses suitable for different assays. This capability, provided by the micro/nano technology based engineering system, offers unprecedented opportunities for future biological research.

A 10×10 microfluidic cell culture array, which could perform 100 different cell-based assays in parallel for long-term cellular monitoring, was presented by Lee and co-workers (Hung *et al.* 2005). The PDMS cell culture microchambers enabled several experimental assessments to be achieved concurrently, such as repeated cell growth/passage cycles, reagent introduction, and real-time optical analysis (Fig. 1.8). Human Hela cells were loaded into individual culture chambers and were studied, both through growing status and their response to fluorescence localization of calcein AM, for a 10 day period. Such a large-scale cell culture array provided a versatile platform for a wide range of assays with applications in drug screening, bioinformatics, and quantitative cell biology.

Recently, Chiou *et al.* (2005) invented optoelectronic tweezers (OET), which use a light intensity pattern to generate virtual energy traps for parallel manipulation of a large number of cells or particles. A typical structure of OET is shown in Fig. 1.9, in which the medium containing cells is sandwiched between an upper ITO-coated glass cover and a bottom photoconductive amorphous silicon-coated glass plate biased with an AC voltage. An optical beam changes the impedance of the amorphous silicon, which then forms a high-resolution patterning of electric fields on the photoconductive surface. The AC driven dielectrophoresis and projected light patterns produce the virtual optoelectric energy traps, which can confine cells into precisely controlled positions. Parallel manipulation of up to 15,000 particles on a $1.3 \times 1.0\text{mm}^2$ area was demonstrated. Since these micron-size energy traps are formed by optical images, each image, including the trapped cell, can be altered or moved to desired locations at a specific time (Fig. 1.9). This capability enables vast possibilities for cellular assays, which are not possible in fixed-in-space cell arrays.

1.7 Complex systems

A complex system (Ottino 2003) generally has very rich information content and can be characterized by the following features:

- large number of building blocks;
- interactions among building blocks and with their environment;
- displays organization without an external organizing principle being applied; and
- adaptability and robustness.

Engineered systems may contain millions of components, yet they are not considered complex. For example, a 747 jet has three million parts, but the system functions precisely as designed in the engineering principle. On the other hand, millions of functional molecules self-organize and form the biological cellular system, which is an example of a complex system.

1.7.1 *Cellular complex system*

A cell is the basic system for performing biological processes, such as growth, apoptosis, and physiological functions. These processes are often under regulatory control through growth factors, hormones, and other external stimulations of a physical or chemical nature, such as light, heat, mechanical force, etc. The cell senses these external factors and responds through a complex, built-in network of signal transduction pathways in order to process the information it receives, and respond accordingly.

DNA and RNA interactions, based upon external stimuli exposure of the cell, serve as the innate mechanisms that drive the process of life. Cells consist of millions of bio-architectures, such as the membrane and cytoplasmic proteins, which provide the framework for cyto-regulatory networks governing receptor-ligand and other protein–protein interactions. These networks govern everything from the onset of cancer, inflammation, mechano-sensation, and differentiation, to a combination of several correlated responses. This is due to the extensive amount of interplay that occurs between intermediate molecules that serve as bridges between multiple pathways, where inflammation, for example, can be affected due to mechano-sensitive regulation of the cell. Medically, these intriguing phenomena result in very apparent translational outcomes. For example, bio-interface design in medical devices produces surfaces that mimic the innate stiffness of biological environments in the form of implant coatings, which aim to mitigate potential immune responses that might result in imminent device fouling.

Studying the cell signal transduction begins with the effort to define the elements involved in the process. In addition to the large molecules, DNA/RNA and proteins, a group of small molecules known as 'second messengers' function as the intermediate molecules in signal transduction, such as cAMP, Ca++, phosphoinositols (PIPs), and diacylglycerols (DAG). The functions of these molecules are universal and are not unique to a specific cell type. For example, the small molecule cAMP functions as a second messenger in many organisms, ranging from prokaryotic *E. coli* to plant phyla such as algae, fungus, and animal phyla—including human cells. Mechanisms involving G protein coupled receptors and protein kinases are also used in most organisms for cell signal transduction. Therefore the basic components, pathways, and regulatory loops defined in one system are most likely conserved in many other systems.

Protein kinases are important regulators in signal transduction and cell function. For example, Rous sarcoma virus (RSV) can transform the host cell to cause abnormal growth. This ability was located to one gene, v-src, which is homologous to a cellular gene, c-src. The transforming ability of v-src is believed to be best described as a hijacking of the cellular gene function. Therefore the virus protein, which was constitutively active, took advantage of the cellular machinery and activated the host cell system to achieve its biological functions. Most of these virus 'oncogenes' encode protein kinases, meaning they function as enzymes to catalyze the addition of a phosphate group to their specific substrates. These kinases are important in normal cell physiology, as well as in disease progression.

There are about 500 genes in the human genome encoding protein kinases, about 2 per cent of the whole genome. They can modify their substrate protein structure by transferring a phosphate group from ATP, thereby regulating its function. Over 30 per cent of cellular proteins can be phosphorylated by protein kinases, resulting in the regulation of a wide range of cell functions.

Another feature of signal transduction is the fast response rate. A cell has to respond to external stimuli promptly. In some cases the time scale could be seconds, such as in response to hormones or mechanical forces. When stimulated by growth factors, the cells respond in minutes. For this reason the components for signal transduction in a cell are often built-in and do not need *de novo* protein synthesis. Modification of protein structure and regulation of their functions are the most common mechanisms.

Moreover, signal transduction has to be under tight regulatory control. We study not only how the signal was initiated and propagated, but also how the signal transduction was dampened and terminated. These regulatory events are the key to understanding biological processes. The signal transduction, cellular response, and regulation are outcomes of the self organized complex system (Weng *et al.* 1999).

1.7.2 *Control a biological complex system*

Exploring the interactions among biomolecules and building the knowledge base of the signaling and regulatory pathways can certainly provide in-depth understanding of the biological complex system. However, completely revealing the network pathways is unachievable due to its laborious nature and time consumption. Even once the knowledge base is accomplished, the information content will be too rich to be deciphered.

The advent of micro/nanotechnology during the past two decades has been able to shift the exploration of life sciences toward a new frontier by enabling direct handling of bioparticles of comparable sizes, such as DNA/RNA or proteins at the nano scale and cells at the micro scale. Length scale matching is a fundamental requirement for efficient transduction of information. For example, via functionalized micro-scale electrochemical sensors, ultra-sensitive bio-marker (nucleic acid) identification with the presence of only tens of molecules has been realized. These diagnostic sensors function by measuring molecular output from cellular secretions or lysates, and may lead to the early detection of diseases, e.g. cancer, resulting in a significant increase in survival rates.

Progress in the therapeutic domain, e.g. implanted devices or drugs, is somewhat limited due to challenges of a bio/bio interface with *live cells.* Interactions between an engineered system and a live biological system face a fundamental difference in their system assembly principles and are, therefore, non-trivial tasks. In engineered systems, the building blocks are assembled based on accurately known design principles and rigid requirements. A biological system is governed by self organization principles far from comprehension. Exploring and understanding the self organized principle of a complex system is *an inverse problem,* which is the core of the challenges in studying biological sciences.

In order to resolve this fundamental difference, we propose a 'system-in-system' approach. Micro/nano scale actuators of this seamless integrated system will stimulate a biological system for therapeutic purposes. Sensors will measure the cellular output responses. A search algorithm will serve as a guiding foundation towards directing biological phenotypical/genetypical outcomes. In this case, both engineering and biological systems are 'fused' into one 'system-in-system'.

We took this system-in-system approach (Wong *et al.* 2008) of trying to direct the cell toward a desired phenotype, through an engineered system control scheme. When stimulants with arbitrarily decided concentrations are applied to cells, the cellular network will react to the inputs and the biological system will change accordingly. If the overall system output does not meet our expectation, a search algorithm will assist us in choosing the next group of concentrations, which are then fed back to alter the system output until the desired phenotype is achieved. With this approach (Fig. 1.10), we do not need to experimentally map out or to analytically simulate the entire unknown network of pathways. After the stimulants are applied, *the cells will do the hard work* and leave us to judge the performance based on the output signal. This system-in-system will 'self-guide' and converge toward an optimized combinatorial drug.

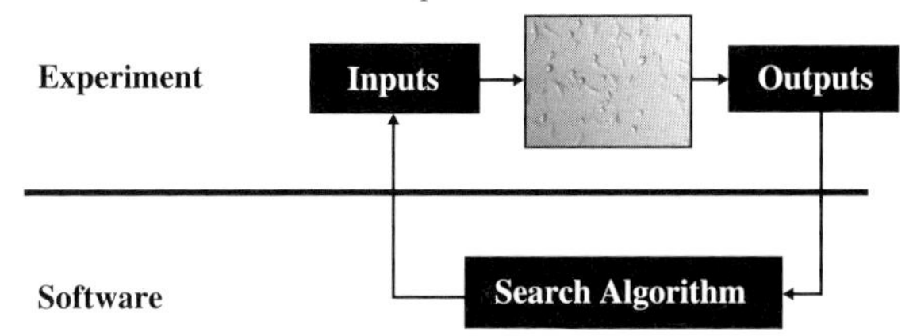

FIG. 1.10: System-in-system approach of controlling a complex biological system (Wong *et al.* 2008).

The catch of this approach is the large search space. For example, six stimulants with ten concentrations each will require 1,000,000 potential trials. A brute force test of all these combinations can be as laborious and time consuming as to explore the cellular network itself. In other words, the success of the feedback search scheme depends upon whether convergence to the desired state can happen within a small number of experimental iterations.

Wong *et al.* (2008) applied this approach to inhibit vesicular stomatitis virus (VSV) infection of NIH 3T3 fibroblasts. VSV was engineered with green florescent protein (GFP) expression capability. The percentage of infected host cells can be determined and used as the output indicator. With a stochastic search algorithm based feedback loop control, only tens of iterations were needed to lead toward one hundred per cent inhibition of viral infection. Furthermore, the dosages of the drug combinations were only 10 per cent of that required when used individually.

It is truly a surprise to see that the rapid converging rate of the feedback search results in several orders of magnitude reduction of a typical testing effort. Many other similar experiments have been performed on different cell systems (Sun *et al.* 2009). The outcomes are consistent in that only tens of iterations can direct the cells toward a desired fate. These results imply that the response functions of cells to the multiple drug treatments are very smooth, such that the search can be accomplished in a small number of iterations. In fact, the smooth cellular responses to external stimulations are due to the robustness and adaptability of complex systems.

The system-in-system approach can also provide an alternative means to study cellular functions. Cells are complex systems with many redundant functions, and it is difficult to predict how a cell will respond to multiple stimulations at one time. This approach overlooks these details and lets the system determine what works best for itself. If researchers are more interested in how the cellular network functions, system control schemes can provide an initial bird's-eye view, and then hone the important molecular activities in the complex network of pathways. The top-down system control approach can bypass the impossible mission of understanding the information rich biological complex system and lead us toward the desired state with much less effort.

References

Ajay, and Murcko, M. A., 1995. 'Computational methods to predict binding free energy in ligand-receptor complexes.' *Journal of Medicinal Chemistry*, **38**, pp. 4953–4967.

Asokan, S. B., Jawerth, L., Carroll, R. L., Cheney, R. E., Washburn, S., and Superfine, R., 2003. 'Two-dimensional manipulation and orientation of actin-myosin systems with dielectrophoresis.' *Nano Letters*, **3**, pp. 431–437.

Atkins, P. W., and De Paula, J., 2002. *Physical Chemistry*, 7th edition. Oxford University Press, Oxford.

Belder, D., and Ludwig, M., 2003. 'Surface modification in microchip electrophoresis.' *Electrophoresis*, **24**, pp. 3595–3606.

Bousse, L., Cohen, C., Nikiforov, T., Chow, A., Kopf-Sill, A. R., Dubrow, R., and Wallace Parce, J., 2000. 'Electrokinetically controlled microfluidic analysis systems.' *Annual Review of Biophysics and Biomolecular Structure*, **29**, pp. 155–181.

Branden, C. I., and Tooze, J., 1999. *Introduction to Protein Structure*, 2nd edition, Garland Publishing, New York..

Brough, B., Northrop, B., Schmidt, J., Tseng, H., Houk, K., Stoddart, F., and Ho, C. M., 2006. 'Evaluation of synthetic linear motor-molecule actuation energetics.' *Proceedings of National Academy of Sciences*, **103**, pp. 8583–8588.

Brough, B., Christman, K. L., Wong, T. S., Kolodziej, C. M., Forbes, J. G., Wang, K., Maynard, H. D., and Ho, C. M., 2007. 'Surface initiated actin polymerization from top-down manufactured nanopatterns.' *Soft Matter*, **3**, pp. 541–546.

Chen, C. S., Jiang, X. Y., and Whitesides, G. M., 2005. 'Microengineering the environment of mammalian cells in culture.' *MRS. Bulletin*, **30**, pp. 194–201.

Chiou, P. Y., Ohta, A. T., and Wu, M. C., 2005. 'Massively parallel manipulation of single cells and microparticles using optical images.' *Nature*, **436**, pp. 370–372.

Cho, S. K., Moon, H., and Kim, C. J., 2003. 'Creating, transporting, cutting, and merging liquid droplets by electrowetting-based actuation for digital microfluidic circuits.' *Journal of Microelectromechanical Systems*, **12**, pp. 70–80.

Chou, S. Y., Krauss, P. R., and Renstrom, P. J., 1995. 'Imprint of sub-25 nm vias and trenches in polymers.' *Applied Physics Letters*, **67**, pp. 3114–3116.

Christman, K. L., Enriquez-Rios, V. D., Maynard, H. D., 2006. 'Nanopatterning proteins and peptides.' *Soft Matter* **2**, 928–939.

Desai, A., Lee, S. W., and Tai, Y. C., 1999. 'A. MEMS electrostatic particle transportation system.' *Sensors & Actuators A: Physical*, **73**, pp. 37–44.

Edmondson, S., Osborne, V. L., and Huck, W. T. S., 2004. 'Polymer brushes via surface-initiated polymerizations.' *Chemical Society Reviews*, **33**, pp. 14–22.

El-Ali, J., Sorger, P. K., Klavs, F., and Jensen, K. F., 2006. 'Cells on chips.' *Nature*, **442**, pp. 403–411.

Elowitz, M. B., Levine, A. J., Siggia, E. D., and Swain, P. S., 2002. 'Stochastic gene expression in a single cell.' *Science*, **297**, pp. 1183–1186.

Fan, R., Karnik, R., Yue, M., Deyu Li, D., Majumdar, A., and Yang, P., 2005. 'DNA translocation in inorganic nanotubes.' *Nano Letter*, **5**, pp. 1633–1637.

Fan, S. K., Hashi, C., and Kim, C. J., Jan 2003. 'Manipulation of multiple droplets on NxM grid by cross-reference EWOD driving scheme and pressure-contact packaging,' in Proceedings of IEEE. Conference on Micro-Electro-Mechanical Systems, Kyoto, Japan, pp. 694–697.

Falconnet, D., Csucs, G., Grandin, H. M., and Textor, M., 2006. 'Surface engineering approaches to micropattern surfaces for cell-based assays.' *Biomaterials*, **27**, pp. 3044–3063.

Gold, L., and Tuerk, C., 1989. *SELEX-NeXstar Pharmaceuticals*. NeXstar Pharmaceuticals, Boulder, CO.

Harrison, D. J., Karl Fluri, K., Kurt Seiler, K., Fan, Z., Effenhauser, C. S., and Manz, A., 1993. 'Micromachining a miniaturized capillary electrophoresis-based chemical analysis system on a chip.' *Science*, **261**, pp. 895–897.

Hermann, T., and Patel, D. J., 2000. 'Adaptive recognition by nucleic acid aptamers.' *Science*, **287**, pp. 820–825.

Ho, C. M., 2001. *Fluidics: The Link Between Micro and Nano Sciences and Technologies*. Technical Digest of the 14th IEEE. International MEMS. Conference (ISBN-0-7803-6251-9), pp. 375–384, Interlaken, Switzerland.

Ho, C. M., and Tai, Y. C., 1998. 'Micro-electro-mechanical-systems and fluid flows.' *Annual Review of Fluid Mechanics*, **30**, pp. 579–612.

Ho, D., Garcia, D., and Ho, C. M., 2006. 'Nanomanufacturing and characterization modalities for bio-nano-informatics systems.' *Journal of Nanoscience and Nanotechnology*, **6**, pp. 875–891.

Huang, B., Wu, H., Devaki Bhaya, D., Grossman, A., Granier, S., Kobilka, B. K., and Zare, R. N., 2007. 'Counting low – copy number proteins in a single cell.' *Science*, **315**, pp. 81–84.

Huang, Y., Joo, S., Duhon, M., Heller, M., Wallace, B., and Xu, X., 2002. 'Dielectrophoretic cell separation and gene expression profiling on microelectronic chip array.' *Analytical Chemistry*, **74**, pp. 3362–3371.

Hughes, M. P., and Morgan, H., 1998. 'Dielectrophoretic trapping of single sub-micrometre scale bioparticles.' *Journal of Physics D—Applied Physics*, **31**, pp. 2205–2210.

Huh, D. *et al.*, 2005. 'Microfluidics for flow cytometric analysis of cells and particles.' *Physiological Measurement*, **26**, pp. R73–R98.

Hung, P. J. *et al.*, 2005. 'Continuous perfusion microfluidic cell culture array for high-throughput cell-based assays.' *Biotechnology and Bioengineering*, **89** (1), pp. 1–8.

Iijima, S., 1991. 'Helical microtubules of graphitic carbon.' *Nature*, **354**, pp. 56–58.

Israelachvili, J. N., 1992. *Intermolecular and Surface Forces*. Academic Press, New York.

Joesten, M. D., and Schaad, L. J., 1974. 'Hydrogen bonding.' *Journal of Molecular Structure*, **30**, pp. 425.

Jung, D. R., Kapur, R., Adams, T., Giuliano, K. A., Mrksich, M., Craighead, H. G., and Taylor, D. L., 2001. 'Topographical and physicochemical modification of material surface to enable patterning of living cells.' *Critical Reviews in Biotechnology*, **21**, pp. 111–154.

Kim, C. J., Pisano, A. P., and Muller, R. S., 1992. 'Silicon-processed overhanging microgripper.' *Journal of Microelectromechanical Systems*, **1** (1), pp. 31–36.

Lee, G. U., Chrisey, L. A., and Colton, R. J., 1994. 'Direct measurement of the forces between complementary strands of DNA.' *Science*, **266**, pp. 771–773.

Lee, J. F., Hesselberth, J. R., Meyers, L. A., and Ellington, A. D., 2004a. 'Aptamer database.' *Nucleic Acids Research*, **32**, pp. D95–D100.

Lee, J. N. *et al.*, 2004b. 'Compatibility of mammalian cells on surfaces of poly (dimethylsiloxane).' *Langmuir*, **20**, pp. 11684–11691.

Lee, L., Hyejin Moon, H., Jesse Fowler, J., Thomas Schoellhammer, T., and Chang-Jin Kim, C. J., 2002. 'Electrowetting and electrowetting-on-dielectric for microscale liquid handling.' *Sensors and Actuators A*, **95**, pp. 259–268.

Lee, Y. K., Shih, C., Tabeling, P., and Ho, C. M., 2007. 'Experimental study and nonlinear dynamic analysis of time-periodic micro chaotic mixers.' *Journal of Fluid Mechanics*, **575**, pp. 425–448.

Li, D., 2004. *Electrokinetics in Microfluidics.* Academic Press, New York.

Li, N., and Ho, C. M., 2008. 'Photolithographic patterning of organosilane monolayer for generating large area two-dimensional B lymphocyte arrays.' *Lab on a Chip*, **8**, p. 2105.

Lide, D. R., 1994. *Handbook of Chemistry and Physics.*CRC., Boca Raton, FL.

Liu, G. Y., Xu, S., and Qian, Y. L., 2000. 'Nanofabrication of self-assembled monolayers using scanning probe lithography.' *Accounts of Chemical Research*, **33**, pp. 457–466.

Manz, A., Harrison, D. J., Verpoorte, E. M. J., Fettinger, J. C., Paulus, A., Ludi, H., and Widmer, H. M., 1992. 'Planar chips technology for miniaturization and integration of separation techniques into monitoring systems: Capillary electrophoresis on a chip.' *Journal of Chromatography*, **593**, pp. 253–258.

Mao, X., Huang, T.J., and Ho, C.M., 2009. The Lab-on-a-Chip Approach for Molecular Diagnostics. Grody, W.W., Nakamura, R.M., Strom, C. (Ed.), *Handbook of Molecular Diagnostics*, Elsevier.

Marcus, J. S., Anderson, W. F., and Quake, S. R., 2006. 'Microfluidic single-cell mRNA isolation and analysis.' *Analytical Chemistry*, **78**, pp. 3084–3089.

Merchant, F. A., and Castleman, K. R., 2002. 'Strategies for automated fetal cell screening.' *Human Reproduction Update*, **8**, pp. 509–521.

Melin, J., and Quake, S. R., 2007. 'Microfluidic large-scale integration: The evolution of design rules for biological automation.' *Annual Review of Biophysics and Biomolecular Structure*, **36**, pp. 213–231.

Morgan, H., and Green, N., 2003. *AC. Electrokinetic: Colloids and Nanoparticles* (ISBN 0 86380 255 9). Research Studies Press Ltd., Baldock, Hertfordshire, UK.

Ohta, A. O., Chiou, P. Y., Han, H. T., Liao, J. C., Bhardwaj, U., McCabe, E. R. B., Yu, F., Sun, R., and Yu, M. C., 2007. 'Dynamic cell and microparticle control via optoelectronic tweezers.' *Journal of Microelectromechanical Systems*, **16**, pp. 491–499.

Ottino, J. M., 2003. 'Complex system.' *AIChE. Journal*, **49**, pp. 292–299.

Pallandre, A., de Lambert, B., Attia, R., Jonas, A. M., and Viovy, J., 2006. 'Surface treatment and characterization: Perspectives to electrophoresis and lab-on-chips.' *Electrophoresis*, **27**, pp. 584–610.

Piner, R. D., Zhu, J., Xu, F., Hong, S., and Mirkin, C. A., 1999. '*Dip-pen nanolithography.*' *Science*, **283**, pp. 661–663.

Pohl, H. A., 1978. *Dielectrophoresis.* Cambridge University Press, Cambridge, UK.

Probstein, R. F., 1994. *Physicochemical Hydrodynamics: An Introduction.* John Wiley & Sons, New York.

Ramos, A., Morgan, H., Green, N. G., and Castellanos, A., 1998. 'AC electrokinetics: A review of forces in microelectrode structures.' *Journal of Physical D: Applied Physics*, **31**, pp. 2338–2353.

Senaratne, W., Andruzzi, L., and Ober, C. K., 2005. 'Self-assembled monolayers and polymer brushes in biotechnology: Current applications and future perspectives.' *Biomacromolecules*, **6** (5), pp. 2427–2448.

Sims, C. E., and Allbritton, N. L., 2007. 'Analysis of single mammalian cells on-chip.' *Lab on a Chip*, **7**, pp. 423–440.

Stone, H. A., Stroock, A. D., and Ajdari, A., 2004. 'Engineering flows in small device, microfluidics toward a lab-on-a-chip.' *Annual Review of Fluid Mechanics*, **36**, pp. 381–411.

Stroock, A. D., Dertinger, S. K. W., Ajdari, A., Mezic, I., Stone, H. A., and Whitesides, G. M., 2002. 'Chaotic mixer for microchannels.' *Science*, **295**, pp. 647–654.

Sun, C. P., Usui, T., Yu, F., Al-Shyoukh, I., Shamma, J., Sun, R., and Ho, C. M., 2009. 'Integrative systems control approach for reactivating Kaposi's Sarcoma-associated herpesvirus (KSHV) with combinatory drugs.' *Integrative Biology*, **1**, pp. 123–130.

Treacy, J., Ebbesen, T. W., and Gibson, J. M., 1996. 'Exceptionally high Young's modulus observed for individual carbon nanotubes.' *Nature*, **381**, pp. 678–680.

Tsutsui, H., and Ho, C. M., 2009. 'Cell separation by non-inertial force fields in microfluidic systems.' *Mechanics Research Communications*, **36**, pp. 92–103.

Ulman, A., 1996. 'Formation and structure of self-assembled monolayers.' *Chemical Reviews*, **96**, pp. 1533–1554.

Ulmanella, U., and Ho, C. M. 2008. 'Molecular effects on boundary condition in micro/nanoliquid flows.' *Physics of Fluids*, **20**, 101512.

Unger, M. A., Chou, H. P., Thorsen, T., Scherer, A., and Quake, S. R., 2000. 'Monolithic microfabricated valves and pumps by multilayer soft lithography.' *Science*, **288**, pp. 113–116.

Vijayaraghavan, A., Blatt, S., Weissenberger, D., Oron-Carl, M., Hennrich, F., Gerthsen, D., Hahn, H., and Krupke, R., 2007. 'Ultra-large-scale directed assembly of single-walled carbon nanotube devices.' *Nano Letters*, **7**, pp. 1556–1560.

Washizu, M., and Kurosawa, O., 1990. Electrostatic manipulation of DNA in microfabricated structures.' *IEEE. Transactions on Industry Applications*, **26**, pp. 1165–1172.

Watson, J. D., and Crick, F. H. C., 1953. 'A structure for deoxyribose nucleic acid.' *Nature*, **171**, pp. 737–738.

Weng, G., Bhalla, U. S., and Iyengar, R., 1999. 'Complexity in biological signaling systems.' *Science*, **284**, pp. 92–96.

Wong, E. W., Sheehan, P. E., and Lieber, C. M., 1997. 'Nanobeam mechanics: Elasticity, strength, and toughness of nanorods and nanotubes.' *Science*, **277**, pp. 1971–1975.

Wong, P. K., Chen, C. Y., Wang, T. H., and Ho, C. M., 2004a. 'Electrokinetic bioprocessor for concentrating cells and molecules.' *Analytical Chemistry*, **76**, pp. 6908–6914.

Wong, P. K., Wang, J. T. H., Deval, J. H., and Ho, C. M., 2004b. 'Electrokinetics in micro devices for biotechnology applications.' *IEEE/ASME. Transactions on Mechatronics*, **9** (2), pp. 366–376.

Wong, P. K., Yu, F., Shahangian, A., Cheng, G., Sun, R., and Ho, C. M., 2008. 'Closed-loop control of cellular functions using combinatory drugs guided by a stochastic search algorithm.' *Proceeding of National Academy of Science*, **105** (13), pp. 5105–5110.

Wong, I. and Ho, C.M., 2009a. 'Surface molecular property modifications for poly (dimethylsiloxane) (PDMS) based microfluidic devices' *Microfluidics and Nanofluidics* **7**, pp. 291–306.

Wong, T. S., Brough, B., and Ho, C. M., 2009b. 'Creation of functional micro/ nano systems through top-down and bottom-up approaches.' *Molecular & Cellular Biomechanics*, **6**, pp. 1–56.

Wong, T. S., Huang, A. P. H., and Ho, C. M., 2009c. 'Wetting behaviors of individual nanostructures.' *Langmuir*, **25**, pp. 6599–6603.

Whitesides, G. M. *et al.*, 2001. 'Soft lithography in biology and biochemistry.' *Annual Review of Biomedical Engineering*, **3**, pp. 335–373.

Xia, Y. N., and Whitesides, G. M., 1998. 'Soft lithography.' *Angewandte Chemie International Edition*, **37**, pp. 550–75.

Xie, J. *et al.*, 2005. 'Microfluidic platform for liquid chromatography-tandem mass spectrometry analyses of complex peptide mixtures.' *Analytical Chemistry*, **77**, pp. 6947–6953.

Yang, X., Grosjean, C., Tai, Y. C., and Ho, C. M., 1998. 'A. MEMS thermopneumatic silicone rubber membrane valve.' *Sensors and Actuators A: Physical*, **64**, pp. 101–108.

Yang, X., Yang, J., Tai, Y. C., and Ho, C. M., 1999. 'Micromachined particle membrane filters.' *Sensors and Actuators A., Physical*, **73**, pp. 184–191.

Yi, C. Q. *et al.*, 2006. 'Microfluidics technology for manipulation and analysis of biological cells.' *Analytica Chimica Acta*, **560**, pp. 1–23.

2

MICRO/NANOFLUIDIC PROCESSES

Patrick Tabeling and Yi-Kuen Lee

This chapter describes a number of chemical engineering processes that have been successfully miniaturized and implemented on microsystems. We hope that this description will be useful for engineers or researchers willing to develop biochemical processes on chip, in particular for biomedical applications.

The chapter is dedicated to microsystems, since the vast majority of the devices existing to date, relevant to the biomedical field, have internal sizes comprised between tens and several hundred micrometers. There are several reasons for this. Although bottlenecks still exist, microscale technologies are reaching a stage where a substantial number of practical problems can be solved in a reliable and inexpensive way, facilitating both laboratory analysis and industrialization. A company willing to manufacture a micrometric biochip has several options for the materials and it will benefit from an important body of experience based on the technological know-how accumulated over the last ten years. On the other hand, the physics at the microscale has specific characteristics, now well described in review and textbooks (Ho and Tai 1998, Stone *et al.* 2004, Tabeling 2005, Bruus 2007), which facilitate fluid control, favor the reduction of the processing time along with considerably reducing the sample size—a feature which thus opens the route to high throughput screening- (Tabeling 2005, Bruus 2007). In contrast, only a few nanosystems dedicated to chemical processes exist and they raise formidable issues concerning the handling of liquids, the development of surface treatments, and the fabrication. The incentive of going to the nano scale is taking advantage of novel phenomena, absent at the micro scale (Eijkel and Van den Berg 2005, Tabeling 2005).

This area of research is wide open at the moment; nonetheless, as far as chemical engineering processes are concerned (which is the subject of the chapter), the number of nanofluidic systems demonstrating processes relevant to biomedical applications is still scarce. To provide the reader with a few examples, we may mention DNA sieves using nanopillars (Baba *et al.* 2003), DNA or virus separation nanodevices based on entropic trapping (Ilic *et al.* 2004; Mannion *et al.* 2006; Fu *et al.* 2006). Applications of nanofluidics have been reviewed in 2005 by J.C. Eijkel and A. Van den Berg. In this chapter, we will concentrate on microfluidic devices and will not describe nanofluidic systems.

Among the processes that can be of interest for biomedical applications, mixing, concentrating, and extracting are certainly important. Separation is also extremely important, but this topic deserves an entire chapter—it appears elsewhere in the book.

In a tentative classification of microfluidic systems, one may distinguish between continuous and discrete devices, i.e between systems where processes take place directly in microchannels or microchambers and systems where reactants are encapsulated and processed in microdroplets thus being promoted to the rank of microreactors. We will use this rough classification in the presentation of the subject. Historically continuous flow processes in microsystems have given rise to an important number of studies and devices over the last ten years, while studies on similar processes using droplet based microfluidics appeared over the last five years and are naturally less abundant. Similarly as in chemical engineering, the two approaches are not competitive but complementary. It is too early at the moment to assess the pros and cons of the two approaches. We hope that the present chapter will provide a few elements for a future review addressing the subject.

The chapter is thus organized as follows: in the first section we discuss continuous microflows, presenting techniques allowing for concentrating and mixing liquids along with extracting species initially diluted in a solvant. The second section is similar to the first, but with microdroplets. In this section, we describe concentration, mixing, and liquid liquid extraction techniques.

2.1 Continuous microflows

2.1.1 *Mixing*

Mixing is one of the most important micro/nanofluidic processes in biochemistry analysis, fine chemical production, polymer chain reaction (PCR), genotyping, sequencing, and synthesis of nucleic acids. This is because many biochemical processes require effective mixing to function properly. Mixing in conventional laboratories is usually performed by either manually shaking, tapping the test tube followed by centrifuging, or mixing using a mechanical or magnetic stirrer (Evensen *et al.* 1998). However, these conventional mixing methods are difficult, if not impossible, to implement in miniaturized systems. Therefore, the pursuit of novel mechanisms for effective mixing is critical for the development of bio-MEMS or lab-on-chip devices.

Traditionally, turbulent enhancement is the most effective process to achieve efficient mixing. For example, Regenfuss *et al.* (1985) and Bokenkamp *et al.* (1998) applied turbulent-jet and multiple T-shaped turbulent flows, respectively, to study fast chemical reactions at millimeter-sized scales. Nonetheless, it is much more difficult to generate turbulence at microscales due to the higher viscous dissipation at small Reynold's numbers (Raynal and Gence 1997). Due to the lack of turbulence in microfluidic devices, mixing largely depends on molecular diffusion. However, molecular diffusion time, especially for the mixing of large biomolecules, can be too large to complete mixing in a reasonable time frame.

Mixing by chaotic advection is considered to be one of the most promising techniques for mixing in microfluidic devices (Ottino 1989; Ottino and Wiggins 2004).

Aref was among the first researchers to use the term chaotic advection (Aref 1984) to describe mixing achieved in a laminar flow at a low Reynolds number by comparing this phenomenon with turbulent mixing (Aref 2002). It is well known that chaotic mixing can occur at any Reynolds number and that it is the degree of the spatial-temporal complexity of the flow that determines whether the advected particle paths are chaotic, not the force balance in the momentum equation (Jones and Aref 1988).

From a chaotic dynamics viewpoint, mixing by chaotic advection can be simplified into two basic flow components: the stretching and folding of fluid elements. For incompressible flow, the evolution of the fluid particle is governed by Navier-Stokes equations. The particle trajectory $\mathbf{x}(t) = (x(t), y(t), z(t))$ can be better described by a set of ordinary differential equations:

$$\dot{\mathbf{x}} = f(t, \mathbf{X})$$

where $\mathbf{X}$ is the initial condition of the fluid particle.
The mixing efficiency can be mathematically described as the time average of the line stretch, i.e. the ratio of the infinitesimal fluid line element after a long period of mixing to the initial line element.

$$\sigma(\mathbf{X}, \mathbf{M_i}) = \lim_{t\to\infty |d\mathbf{X}|\to 0} \left[\frac{1}{t} \ln\left(\frac{d\mathbf{x}}{d\mathbf{X}}\right)\right]$$

where $d\boldsymbol{x}$, $d\boldsymbol{X}$ are the differential line elements at time $t \to \infty$ and $t=0$, respectively. It has been shown that Lyapunov exponent is equivalent to the long time average of the specific rate of line stretch in the fluid (Ottino 1989):

$$\sigma(\mathbf{X}, \mathbf{M_i}) = \lim_{t\to\infty} \left(\frac{1}{t} \ln \lambda\, (\mathbf{X}, \mathbf{M_i}, t)\right)$$

Therefore, Lyapunov exponent can be used as a mixing index to evaluate the performance of a micromixer. Since most flow in microfluidic devices is laminar, the decrease of the *striation thickness* for the lamellar fluid structure is proportional to the exponential rate of decrease for the maximum Lyapunov exponent (Tang and Boozer 1999). The striation thickness decrease means that the interfacial area in fluid will increase dratically. Eventually, the mixing will be completed by molecular diffusion. Experimentally, it is very common to evaluate mixing by considering the intensity uniformity of the fluid flow inside the micromixer. The uniformity was quantified by calculating the experimental mixing index I_{mix} by considering the deviation of the pixel intensity C from its maximum intensity value in an image field:

$$I_{mix} = 1 - \frac{1}{\bar{C}}\sqrt{\frac{\sum (C - \bar{C})^2}{N}}$$

where N is the total number of pixels and $\overline{C}$ is the averaged pixel intensity of the concentration function.

Recently, a large number of different types of new micro mixers have been widely published in research journals and microsystem-related conferences, such as IEEE MEMS, MicroTAS (Micro Total Analysis System), IMRET (International Conferences on MicroREaction Technology), etc. Kakuta *et al.* gave an early review on micromixers developed before 2001. A detailed review of micromixers designed during recent years was undertaken by Nguyen and Wu (2005) and Hessel *et al.* (2005).

In general, micromixers can be classified into passive micromixers and active micromixers. Passive micromixers do not require external energy to generate secondary flow for mixing, but rely on pumping fluids in microchannels with different types of bifurcation (T-shaped mixer, Y-shaped mixer, etc.) and combination. Passive micromixers can be categorized into: parallel lamination (Ehrfeld *et al.* 1999; Kakuta *et al.* 2001); serial lamination (Branebjerg *et al.* 1996; He *et al.* 2001); injection; jet collision; hydrodynamic focusing (Knights *et al.* 1998; Drese, 2004); chaotic advection channels with different geometry (Stroock *et al.* 2002; Wang *et al.*, 2003), such as patterned groove on channel walls: slanted grooves, staggered herringbone and other derivative structures; zig-zag shaped; Tesla structure; and 2-D and 3-D serpentine channels (Liu *et al.* 2000). Typical strategies to achieve passive and active micromixing are illustrated in Fig. 2.1.

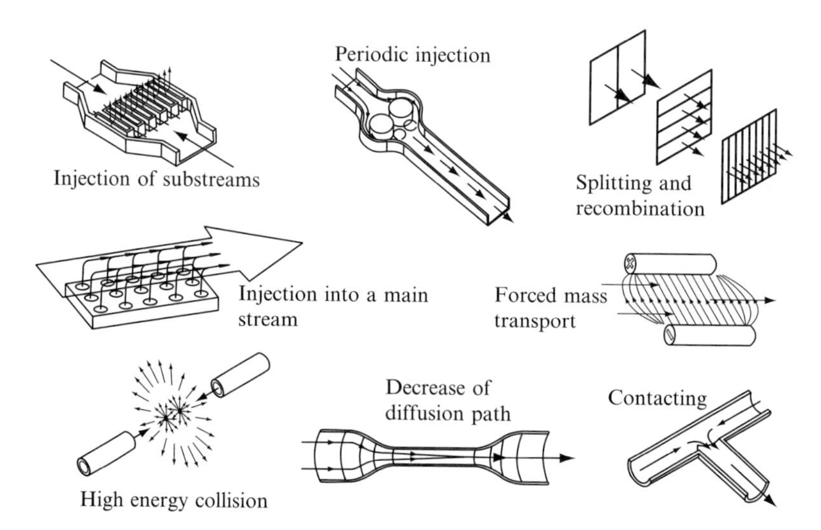

Fig. 2.1: Schematic diagrams of typical strategies to achieve passive and active micromixing. Hessel *et al.* (2005).

Active micromixers depend on external energy sources to induce secondary flow for mixing enhancement. The external energy sources for active micromixers reported in the literature include: periodic pressure forcing (D'Allessandro *et al.* 1999, Deshmukh *et al.* 2000, Lee *et al.* 2002, Okkels *et al.* 2004, Dodge *et al.* 2005), integrated micromixer/micropump (Voldman *et al.* 2000), piezoelectric (Woias *et al.* 2000), thermal bubble driven (Tsai and Lin, 2002), electrohydrodynamic, electroosmotic, electrokinetic instability driven, magnetohydrodynamic, acoustic, electrorheological, laser induced, ferrofluid based, dielectrophoretic, small impellers, and others.

It is instructive to mention in more detail two micromixers that we referred to above. The first micromixer functions using the stretching and folding of the material lines (Lee *et al.* 2002, Okkels *et al.* 2004, Dodge *et al.* 2005) This mixer consists of a principal canal, along which a stationary flow is imposed (see Fig. 2.2). This canal intersects adjacent canals. Along these canals, a non-stationary flow is imposed (for example, periodic with time). The two flows (principal and secondary) merge in the intersection region. We are now dealing with a case where the velocity field is not stationary, which implies that chaotic trajectories can form. It has been shown in this system that, at the intersection of the canals, a given material line is subjected to a succession of folding and stretching characteristic of chaotic regimes. This stretching and folding intervenes a finite number of times (one, two, or three times in the intersection, depending on the values of the parameters controlling the principal flow and the periodic perturbation). We observe that, despite the small amount of stretching and folding carried out at the intersection, the system mixes effectively (Fig. 2.2). In some range of parameters, the system also gives rise to unusual spatio temporal resonance phenomena that inhibit mixing and favor separation processes (Okkels *et al.* 2004, Dodge *et al.* 2005). If identical side channels are periodically placed along the streamwise direction of the main channel, the Poincare sections of the system at

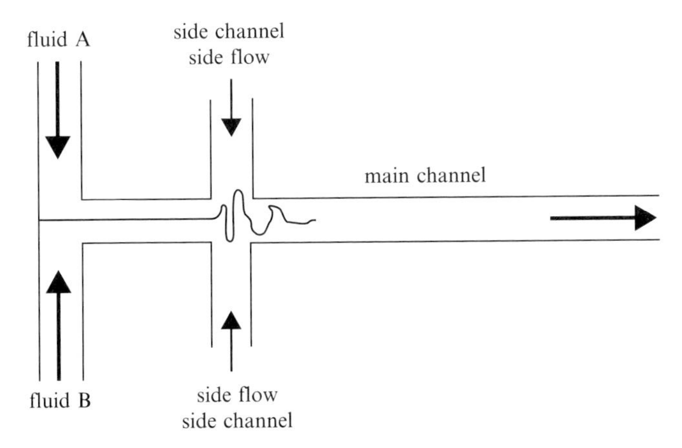

Fig. 2.2: Stretching/folding micromixer proposed by Lee *et al.* (2002).

certain perturbation amplitudes show Kolmogorov Arnold Moser (KAM) curves (Niu and Lee 2003). The route from the quasi-period to chaos is revealed to be destruction of KAM curves and shrinkage of the quasi-periodic areas. Lyapunov exponents (LE) can be used as the mixing index and the criteria to evaluate the chaotic behavior of the system. LE is closely related to the amplitude and frequency of side channel perturbation (Lee *et al.* 2007).

The second mixer, developed by Stroock *et al.* 2002 is a passive chaotic mixer. One obvious advantage is that it does not require an external source to perturb the flow and produce mixing (but we need some source of energy to drive the flow). The mixer consists of a canal along which a series of herringbone-shaped grooves have been placed. These hollows force the fluid to circulate obliquely with respect to the direction of the principal flow. Due to conservation of mass, return flows are developed. In the end, we obtain a helix-shaped movement of the fluid. Figure 2.3 represents such a system, which is made in PDMS using soft lithography. The herringbones do not form a homogenous motif along the canal. The pattern changes drastically every five herringbones. This translates, kinematically, to a displacement of the centers of the fluid helices. This variation of the helix structure along the canal can be seen as a time-dependent modification of the flow, passing from a stationary regime to a non-stationary regime, which is conducive to chaos. Chaos is created in Fig. 2.3. Here, the existence of chaos is revealed by a succession of folding and stretching of the dye placed in the system. The mixing length L_m was estimated in the article,

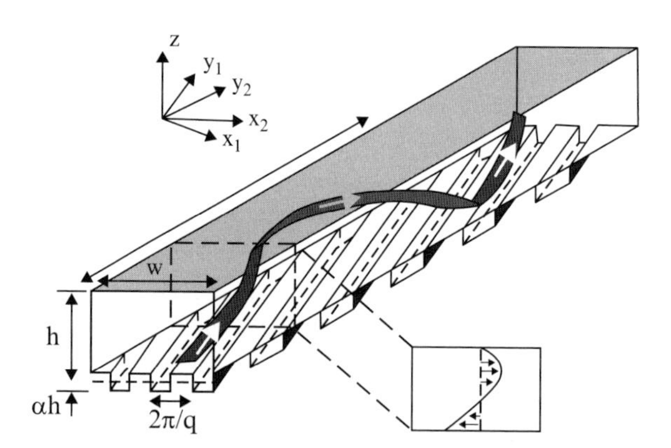

Fig. 2.3: Pressure-driven flow in a channel bearing a topographic pattern of oblique grooves on one wall. Close to this wall, the flow tends to orient along the grooves leading to a lateral motion to the left-wall, transverse recirculation then occurs in the middle of the channel (see the schematic representation of the transverse flow in a cross-section). The overall resulting flow pattern consists of helical flow lines, one of which is schematically depicted.

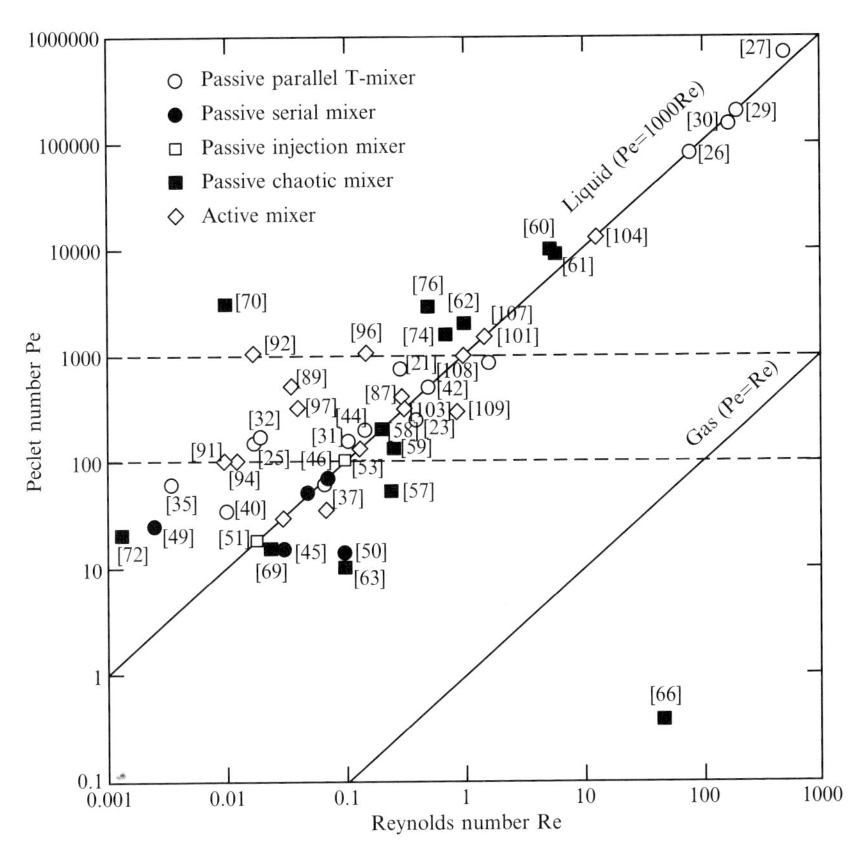

Fig. 2.4: The Pe-Re diagram of typical micromixers summarized by Nguyen and Wu (2005).

$$L_m \sim w \, \log(Pe)$$

where w is the channel width, and Pe is the Peclet number calculated for the width of the canal and the average flow. In practice, this estimation, confirmed by experiment, leads to mixing lengths several times larger than the distance w, for Peclet numbers on the order of 10^5 (Stroock *et al.* 2002)

Generally speaking, to design a proper micromixer for one specific application, there are two important key parameters to take into account: the Reynold's number (Re) and Peclet number (Pe). In most microfluidic devices, the Reynold's number is usually quite small because of the dominance of viscous force. On the other hand, the Peclet number is defined to be the ratio of the mass transport due to convection and that of molecular diffusion:

$$\text{Pe} = \frac{UL}{D}$$

where U is the average velocity of flow, L is a characteristic length of the micro mixer and D is the diffusion constant of the fluid particle of interest. The Peclet number can be used to determine whether mixing is convection-dominated or diffusion-dominated. In general, the Peclet number of gas is at least larger than that of liquid by three orders of magnitude due to the difference of their diffusion constants. Nguyen and Wu (2005) have summarized a large number of passive and active micromixers both in table format and in the Pe-Re diagram as shown in Fig. 2.4. Obviously, Re in most micromixers is less than 1.

In summary, a large number of micromixer research works (more than 250), including design, fabrication, and characterization, have been reported in the literature. This can provide a good foundation for practical applications in the biotechnology, pharmaceutical, and chemical engineering industries. One successful application of micro mixer is the mass production of nitroglycerin in the range of 10 kg per hour at a plant in Xi'an Huian Chemical, Xi'an, China with the help of collaborators from Germany's Institute for Microtechnology Mainz (Thayler 2005).

Although the previous micromixer review papers successfully compare several key design parameters of different types of micromixers, channel width, height, velocity, Re, Pe, and materials, one of the most important performance indices, mixing efficiency, is still missing. It is suggested that more future research should be focused on the development of computer-aided design tools, not simply CFD, for the prediction of mixing efficiency.

2.1.2 *Concentrating/diluting*

Concentrating an analyte prior to being detected and quantified is an important task in analytical chemistry in general and particularly when the sample is embedded in a microfluidic environment. Let us stress that for a nanomolar (per liter) solution—a concentration that is not particularly low—one cube of ten micron size hosts 600 molecules only. Concentrating the analyte obviously leads to increasing the signal over noise ratio and facilitates detection.

The number of specific microfluidic systems dedicated to concentrate and dilute is modest at the moment. Concentrating at the micro scale can be done in various ways: the most common approach is still based on the use of functionalized particles. We will not mention this technique, which would require an entire chapter. Typically the particles are injected and trapped on the chip, often close to a restriction, and the protocol is essentially the same as the one used classically. This approach can be illustrated by ELISA test, successfully miniaturized by several investigators (Sato *et al* 2003; Rossier *et al.* 2002). One issue is the handling of the particles, in particular the avoidance of clogging. We do not describe this topics here and instead mention two techniques which do not use particles: the first is based on geometry and the second is based on pervaporation.

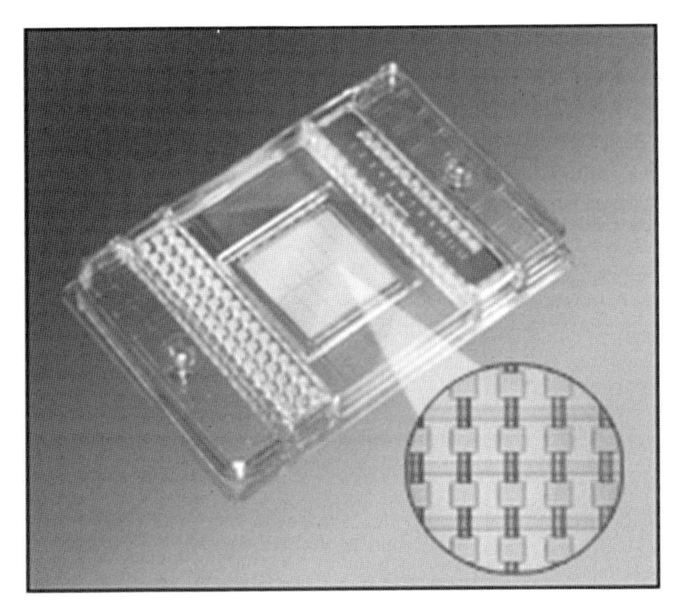

Fig. 2.5: Digital microarray produced by Fluidigm Corp., USA. It consists of integrated channels, chambers, and valves that can be used to detect somatic mutations in a high background of unmutated cells. After sample and reagent mixtures are transferred to wells on the input frame, the mixtures are automatically pressurized into their respective partitioning domains.

The first method, called 'digital microarray', is based on the remark that if one divides a sample that contains a small number of molecules of interest into a large number of microwells, obviously a few wells will contain one molecule or more and the others nothing. In the first set of microwells, the concentration is considerably enhanced compared to the initial sample.

Modern instrumentation uses this idea to perform detections of molecules present, in small concentrations, in a sample. Some systems are now available commercially. For example, Fig. 2.5 shows a device produced by Fluidigm Corp., USA, which concentrates DNA solutions by using a network of microwells. In this particular case, the fluids are manipulated on chip thanks to the use of integrated microvalves, a powerful innovation that requires the use of elastomeric materials (Unger *et al.* 2000).

A novel approach is based on pervaporation: the idea here is to bring a solution in a microwell separated from a microchannel by a membrane. In the microchannel, an air flow is driven. If the membrane is selectively permeable to the solvent, the solvent will migrate gently through the membrane, become exposed to the air flow and eventually evaporate. This process therefore works at concentrating the solute initially contained in the sample. A similar approach, using membranes

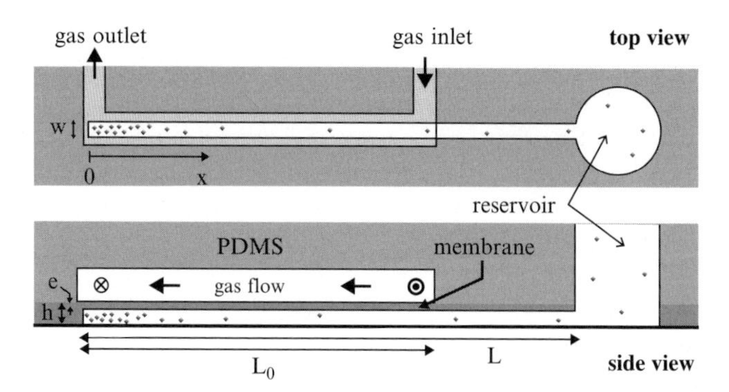

Fig. 2.6: Schematics of the experiment of Leng *et al.* (2006) (top and side views). A microchannel contains a solution; it is separated by a membrane from another microchannel in which an air flow is driven. If the membrane is selectively permeable to the solvant, the solvant will migrate gently through the membrane, become exposed to the air flow and eventually evaporates. This process therefore works at concentrating the solute initially contained in the sample.

again, but implemented differently, has also been developed (Foote *et al.* 2005). The concept is old—membranes have been used for a long time to concentrate solutions—but its microfluidic substantiation is new. In this topic, one may mention a work published recently, based on the properties of PDMS (Polydimethylsiloxane) (Leng *et al.* 2006) the most important material used in microfluidic research at the moment. PDMS is permeable to water and therefore it can be used for concentrating aqueous solutions. In the work of Leng *et al.* 2006, and represented in Fig. 2.6, pervaporation velocities are on the order of 20 nm/s for a 20 micron thick membrane. By this mechanism, the solution brought inside the microchannel concentrates. For a parallelepipedic chamber of size $h \times l \times L$ (h is the height, l the width and L the length) separated from an air channel by a membrane, the rate of increase of the solute concentration is given the following expression:

$$\frac{dC}{dt} = \frac{\rho v_e}{h}$$

where C is the aqueous solution concentration and ρ is the solute density. The expression shows that miniaturization speeds up the concentration process. In practice, it takes a few minutes to obtain a twofold increase in the solute concentration. This system is used to analyze complex systems, such as surfactant solutions.

Another system which performs a similar task with droplets was fabricated recently (Shim *et al.* 2007). As in the previous system, to perform reversible dialysis, the bottom of the wells are constructed from a thin PDMS membrane that is slightly permeable to water, but impermeable to a number of solutes, such as proteins and salts. The other side of the membrane contains a 100 nL reservoir, through which flows either dry air or an aqueous salt solution. Again, this produces a chemical potential gradient between the protein solution stored in the well and the reservoir. When dry air, or a concentrated salt solution, is introduced into the reservoir, water permeates from the protein drops through the membrane into the reservoir. In practice a drop in a well will shrink and completely evaporate in about one hour. If pure water is introduced in the reservoir, then the chemical potential gradient is reversed and water flows from the reservoir to the drop, thereby diluting each component as the drops swell. This system has been used for crystallizing proteins. Controlled protein crystallization requires nucleation of a germ followed by a sequence of concentration and dilution steps. It is interesting to note that this task was successfully performed with microfluidic technology, at the single drop level.

2.1.3 *Liquid liquid extraction*

Liquid liquid extraction is commonly used in chemical engineering with important applications in the field of environmental sciences. The liquid liquid extraction method is based on the exploitation of a solubility contrast between two liquids. The idea is to bring the sample in contact with an immiscible liquid, and wait for equilibrium to be completed. If the partition factor is favorable most of the solute will be transferred into the second phase at the equilibrium state. This process was miniaturized by Kitamori's group in the last few years (Tokeshi *et al.* 2000). The sketch of the device that was made is shown in Fig. 2.7. The two liquids—one

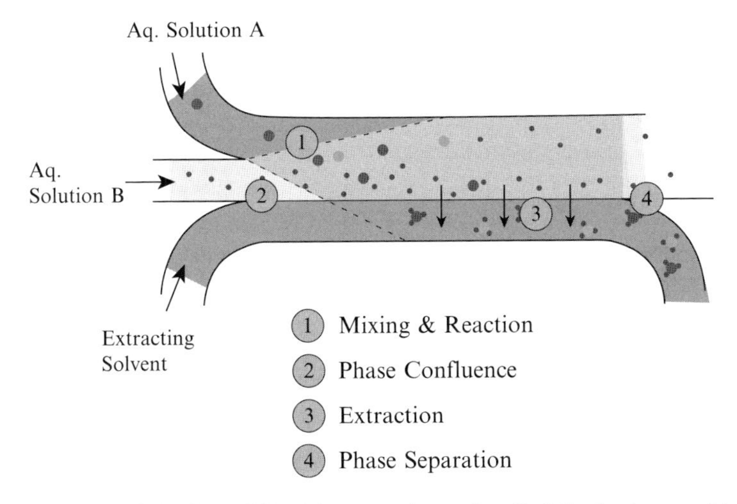

Fig. 2.7: Sketch of the liquid liquid extraction microfluidic device used by Tokeshi *et al.* (2000).

including the analyte, the other being the extracting phase—are driven into a microfluidic system into a Y junction; after merging, they flow side by side and separate again at a second Y junction. One of the two phases can be gaseous When the fluids at hand are immiscible it is necessary to pattern the surfaces exposed to the fluids in the active region so as to stabilize the interface between them. If the extracting liquid is appropriately chosen, a substantial fraction of the analyte will be transferred from the sample to the extracting phase during the time the two liquids are flowing side by side. As the two liquids are brought along each other, boundary layers develop along each side of the interface, one enriched, the other depleted; their thicknesses are given by the solution to the so called Leveque problem. Ideally, the length of the main section is such that the two boundary layers occupy the entire width of the main section. This 'optimal' length is given by:

$$L_0 = \frac{Uw^2}{D}$$

where w is the channel width in the main part, and D is the smaller coefficient of diffusion of the analyte of the two liquids. If this condition is satisfied, equilibrium is reached just before the two streams separate. The device thus allows for extracting a part equal to $1/(1+k)$ of the analyte initially present in the sample, where k is the partition coefficient. In practice, it is important to work with small width w so as the length L_o remains in acceptable ranges, i.e less than one centimeter or so. As previously mentioned a delicate issue is the stability of the interface between the two fluids, which requires using *in situ* surface treatments, as developed in Kitamori's group.

2.1.4 *Distributing concentration levels in many wells*

It is common to desire screening solutions hosting physico-chemical processes at different concentrations. In order to perform such a task, one needs to distribute the analytes and the solvant among different wells so that each of them contains a specific composition. Several devices dedicated to this task have been proposed over the last few years. A passive system was devised in 2002 in Whitesides' group (Jeon *et al.* 2002). The chip is shown in Fig. 2.8. Two inlets feed the system with two buffers, one including an analyte at a certain concentration and the other not. The pyramidal branch array of microchannels serves to split, combine and mix fluid streams as they flow through the network of microfluidic channels. Each resulting microchannel contains a different proportion of the analyte. The system eventually feeds different chambers at different concentrations that vary stepwise from one chamber to the other.

Another system, including active elements (valves), was devised in Quake's group (Thordsen *et al.* 2002). The system operates as a multiplexer. With several tens of valves, it allows hundreds of microchambers. This system is used to screen

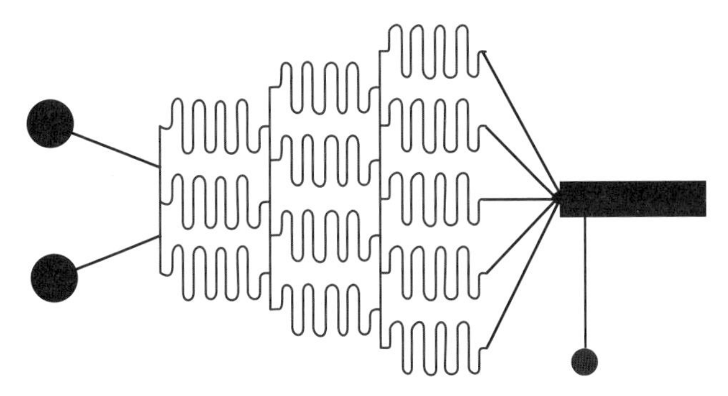

FIG. 2.8: System allowing five chambers (on the right side) incorporating solutions at five different concentrations. The system is fed by the two entries shown on the left side (Jeon *et al.* 2002).

the physico-chemical properties of complex mixtures. An example is the determination of 'optimum' conditions for the crystallization of proteins.

2.2 Processes using droplet based microfluidics (digital microfluidics)

2.2.1 *Screening and mixing*

A digital microfluidic system dedicated to carrying out microreactions typically includes several steps that are well illustrated in Fig. 2.9. This cartoon formed the cover page of an issue of *Angewandte Chimie* (Song *et al.* 2003); it is now often used to discuss the potentialities of digital microfluidics, in particular for combinatorial analysis. Another form of digital microfluidics, where droplets are driven over a plate by electrowetting/dielectrophoresis phenomena can also be used to carry out chemical processes. We will not cover this aspect here and refer the reader to the chapter written by C.J. Kim *et al.* in this volume. We concentrate here on digital microfluidic systems in microchannels, i.e. where droplets or bubbles are conveyed in a liquid continous phase in a network of microchannels. We refer again to Fig. 2.9 to describe in more detail this type of system. In this figure, droplets are formed at a T-junction. Each droplet is filled with reactants at different concentrations. The filling method is based on the particular design of the droplet emitter. This emitter is fed by three liquids of different natures. Each separate flow-rate varies in time in such a way that the total flow-rate is maintained at a constant. In such circumstances, provided that the overall viscosity and the interfacial tension of the reactant mixture does not significantly change with the composition, the hydrodynamic conditions that control the droplet formation process remain the same and the droplet characteristics do not change with time. One then has a

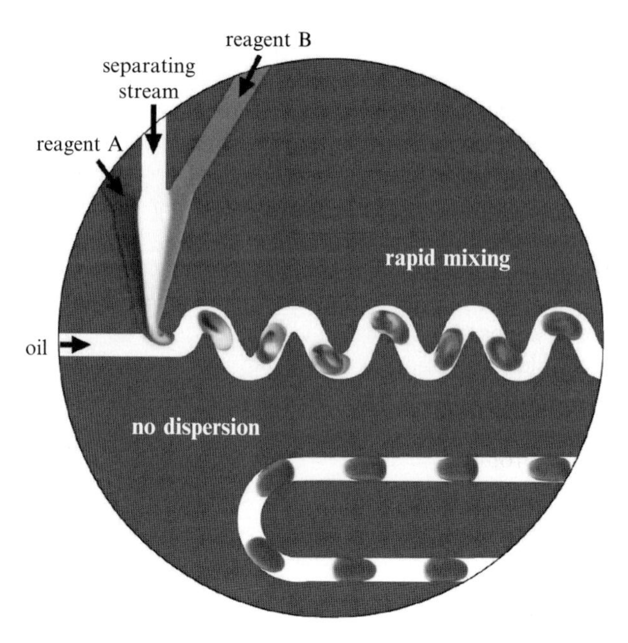

FIG. 2.9: A typical droplet-based microfluidic system to carry out microreactions (Song *et al.* 2003).

system which produces similar droplets incorporating mixtures with different compositions. As the droplet is formed, they move downstream, and internal recirculations develop; these recirculations are unavoidable. They result from the influence of the walls, which force the fluid layers located close to them to flow at lower speed. As a consequence, in order to ensure mass conservation, internal recirculations must develop. These recirculations are extremely useful for reducing the mixing time of the reactants within each droplet. Experiments indicate that the resulting mixing time lies in the millisecond range (Song *et al.* 2003). Nonetheless, the physical group of parameters that control this process is still unclear. According to current dispersion theories the mixing time should be on the order of a fraction of L^2/D (where L is the droplet size and D is the diffusion coefficient of the solute in the solution) while experiments and numerical simulations indicate a dependence with the velocity (Sarrazin *et al.* 2006). Whether there is a velocity dependence or not, it is clear that digital microfluidics offers a context where reactants can be isolated in microdroplets, which thus play the role of isolated microreactors, with mixing times in the millisecond range and volumes in the nanoliter range. Moreover, as pointed out in Song *et al.* 2003, fast screening of chemical reactions becomes feasible (in such systems, emission droplets frequencies are typically kHz). Droplet based microfluidics is therefore a formidable tool which has the potential of revolutionizing traditional approaches developed in biology and chemistry. The future will tell whether this potentiality will be substantiated.

2.2.2 *Extracting*

We concentrate here on the process of droplet extraction/purification in microfluidic systems. The process is exceedingly important in chemical engineering and its microfluidic version should have great promise owing to the large surface/volume ratios achieved in miniaturized devices. Concerning microfluidic droplet-based extraction, we are just at the beginning and at the moment there is a poor documentation on the topics. Here we present a recent contribution that provides a first analysis of the physical process that drives the exchanges between the droplets and the external phase (Mary *et al.* 2007, 2008). We are still far from the end of the story since the analysis is carried out in a restricted framework (essentially the droplets are assumed to occupy the entire width of the microchannels) and perhaps more importantly, there is no chemical process involved. The experiments of Mary *et al.* (2007, 2008) are sketched in Fig. 2.10. Water droplets are formed in a hydrofocusing geometry, with octanol as the external phase. Droplets move downstream along either straight or zigzag PDMS microchannels. The materials transferred across the droplet interfaces are fluorescein or rhodamine, whose partition coefficients—measured independently—are 0.01 and 55 respectively. The tracer is initially introduced either into the dispersed phase (we then refer to purification) or into the external phase (extraction). The measurements are performed using a sensitive camera.

Figure 2.11 shows how droplets exchange matter with the continuous phase (octanol) as they move downstream. In the first rapid phase, internal recirculations homogenize the material along the streamlines (Step I). In the second phase (Step II), and much slower, the droplet completes the homogenization of the tracer; this process is driven by diffusion.

How long does it take to complete the extraction process? Measurements of the time it takes to reach equilibrium provide estimates on the order of fractions of seconds. This is much faster than standard technologies. It is interesting to note

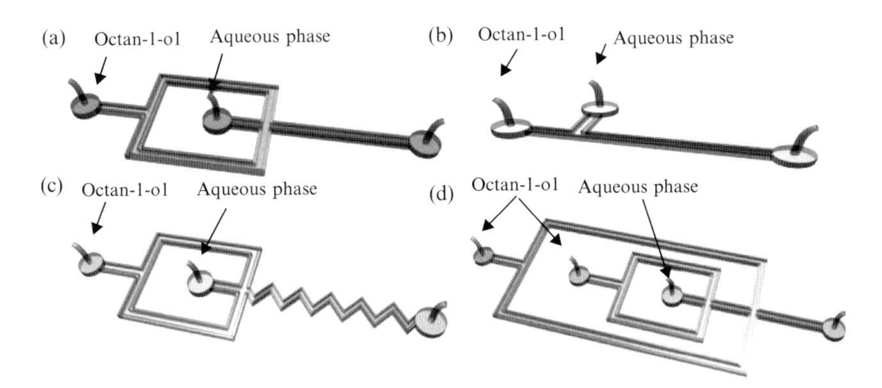

Fig. 2.10: Schematic views of the microfluidic chips used in Mary *et al.* (2008). (a) Hydrofocusing device. (b) T geometry device. (c) Flow-focusing geometry device with winding channels. (d) Double flow-focusing geometry device.

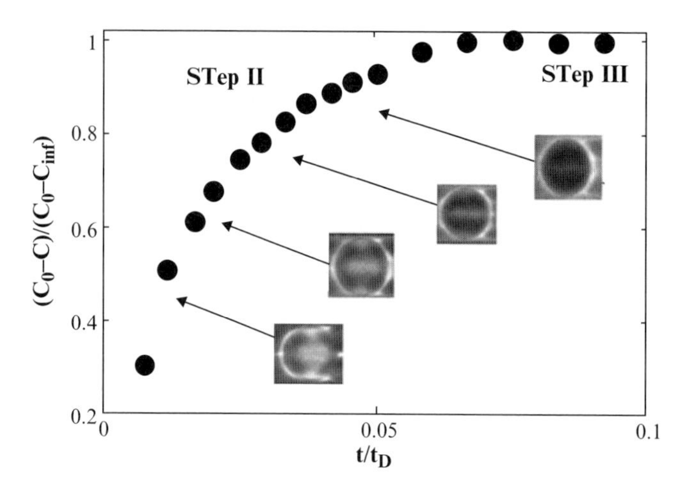

FIG. 2.11: Intensity of fluorescence within the droplets (water/glycerol mixture, 60 wt% in glycerol) as a function of time for a droplet, 185 μm in diameter, moving at 8 mm/s in a straight channel, with height, $h = 70\,\mu$m. The droplet gradually extracts fluorescein from the external phase as it moves downstream; it eventually reaches a steady state for which an equilibrium, determined by the partition factor between the dispersed and continous phase, is achieved.

that naive arguments indicate that, similarly as in the mixing problem mentioned above, this time should be essentially diffusive and thus be independent of the droplet speed. Surprisingly, experiment reveal an unescapable dependence of this time with the droplet velocity. The physical origin of this feature is quite subtle; it is detailed in Mary *et al.* 2008.

2.2.3 *Storing droplets*

Droplets encapsulating reactants must be stored on the chip when the kinetics of the reaction is slow, which often happens in biochemistry. An example is protein crystallization, which often requires times in the range of hours or tens of hours to be achieved. In this context, S. Fraden's group has proposed a novel chip, called 'phase chip', which exploit surface tension forces to guide each drop to a storage chamber (Shim *et al.* 2007). After the drops are formed, they are confined in a flow channel, which has a rectangular cross-section. The device is designed such that the channels flatten and elongate the drop. Wells, located to the side of the channel, are deeper than the flow channel. A drop in a well can adopt a spherical shape, minimizing its surface area and its surface energy. A drop that partially occupies both a channel and well will experience a gradient in surface energy, with the resulting force acting to drive and store the drop inside the well. As the wells exist as pockets on the sides of the channel, the enclosed, stored droplets are outside the flow stream and shielded from dislodgment by hydrodynamic forces.

Drops sequentially fill the wells, with the first drop going into the first well. Subsequent drops pass over all filled wells, entering the first empty well. To prevent coalescence of the drops during the loading process, surfactants must be added to the bulk.

2.3 Conclusion

As a conclusion, one may say that microfluidic technology allows us to invent formidable tools for chemical/biological analysis and production. Microfluidic technology renews in depth well established methodologies in chemistry and biology. The current advantages brought by this technology are low sample consumption, cost reduction, possibility of integration, and parallelization. On top of this, the physics of miniaturization ensures that the time it takes for the processes to complete are considerably reduced in comparison to traditional approaches. The future will tell to what extent microfluidic technology will impact on domains such as biomedical science. We may probably expect that the impact will be considerable.

2.4 Acknowledgments

This work is partially supported by the Hong Kong Research Grants Council (Project Reference No. 616106) and by a grant from the PROCORE France/ Hong Kong Joint Research Scheme (Reference No. F-HK21/07T-II).

References

Aref, H., 1984. 'Stirring by chaotic advection.' *Journal of Fluid Mechanics*, **143**, pp. 1–21.

Aref, H., 2002. 'The development of chaotic advection.' *Physics of Fluids*, **14** (4), pp. 1315–1325.

Baba, M. *et al.*, 2003. 'DNA size separation using artificially nanostructured matrix.' *Applied Physics Letters*, **83** (7), pp. 1468–1470.

Bokenkamp, D. *et al.*, 1998. 'Microfabricated silicon mixers for submillisecond quench-flow analysis.' *Analytical Chemistry*, **70** (2), pp. 232–236.

Branebjerg, J. *et al.*, 1996. 'Fast mixing by lamination.' *The 9th Annual International Workshop on MEMS* (IEE. M.,EMS 96), pp. 441–446. IEEE, San Diego.

Bruus, H., 2007. *Theoretical Microfluidics*. Oxford University Press, Oxford.

D'Alessandro, D., Dahleh, M., and Mezic, I., 1999. 'Control of mixing in fluid flow: A maximum entropy approach.' *IEEE Transactions on Automatic Control*, **44** (10), pp. 1852–1863.

Deshmukh, A. A., Liepmann, D., and Pisano, A. P., 2000. 'Continuous micromixer with pulsatile micropumps.' *Technical Digest of the IEEE Solid State Sensor and Actuator Workshop*, pp. 73–76. Hilton Head Island, SC.

Dodge, A., *et al.*, 2004. 'An example of a chaotic micromixer: The cross-channel micromixer.' *Comptes Rendus Physique*, **5** (5), pp. 557–563.

Dodge, A., *et al.*, 2005. 'Spatiotemporal resonances in a microfluidic system.' *Physical Review E*, **72**, p. 056312.

Drese, K. S., 2004. 'Optimization of interdigital micromixers via analytical modeling – exemplified with the SuperFocus mixer.' *Chemical Engineering Journal*, **101** (1–3), pp. 403–407.

Ehrfeld, W. *et al.*, 1999. 'Characterization of mixing in micromixers by a test reaction: Single mixing units and mixer arrays.' *Industrial & Engineering Chemistry Research*, **38** (3), pp. 1075–1082.

Eijkel, J. C. T., and Van den Berg, A., 2005. 'Nanofluidics: What is it and what can we expect from it?' *Microfluidics and Nanofluidics*, **1** (3), pp. 249–267.

Evensen, H. T., Meldrum, D. R., and Cunningham, D. L., 1998. 'Automated fluid mixing in glass capillaries.' *Review of Scientific Instruments*, **69** (2), pp. 519–526.

Foote, R. S. *et al.*, 2005. 'Preconcentration of proteins on microfluidic devices using porous silica membranes.' *Analytical Chemistry*, **77** (1), pp. 57–63.

Fu, J., Yoo, J., and Han, J., 2006. 'Molecular Sieving in Periodic Free-Energy Landscapes Created by Patterned Nanofilter Arrays.' *Physical Review Letters*, **97**, pp. 018103–4.

He, B. *et al.*, 2001. 'A picoliter-volume mixer for microfluidic analytical systems.' *Analytical Chemistry*, **73** (9), pp. 1942–1947.

Hessel, V., Lowe, H., and Schonfeld, F., 2005. 'Micromixers: A review on passive and active mixing principles.' *Chemical Engineering Science*, **60** (8–9), pp. 2479–2501.

Ho, C. M., and Tai, Y. C., 1998. 'Micro-electro-mechanical-systems (MEMS) and fluid flows.' *Annual Review of Fluid Mechanics*, **30** (1), pp. 579–612.

Ilic, B., Yang, Y., and Craighead, H. G., 2004. 'Virus detection using nanoelectromechanical devices.' *Applied Physics Letters*, **85** (13), pp. 2604–2606.

Jeon, N. L. *et al.*, 2002. 'Neutrophil chemotaxis in linear and complex gradients of interleukin-8 formed in a microfabricated device.' *Nature Biotechnology*, **20** (8), pp. 826–830.

Jones, S. W., and Aref, H., 1988. 'Chaotic advection in pulsed-source sink systems.' *Physics of Fluids*, **31** (3), pp. 469–485.

Kakuta, M., Bessoth, F. G., and Manz, A., 2001. 'Microfabricated devices for fluid mixing and their application for chemical synthesis.' *Chemical Record*, **1** (5), pp. 395–405.

Knight, J. B. *et al.*, 1998. 'Hydrodynamic focusing on a silicon chip: Mixing nanoliters in microseconds.' *Physical Review Letters*, **80** (17), pp. 3863–3866.

Lee, Y. K., Tabeling, P., Shih, C., and Ho, C. M., 2002. 'Characterization of a MEMS-fabricated mixing device.' *Proceedings of MEMS, ASME International Mechanical Engineering Congress and Exposition*, pp. 505–511. Orlando, FL.

Lee, Y. K., Tabeling, P., Shih, C., and Ho, C. M., 2007. 'Experimental study and nonlinear dynamic analysis of time-periodic micro chaotic mixers.' *Journal of Fluid Mechanics*, **575**, pp. 425–448.

Leng, J., Lonetti, B., Tabeling, P., Joanicot, M., and Ajdari, A., 2006. 'Micro-evaporators for kinetic exploration of phase diagrams.' *Physical Review Letters*, **96** (8), pp. 84503–84504.

Liu, R. H. *et al.*, 2000. 'Passive mixing in a three-dimensional serpentine microchannel.' *Journal of Microelectromechanical Systems*, **9** (2), pp. 190–197.

Mannion, J. T. *et al.*, 2006. 'Conformational analysis of single DNA molecules undergoing entropically induced motion in nanochannels.' *Biophysical Journal*, **90** (12), pp. 4538–4545.

Mary, P., Studer, V., and Tabeling, P., 2007. 'Microfluidic quantitative extraction in droplets.' *The 11th International Conference on Miniaturized Systems for Chemistry and Life Sciences.* MicroTAS 2007, Paris. The Chemical and Biological Microsystems Society.

Mary, P., Studer, V., and Tabeling, P., 2008. 'Microfluidic droplet-based liquid-liquid extraction.' *Analytical Chemistry*, **80** (8), pp. 2680–2687.

Nguyen, N. T., and Wu, Z. G., 2005. 'Micromixers – a review.' *Journal of Micromechanics and Microengineering*, **15** (2), pp. R1–R16.

Niu, X. Z., and Lee, Y. K., 2003. 'Efficient spatial-temporal chaotic mixing in microchannels.' *Journal of Micromechanics and Microengineering*, **13** (3), pp. 454–462.

Okkels, F., and Tabeling, P., 2004. 'Spatiotemporal resonances in mixing of open viscous fluids.' *Physical Review Letters*, **92** (3), p. 038301.

Ottino, J. M., 1989. *The Kinematics of Mixing: Stretching, Chaos, and Transport.* Cambridge University Press, Cambridge.

Ottino, J. M., and Wiggins, S., 2004. 'Introduction: Mixing in microfluidics.' *Philosophical Transactions of the Royal Society A: Mathematical, Physical and Engineering Sciences*, **362** (1818), pp. 923–935.

Raynal, F., and Gence, J. N., 1997. 'Energy saving in chaotic laminar mixing.' *International Journal of Heat and Mass Transfer*, **40** (14), pp. 3267–3273.

Regenfuss, P. *et al.*, 1985. 'Mixing liquids in microseconds.' *Review of Scientific Instruments*, **56** (2), pp. 283–290.

Rossier, J., Reymond, F., and Michel, P. E., 2002. 'Polymer microfluidic chips for electrochemical and biochemical analyses.' *Electrophoresis*, **23** (6), pp. 858–867.

Sarrazin, F. *et al.*, 2006. 'Experimental and numerical study of droplets hydrodynamics in microchannels.' *AIChE Journal*, **52** (12), pp. 4061–4070.

Sato, K. *et al.*, 2003. 'Microchip-based chemical and biochemical analysis systems.' *Advanced Drug Delivery Reviews*, **55** (3), pp. 379–391.

Shim, J. U. *et al.*, 2007. 'Control and measurement of the phase behavior of aqueous solutions using microfluidics.' *Journal of the American Chemical Society*, **129** (28), pp. 8825–8835.

Song, H., Tice, J. D., and Ismagilov, R. F., 2003. 'A microfluidic system for controlling reaction networks in time.' *Angewandte Chemie International Edition*, **42** (7), pp. 768–772.

Stone, H. A., Stroock, A. D., and Ajdari, A., 2004. 'Engineering flows in small device: Microfluidics toward a lab-on-a-chip.' *Annual Review of Fluid Mechanics*, **36** (1), pp. 381–411.

Stroock, A. D. *et al.*, 2002. 'Chaotic mixer for microchannels.' *Science*, **295** (5555), pp. 647–651.

Tabeling, P., 2005. *Introduction to Microfluidics.* Oxford University Press, Oxford.

Tabeling, P. *et al.*, 2004. 'Chaotic mixing in cross-channel micromixers.' *Philosophical Transactions of the Royal Society of London Series A-Mathematical Physical and Engineering Sciences*, **362** (1818), pp. 987–1000.

Tang, X. Z., and Boozer, A. H., 1996. 'Finite time Lyapunov exponent and advection-diffusion equation.' *Physica D*, **95** (3–4), pp. 283–305.

Thayler, A. M., 2005. 'Harnessing microreactions.' *Chemical and Engineering News*, **83** (22), p. 3.

Thorsen, T., Maerkl, S. J., and Quake, S. R., 2002. 'Microfluidic Large-Scale Integration.' *Science*, **298**, pp. 580–584.

Tokeshi, M., Minagawa, T., and Kitamori, T., 2000. 'Integration of a microextraction system on a glass chip: Ion-pair solvent extraction of Fe(II) with 4,7-diphenyl-1,10-phenanthrolinedisulfonic acid and tri-n-octylmethylammonium chloride.' *Analytical Chemistry*, **72** (7), pp. 1711–1714.

Tsai, J. H., and Lin, L., 2002. 'Active microfluidic mixer and gas bubble filter driven by thermal bubble micropump.' *Sensors and Actuators A-Physical*, **97–98**, pp. 665–671.

Unger, M. A. *et al.*, 2000. 'Monolithic microfabricated valves and pumps by multilayer soft lithography.' *Science*, **288** (5463), pp. 113–116.

Voldman, J., Gray, M. L., and Schmidt, M. A., 2000. 'An integrated liquid mixer/valve.' *Journal of Microelectromechanical Systems*, **9**, pp. 295–302.

Wang, H. Z. *et al.*, 2003. 'Numerical investigation of mixing in microchannels with patterned grooves.' *Journal of Micromechanics and Microengineering*, **13** (6), pp. 801–808.

Woias, P., Hauser, K., and Yacoub-George, E., 2000. 'An active silicon micromixer for mTAS applications,' in The 11th International Conference on Miniaturized Systems for Chemistry and Life Sciences (MicroTAS 2000), Enschede, Netherlands.

Yamada, M. *et al.*, 2006. 'A microfluidic flow distributor generating stepwise concentrations for high-throughput biochemical processing.' *Lab on a Chip*, **6** (2), pp. 179–184.

3

NUMERICAL SIMULATION OF MICROFLOWS AND NANOFLOWS

Narayan R. Aluru and George Em Karniadakis

Computational fluid dynamics (CFD) has enabled revolutionary advances in many areas of classical fluids for several decades now. Traditional CFD methods, however, find limited applications in emerging areas of biological micro and nanofluids. A number of computational techniques ranging from quantum to molecular to mesoscopic methods have been developed to understand fundamental issues governing fluid mechanics at small scales. In addition, to deal with the disparate length and time scales involved in biofluids, the development of efficient and accurate multiscale methods becomes critical. In this chapter we review recent advances in numerical simulation of micro and nanoflows. For coarse-grained simulation of microfluidics, we present an overview of Lattice Boltzmann, Brownian dynamics, stochastic rotation dynamics, and smoothed particle hydrodynamics methods and discuss the dissipative particle dynamics method in detail as it shares many features with the other methods. In the area of nanoflows, we review recent advances in non-equilibrium molecular dynamics methods focusing on the development of self-consistent and grand canonical methods for electric-field mediated transport. We present examples showing the significance of quantum effects in nanoflows. Finally, we discuss multiscale modeling focusing on direct coupling of molecular dynamics with Navier–Stokes equations and hierarchical coupling of quantum, molecular dynamics, and classical fluid equations.

3.1 Introduction

Modeling biological fluids in micro- and nanodomains requires new approaches beyond the traditional CFD methods. There are new phenomena to capture at the submicron scale, including quantum effects, but the main computational challenge in biological systems is the disparate range of spatio-temporal scales. In particular, atomistic methods such as non-equilibrium molecular dynamics (MD) simulation as well as mesoscopic methods, such as Lattice Boltzmann and similar approaches, are required to deal with active particulates (e.g. cells) in the flow or active wall surfaces in biological devices (e.g. protein coated surfaces). More often, it is the coupling of atomistic, mesoscopic, and continuum methods which is required in emerging biological applications, the so-called *multiscale modeling* of such systems, to bridge the aforementioned gap in scales. A detailed presentation of

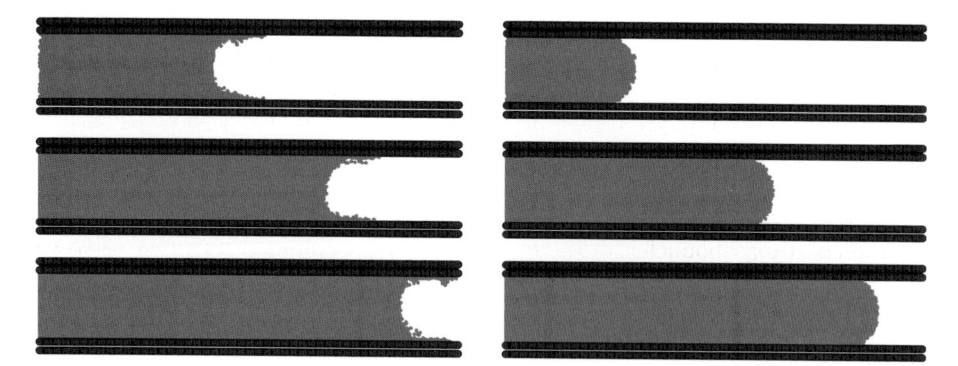

Fig. 3.1: Microchannel flow at different times with (Left) hydrophilic walls and (Right) hydrophobic walls. A multibody dissipative particle dynamics (DPD) method is employed in the simulations; see Section 3.2.5. (Courtesy of Wenxiao Pan).

suitable methods is given in the book by the authors, see (Karniadakis, *et al.* 2005). Here, we focus our attention on a review and recent developments of mesoscopic methods for microfluidics and on new molecular dynamics formulations suitable for charged nanoflows; a review and more references on particle-based methods can also be found in Koumoutsakos (2005).

Wetting of the solid surfaces by liquids is important in applications with walls which are hydrophobic or hydrophilic, see an example in Fig. 3.1. While there exist macroscopic descriptions of wetting, they do not fully incorporate effects such as hysteresis, layering, and monolayer spreading. Hysteretic effects can have a dramatic effect on the value of the contact angle and the strength of surface tension. In modeling wetting phenomena it is also important to incorporate the possible deformation of the solid in wetting processes, since local regions of high stress can produce distortions of the elastic solid (Shanahan, 1988), e.g. the elastomeric walls used in some patterned drug delivery applications.

Fluid flows in nanometer scale channels and pores, referred to as *nanoflows*, play an important role in determining the functional characteristics of many biological and engineering devices and systems. For example, ionic channels are naturally occurring nanotubes found in the membranes of all biological cells (Hille 2001). They are defined as a class of proteins, and each channel consists of a chain of amino acids folded in such a way that the protein forms a nanoscopic water-filled pore controlling the transport of ions in and out of the cell, and in and out of compartments inside cells like mitochondria and nuclei, thereby maintaining the correct internal ion composition that is crucial for cell survival and function. Each channel carries a strong and steeply varying distribution of permanent charge, which depends on the particular combination of channel and prevalent physiological conditions. The narrowest diameter of an ion channel can vary from a few angstroms to tens of angstroms. Many ion channels have the ability to selectively transmit or block a particular ion species, and most exhibit switching properties similar to electronic devices. Malfunctioning channels cause or are associated with many diseases, and a large number of drugs act directly or indirectly on ion channels. Given the physiological importance of ion channels, it is important to understand the flow of water and electrolytes in naturally existing nanoscopic pores in the presence of a strong permanent charge.

Another application in which modeling of nanoflows is required is the translocation of deoxyribonucleic acid (DNA) through a nanopore. Functional analysis of the genome will require sequencing the DNA of many organisms, and nanometer-scale pores are being explored for DNA sequencing and analysis. The characteristic that makes nanopores useful for analysis of DNA or other individual macromolecules is that the scale of the pores is the same as that of the molecules of interest. There are a number of other applications such as drug delivery systems, chemical and biological sensing, and energy conversion devices, and a number of other nanodevices in which nanoflows are important and need to be modeled in sufficient detail.

In this chapter we present first *particle-based* approaches with an emphasis on methods that are suitable in simulating complex biological flows, such as the dissipative particle dynamics (DPD), stochastic rotation dynamics (SRD), smoothed particle hydrodynamics (SPH), Brownian dynamics (BD), and Lattice Boltzmann method (LBM). All these methods, except LBM, follow a Lagrangian description and hence suspensions of particles, cells, or DNA can be modeled relatively easily. They are mesoscopic methods, which implies that they can model accurately spatial scales above 10 nm to 100 nm. They also exhibit hydrodynamic behavior (except the basic BD method) and can be used in larger domains as well and at finite Reynolds numbers, or alternatively they can be interfaced with continuum-based approaches (e.g. the Navier–Stokes equations) to extend their applicability. This important subject involving multiscaling is presented in the third part of this chapter.

In the second part of the chapter, we first present advances in non-equilibrium molecular dynamics (MD) simulations. MD simulations are routinely used to understand fundamental issues governing structure and transport of fluids in nanometer channels or pores. External electric fields are used to transport electrolytes and biological macromolecules through nanopores for various applications. An outstanding issue in MD simulations is the treatment of the external electric field. We present a self-consistent MD method that can be used to accurately describe transport and selectivity of ions through nanopores. Second, the significance of quantum-mechanical effects in nanoflows is not clear. We present recent results on carbon nanotubes where quantum effects on water structure and transport are clearly seen. Finally, there is a growing interest towards discovering new materials that cannot only be effective transporters of water, but can also closely mimic the structure and dynamics of water observed in biological channels. It was recently observed that boron nitride nanotubes exhibit interesting physical properties; and we present simulation results to show that boron nitride nanotubes posses superior water transport properties compared to many existing materials, see a visualization in Fig. 3.2.

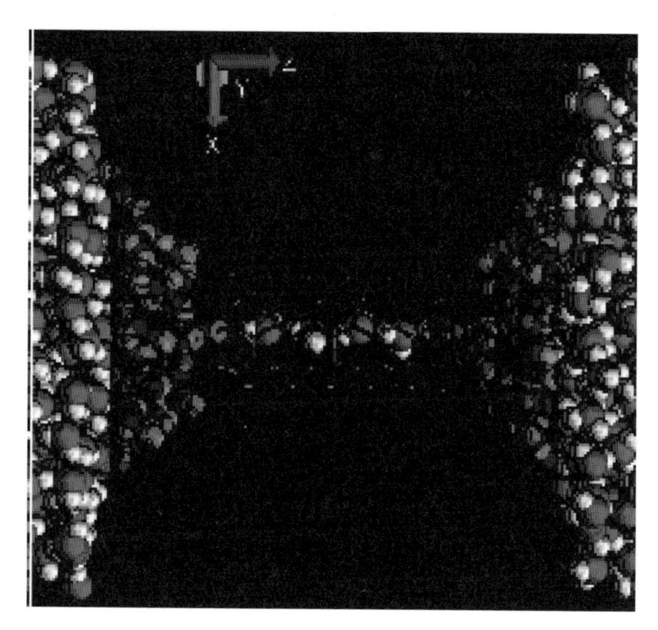

FIG. **3.2:** Visualization of a boron nitride nanotube in a water bath, see Section 3.3.3.

In the third part of the chapter, we present a discussion on multiscale methods. First, we review methods to directly couple molecular dynamics with Navier–Stokes equations. Next, we discuss hierarchical multiscale methods. In hierarchical multiscale methods information is passed from one length scale to the next, larger scale. By passing information from the quantum to the molecular and meso and macro length scales, we present simulation results to understand the structure, dynamics, and transport of water and electrolytes in silica nanochannels.

3.2 Microfluidics modeling: coarse-grained simulations

The dynamics and energetics of complex biological fluids, e.g. DNA molecules, platelets and red blood cells, large proteins, etc., is affected and often dominated by hydrodynamic interactions. Both the solute as well as the solvent have to be described accurately to capture such interactions, which may take place across many orders of magnitude in spatial and temporal scales. Efficient computer simulation of such systems requires a coarse-grained and simplified description, especially of the solvent dynamics. In this section, we present several mesoscopic techniques that coarse-grain out the *fast modes* of the solvent and hence allow us to simulate much larger time scales and bigger systems than in a molecular dynamics simulation. A great challenge for these mesoscopic methods is the resolution of the correlated motion of many solvent particles, leading to long-range hydrodynamic interactions that dominate their thermal fluctuations. In this Stokes regime, it is

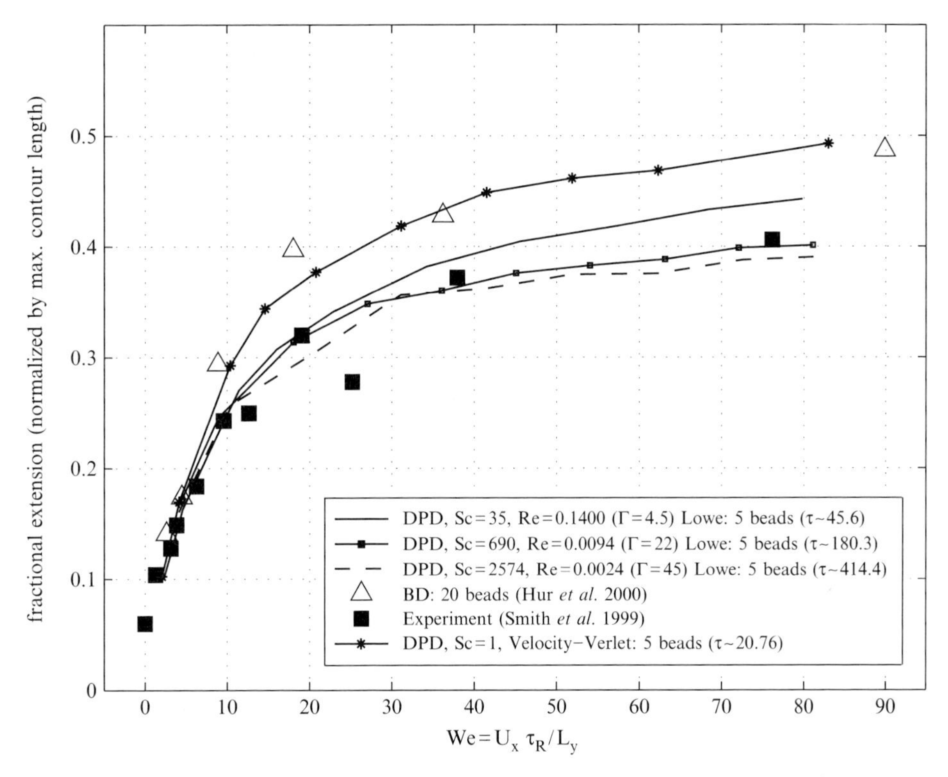

FIG. 3.3: Simulating the dynamics of DNA in microchannels: mean fractional extension for wormlike chains versus the Weissenmber number We. Comparisons are made between dissipative particle dynamics (DPD), Brownian dynamics (BD) (Hur *et al.*, 2000), and DNA experiments (Smith *et al.*, 1999).

customary to employ the Oseen tensor in analytical theories but also in mesoscopic techniques, e.g. in Brownian dynamics (Doi and Edwards 1986). However, in most of the methods that we present here, such long-range interactions are approximated by *local* particle interactions leading to huge computational savings. This leads to simulation approaches that can realistically simulate complex fluids in microchannels.

An example is the simulation of λ-phage DNA in microchannels. DNA molecules under steady shear have been extensively studied in experimental (Smith *et al.* 1999) and computational (Hur, *et al.* 2000; Jendre-jack, *et al.* 2002; Symeonidis, *et al.* 2006) works. Figure 3.3 shows results from two different mesoscopic simulations: Brownian dynamics (BD) and dissipative particle dynamics (DPD). A wormlike model is used to capture the DNA behavior, see Section 3.2.5.2 and (Symeonidis, *et al.* 2006). The simulation results show good agreement with experimental results, which involve spatial scales of a few *microns*, temporal scales of many *seconds*, and shear velocity scales of 10 to 200 *microns per second*. Clearly, such spatio-temporal scales are beyond the reach of MD simulations.

In the following, we present an overview of the following methods: Lattice Boltzman (LBM), Brownian dynamics (BD), stochastic rotation dynamics (SRD), smoothed particled hydrodynamics (SPH), and dissipative particle dynamics (DPD). We present more details for the last method (DPD) as it shares many features with the other methods, e.g. SPH, BD, and SRD.

3.2.1 *Lattice Boltzmann method (LBM)*

The lattice–Boltzmann method (LBM) has been used extensively for simulating electrokinetic flows in microchannels (Li and Kwok 2004), for mixing (Monaco *et al.*, 2007), for complex fluids (Usta, *et al.* 2007), and also for biological applications (Sun and Munn 2005; Ding and Aidun 2006; Zhu, *et al.* 2007). For example, in (Li and Kwok 2004) 2-D LBM was employed in conjunction with a modified Poisson–Boltzmann equation and good agreement with experimental results was obtained for KCl and LiCl electrolyte solutions in pressure-driven microchannel flows. In Sun and Munn (2005), also 2-D LBM was used to simulate the red blood cells and estimate the flow resistance at different levels of hematocrit; white blood cells were also included in the model to predict the increase in resistance due to white cell rolling and adhesion. Similarly, in (Ding and Aidun 2006) 3-D LBM was used to investigate the clustering of red blood cells in small vessels; it was found that the cluster size distribution is insensitive to inertia up to particle Reynold's number $Re = 3$. In Zhu, *et al.* (2007) LBM was used in conjunction with a lattice spring model (LSM) that models the micromechanics of elastic solids in a study of compliant microcapsules and pillars in microchannels; such fluid-filled capsules model biological cells, e.g. leukocytes.

LBM represents a 'minimal' form of the Boltzmann equation, solved on discrete lattices, and can be used for gas or liquid as well as for particulate microflows. Progress has been made for handling accurately complex geometries (Mei and Shyy 1998) and it seems to be particularly effective in the regime in which microdevices operate. A schematic representation of LBM with respect to other methods and its history of development is shown in Fig. 3.4, adopted from the book by Succi (2001). The left column shows the classical BBGKY (Bogoliubov–Born–Green–Kirwood–Yvon) hierarchy leading from atomistic to continuum flow equations. The right column shows the corresponding approximations in the framework of lattice methods. The LBM formulation is presented in many papers but we point to the work Chen and Doolen (1998) who employ a square lattice in two dimensions with three speeds and nine velocities; a popular model for multiphase in microfluidics is the Shan–Chen model (Shan and Chen, 1994). The book by Succi (2001) extensively covers many topics related to theoretical foundations of the method.

While successful for simple isothermal fluid flows, the standard LBM formulation is not Galilean invariant for non-isothermal and multiphase flows; for isothermal flows it is Galilean invariant up to order $\mathcal{O}(M^4)$, where M is the Mach number. It has also been shown to be unstable for small values of the viscosity. Such considerations have led to the development of the *entropic* lattice Boltzmann

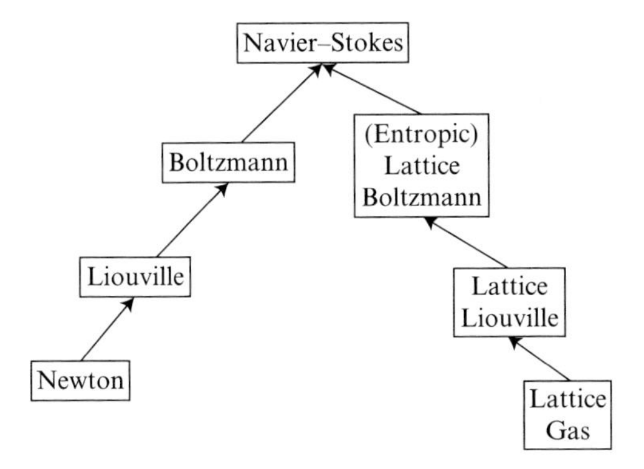

Fig. 3.4: Analogue between the BBGKY hierarchy and its lattice counterpart. Adopted from Succi (2001).

method in Karlin, *et al.* (1999) and Boghosian, *et al.* (2003). At the heart of this formulation is the use of the H-theorem of Boltzmann, which measures irreversibility. It has been shown rigorously by (Yong and Luo 2003) that the H-theorem does not exist for the lattice Boltzmann equation with *polynomial* equilibria. Enforcing the H-theorem in the lattice Boltzmann formulation guarantees an asymptotically homogeneous spatial distribution of particles as $t \to \infty$, which in turn translates into numerical stability. In the continuum case, the Boltzmann inequality produces the Maxwellian distribution (used as equality in equilibrium). To this end, in entropic lattice Boltzmann, use of different convex H functions is made that can be minimized relatively easily. The requirements are that Galilean invariance be preserved, as well as realizability and solvability. The latter condition implies that local equilibrium can be expressed in terms of the density and velocity, i.e. the continuum variables. It seems that the new entropic LB method offers many possibilities and potentially can overcome the aforementioned short-comings of the more traditional LBM. However, its extensions to non-isothermal flows as well as complex fluid flows has not been established yet. We also note that the limit of LBM to the Navier–Stokes equations has been obtained so far only for incompressible flows and not for compressible flows. In a sense, we have a quasi-compressible formulation that can be used for both incompressible and compressible flows. In this limit the Mach number scales with the Knudsen number, and the fluctuation of density around its mean value scales with the Knudsen number squared (Boghosian, *et al.* 2003).

With regard to boundary conditions, the so-called *bounce-back* scheme has been used to simulate wall boundary conditions. In the bounce-back scheme, when a particle distribution streams to a wall node, the particle distribution scatters back to the node it came from (Chen and Doolen 1998). However, this approach leads to relatively low-order accuracy, and more recent work has attempted to include corrections in the distribution function by including velocity gradients at the wall nodes (Skordos 1993), explicitly imposing a pressure constraint (Noble *et al.*, 1995) or employing extrapolation techniques on staggered meshes (nodes at midpoints of lattice), similarly to the classical finite difference discretizations (Chen, *et al.* 1996). An analysis of the accuracy of the boundary conditions as a function of the relaxation parameter was performed in Holdych *et al.* (2004). Also, new boundary conditions with error estimates have been formulated in Ansumali and Karlin (2002), and have been applied to slip flows as well as to no-slip flows. Particular care has to be exercised with the *slip* boundary condition as some technquies, e.g. bounce-back, may introduce erroneous values of slip velocity.

Next, we present an interesting application of LBM of micromixing in biofluidics (J.G. Brasseur and Y. Wang, unpublished), see Fig. 3.5. A *multi-grid* lattice-Boltzmann algorithm has been applied to analyze fundamental fluid and transport dynamics of nutrient uptake in the small intestine. Of interest are the mechanisms

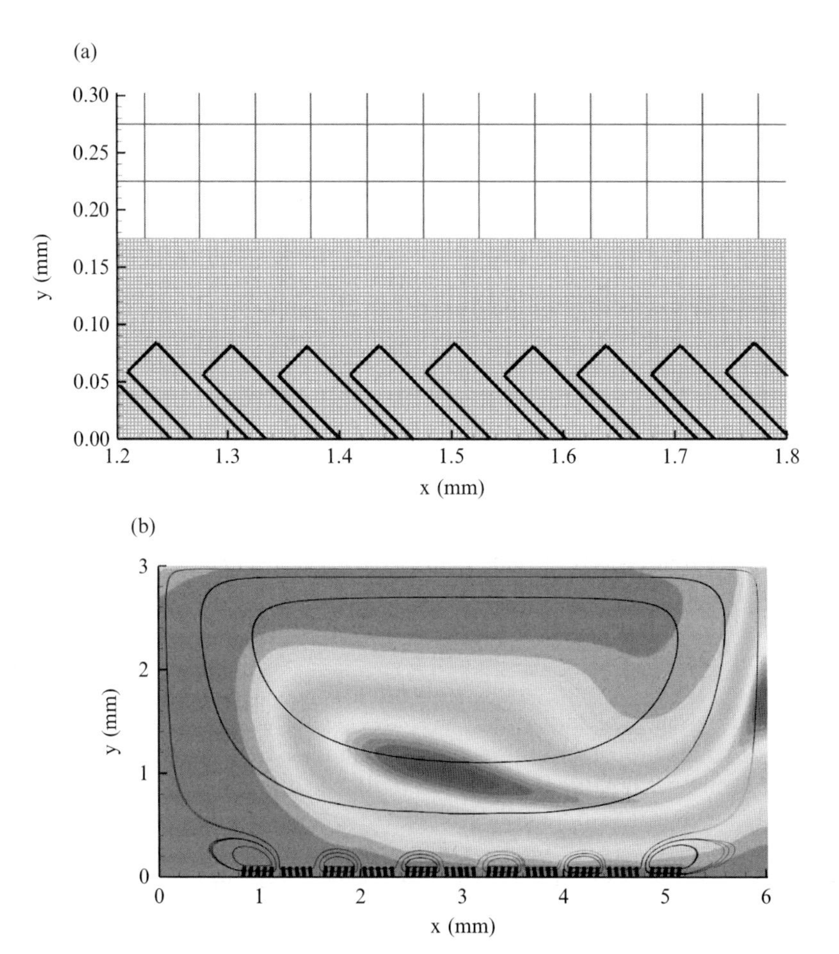

FIG. 3.5: (a) Coarse–fine grids for LBM simulations. (b) Flow streamlines and concentration field. (Courtesy of James G. Brasseur and Yanxing Wang.)

that control that rate of transport of nutrient molecules to the epithelium across which they are transported and taken up into the bloodstream. The process involves three interacting scales: (1) the centimeter scale of lumen contractions induced by muscles in the wall of the intestines, (2) the hundred micron scale of villi line the mucosa of the gut and which contain smooth muscle fibers that control their motions, and (3) the scale of individual molecular motion. It is hypothesized that nutrient absorption depends on a tuned coupling between the macrofluid motions driven by lumen contractions and microfluid motions induced by moving groups of villi. To explore this hypothesis, Brasseur and Wang developed a two-dimensional multi-grid model within the framework of LBM in which the discretized Boltzmann equation for particle distribution functions is solved with a low Mach number approximation for the collision operator. Scalar concentration is carried by a separate variable to model nutrient transport from the bulk flow to the surface of modeled moving villi. As illustrated in Fig. 3.5(a), the discretized distribution function is solved simultaneously on a 'coarse' and 'fine' lattice while conserving mass, momentum, and scalar, and by adjusting eddy-diffusivity time scales appropriately to guarantee continuity of molecular flux. Second-order boundary conditions are applied on surfaces with special requirements made for the moving boundaries. Zero scalar flux boundary conditions are applied on all surfaces except for the villi, where the scalar is fixed at zero to model instantaneous uptake. The modeled villi in Fig. 3.5(a) move backwards and forwards in periodic fashion. As illustrated in Fig. 3.5(b) the large-scale eddying motion is introduced by a moving lid over a cavity. The concentration field is shown by the color isocontours in the figure at one instant in time. As shown by the instantaneous streamlines in Fig. 3.5(b), the villi motion creates a 'micro-mixing layer'. Brasseur and Wang found that, when properly coupled with the large eddy, the local mixing and transport induced by the micromixing layer enhances uptake of scalar at the villi surface.

3.2.2 *Brownian dynamics (BD)*

Brownian dynamics can be used to simulate particles subject to Brownian motion and is very popular in simulating the dynamics of polymers and DNA molecules; it has also been used in simulations of protein folding (Rojnuckarin, *et al.* 1998) through a weighted ensemble procedure. In Brownian dynamics explicit solvent molecules are replaced by a stochastic force, which gives rise to a time-dependent stochastic differential equation to be solved. This allows the study of the temporal dynamics of complex fluids, e.g. biopolymers, large proteins, DNA molecules, and colloids. Both equilibrium and non-equilibrium flows can be simulated but in the basic BD version, the flow is prescribed and there is no feedback by the solution to the solvent. If hydrodynamic interactions are included then there is a two-way coupling but the resulting system may be somewhat diffcult to solve. In either case with BD we can simulate inertia-free flows only. We also note that BD should be used carefully as it does not conserve momentum.

The total force on a particle consists of a drag force $\mathbf{F}_i^D$, a Brownian force $\mathbf{F}_i^R$ and all other non-hydrodynamic forces $\mathbf{F}_i^E$. Due to the inertia-free assumption, the total force is zero, and hence:

$$\mathbf{F}_i^D + \mathbf{F}_i^R + \mathbf{F}_i^E = 0.$$

In the **basic BD version** we neglect hydrodynamic interactions and hence the drag force on the particle is given by the Stokes law, i.e.

$$\mathbf{F}_i^D = -\gamma\left(\frac{d\mathbf{r}_i}{dt} - \boldsymbol{v}^{\infty}(\mathbf{r}_i)\right),$$

where γ is the drag coeffcient and we denote by $\boldsymbol{v}^{\infty}$ the unperturbed velocity of the solvent computed at the particle position $\boldsymbol{r}_i$. Combining the above two equations, we obtain the *Langevin equation*:

$$\frac{d\mathbf{r}_i}{dt} = \boldsymbol{v}^{\infty}(\mathbf{r}_i) + \frac{1}{\gamma}\left(\mathbf{F}_i^R + \mathbf{F}_i^E\right). \tag{3.1}$$

In order to enforce thermal equilibrium, the fluctuation-dissipation theorem is invoked and this specifies a relationship between the drag coefficient and the amplitude of the Brownian motion, i.e. we have that

$$< \mathbf{F}_i^R(t)\mathbf{F}_j^R(t') >= \gamma 2k_B T\delta_{ij}\delta(t - t')\boldsymbol{\Delta},$$

where the brackets denote averaging and we also have that $< \mathbf{F}_i^R >= 0$. Here k_B is the Boltzmann constant, and $\boldsymbol{\Delta}$ is the unit second-order tensor.

The Langevin equation is integrated in time using an explicit Euler time scheme, i.e.

$$\mathbf{r}_i(t + \Delta t) = \mathbf{r}_i(t) + \frac{d\mathbf{r}_i}{dt}\Delta t.$$

Since the Langevin equation is a stochastic ODE it requires solving it many times to obtain a suffcient number of *sample trajectories* for accurate estimate in the statistics of the dynamics. The accuracy is of the order of $\sqrt{N}$, where N denotes the number of computed trajectories, but *variance reduction* techniques can be employed to improve further the accurcy, see (Ottinger 1996) for details. The most expensive component in the basic version of BD is the calculation of random numbers, which are drawn from a Gaussian distribution.

The **hydrodynamic interactions** can be added to the basic BD version through an *interaction tensor* $\boldsymbol{\Omega}_{ij}$, with $\boldsymbol{\Omega}_{ii} = 0$. It is typical to add this term to the diffusion tensor (Ottinger 1996), in the form

$$\mathbf{D}_{ij}(\mathbf{r}_i, \mathbf{r}_j) = \frac{k_B T}{\gamma}\left(\delta_{ij}\boldsymbol{\Delta} + \gamma\boldsymbol{\Omega}_{ij}(\mathbf{r}_i, \mathbf{r}_j)\right).$$

The modified Langevin equation is then written as:

$$\frac{d\mathbf{r}_i}{dt} = \boldsymbol{\nu}^{\infty}(\mathbf{r}_i) + \frac{1}{k_B T}\sum_{j=1}^{N}\left[\mathbf{D}_{ij}\cdot\mathbf{F}_j^E(\mathbf{r}_k) + \sqrt{2}\mathbf{R}_{ij}(\mathbf{r}_k)\cdot\mathbf{n}_j(t)\right], \tag{3.2}$$

where the terms $\mathbf{R}_{ij}$ are computed so that the fluctuation-dissipation theorem is satisfied, which leads to

$$\mathbf{D}_{ij}(\mathbf{r}_i, \mathbf{r}_j) = \sum_{n=1}^{N}\mathbf{R}_{in}(\mathbf{r}_k)\cdot\mathbf{R}_{jn}^T(\mathbf{r}_k).$$

Also, the random vectors $\mathbf{n}_i$ have zero mean and they satisfy

$$< \mathbf{n}_i(t)\mathbf{n}_j(t^{'}) >= \delta_{ij}\delta(t-t^{'})\boldsymbol{\Delta}.$$

This equation should be inverted to compute the $\mathbf{R}_{ij}$ terms, see (Fixman 1986) for an effcient procedure. An important step in dealing with hydrodynamic interactions is the choice of the interaction tensor $\boldsymbol{\Omega}_{ij}$, which is typically a modified version of the Oseen–Burgers tensor that should produce a positive-definite diffusion tensor.

The modified Langevin equation (3.2) is difficult to solve, so in simulation practice it is customary to simply vary the drag coeffient γ to fit experimental data in polymer solutions or in DNA molecules (Doyle *et al.* 2000). However, in many cases such adjustments are not sufficient especially in trying to match stress-strain behavior or in strong interactions with the microchannel walls. Hydrodynamic interactions in microchannels with biopolymers cause a significant depletion layer at the walls, which is dependent on molecular weight. In fact this finding has been exploited in order to develop separation processes for DNA (Agarwal, Dutta, and Mashelkar 1994). To this end, in Jenderjack, *et al.* (2004) a solution procedure is developed that employs Green's functions for the interaction tensor which are added to the free-space interaction tensor. These extra Green's functions impose the no-slip boundary condition and can be computed in a pre-processing stage for a microchannel of arbitrary cross-section using standard finite element solvers. The study of Jenderjack *et al.* (2004) on microchannels from 1 to 10 μm and DNA with contour length of 10 to 126 μm showed that higher molecular weight chains are more concentrated in the center of the channel. This, in turn, causes higher molecular weight chains to travel faster compared to lower molecular weight chains. While this is somewhat surpising, such results suggest an effective separation process of DNA based on molecular weight. It is also worth mentioning that for this problem, Brownian motion is in competition with the wall hydrodynamic effect, and causes a drift of DNA away from the centerline, so the maximum of the concentration profile is not exactly at the centerline of the microchannel but close to it.

3.2.3 *Stochastic Rotation Dynamics (SRD)*

Stochastic Rotation Dynamics (SRD) (Ihle and Kroll 2001) or Multi-Particle-Collision (MPCD) (Malevanets and Karpal 1999; Malevanets and Karpal 2000;

Lamura *et al.* 2001) is an off-lattice Lagrangian method similar to the dissipative particle (DPD) method but it employs rotations of the velocities of the particles relative to the center of mass velocity. It conserves mass, momentum, and energy at each cell. Specifically, it involves two steps: the *streaming step*, where particles move balistically with their positions changing according to

$$\mathbf{r}_i(t + \Delta t) = \mathbf{r}_i(t) + \Delta t \boldsymbol{v}_i(t),$$

and a *collision step*, where the particles are sorted out into cubes of side length a. In this step, the rotation takes place according to

$$\boldsymbol{v}_i(t + \Delta t) = \boldsymbol{v}_{cm}(t) + \mathbf{R}_p[\boldsymbol{v}_i(t) - \boldsymbol{v}_{cm}(t)],$$

where $\boldsymbol{v}_{cm}$ is the velocity of the center of mass for each cell, and $\mathbf{R}_p$ is the rotation matrix for rotations by a fixed angle α. The orientation of the rotation axis is random for each cell and at each time step. In order to ensure Galilean invariance, a random shift is applied before the collision step.

SRD simulates a 'dissipative' gas as many of the mesoscopic methods do, with a Schmidt number of order one. Analytical expressions for the transport coefficients have been derived in terms of the parameters of the model (Malevanets and Karpal 1999; Ihle and Kroll 2003). For example, the dynamic viscosity μ can be computed analytically as the sum of a kinetic contribution μ_k and a collisional contribution μ_c, where

$$\mu_k = \frac{k_B T \Delta t \rho}{a^3}\left[\frac{5\rho}{(4 - 2\cos\alpha - 2\cos 2\alpha)(\rho - 1)} - \frac{1}{2}\right]$$

and

$$\mu_c = \frac{m(1 - \cos\alpha}{18 \Delta t a}(\rho - 1),$$

where m is the mass of the particle. The mean free path $\lambda = \Delta t \sqrt{k_B T/m}$ is very important in determining the viscosity of the simulated fluid and thereby its dynamics. At large values of λ the kinetic contribution dominates while at small values of λ the collisional contribution to the total viscosity dominates. It was found in Ripoli *et al.* (2004) that in order to simulate liquid-like behavior with large Schmidt numbers we need to take small time step since $\mu_c \propto 1/\Delta t$. On the other hand, an analytical expession for the diffusion coefficient is

$$D = \frac{k_B T \Delta t}{m}\left[\frac{1}{\rho_1} - \frac{1}{2}\right],$$

where $\rho_1 = 2(\rho - 1)(1 - \cos\alpha)/(3\rho)$ and hence $D \propto \Delta t$. These estimates suggest that $Sc \propto (\Delta t)^{-2}$, and hence small time steps lead to large values of Schmidt number, e.g. $\lambda \leq 0.1$, which provides the appropriate time step. In addition, in Ripoli *et al.* (2004) it is recommended that the rotation angles should also satisfy $\alpha \geq 90^\circ$. For such parameters, liquid-like behavior is observed with the Schmidt number in the range of 10 to 100, which is still more than an order of magnitude lower than the realsitic values of the Schmidt number for liquids of interest to biological applications. A more systematic procedure in obtaining the parameters of the SRD model is described in Hecht *et al.* (2003).

SRD has been applied to flows in channels and around objects even at finite Reynold's number but one of the most interesting applications so far has been the modeling of red blood cells (RBCs) in capillary flows (Noguchi and Gompper 2005). It was found that as the flowrate through the capillary increases, RBCs transition from a non-axisymmetric discocyte to an axisymmetric parachute shape, along the flow direction, thereby reducing the flow resistance.

3.2.4 *Smoothed Particle Hydrodynamics (SPH)*

Smoothed Particle Hydrodynamics (SPH) was originally developed for astrophysical compressible flow (Gingold and Monaghan 1977) but recent progress has made the method suitable for simulating microchannel complex flow physics as well as reactive transport (Tartakovsky *et al.* 2007). An advantage of SPH is that it is relatively easy to incorporate new physics into the formulation (Morris, *et al.* 1997). It is a Lagrangian (grid-free) method so moving particles or deformable boundaries can be modeled readily. Similarly to SRD and DPD, SPH is Galilean invariant but unlike DPD it does not involve any random force in the equation of motion.

In the SPH formulation both fluid and solid boundaries are represented by point particles. The key concept of the method is that continuous fields can be approximated by a linear combination of smooth bell-shaped functions (or kernels), $W(\mathbf{r}-\mathbf{r}_i)$, centered on the particles. A scalar field and its gradient are expressed as functions of a set of extensive variables a_i for a particle located at $\mathbf{r}_i$ as

$$\Phi(r) = \sum_i \frac{a_i}{n_i} W(\mathbf{r} - \mathbf{r}_i) \tag{3.3}$$

$$\nabla\Phi(r) = \sum_i \frac{a_i}{n_i} \nabla_r W(\mathbf{r} - \mathbf{r}_i). \tag{3.4}$$

Based on these approximations, we can write the mass m_i (or number density n_i) and momentum equations for each particle as

$$n_i = \sum_j W(\mathbf{r}_j - \mathbf{r}_i), \quad i, j \in \text{ fluid + solid}$$

and

$$\frac{d\boldsymbol{v}_i}{dt} = -\frac{1}{m_i} \sum_{j\in\text{fluid+solid}} \left(P_j/n_j^2 + P_i/n_i^2\right) \nabla_i W(\mathbf{r}_i - \mathbf{r}_j) + \tag{3.5}$$

$$\frac{1}{m_i} \sum_{j\in\text{fluid+solid}} \frac{(\mu_i + \mu_j)(\boldsymbol{v}_i - \boldsymbol{v}_j)}{n_i n_j (\mathbf{r}_i - \mathbf{r}_j)^2} (\mathbf{r}_i - \mathbf{r}_j) \cdot \nabla_i W(\mathbf{r}_i - \mathbf{r}_j, h) + \mathbf{F}_i^E, \; i \in \text{ fluid.}$$

Here h is a smoothing parameter with $2h$ the width of the kernel W. Also, μ is the dynamic viscosity and P_i the pressure. For incompressible flows the pressure P_i is interpreted as the dynamic component, i.e. the hydrostatic componenent should be subtracted out (Morris, *et al.* 1997). The last term in the above equation represents the viscous contributions, which may appear in different forms in the literature. For example, a symmetrized form derived in Sigaloti *et al.* (2003) leads to a stable formulation even for low-order polynomial kernels W. The latter are typically chosen as quintic splines to avoid numerical instabilities but with the symmetrized formulation cubic splines seem to work as well. The pressure is obtained from an equation of state that has a form

$$P = P_0\left[\left(\frac{\rho}{\rho_0}\right)^{\gamma} - 1\right],$$

with $\gamma = 7$ so that small variations in density cause large pressure changes. At the low Reynold's numbers encountered in microfluidics it is more accurate to take $\gamma = 1$. An alternative equation of state is the one often used in the artificial compressibility method, i.e.

$$P = c^2\rho,$$

where c is the speed of sound; its value has to be sufficiently large in order to simulate a liquid flow but not too large so that it makes the time step prohibitively small (Morris *et al.* 1997).

The boundary conditions are complicated, especially at curved walls. A simple bounce-back condition or the use of images may work in most cases but a more elaborate formulation was presented in Takeda, *et al.* (1994), where extra terms are included to account for the defficiency of the kernel right at the boundary; this procedure was extended further in Morris, *et al.* (1997).

To integrate the momentum equation, an explicit Verlet algorithm is typically employed in the form

$$\mathbf{r}_i(t + \Delta t) = \mathbf{r}_i(t) + \Delta t \boldsymbol{v}_i(t) + 0.5\Delta t^2 \mathbf{a}_i(t)$$

where $\mathbf{a}_i$ denotes acceleration, and

$$\boldsymbol{v}_i(t + \Delta t) = \boldsymbol{v}_i(t) + 0.5\Delta t[\mathbf{a}_i(t) + \mathbf{a}_i(t + \Delta t)].$$

Although a Lagrangian method, SPH is only conditionally stable because of the finite-speed sound waves, so

$$\Delta t \leq 0.25h/(3c),$$

and also due to constraints in particle acceleration (Monaghan 1992):

$$\Delta t \leq \min_i \left(h/(3|\mathbf{a}_i|)\right)^{1/2},$$

and due to explicit diffusion (Morris, Fox, and Zhu 1997):

$$\Delta t \leq \min_i \left(\rho_i h^2/(9\mu_i)\right).$$

SPH has been validated for channel and pipe flow at small Reynolds number (Morris *et al.* 1997; Sigaloti *et al.* 2003). Even for first-order boundary conditions, the simulation error for a flow in a channel of 1 mm with 50 particles across is less than 1 per cent while with 20 particles is about 2 per cent, indicating first-order accuracy, which is typical for most particle-based schemes. However, for compex-geometry domains, the pressure exhibits noticeable perturbations in the vicinity of the boundaries, with errors of 5 percent to 10 per cent locally.

An example of SPH application relevant to biological applications is the simulation of droplet dynamics inside an inverted Y-shaped channel presented in Tartakovsky and Meakin (2005). Three snapshots of the flow are shown in Fig. 3.6 for a specific value of the gravitational force that sets the corresponding Bond number $Bo = \rho g A/\sigma = 3.21$, where σ is the surface tension and A the cross-sectional area. For slightly smaller values of the Bond number the droplet stops at the junction and flow continues in the form of films on the outer walls of the channels. However at $Bo = 3.21$, the droplet is able to overcome the capillary barrier and enter the intersection. At even larger Bo values, the droplet flows quickly through the junction without film development. This complicated flow pattern was computed with particles placed initially randomly on a 64×64 box. The action of surface tension was modeled simply with a combination of short-range repulsive and (relatively) long-range attractive particle–particle interactions, with the range of attractive interactions equal to the range of the SPH kernel. The effect of surface tension is then simulated by allowing particle-particle interactions subject to these potentials. In addition, the equation of state was of the van der Waals type, i.e.

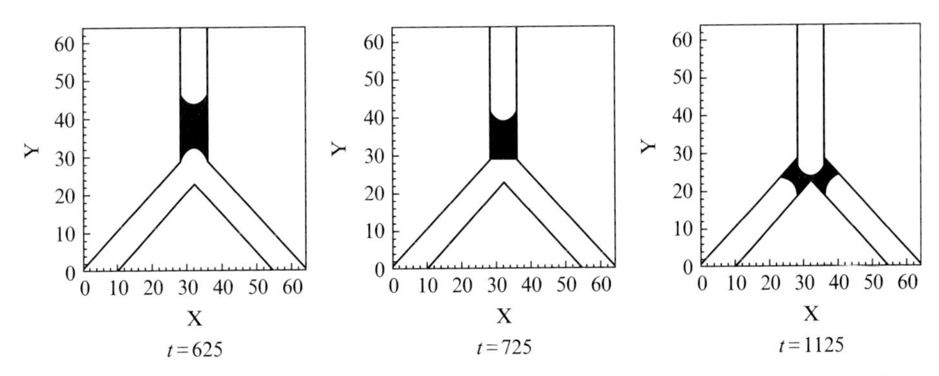

FIG. 3.6: SPH modeling of surface tension and contact angle: droplet flow through a Y-shaped junction at three different times corresponding to Bond number $Bo = 3.21$. The size of the flow domain is in units of the kernel half-width h. (Courtesy of Alexandre Tartakovsky.)

$$P = \frac{\rho(k_B/m)T}{1 - \rho(C_1/m)} - (C_2/m)\rho^2,$$

where C_1 and C_2 are the van der Waals constants.

3.2.5 Dissipative Particle Dynamics (DPD)

The dissipative particle dynamics (DPD) method combines features from both MD and LBM and it is closely related to SRD and SPH. It can also be thought of as momentum-preserving Brownian dynamics with hydrodynamic interactions. Today the method has been used in microchannel flow of DNA suspensions (Fan *et al.* 2003; Symeonidis *et al.* 2006), in modeling colloidal micropumps (Palma *et al.* 2006), and in multiphase flow in microchannel networks (Liu, *et al.* 2006).

The initial DPD model was proposed by Hoogerburgge and Koelman as a polymer simulation method to avoid the artifacts associated with traditional LBM simulations while capturing spatio-temporal hydrodynamic scales much larger than those achievable with MD (Hoogerburgge and Koelman 1992). The DPD model consists of particles that correspond to *coarse-grained* entities, thus representing molecular clusters rather than individual atoms. The particles move off-lattice, interacting with each other through a set of prescribed and velocity-dependent forces (Hoogerburgge and Koelman 1992; Espanol and Warren 1995). Specifically, there are three types of forces acting on each dissipative particle:

- A purely repulsive conservative force.
- A dissipative force that reduces velocity differences between the particles.
- A stochastic force directed along the line connecting the center of the particles.

These forces can be interpreted as follows: the conservative forces cause the fluid particles to be as evenly distributed in space as possible as a result of certain 'pressures' among them. The conservative forces are given by *soft potentials*, see Fig. 3.7, and this eliminates some of the computational difficulties associated with the hard potentials used in MD simulations. The frictional forces represent viscous resistances between different parts of the fluid. Finally, the stochastic forces represent degrees of freedom that have been eliminated during the coarse-graining process. The last two forces effectively implement a thermostat, so that thermal equilibrium is achieved. Correspondingly, the amplitude of these forces is dictated by the fluctuation–dissipation theorem (Espanol and Warren 1995), which ensures that in thermodynamic equilibrium the system will have a canonical distribution. All three forces are modulated by a weight function that specifies the range of interaction between the particles and renders the interaction local.

The distinguishing feature of the DPD forces is that they conserve momentum, and therefore the DPD model satisfies mass and momentum conservation, which are responsible for the hydrodynamic behavior of a fluid at large scales (Espanol 1995). Also, by changing the conservative interactions between the fluid particles,

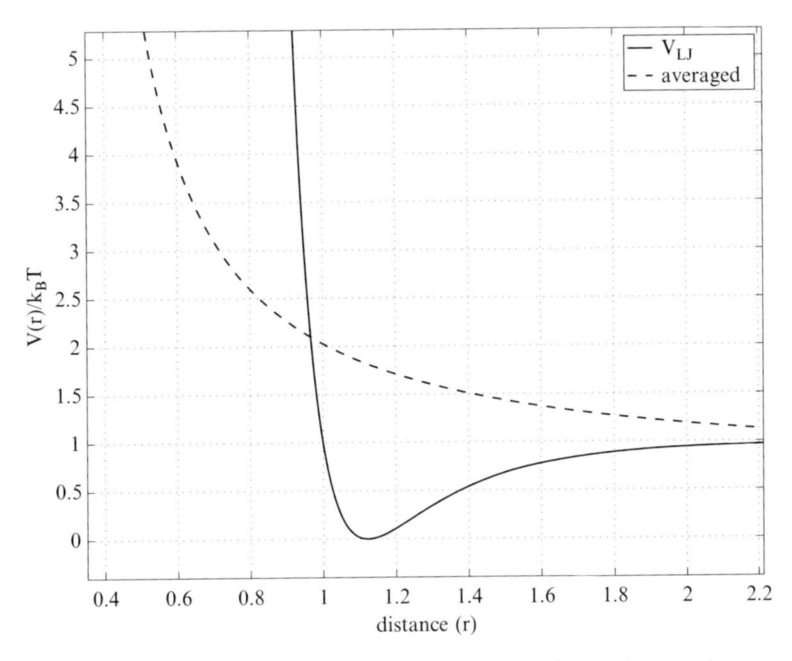

Fig. 3.7: Lennard–Jones potential and its averaged soft repulsive-only potential.

one can easily construct 'complex' fluids, such as polymers, colloids, amphiphiles, and mixtures.

In summary, the DPD method is characterized by the following conditions:

- Position and velocity variables are continuous, as in MD, but the time step is updated in discrete steps, as in LBM.
- The conservative forces between DPD particles are soft-repulsive, which makes it possible to extend the simulations to longer time scales compared to MD.
- Hydrodynamic behavior is expected at much smaller particle numbers than in classical MD.
- Mass and momentum are locally conserved, which results in hydrodynamic flow effects on the macroscopic scale.
- The characteristic kinetic time in DPD is large compared to MD time scales.

A conceptual picture of DPD is that of soft microspheres randomly moving around but following a preferred direction dictated by the conservative forces. DPD can be interpreted as a Lagrangian discretization of the equations of fluctuating hydrodynamics as the particles follow the classical hydrodynamic flow but they display thermal fluctuations. These fluctuations are included *consistently* based on the principles of statistical mechanics. The DPD method is particularly attractive for the computer simulation of biopolymer solutions, since by employing

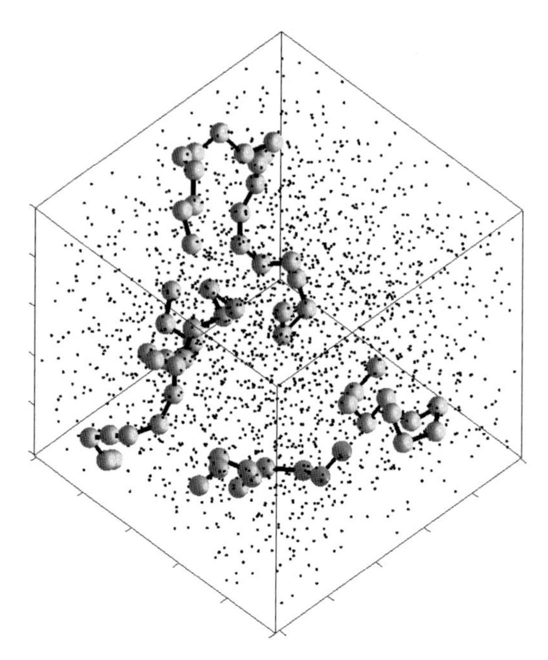

FIG. 3.8: Polymer chains flowing in a sea of solvent DPD particles.

bead-spring representations of the polymer chains we can formulate and compare a variety of realistic conservative inter-monomer forces, see Section 3.2.5. A schematic of dilute polymer solution in a DPD solvent is shown Fig. 3.8.

In order to appreciate the potential and computational complexity of DPD, in the following section we summarize the governing equations for simple and complex fluids, and subsequently we present a time-staggered algorithm and some results.

3.2.5.1 *The DPD Equations* We consider a system consisted of N particles having equal mass (for simplicity in the presentation) m, position $\mathbf{r}_i$, and velocities $\boldsymbol{v}_i$. The aforementioned three types of forces exerted on a particle i by particle j are given by

$$\mathbf{F}_{ij}^{C} = F^{(C)}(r_{ij})\mathbf{e}_{ij}, \qquad \mathbf{F}_{ij}^{D} = -\gamma\omega^{D}(r_{ij})(\boldsymbol{v}_{ij}\cdot\mathbf{e}_{ij})\mathbf{e}_{ij}, \qquad \mathbf{F}_{ij}^{r} = \sigma\omega^{r}(r_{ij})\xi_{ij}\mathbf{e}_{ij},$$

where $\mathbf{r}_{ij} = \mathbf{r}_i - \mathbf{r}_j$, $\boldsymbol{v}_{ij} = \boldsymbol{v}_i - \boldsymbol{v}_j$, $r_{ij} = |\mathbf{r}_{ij}|$ and the unit vector $\mathbf{e}_{ij} = \frac{\mathbf{r}_{ij}}{r_{ij}}$. The variables γ and σ determine the strength of the dissipative and random forces, respectively, ξ_{ij} are symmetric Gaussian random variables with zero mean and unit variance, and ω^D and ω^R are weight functions, e.g.

$$\omega^{R}(r_{ij}) = \left(1 - r_{ij}/r_c\right)^{p} \text{ for } r_{ij} \leq r_c, \tag{3.6}$$

where $p = 1$ for the standard DPD method. However, other choices for these envelopes can increase the Schmidt number of the DPD fluid, e.g. $p = 0.25$.

The conservative force $\mathbf{F}_{ij}^C$ is similar to that in the MD formulation. This force as well as the other two act within a sphere of radius r_c, which defines the length scale of the system; it corresponds to a *soft repulsive-only* interaction potential. By averaging the Lennard–Jones potentials or the corresponding molecular field over the *rapidly* fluctuating motions of atoms over short time intervals, an effective average potential is obtained of the form shown in Fig. 3.7. A linear approximation of this is as follows (Groot and Warren 1997): $\mathbf{F}_{ij}^C = a_{ij}(1 - r_{ij})\mathbf{e}_{ij}$ for $r_{ij} \leq r_c = 1$ and is otherwise zero. Unlike the hard Lennard–Jones potential which is unbounded at $r = 0$, the *soft potential* employed in DPD has the finite value a_{ij} at $r = 0$. This parameter is obtained (Groot and Warren 1997) by enforcing the proper compressibility of the system defined by

$$\kappa^{-1} = \frac{1}{k_B T}\frac{\partial p}{\partial n}|_{\text{sim}} = \frac{N_m}{k_B T}\frac{\partial p}{\partial n}|_{\text{MD}} \tag{3.7}$$

where n is the number density, N_m is the coarse-graining parameter, k_B is the Boltzmann constant and T the temperature of the system. (We note that 'sim' refers to simulation and that in MD we have that $N_m = 1$.) By matching the diffusion constant (D_{sim}) in DPD simulation with that of water (D_{water}) we find the DPD time scale as

$$\tau = \frac{N_m D_{\text{sim}} r_c^2}{D_{\text{water}}} \propto N_m^{5/3}.$$

This time scale and the soft potential explain why the DPD method is several orders of magnitude faster than straightforward MD. With respect to the latter, the soft potential removes the 'caging effect' of an atom so that the diffusivity of atoms is increased by a factor of 1000. The effect of the time scale is to decrease the corresponding CPU time in proportion to the coarse-graining parameter N_m; hence the *total speed-up* with respect to MD is $1000 \times N_m \times N_m^{\frac{2}{3}}$ for a given system volume. Thus, for $N_m = 5$ and 10 the speed-up factor is 73,000 and 464,000, respectively!

The time evolution of DPD particles is described by Newton's law

$$d\mathbf{r}_i = \mathbf{v}_i \delta t; \quad d\mathbf{v}_i = \frac{\mathbf{F}_i^C \delta t + \mathbf{F}_i^D \Delta t + \mathbf{F}_i^R \sqrt{\delta t}}{m_i},$$

where $\mathbf{F}_i^C = \sum_{i \neq j} \mathbf{F}_{ij}^C$ is the total conservative force acting on particle i; $\mathbf{F}_i^D$ and $\mathbf{F}_i^R$ are defined similarly. The random force, which represents Brownian motion, appears with a factor of $\sqrt{\Delta t}$ in the velocity increment. The dissipative and random forces are characterized by strengths $\omega^D(r_{ij})$ and $\omega^R(r_{ij})$ coupled due to the *fluctuation–dissipation* theorem (Espanol and Warren 1995) as follows:

$$\omega^D(r_{ij}) = \left[\omega^R(r_{ij})\right]^2 = \max\left\{\left(1 - \frac{r_{ij}}{r_c}\right)^2, 0\right\}, \quad \sigma^2 = 2\gamma k_B T.$$

Imposing wall or inflow/outflow **boundary conditions** in DPD is not a trivial issue but similar approaches used in LBM or SPH can be adopted. A broad classification of the three main approaches to impose boundary conditions in DPD was provided in (Revenga, *et al.* 1999) as follows:

1. The Lees–Edwards method to impose planar shear, also used in LBM (Wagner and Pagonabarraga 2002), which is essentially a way to avoid modeling directly the physical boundary (Lees and Edwards 1972; Boek *et al.* 1997).
2. Freezing regions of the fluid to create a rigid wall or a rigid body, e.g. in particulate flows, see (Hoogerburgge and Koelman 1992; Boek *et al.* 1997).
3. Combine different types of particle-layers with proper reflections, namely specular reflection, bounce-back reflection, or Maxwellian reflection (Revenga, Zuniga, and Espanol 1998; Willemsen, *et al.* 2000; Fan *et al.* 2003).

The most accurate boundary conditions are based on *combinations* of the above concepts and use of modified repulsive and dissipative forces at the boundaries in an adaptive manner; the interested reader should consult Pivkin and Karniadakis (2005) and Pivkin and Karniadakis (2006).

An important extension of DPD is the *multibody* dissipative particle dynamics method or **M-DPD**, which allows simulation of multiphase flows and hydrophilic and hydrophobic walls, see Fig. 3.1. It also simulates realistic equation of states unlike almost all other mesoscopic methods which do not simulate true liquids. To this end, the weight function in the conservative force, which plays a key role in establishing the correct pressure, was directly derived by the equation of state (Trofimov, *et al.* 2002). Moreover, it was proposed in Warren (2003) to take a negative conservative coefficient for the standard DPD model to make the soft conservative potential attractive and add a repulsive multibody contribution with a different cut-off radius. This combination of the long-range attraction with the short-range repulsion can be achieved together with the true equation of state of the liquid.

3.2.5.2 *Models for polymers* Unlike the MD equations, the DPD equations are stochastic due to random force and non-linear since the dissipative force depends on the velocity. In particular, for complex fluids the presence of both soft and hard potentials suggests the use of time-staggered algorithms for integrating the DPD equations of motion (Symeonidis and Karniadakis 2006). This allows the efficient study of polymeric physical quantities, such as the radius of gyration of the polymeric chain. The conservative forces present in the usual DPD equations can be taylored in such a way so as to describe a variety of interactions—e.g. Lennard–Jones (LJ), Hookean dumbells, Finitely Extensible Non-Linear Elastic (FENE) springs and van der Waals—as long as they are derivable from a given potential $V(r_{ij})$. Figure 3.7 illustrates the need for two different temporal resolutions: the Lennard–Jones (LJ) potential (for monomer–monomer pairs) is a hard repulsion that requires a time-step much smaller than the soft interaction forces of a typical DPD particle pair (which can be thought of as an *averaged* soft potential).

Figure 3.8 shows polymeric chains moving freely in a DPD solvent of N particles. These chains consist of beads (DPD particles) subject to the standard DPD forces: soft repulsive (conservative), dissipative and random. *In addition* to these forces, they are subject to intra-polymer forces of three types:

- **Hookean Spring + LJ:** Within a chain of M monomers each bead is subject to a pairwise harmonic potential

$$U_{\text{HOOKE}} = \frac{\kappa}{2}|\mathbf{r}_i - \mathbf{r}_{i-1}|^2, \text{ where } i = 2, 3, 4, \ldots, M$$

  and κ is the spring constant. The repulsion for each pair of monomer particles is given by the shifted LJ potential

$$U_{\text{LJ}} = 4\epsilon\left[\left(\frac{L}{r_{ij}}\right)^{12} - \left(\frac{L}{r_{ij}}\right)^{6} + \frac{1}{4}\right]$$

  truncated to act *only* for pairs with $r_{ij} < r_c$. For the above formula we pick $\epsilon = k_B T$, $L = 2^{-1/6}$ and $r_c = L2^{1/6} = 1$. We also fix $\kappa = 7$. Therefore, the monomer-monomer conservative interactive forces are given by the gradient $\mathbf{F}^C = -\nabla(U_{\text{HOOKE}} + U_{\text{LJ}})$.

- **Fraenkel (stiff) Spring:** Within a chain of M monomers each bead is subject to a pairwise Hookean spring. The spring is considered to be *stiff* since it has a finite equlibrium length r_{eq}. Stretched to a length greater than r_{eq} the spring exerts an attractive force, while pushed to one smaller than r_{eq} it exerts a repulsive one. Its potential follows the formula

$$U_{\text{STIFF}} = \frac{\kappa}{2}\left(|\mathbf{r}_i - \mathbf{r}_{i-1}| - r_{\text{eq}}\right)^2, \quad \text{where } i = 2, 3, 4, \ldots, M$$

  and κ is the spring constant. Again, we pick $\kappa = 7$ and $r_{\text{eq}} = r_c$ and the monomer–monomer conservative interactive forces are given by $\mathbf{F}^C = -\nabla U_{\text{STIFF}}$.

- **FENE Spring + LJ:** This model is similar to the Hookean spring with LJ forcing. Within a chain of M monomers each bead is subject to a pairwise nonlinear spring force. The spring has a maximum extensibility $r_{\max}$ beyond which the force becomes infinite, and hence any length greater than $r_{\max}$ is not allowed. The FENE potential is described by

$$U_{\text{FENE}} = -\frac{\kappa}{2}r_{\max}^2 \log\left[1 - \frac{|\mathbf{r}_i - \mathbf{r}_{i-1}|^2}{r_{\max}^2}\right], \quad \text{where } i = 2, 3, 4, \ldots, M$$

and κ is the spring constant. In addition to the FENE force, a hard repulsive LJ potential U_{LJ} (identical to the first case above) is present. For the cases presented here we choose $r_{\max} = 2r_c$, $\kappa = 7$. The monomer forces are again $\mathbf{F}^C = -\nabla(U_{\mathrm{FENE}} + U_{\mathrm{LJ}})$.

- **Wormlike Chain:** Polymer models of biological importance (DNA, proteins) have been known to be governed by stiff interactions. The worm-like chain (Kratky and Porod 1949; Yamakawa 1971; Sun 1994) can be thought of as a continuous curve in three-dimensional space. Of importance is the *persistence length* λ_p, which is a measure of the chain's stiffness and is the average length over which the orientation of a curve segment does not change ('persists'). We will focus on the bead-spring representation of the model, which approximates a portion of the worm-like chain with a force law given by the Marko–Siggia (Marko and Siggia 1995) expression

$$F^{(C)} = \frac{k_B T}{\lambda_p}\left[\frac{1}{4(1-R)^2} - \frac{1}{4} + R\right],$$

where $R = \frac{|\vec{r}_i - \vec{r}_{i-1}|}{L_{\text{spring}}} = \frac{r}{L_{\text{spring}}}$ $i = 2, 3, 4, \ldots, M$

and L_{spring} is the maximum allowed length for each chain (spring) segment. The expression is accurate for large values of the ratio $\frac{L_{\text{spring}}}{\lambda_p}$ and exact as $r \to 0$ or $r \to L_{\text{spring}}$.

The Marko–Siggia spring law is an averaged quantity, locally approximating flexible rods. The derivation of the formula accounts for coarse-graining microscopic elements of a long chain (such as bead-rod), by use of statistical mechanics. However, in order to use the Marko–Siggia law in molecules with more than two beads (dumbbells), some authors (Larson *et al.* 1997) account for the different stiffness of the beaded counterparts by altering the persistence length λ_p of the sub-chains. Detailed analysis of such arguments (Underhill and Doyle 2004) has shown that it is possible to minimize the errors arising by the introduction of beads and sub-chains. Here we will adopt the analysis presented in Underhill and Doyle (2004) for stained λ-phage DNA molecules assumed to have $L = 21.1\,\mu m$ (fully extended length) and $\lambda_p = 0.053\,\mu m$ (persistence length). The correction we will apply will linearly approximate the ratio of effective to true persistence length, for three different regions of the extension: low force, half-extended spring, and high-force regimes. More specifically, we define the ratio

$$\lambda^* = \frac{\lambda_p[\text{Eeffective}]}{\lambda_p[\text{True}]}$$

so that when $\lambda^* = 1$ no correction is applied. The tables in Underhill and Doyle (2004) suggest a high, medium, and zero correction for the low-force, half-extension, and high-force regions respectively. We go one step further to introduce a linear fit to the suggested correction values for N-bead chains:

$$\lambda^* \approx (1.0 - \hat{z}) \times 0.022 \times (N - 1) + 1, \text{ if } N \leq 20$$
$$\lambda^* \approx (1.0 - \hat{z}) \times 0.025 \times (N - 1) + 1, \text{ if } N > 20,$$

where $0 \leq \hat{z} \leq 1$ is the instantaneous fractional extension of the whole molecule in the stretching direction. The above expressions approximate fairly accurately the values given in Underhill and Doyle (2004) and are implemented in all instances of $N > 2$ for the Marko–Siggia spring force in this work.

3.2.5.3 *Time-staggered velocity-Verlet algorithm* The most popular DPD integrator is a modified version of the classical velocity-Verlet (first proposed by Groot and Warren (Groot and Warren 1997)). The velocity-Verlet scheme is characterized by explicit calculation of all forces $\mathbf{F}^c$, $\mathbf{F}^d$, $\mathbf{F}^r$ (conservative, dissipative, and random) and is known to be time-step dependent, but at the same time is straightforward and relatively accurate. Below we outline the modified velocity-Verlet scheme with parameter λ. The latter is often set to 0.5 but Groot and Warren (Groot and Warren 1997) have shown that for a certain range of timesteps Δt the optimal value is closer to 0.65. Denoting the total forces by $\mathbf{F}_i = \sum_{j \neq i} \left[\mathbf{F}_{ij}^c + \mathbf{F}_{ij}^d + \frac{\mathbf{F}_{ij}^r}{\sqrt{\Delta t}} \right]$ and the extra polymeric forces by $\mathbf{F}_i^p = \sum_{j \neq i} \mathbf{F}_{ij}^p$ the basic (classical thermostat) velocity-Verlet scheme is outlined in Table 3.1.

To extend this algorithm for the simulation of complex fluids with *soft/hard* potentials, a large time step, Δt, is employed for solvent particles and a smaller one, δt, for polymer particles belonging to a chain. We integrate the polymer particles $\frac{\Delta t}{\delta t}$ times in a separate sub-cycle (using δt for the time step). During the sub-cycle we update the intra-polymer forces, but not the inter-particle (total) ones. This would require CPU time for each sub-cycle equivalent to a standard one. Hence, we cannot expect exact agreement of the new scheme with the classical

Table 3.1: Overview of the traditional velocity-Verlet approach.

▶ $\mathbf{F}_i \leftarrow \mathbf{F}_i + \mathbf{F}_i^p$	: POLYMER
▶ $\mathbf{r}_i \leftarrow \mathbf{r}_i + (\Delta t)\mathbf{u}_i + \frac{(\Delta t)^2}{2m}\mathbf{F}_i$	: SOLVENT, POLYMER
▶ $\widehat{\mathbf{u}}_i \leftarrow \mathbf{u}_i + (\Delta t)\mathbf{F}_i$	: SOLVENT, POLYMER
▶ compute $\widehat{\mathbf{F}}_i(r_{ij}, \widehat{u}_{ij})$	: SOLVENT, POLYMER
compute $\widehat{\mathbf{F}^p}_i(r_{ij})$	: POLYMER
▶ $\widehat{\mathbf{F}}_i \leftarrow \widehat{\mathbf{F}}_i + \widehat{\mathbf{F}}_i^p$	: POLYMER
▶ $\mathbf{u}_i \leftarrow \mathbf{u}_i + \frac{\Delta t}{2m}[\widehat{\mathbf{F}}_i + \mathbf{F}_i]$	: SOLVENT, POLYMER
▶ update $\mathbf{F}_i \leftarrow \widehat{\mathbf{F}}_i$	: SOLVENT, POLYMER
update $\mathbf{F}_\mathbf{i}^\mathbf{P} \leftarrow \widehat{\mathbf{F}}_i^p$	: POLYMER

▷Analyzer

TABLE 3.2: Overview of the time-staggered velocity-Verlet approach.

Step	Polymer	Solvent
▶ $\mathbf{r}_i \leftarrow \mathbf{r}_i + (\Delta t)\mathbf{u}_i + \frac{(\Delta t)^2}{2m}\mathbf{F}_i$		: SOLVENT
$\mathbf{F}_i \leftarrow \mathbf{F}_i + \mathbf{F}_i^p$	: POLYMER	
$\mathbf{r}_i \leftarrow \mathbf{r}_i + (t)\mathbf{u}_i + \frac{(t)^2}{2m}\frac{\Delta t}{t}\mathbf{F}_i$	: POLYMER	
$\widehat{\mathbf{u}}_i \leftarrow \mathbf{u}_i + (t)\mathbf{F}_i$	: POLYMER	
compute $\widehat{\mathbf{F}}_i^p(r_{ij})$	: POLYMER	
$\times \frac{\Delta t}{t}$ times		
▶ $\widehat{\mathbf{u}}_i \leftarrow \mathbf{u}_i + (\Delta t)\mathbf{F}_i$		: SOLVENT
▶ compute $\widehat{\mathbf{F}}_i(r_{ij}, \widehat{u}_{ij})$		: SOLVENT, POLYMER
compute $\widehat{\mathbf{F}}_i^p(r_{ij})$	: POLYMER	
$\widehat{\mathbf{F}}_i \leftarrow \widehat{\mathbf{F}}_i + \widehat{\mathbf{F}}_i^p$	: POLYMER	
$\mathbf{u}_i \leftarrow \mathbf{u}_i + \frac{t}{2m}[\widehat{\mathbf{F}}_i + \mathbf{F}_i]$	: POLYMER	
$\times \frac{\Delta t}{t}$ times		
▶ $\mathbf{u}_i \leftarrow \mathbf{u}_i + \frac{\Delta t}{2m}[\widehat{\mathbf{F}}_i + \mathbf{F}_i]$		: SOLVENT, POLYMER
▶ update $\mathbf{F}_i \leftarrow \widehat{\mathbf{F}}_i$		: SOLVENT, POLYMER
update $\mathbf{F}_i^p \leftarrow \widehat{\mathbf{F}}_i^p$		: POLYMER

▷Analyzer

one, but we can anticipate small differences if the ratio $\frac{\Delta t}{\delta t}$ is not too large and if the (outdated) forces are applied in the correct manner during the δt cycle. The algorithm is summarized in Table 3.2, (see Symeonidis and Karniadakis, 2006 for details).

3.2.5.4 *Polymer migration in microchannels and nanochannels*

Polymer depletion and cross-stream migration phenomena in micro- and nanochannels are important in microfluidic devices and a variety of biological systems. These effects might be relevant in physical processes such as adsorption, lubrication, wall-slip, and polymer transport. Depletion layers arise from steric wall repulsion (de Gennes 1979), when a polymer solution is placed in confined geometries. Depletion was observed in the region next to the fluid–solid interface in several experiments (Fang, Hu, and Larson 2005; Chen *et al.* 2005), and was simulated using various methods including: lattice Boltzmann (Usta, Butler, and Ladd 2006), Brownian Dynamics (Khare, Graham, and de Pablo 2006) and Dissipative Particle Dynamics (Millan *et al.* 2007). An asymptotic analytical solution (Eisenriegler and Maassen 2002) of the depletion layer for an ideal chain in the presence of purely repulsive wall predicts depletion to be effective at about one radius of gyration from the confining surface. Therefore, in micro- and nanochannels the layer is often of the same order as the channel width, and greatly affects the polymer distribution across the channel.

In presence of flow (e.g. Poiseuille, Couette), the polymer migration phenomena changes the polymer distribution across the channel. Several experimental observations (Fang *et al.* 2005; Chen *et al.* 2005) show polymer migration from the walls towards the channel centerline. However, simulations (Usta *et al.* 2006;

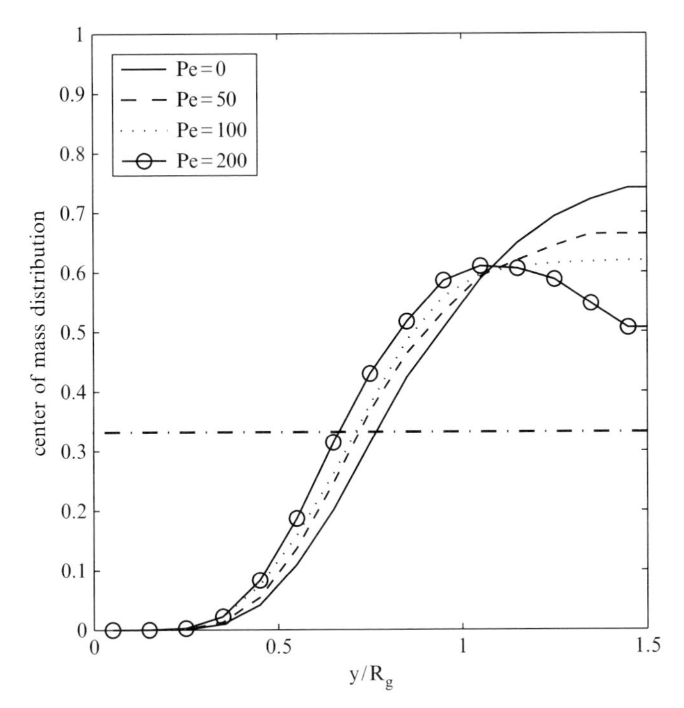

FIG. 3.9: Polymer center-of-mass and conformation distributions in Poiseuille flow. The polymer is modeled with 16 beads and the channel height is three times the radius of gyration. (Courtesy of Dmitry Fedosov.)

Khare *et al.* 2006; Millan *et al.* 2007) showed that polymer migration might proceed towards the walls as well as to the channel centerline. (Fedosov 2010) employed DPD simulations to investigate systematically the depletion layer and polymer migration in microchannels and nanochannels. The simulations were performed at Peclet number $Pe = 50$, 100, and 200 for several channels. Figure 3.9 displays the effect of the Pe number on the results for the center-of-mass distributions; the gap width is three times the radius of gyration of the polymer. The center-of-mass distributions of Fig. 3.9 show the dynamic depletion layer to be steadily reduced relative to the hydrostatic case ($Pe = 0$) as Pe is increased. Results not shown here also suggest that the effect is strongest for the smallest gap. In contrast, migration of polymer from the centerline towards the walls becomes more pronounced as the gap size increases, and between $Pe = 100$ and 200 the significant development is a distribution with two off-center peaks.

3.2.5.5 *Red Blood Cells in Microcirculation*

The human red blood cell (RBC) has a biconcave shape with the diameter of about $8\,\mu m$. When RBCs pass through capillaries, whose diameters are comparable or

Fig. 3.10: Successive snapshots of deformation of coarse-grained model from the DPD simulations of the RBC flow in a microchannel. From left to right: a) The model is placed in the channel with the fluid at rest. b,c) The deformation of the model 0.008 and 0.016 seconds after the body force driving the flow is applied. d) The shape of the model at steady flow. e) The model recovers it equilibrium biconcave shape 0.2 seconds after the body force is turned off. Only a portion of the microchannel is shown for or clarity. (Courtesy of Igor Pivkin.)

smaller than that of RBC, they deform under the flow resistance to a parachute-type shape and return back to their original biconcave shape once the hydrodynamic forces cease. The ability of RBC to deform—which originates in the membrane properties—influences, in turn, the blood flow resistance. The RBC membrane is composed of a lipid bilayer and an attached cytoskeleton. The cytoskeleton consists primarily of spectrin proteins, which form the network by linking short actin filaments, resulting in junction complexes. There are about 3×10^4 such junction complexes per RBC. In (Pivkin 2005) a coarse-grained model based on the spectrin-level RBC model of (Discher, Boal, and Boey 1998) was developed using mean-field theory and was applied to DPD simulations in capillaries of 10 microns in diameter. The RBC was modeled as a collection of DPD particles and this model was immersed in DPD fluid. In the capillaries of this diameter, the blood velocity is typically about 1 mm/s. Initially, the fluid is at rest and the RBC is placed in the middle of the channel while a body force is applied in the axial direction to drive the flow. The RBC deforms under the flow conditions and after some transition period assumes the parachute type shape shown in Fig. 3.10, which is commonly observed in experiments (Tomaiuolo, *et al.* 2007). After the body force is *turned off* the flow slows down and eventually the DPD fluid returns to rest, while the RBC recovers its equilibrium biconcave shape.

3.3 Nanofluidics modeling: molecular simulations

Significant advances have been made to understand fluid flow and chemical reactivity issues in micron-scale devices. While some critical issues arise when considering transport through micron-scale structures, a significant portion of micro scale devices can be treated as though it was mesoscopic or even macroscopic. In contrast, fluid transport through nano scale devices is fundamentally different from transport in micron-scale devices. Specifically, the fundamental issues encountered in nanofluidics can be different from microfluidics because of the following reasons: (i) the surface-to-volume ratio is very high in nanofluidic systems (ii) the channel critical dimension can be comparable to the size of the fluid molecules under investigation (iii) the Debye length (typically ranging from several Angstroms to several tens of Angstroms) characterizing the length scale of ionic interactions in a solution can be comparable to the channel critical dimension. In microfluidic devices, the Debye length is negligible compared to the channel width (iv) the composition of the channel wall as well as its roughness does not significantly affect the flow at microscales. For nanoscale flows, these are important parameters (v) macroscopic definitions and assumptions that transport properties (diffusivity, viscosity, etc.) are constant may not be accurate at nanoscale dimensions (vi) the validity of the classical continuum theory can be questionable for confined nanochannels (vii) the issue of boundary conditions at solid–liquid interfaces at nano scales is not very well understood (viii) anomalous behavior such as immobilization of water, charge inversion, flow reversal, and finite transport for no surface charge density have been observed in confined nanopores, while these are not issues in micro-scale fluid and ion transport.

A popular approach to understand nanoflows is to employ molecular dynamics simulations. A good introduction of molecular dynamics method is provided in (Karniadakis, *et al.* 2005). Equilibrium and non-equilibrium molecular dynamics simulations have been performed to understand structure and dynamics of water and electrolytes. The static behavior includes density distribution, dipole orientation, hydrogen-bonding and clustering, and contact angle of water to the surface in nano-confinement environment. These properties are of fundamental interest and of critical importance. For example, understanding of the density distribution and dipole orientation of the water molecules near a surface is crucial for understanding the electrochemical reaction at a surface (Henderson, *et al.* 2001). There is an extensive literature on the density distribution and dipole orientation of water near a one dimensional surface (Henderson, *et al.* 2001; Yeh and Berkowitz 2000; Spohr, *et al.* 1998; Galle and Vortler 1999; Gordillo and Marti 2003; Puibasset and Pellenq 2003; Muller and Bubbins 1998) (1-D confinement), inside a cylindrical pore (Henderson, *et al.* 2001; Allen, *et al.* 1999; Rovere and Gallo 2003; Walther, *et al.* 2001; Allen, *et al.* 2002; Gallo, *et al.* 2000; Lynden-Bell and Rasaiah 1996b; Leo and Maranon 2003; Gallo, *et al.* 2002a; Gallo, *et al.* 2002b; Green and Lu, 1997) (2-D confinement) and inside a cavity (Levinger 2002; Brovchenko, *et al.* 2001; Egorov and Brodskaya, 2003) (3-D confinement). In recent years, experimental techniques have improved dramatically and it is now possible to probe the structure of a liquid at atomistic detail (Toney, *et al.* 1994; Cheng, *et al.* 2001). For example, based on x-ray scattering measurements, Toney and co-workers (Toney, *et al.* 1994) have proposed that water is ordered in layers extended to about three molecular diameters from an electrode surface and that water density near a charged electrode is very high. Though these experiments can provide a good insight into water density distribution and dipole orientation, they cannot provide detailed and direct information of these functions—rather, they provide information about the integrals involving these functions. In addition, these experiments can probe only relatively simple geometry and cannot be used easily to study how the various parameters (e.g. surface characteristics) influence the static behavior of water in confined states. As a result, molecular dynamics simulations have been used extensively to understand static behavior of water.

Understanding the dynamic behavior of water is critical to many biological and engineering applications. For example, the study of the diffusion of water molecules through nanochannels can help explain the operating mechanism of the water channels, which are responsible for many important biological processes in the cell (Sui *et al.* 2001; Hummer *et al.* 2001b; Beckstein and Sansom 2003). Since diffusion is usually the dominant transport mechanism at small scales, the diffusion of water in nanochannels using molecular dynamics has been explored in detail in the past. Several researchers have investigated the diffusion of water inside artificial cylindrical nanopores (Allen *et al.* 1999; Lynden-Bell and Rasaiah 1996b; Zhou *et al.* 2003), inside silica nanopores (Rovere and Gallo 2003; Spohr *et al.* 1999; Gallo *et al.* 1999), inside carbon nanotubes (Leo and Maranon 2003; Mashl *et al.* 2003; Paul and Chandra 2003), and inside slit nanopores (Brovchenko *et al.* 2001; Zhang *et al.* 2002a). Water in confined nanochannels can exhibit very interesting and different physical characteristics compared to that of bulk water. The properties of water in confined nanochannels can depend strongly on the type of the surface (hydrophilic *vs* hydrophobic, channel wall structure) and on whether or not the nanopore surface is charged.

Molecular dynamics simulations have also been used to understand structure and transport of electrolytes in nanochannels (Freund 2002; Marry *et al.* 2003; Qiao and Aluru 2003; Qiao and Aluru 2004; Zhu *et al.* 2005). In particular, by comparing ion concentrations and velocity profiles obtained from molecular dynamics simulations and classical theories, the limitations of the classical theory for nanochannels have been revealed. In addition, molecular dynamics simulations have provided unique insights into the composition of the electrical double layer, shear viscosity, and hydrogen bonding in the interfacial layer, and charge inversion and flow reversal phenomena.

Here, in Section 3.3.1, we discuss recent advances in molecular dynamics simulations—specifically, in accurately treating the external electric field which is central to understanding electrokinetic flows at nano scale. In addition to molecular effects, quantum effects can also be important for nanoflows. Carbon nanotubes, because of their unique physical properties, are good test beds to understand quantum effects in nanoflows. In Section 3.3.2, we discuss the significance of quantum effects by considering the case of water in carbon nanotubes. Boron nitride nanotubes have recently received significant attention because of their exciting physical properties. In Section 3.3.3 we present results to show that boron nitride nanotubes have superior water permeation properties compared to nanopores made of many other existing materials.

3.3.1 *Recent developments in molecular dynamics*

Recent developments in molecular dynamics (MD) simulation of biological membranes and artificial nanochannels have made it possible to understand ion and water transport and selectivity in nano-confined geometries (Tuckerman and Martyna 2000). Atomistic models of infinitely long nanochannels have been studied in great detail to analyze the structure and diffusion of water molecules inside the channel (Shrivastava and Sansom 2000; Hu, *et al.* 2000). Non-equilibrium MD studies on membranes with an externally applied field have helped in understanding transport, dynamical properties, and energetics of ions in nanochannels (Lynden-Bell and Rasaiah 1996a). To investigate the entry of ions and water into the nanochannel/nanopore due to an applied external electric field, a nanochannel/nanopore with finite length immersed in a bath was investigated using Brownian dynamics (Li *et al.* 1998) and MD simulations (Crozier *et al.* 2001). The external electrical potentials are applied at either end of the bath–nanochannel–bath system (hereafter, simply referred to as the nanochannel–bath system) and a uniform external electric field was assumed to be applied throughout the nanochannel–bath system. This is referred to as the 'uniform field' approximation in this chapter. Typically, the size of the bath is much bigger compared to the critical size (width or diameter) of the nanochannel and assuming a uniform applied electric field in the entire nanochannel–bath system can be a gross simplification. Ramakrishnan *et al.* (2004) modified the uniform field approach, and applied a higher external electric field inside the nanochannel and a lower electric field in the bath, accounting for the water polarization near the channel wall. In this section, we present the self-consistent MD (SCMD) methodology (Raghunathan and Aluru 2007) to accurately account for the applied external electric field in an MD simulation due to immersed electrodes.

3.3.1.1 *Self-consistent formulation* A nanochannel–bath system filled with an electrolyte solution is shown in Fig. 3.11. The Poisson equation, solved in the electrostatic domain Ω (see Fig. 3.11) for this system is given by:

$$\nabla \cdot (\epsilon_i \nabla \phi) = -\rho_{ions} = -F \sum_{i=1}^{N} z_i c_i \text{ in } \Omega \tag{3.8}$$

where ϕ is the electrostatic potential, ϵ_i $(= \epsilon_r \epsilon_0)$ is the dielectric permittivity of the fluid medium, ϵ_r is the relative dielectric permittivity, ϵ_0 is the dielectric constant of vacuum, ρ_{ions} is the charge density of the ions, N is the number of ionic species, c_i is the concentration of the ith ionic species, z_i is the valence of ion i, and F is the Faraday's constant.

The externally applied electrostatic potential on the electrodes at Γ_1 and Γ_2 will enforce a Dirichlet boundary condition, i.e.

$$\phi = \phi_1 \text{ on } \Gamma_1 \tag{3.9}$$

$$\phi = \phi_2 \text{ on } \Gamma_2 \tag{3.10}$$

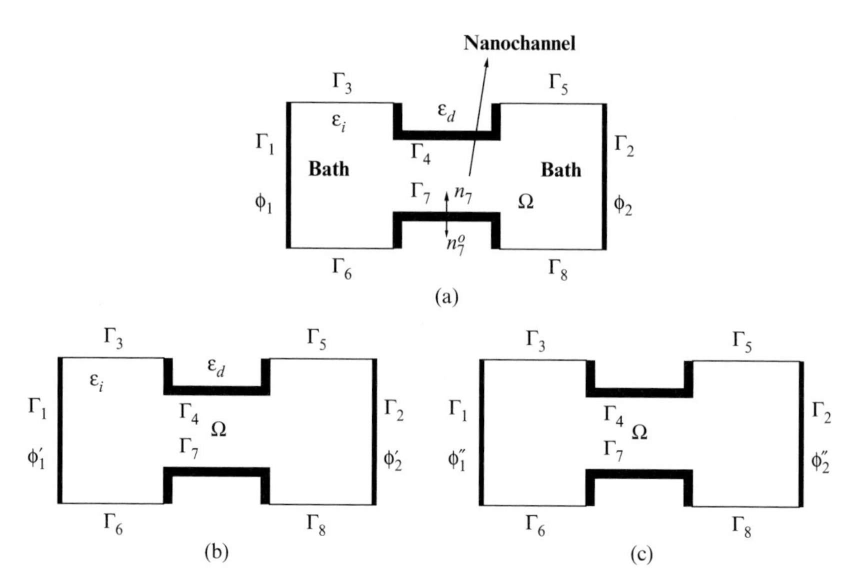

Fig. 3.11: The domain and boundaries of (a) the electrostatic problem to be solved and its decomposition into (b) the particular solution and (c) the homogeneous solution.

The boundary conditions on the top and bottom surfaces Γ_3, Γ_5, Γ_6 and Γ_8 of the baths are given by (symmetric boundary conditions):

$$\frac{\partial \phi}{\partial n} = 0 \text{ on } \Gamma_j, \text{ for } j = 3, 5, 6, 8 \tag{3.11}$$

The boundary conditions on the channel wall interfaces Γ_4 and Γ_7 are:

$$\epsilon_i \frac{\partial \phi}{\partial n_j} = \epsilon_d \frac{\partial \phi}{\partial n_j^d} \text{ on } \Gamma_{\mathrm{j}}, \text{ for j} = 4, 7 \tag{3.12}$$

where n_j and n_j^d denote the inward and outward normals on the boundary Γ_j. ϵ_d is the dielectric constant of the exterior (or outward) medium. Note that in the approach described below, the value of ϵ_i and ϵ_d are not needed to satisfy this boundary condition.

Due to the 'layering' of water molecules in the interfacial region (Ballenegger and Hansen 2005), the dielectric constant of water at the interface and inside the nanochannel is typically reduced (Stern and Feller 2003) and cannot be estimated accurately without a molecular description of water. Therefore in solving for electrostatics in MD, the charge density of water molecules, ρ_{water}, is included explicitly in the Poisson equation, i.e. $\rho_{total} = \rho_{ions} + \rho_{water}$. Assuming a dielectric constant of vacuum ($\epsilon_r = 1$), eqn (3.8) becomes,

$$\nabla \cdot \nabla(\phi) = -\frac{\rho_{total}}{\epsilon_0} \text{ in } \Omega \tag{3.13}$$

There are two major issues while trying to solve eqn. (3.13) directly. First, as water molecules are accounted for in the charge density explicitly, the domain Ω has to be resolved into a fine grid with a grid-size comparable to the radius of a single water molecule (≈1 Å). Second, the charge density is assumed to have a continuous distribution, which can lead to an error in the calculated electrostatic potential, as the ionic charges are discrete in nature. Both these limitations are resolved by decomposing eqn. (3.13) into a *particular solution* for the charge distribution in an unbounded domain and a *homogeneous solution* within the given domain, with appropriate boundary conditions (Tikhonov and Samarskii 1990). Therefore, the solution to the Poisson problem ϕ, is now the sum of two solutions:

$$\phi = \phi' + \phi'' \tag{3.14}$$

$$\frac{\partial \phi}{\partial n} = \frac{\partial \phi'}{\partial n} + \frac{\partial \phi''}{\partial n} \tag{3.15}$$

where ϕ' is the particular solution and ϕ'' is the homogeneous solution. The particular solution ϕ' is computed by solving the following equation in the unbounded domain:

$$\nabla \cdot \nabla(\phi') = -\frac{\rho_{total}}{\epsilon_0} \text{ in } \Omega \cup \Omega' \tag{3.16}$$

where Ω' is the domain exterior to Ω. The solution to eqn. (3.16) is given by (Tsang 1997):

$$\phi'(R) = \int_{\Omega \cup \Omega'} \frac{\rho_{total}(r)dv}{4\pi\epsilon_0 |R - r|} \tag{3.17}$$

$$= \sum_{i=1}^{n_q} \frac{q_i}{4\pi\epsilon_0 |R - r|} \tag{3.18}$$

Here, R is the point in space where the potential ϕ' is calculated, $\rho_{total}(r)$ is the charge density at position r, N_q is the charge on the ith atom in the system, n_q is the total number of charged atoms in the system, which includes both ions and water, and r_i is the coordinate of the charged atom i in the system. The particular solution can be obtained by using a direct summation calculation of the Coulomb interactions or by using particle methods for fast summations (Greengard and Rokhlin 1987). Here, the Particle Mesh Ewald (PME) (Darden *et al.* 1993) method was used with a correction of the ewald summation given by Yeh *et al.* (1999) to remove the periodicity in the x-direction and calculate ϕ' from eqn. (3.18). The solution to eqn. (3.18) gives $\phi' = \phi'_i$ on Γ_i, $i = 1, 2$ and $\frac{\partial \phi'}{\partial n} = \left(\frac{\partial \phi'}{\partial n}\right)_i$ on Γ_i, $i = 3, \cdots, 8$.

To obtain the homogeneous solution, appropriate boundary conditions need to be applied such that the decomposed solution retains the same boundary conditions as applied in the original problem. The homogeneous equation along with the corresponding boundary conditions are given by:

$$\nabla \cdot \nabla(\phi^{''}) = 0 \text{ in } \Omega \tag{3.19}$$

$$\phi^{''} = \phi_1 - \phi^{'} \text{ on } \Gamma_1 \tag{3.20}$$

$$\phi^{''} = \phi_2 - \phi^{'} \text{ on } \Gamma_2 \tag{3.21}$$

$$\frac{\partial \phi^{''}}{\partial n} = \frac{\partial \phi}{\partial n} - \frac{\partial \phi^{'}}{\partial n} \text{ on } \Gamma_i,\ i = 3,\ 5,\ 6,\ 8 \tag{3.22}$$

$$\frac{\partial \phi^{''}}{\partial n} = 0 \text{ on } \Gamma_i,\ i = 4,\ 7 \tag{3.23}$$

Since the water and wall dielectric properties are explicitly taken into account in the particular solution, we impose the boundary condition given in eqn (3.23) on $\mathit{\Gamma}_4$ and $\mathit{\Gamma}_7$ for the homogeneous equation. Other boundary conditions (e.g. $\frac{\partial \phi}{\partial n} = 0$) on $\mathit{\Gamma}_4$ and $\mathit{\Gamma}_7$ can also be easily implemented in the approach presented here. The implementation of the decomposition discussed above is discussed in detail in (Raghunathan and Aluru 2007).

3.3.1.2 *Dual control volume Grand Canonical approach*

In the nanochannelbath setup (see Fig. 3.11 (a)), K^+ and Cl^- ions migrate from one bath to the other through the nanochannel in the presence of an external electric field. This causes the depletion or the accumulation of the ions in the bath. The concentration of the K^+ and Cl^- ions needs to be maintained at a constant in the bath. For this purpose, a Grand Canonical (μVT) MD simulation is needed to maintain the chemical potential (and concentration), volume and temperature in the MD simulation (Zheng *et al.* 2005; Attard 2004; Zhang *et al.* 2002*b*; Boinepalli and Attard 2003). The two baths in the nanochannel–bath setup are considered to be the two control volumes where the concentration needs to be maintained. Using the DCV-GCMD technique proposed by Heffelfinger *et al.* (1994), the chemical potential (and concentration) within the control volumes is maintained. The original DCV-GCMD technique was developed for Lennard–Jones (LJ) fluids and recently (Raghunathan and Aluru 2007) it has been extended to electrolytic solutions. Though the DCV-GCMD technique is necessary for the model presented here, it is computationally more expensive than standard NVT or NVE MD simulations due to the extra effort required in the insertion/deletion of particles into the system.

For the DCV-GCMD simulation of the nanochannel–bath system, a control volume is placed in each of the two baths (see Fig. 3.12). The control volumes are provided with the chemical potential of the ionic species (K^+/Cl^- ions) that needs to be maintained during the MD simulation. During the simulation, the stochastic

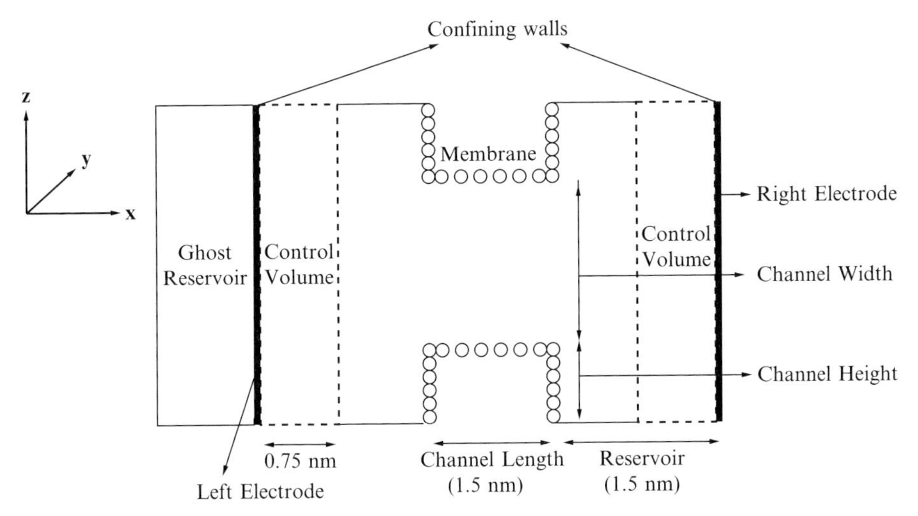

FIG. 3.12: A sketch of a nanochannel–bath system to investigate electric field-mediated transport. The two control volumes on either side are maintained at a constant chemical potential using DCV-GCMD.

insertion/deletion trials are performed every n_{MD} time steps in MD. The MD simulation is 'frozen' in time after every n_{MD} time steps. This is followed by a number of stochastic insertion/deletion trials, denoted by n_{trial}, performed in each control volume, using the Metropolis algorithm (Hammersley and Handscomb 1964). This 'stochastic' step maintains the chemical potential and concentration of the K^+ and Cl^- ions in the control volume. A 'ghost' reservoir containing LJ atoms is attached to the simulation setup (see Fig. 3.12). The 'ghost' reservoir contains LJ atoms with no charge and these atoms do not enter the nanochannel–bath system due to the presence of the confining walls. Whenever an ion is deleted from the control volume, it is placed into the 'ghost' reservoir and if an atom needs to be inserted into the control volume, it is taken from the 'ghost' reservoir. This reservoir thus ensures that the total number of atoms in the MD simulation is constant during DCV-GCMD. It should be noted that the 'ghost' reservoir is used only to overcome a software limitation and is not critical for the implementation of DCV-GCMD. During the trial insertions/deletions, the change in potential energy U, for inserting or deleting an ion in the control volume is calculated by summing the electrostatic and LJ interactions. U is used in the insertion/deletion acceptance criteria in DCV-GCMD, as well as to calculate the chemical potential of the control volume using the Widom's particle insertion method (Ding and Valleau 1994). More details on the DCV-GCMD simulation procedure can be found in Raghunathan and Aluru (2007).

During the stochastic step, the probability of a trial insertion or deletion (of K^+ or Cl^- ions) in the control volumes is equal. The probability of adjusting the number of any one of the K^+ or Cl^- ions in the control volume is also the same. Trial insertion of an ionic species i into the control volume is accepted if:

$$\left[\frac{1}{N_i(c)+1}\right]\exp\left[\beta\mu_i(c)+\ln\frac{V(c)}{\Lambda_i^3}-\beta\Delta U_i\right]\geq\xi \tag{3.24}$$

where $V(c)$, $N_i(c)$ and $\mu_i(c)$ are the volume, number of ions of type i, and the input chemical potential of type i in the control volume c, respectively. $\beta=\frac{1}{k_BT}$ where k_B is Boltzmann's constant and T is the temperature, ΔU_i is the change in the potential energy of creating a particle of type i, Λ_i is the De Broglie wavelength, and ξ is a random number generated uniformly in (0,1). If accepted, the ion inserted into the control volume is given a velocity from the Gaussian distribution (Lísal *et al.* 2004) at the temperature at which the MD simulation is performed. A deletion attempt in the control volume is successful if:

$$N_i(c)\exp\left[-\beta\mu_i(c)-\ln\frac{V(c)}{\Lambda_i^3}-\beta\Delta U_i\right]\geq\xi \tag{3.25}$$

In order to verify that the chemical potential in the control volume is maintained, we also calculated the chemical potential of each ionic species in the control volumes during the DCV-GCMD simulation using the Widom's Particle insertion method (Ding and Valleau 1994; Henderson 1983). In this method the 'test' atoms (ions), are inserted at random positions into the control volume at regular intervals. These inserted 'test' ions are assumed to have no influence on the existing ions and water molecules present in the system. If u_i denotes the energy of the 'test' ion i with respect to N atoms present in the control volume, the chemical potential of that ionic species μ' in the control volume is given by

$$\beta\mu'=\ln\left(\frac{N_i(c)\sigma^3}{V(c)}\right)-\ln\langle\exp(-\beta\Delta U_i)\rangle \tag{3.26}$$

where $\langle\rangle$ denotes the grand canonical average over the test atoms (ions) of type i in the control volume and σ is the LJ ion–ion distance parameter. The average is calculated over statistical data from randomly inserted test particles of ionic species i into the control volume.

3.3.1.3 *Results*

We present a few examples comparing the self-consistent molecular dynamics (SCMD) simulations, uniform applied electric field MD simulations, and classical continuum theory simulations. The classical continuum theory for electric field-mediated ion transport is given by the Poisson equation (3.8) and the Nernst–Planck equation,

$$\nabla\cdot J_i=0\text{ in }\Omega \tag{3.27}$$

$$J_i = -D_i \nabla c_i - \mu_i z_i F c_i \nabla \phi \text{ in } \Omega, \ i = 1 \text{ to } N \tag{3.28}$$

In the above equations, ϕ is the electrostatic potential obtained by solving eqn. (3.8), J_i is the flux of ionic species i, c_i is the concentration of ith ionic species, z_i is the valency of the species i, D_i and μ_i are the diffusion coefficient and mobility of species i, respectively, and N is the total number of species. For continuum calculations, the relative permittivity of water, ϵ_r, in eqn (3.8), is assumed to be 80, its bulk value.

Electrical potential and electric field calculations in MD simulations of nanochannels are important as they effect the water and ion transport, ion selectivity, and current-voltage (I-V) characteristics in the channel. The electrical potential ϕ (ϕ is the total potential including the particular solution and the homogeneous solution) is calculated as described above. Figure 3.13 shows the SCMD and uniform field MD results for the electrical potential and the x-component of the electric field in the axial direction (x-direction) along the centerline of the 3.5 nm nanochannel for channel heights of 0.5 and 1.5 nm, respectively. For the potential and electric field profiles obtained in the large channel height case of 1.5 nm, SCMD simulation shows a higher potential drop in the channel region and a corresponding reduction in the potential drop in the reservoir region when compared to the 0.5 nm channel height case. In the uniform field MD formulation, the

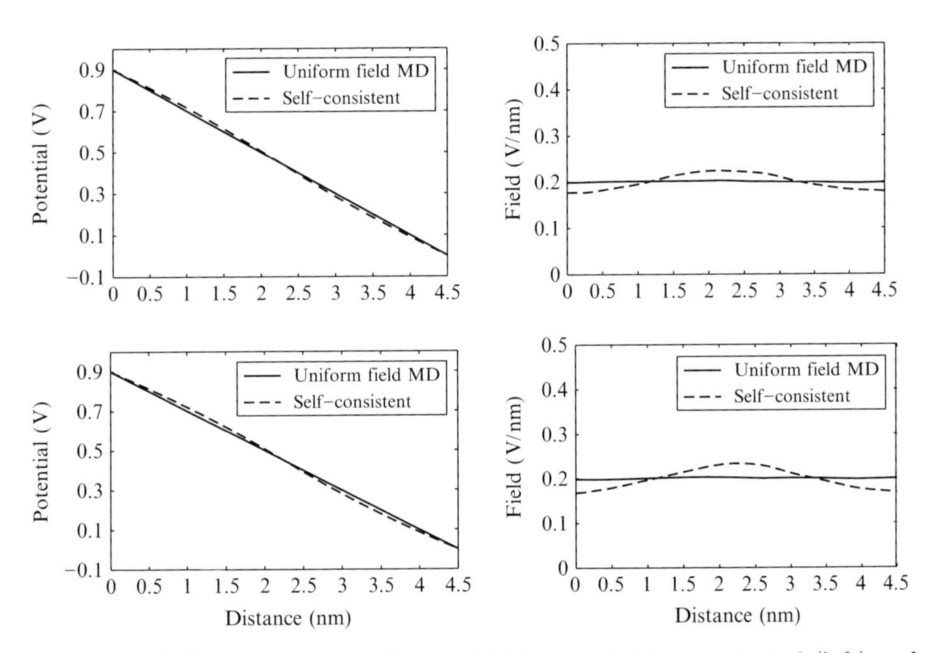

Fig. 3.13: Self-consistent *vs* uniform field MD calculations: potential (left) and electric field (right) calculations in the axial direction along the center of the 3.5 nm wide nanochannel; channel height = 0.5 nm (top), 1.5 nm (bottom).

potential drop is lower across the nanochannel compared to SCMD. Note that in the uniform field MD formulation, though the applied electric field is constant, the total electric field is not necessarily constant in the entire nanochannel–bath system due to the electric field from the particular solution. In the case of 1.5 nm channel height, a difference of 8 per cent in the electric field along the channel centerline is observed in the middle of the channel between the two formulations. This difference in the electric field can result in different ion migration rates and transport properties in the nanochannel system from SCMD and uniform field MD simulations. SCMD results for the electrical potential and electric field are also compared with the results from the classical theory (see eqn. (3.28)). We do not expect the results from the classical theory to match with SCMD in the entire system, but along the centerline of the channel, the classical theory can be a good approximation because of the bulk-like nature of the electrolyte along the centerline. Figure 3.14 compares the electrical potential and field obtained from continuum theory with SCMD for nanochannels with channel heights of 0.5 and 1.5 nm. The potential profiles at the centerline region along the x-direction are in good agreement.

To understand the effect of the applied electric field on the channel selectivity, SCMD and uniform field MD simulations were compared for a 1 nm wide channel

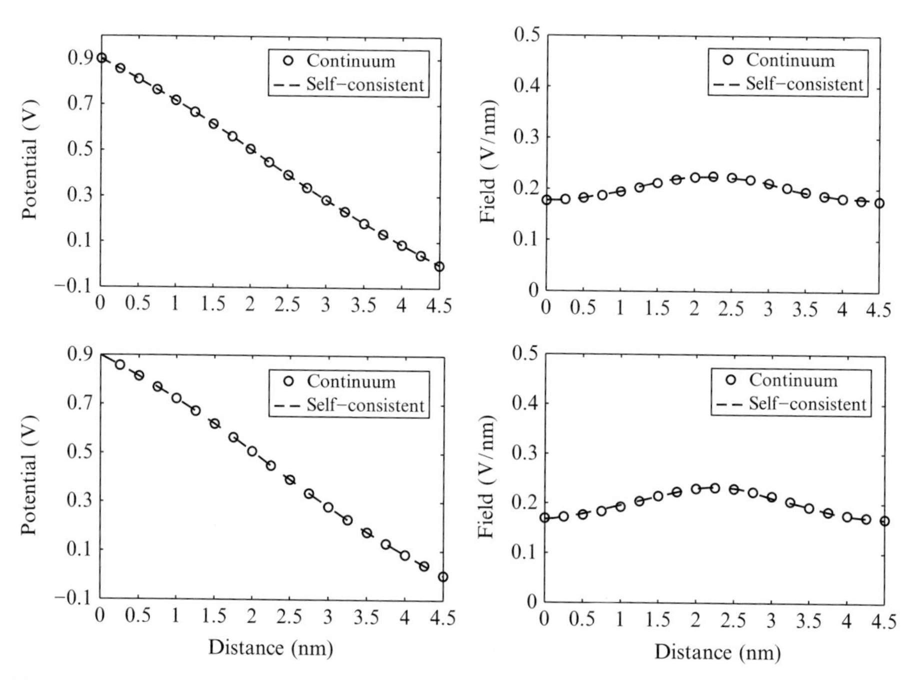

Fig. 3.14: Self-consistent *vs* continuum calculations: potential (left) and electric field (right) in the axial direction along the center of the 3.5 nm wide channel; channel height = 0.5 nm.

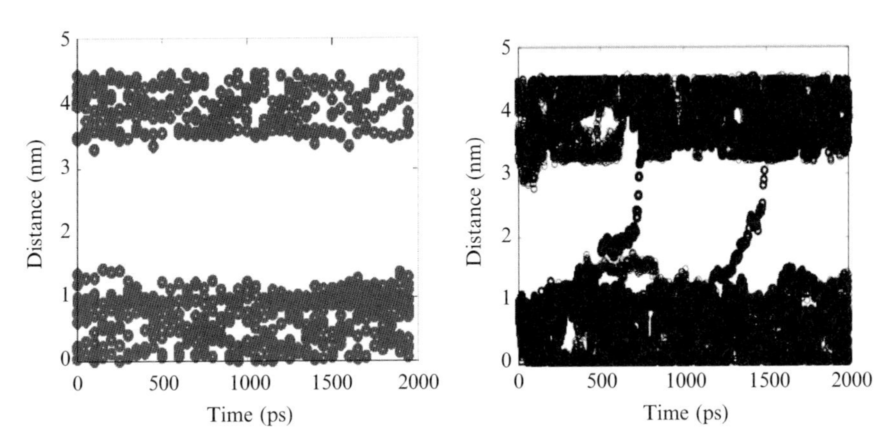

FIG. 3.15: Trajectory of K^+ ions in the uniform field MD (left) and SCMD (right) simulations of a 1 nm wide channel.

with a channel height of 1.5 nm. The electrode potentials were kept the same as in the previous examples. It is found that during a 2 ns simulation time, the K^+ ions go through the channel in SCMD simulations, while in the uniform field MD simulations no ion gets through the nanochannel. Figure 3.15 shows the trajectory of a K^+ ion transport in the SCMD and uniform field MD simulations in 2 ns. This dramatic difference in the transport of ions observed between the two simulations implies that the selectivity of a nanochannel can be accurately studied only with a proper treatment of the applied electric field in molecular dynamics simulations.

3.3.2 *Quantum effects in nanoflows*

To understand quantum effects in nanoflows, we consider the example of water in carbon nanotubes. An extensive review of water in carbon nanotubes can be found in Sinnott and Aluru (2006). In this section, we will review a recent result (Won *et al.* 2006) where it has been shown that water in two different carbon nanotubes with similar diameters can exhibit different transport properties. To account for chirality effects, quantum-mechanical calculations have been performed to compute the partial charges. Using quantum partial charges, computed from 6-31G**/B3LYP density functional theory (DFT), in molecular dynamics (MD) simulations, it was found that water inside (6,6) and (10,0) single-walled carbon nanotubes (SWCNTs) with similar diameters but with different chiralities has remarkably different structural and dynamical properties.

The system that has been simulated consisted of the SWCNT, water, and a slab. The SWCNT was fixed in the slab as shown in Fig. 3.16. The simulation box was 3.36 nm × 3.234 nm × 5.82 nm. The reservoirs were initially filled with water. The extended simple point charge (SPC/E) model was used in the MD simula-

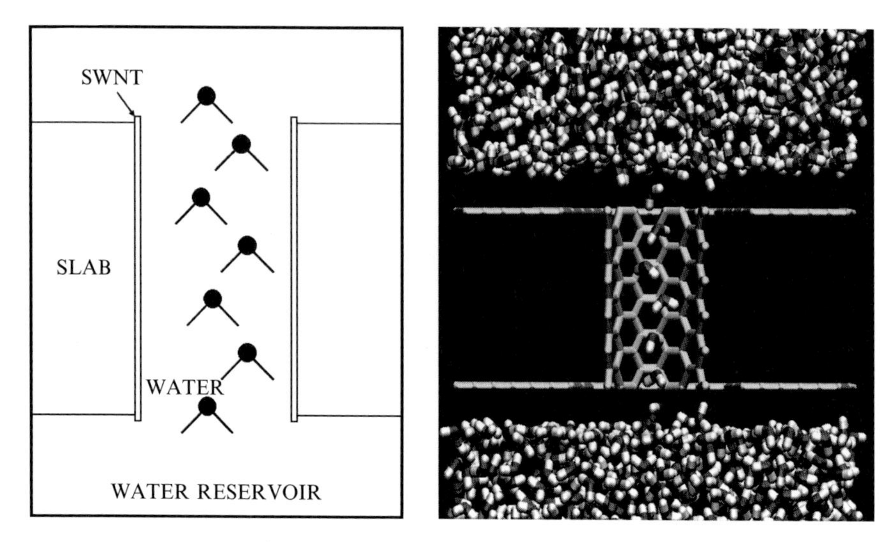

Fig. 3.16: Schematic (left) and visualization (right) of a single-walled carbon nanotube in a water bath.

tions. Neutral carbon atoms and carbon atoms with the partial charges were employed in separate simulations to investigate the effect of partial charges.

The CHelpG partial charges for the (6,6) armchair and the (10,0) zig-zag SWCNTs along the axis of the tube with water inside the tube were calculated based on four different representative configurations from equilibrium MD simulations performed without partial charges on the carbon atoms. Figure 3.17 shows the CHelpG partial charges for the tubes. The effect of the tube chirality on the nanotube electrostatics can be understood by comparing the top and bottom figures in Fig. 3.17. The magnitude of the average partial charge (when water is included) at the ends of the zigzag SWCNT is approximately 4.5 times greater than that of the armchair SWCNT. This suggests that the electrostatic potential at the ends of the nanotube depends strongly on chirality.

Using MD simulations, we investigated the dipole orientation and transport properties of water confined to a nanotube for over 28 ns of simulation time. All the water molecules in the uncharged and charged (6,6) tube are oriented such that the dipole vectors of the water molecules point either towards the top water reservoir or towards the bottom water reservoir at any instant. Once a water molecule flips and reverses its orientation, all other water molecules flip simultaneously. Unlike in the (6,6) tube, since the magnitude of the partial charges near the ends of the (10,0) SWCNT is high, a strong electrostatic field is generated at the ends of the tube. The direction and magnitude of the electrostatic field cause the water molecules at both the ends to be reoriented such that hydrogen atoms point towards the bath region (Fig. 3.18). Since the water dipoles at both ends point

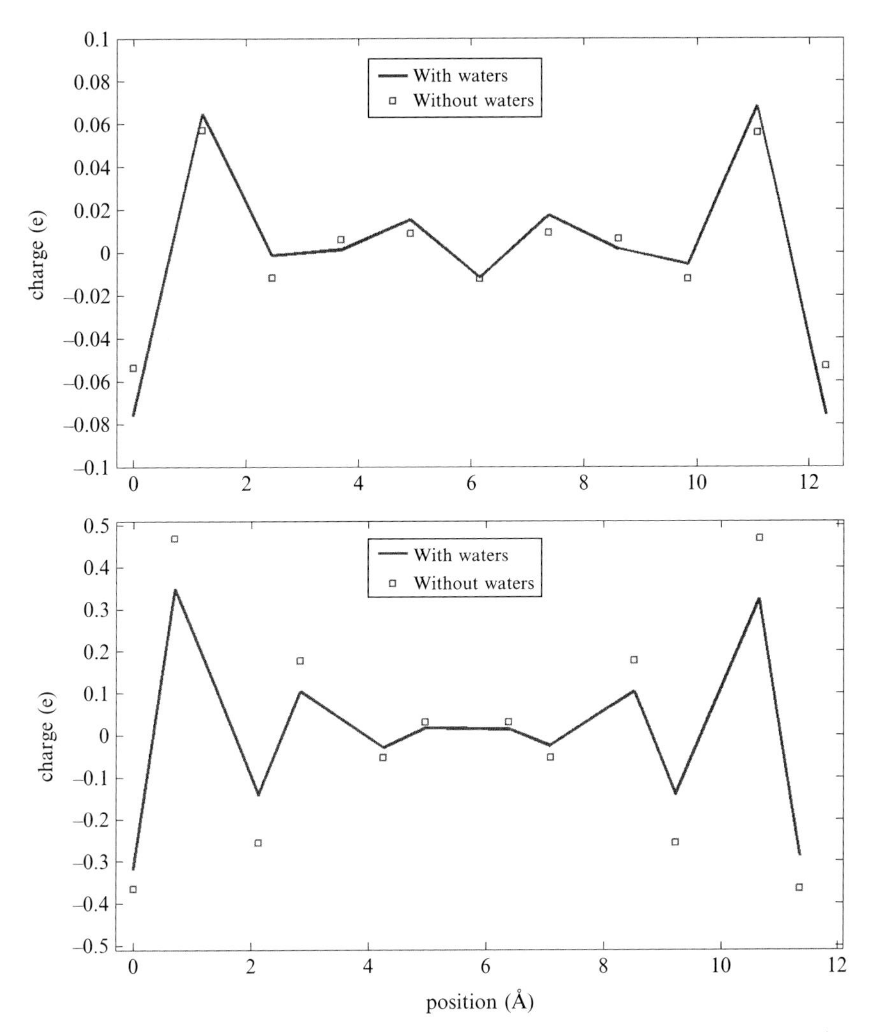

Fig. 3.17: Partial charge distribution on a (6,6) SWCNT (top) and a (10,0) SWCNT (bottom) along the tube axial direction. Solid line: Water is included in the calculation, Squares: No water is included in the calculation. A tube starts at 0 Å in the axial direction.

away from each other, the central water molecule forms an L-defect which is similar to the water orientation in an aquaporin-1 channel.

The effect of the partial charges on water transport is evaluated by the self-diffusion coefficient and translocation time, which is defined as the time taken for a water molecule to travel from one end of the tube to the other end of tube. Table 3.3 shows that the water transport rate is enhanced when partial charges are introduced into the system.

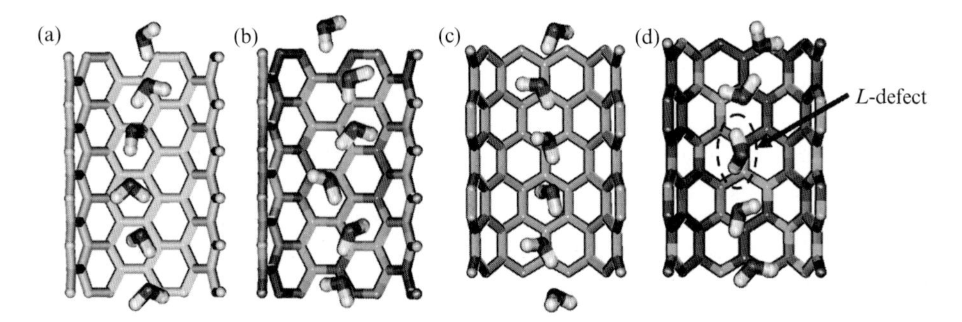

Fig. 3.18: A snapshot of water dipole orientation in (a) an uncharged (6,6) SWCNT, (b) a partially charged (6,6) SW CNT, (c) an uncharged (10,0) SWCNT, and (d) a partially charged (10,0) SWCNT. Color of the SWCNTs in (b) and (d) represents the partial charge magnitude of each carbon atom.

Table 3.3: Single-file transport properties of water in uncharged and partially charged SWCNTs.

Chirality	Charge type	Diffusion coefficient $[10^{-5}\ \mathrm{cm}^2/\mathrm{s}]$	No. of translocation events [#/us]	Translocation time [ps]
(6,6)	No partial charge	1.16 ± 0.08	4.94	379.54
(6,6)	Partial charge	1.41 ± 0.04	9.13	248.93
(10,0)	No partial charge	1.10 ± 0.05	5.31	353.96
(10,0)	Partial charge	1.28 ± 0.13	6.69	263.99

The effect of the partial charges on the water orientation and transport properties can be understood in more detail by evaluating the potential of mean force (PMF) of the water molecules inside the tubes. The energy barriers of the tubes are found to be 4.80 k_BT and 5.11 k_BT for the partially charged (6,6) and (10,0) SWCNTs. These water permeation barriers are of the same order of magnitude as that of an aquaporin-1 water channel which is about 5 k_BT. The PMF analysis shows that a larger energy fluctuation inside the partially charged (10,0) tube induces a slower diffusion coefficient when compared with the partially charged (6,6) tube. The substantial water-nanotube electrostatic interaction in the partially charged (10,0) tube changes the single-file water structure and gives rise to the formation of an *L*-defect in the center of the nanotube.

3.3.3 *Water conduction in boron nitride nanotubes*

Molecular dynamics (MD) simulations by Hummer and his coworkers (Hummer *et al.* 2001a) have indicated that a (6,6) CNT with a diameter of approximately 8 Å can conduct water at 300 K. The wetting behavior of the carbon nanotube was confirmed by experimental study (Majumder *et al.* 2005). Boron nitride nanotubes (BNNTs) possess many of the superior properties of CNTs such as a high Young's modulus (Chopra and Zettl 1998) and thermal conductivity (Chang *et al.* 2005), but unlike CNTs, BNNTs exhibit high resistance to oxidation, and a wide band-gap regardless of its chirality. These exciting properties allow BNNTs to act as complementary materials to CNTs or even replace the CNTs for applications requiring chemical stability, high-temperature resistance or electrical insulation. Recently, the superior water permeation properties of BNNTs have also been reported (Won and Aluru 2007). Specifically, we have reported that a (5,5) BNNT with a diameter of 6.9 Å and a finite length of 14.2 Å can conduct water while a CNT with a similar diameter and length has only intermittent filling of water.

To gain fundamental insights into the water permeability of BNNTs and to compare the results with those in CNTs, we performed molecular dynamics simulations on two finite length (5, 5) BNNT and (5,5) CNT, with diameter of 6.9 Å and length of 14.2 Å. Both tubes are saturated at the ends with hydrogen atoms. The MD simulation domain consists of the nanotube, water, and a slab. The nanotube is fixed in slab as shown as in Fig. 3.2. The boron and nitride atoms in the BNNT and carbon atoms in CNT are modeled as uncharged Lennard–Jones particles. The extended simple point charge (SPC/E) model was used in the simulations. The simulations were performed for 40 ns with 1.0 fs time step, with a constant pressure of 1 bar and a constant temperature of 300 K.

The MD simulation was started with an empty (5,5) BNNT. The water from the water reservoir filled the empty (5,5) BNNT within 50 ps of simulation time (Fig. 3.20a). There was a small fluctuation in the number of water molecules occupying the BNNT, but during the simulation time of 40 ns, the BNNT is occupied by approximately 5 water molecules forming a single file chain. In addition, water molecules traversed the tube at a rate of about 5.1 molecules per nanosecond. Despite being the same size as the BNNT, the initially empty (5,5) CNT was barely filled by water. A few water molecules enter the CNT during the simulation time. The water molecules inside the CNT formed a single-file chain only a few times but it didn't last longer than 1 ns (Fig. 3.19a). The water structure inside the nanotube can be best understood by examining the water density distribution in the tube axial and radial directions (Fig. 3.20b, Fig. 3.20c, Fig. 3.19b, and Fig. 3.19c). Figure 3.20b shows the water density averaged within 0.8 Å from the tube axis along the z-axis. The five peaks indicate that there are five favorable locations of water inside the tube. Unlike in (5,5) BNNT, the axial density distribution of water inside (5,5) CNT (Fig. 3.19b) indicates that the water molecules like to reside at the ends of the nanotube.

The different wetting behavior of the (5,5) BNNT and CNT can be explained by the potential of mean force (Kjenllander and Greberg 1998)(PMF) analysis. The mean force distribution was obtained by sampling the force experienced by the water molecules in each bin. The energy barrier (E_b) was found to be 5.29 K_BT for the (5,5) BNNT and 9.83 K_BT for the (5,5) CNT. An energy barrier of around 5 K_BT is considered small for water permeation (de Groot and Grubmller 2005; Borgnia and Agre 2001). The probability for the water to overcome the energy

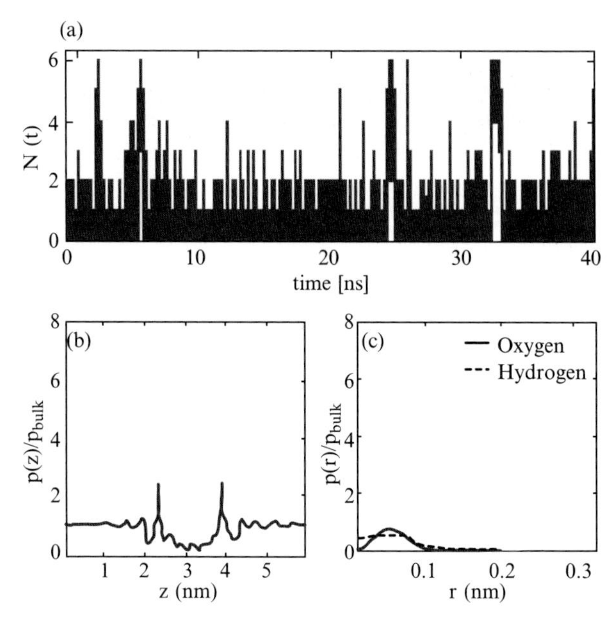

Fig. 3.19: Number N(t) of water molecules inside the (5,5) single-walled carbon nanotube (a), density of water along tube axial direction, $\rho(z)$, within 0.8 Å from the tube axis (r=0), The CNT is located from 2.43 to 3.85 nm (b), Radial density profiles ρ(r) of water inside the tube (c).

barrier, $K_r \propto \exp\left(-E_b/K_BT\right)$ indicates that the water molecules in the (5,5) CNT case have approximately 90 times lower chance to overcome the barrier compared to the (5,5) BNNT case.

To further understand the wetting behavior of BNNT, we performed an MD simulation where the nitride atoms in the BNNT were assigned the same Lennard–Jones (LJ) parameters as the boron atom and another MD simulation where the boron atoms in the BNNT were assigned the same LJ parameters as the nitride atom. We found that in the first case (when all the atoms in the BNNT had boron LJ parameters) the BNNT was almost empty of water and in the second case (when all the atoms in the BNNT had nitride LJ parameters) the BNNT was completely filled with water. From these results, we can conclude that even though the water molecule–boron atom van der Waals attractions ($\epsilon_{B\text{-}O}$= 0.5082 KJ/mol) are stronger than the water molecule–carbon atom van der Waals attractions ($\epsilon_{C\text{-}O}$= 0.4340 KJ/mol), they are not strong enough for water molecules to enter the BNNT and water conduction in BNNT is primarily due to the water molecule-nitride atom van der Waals interactions ($\epsilon_{N\text{-}O}$= 0.6277 KJ/mol).

Table 3.4 presents a comparison between the properties of water in the (5,5) BNNT (our simulations indicate that this is the smallest diameter finite length BNNT that will conduct water) and (6,6) CNT (our simulations also indicate that

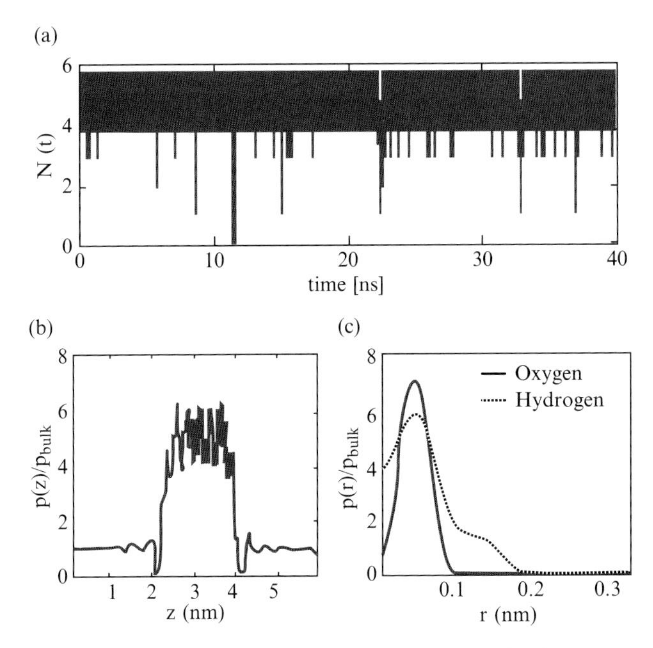

Fig. 3.20: Number N(t) of water molecules inside the (5,5) boron nitride nanotube (a), density of water along tube axial direction, $\rho(z)$, within 0.8 Å from the tube axis (r=0), the CNT is located from 2.43 to 3.85 nm (b), radial density profiles ρ(r) of water inside the tube (c).

Table 3.4: Comparison of water properties in the (5,5) BNNT and (6,6) CNT.

Tube type	Diameter [Å]	Flipping frequency	D_z [10^{-5} cm^2/s]	Average # of H- bonds per water
(5,5) BNNT	6.9	0.05	1.18 ± 0.06	0.84
(6,6) CNT	8.2	0.375	1.16 ± 0.08	0.86

this is the smallest diameter finite length CNT that will conduct water). The dipole orientation of water in the (5,5) BNNT is similar to that in the (6,6) CNT. All the water molecules in both tubes orient such that the dipole vectors of the water molecules point either towards the top water reservoir or towards the bottom water reservoir at any instant. Once a water molecule flips and reverses its orientation, all other water molecules flip simultaneously. During 40 ns of simulation time, the flipping occurs only twice in the BNNT, while water molecules in the (6,6) CNT flipped 6 times over a 16 ns sampling time. The axial diffusion coefficient of water in BNNT was found to be 1.18×10^{-5} cm^2/s which is close to the diffusion coefficient in the (6,6) CNT. The average number of hydrogen bonds per water molecule in both tubes is also very similar.

3.4 Multiscale modeling

In many microfluidic applications we often need to model accurate molecular details of a surface, e.g. flow over nanotubes with hydrophobic surface or nano-patterned channel walls or proteins adhered to the wall. MD simulations can be employed to capture the molecular details but they can only be used for a short time and very small length scales due to their large computational requirements compared to the computational complexity of continuum discretizations. Multi-scale approaches both in time and space can overcome this difficulty. They may involve MD simulations coupled directly to continuum (e.g. Navier–Stokes equations) or coupled to mesoscopic descriptions (e.g. DPD or LBM). The multiscale approach would potentially extend the range of applicability of each specific method and provide a unifying description of liquid flows from nano scales to larger scales. The *incompressible* Navier–Stokes equations are involved in the coupling, although some authors have attempted to couple MD to compressible Navier– Stokes equations; see, for example, Flekkoy *et al.* (2000).

There are two main approaches in coupling MD to Navier–Stokes equations:

- The state-exchange method of O'Connell and Thompson (1995).
- The flux-exchange method of Flekkoy *et al.* (2000).

In the *state-exchange* method the state information bewtween the MD simulation and the Navier–Stokes equations is transferred through an overlap region where the particles' dynamics is constrained. In O'Connell and Thompson (1995) the constrained dynamics is imposed via a relaxation technique with a constant *ad hoc* relaxation parameter. This was improved in the work of Wang and He (2007) where the relaxation (coupling) parameter was computed dynamically invoking the momentum consistency condition, and also in Nie *et al.* (2004) where an external force was used. Also, in Hadjiconstantinou and Patera (1997) and in Werder *et al.* (2005) an alternating Schwarz method is used to solve sequentially the problems in the continuum and atomistic domains and state exchange is performed until convergence is achieved. In addition, in Werder *et al.* (2005) a boundary force is applied that takes into account the physical properties of the fluid being simulated.

In the MD-continuum coupling the use of an overlap region is necessary, since MD induces local structure in the fluid at interfaces and especially for density (Koplik and Banavar 1995). The two-domain coupling would therefore have the form shown in the sketch of Fig. 3.21, consisting of a region (MD) where the molecular dynamics simulation is performed, the region where the Navier–Stokes equations will be solved, and an overlap region, where both descriptions are valid. However, in order to terminate the MD region, in addition to standard particle motion in the overlap region a constraint should be imposed of the form

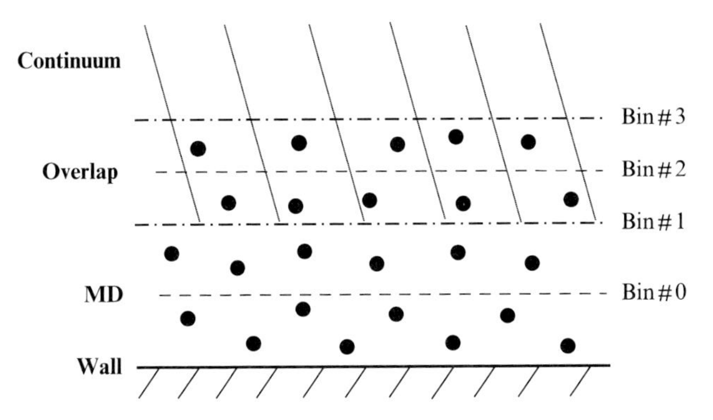

Fig. 3.21: Domain for MD-continuum coupling.

$$\sum_{n=1}^{N_i} p_n - M_i v_i = 0,$$

where N_i is the total number of particles and M_i is the mass of the continuum fluid element in the i-bin, and p_n is the momentum of the nth particle in the v-direction. This constraint can be integrated into standard Lagrange's equations governing the motion of the rest of the molecules in the MD region. This approach was successfully implemented by O'Connell and Thompson (1995), who used an overlap region of 14σ. A free parameter in this approach is the strength of the constraint ξ in relation to the extent of the overlap region. Specifically, the equations of motion in the i-bin are

$$\dot{\mathbf{x}}_i = \frac{p_i}{m} + \xi \left[\frac{M_i}{mN_i} - \frac{1}{N_i} \sum_{n=1}^{N_i} \frac{p_n}{m} \right], \tag{3.29}$$

$$\dot{\mathbf{P}}_i = -\frac{\partial V}{\partial \mathbf{x}}, \tag{3.30}$$

where m is the particle mass. Small values of the parameter will provide an inadequate coupling between MD and Navier–Stokes, while large values will lead to excessive damping of particle fluctuations, which in turn will lead to divergence in the solution. In O'Connell and Thompson (1995), a value of $\xi = 0.01$ was used for simulating a slow-startup Couette flow. It was shown that the best choice of ξ is to have $\Delta t_{\mathrm{MD}}/\xi$ greater than the autocorrelation time t_{vv}, where Δ_{MD} is the time step in the MD simulation. This relaxation approach does not handle correctly the mass flux at the MD-continuum interface, and this is an important limitation of the method. Improvements of this method have been presented in Nie *et al.* and He 2007; Werder *et al.* (2005).

The *flux-exchange* method developed by Flekkoy *et al.* (2000) is *conservative*, since it relies in the matching of fluxes of mass and momentum between the MD and Navier–Stokes domains. In particular, the mass flux continuity is enforced by the equation

$$ms(\boldsymbol{x}, t) = A\rho\boldsymbol{v}\cdot\boldsymbol{n},$$

where $\boldsymbol{A}$ is the area and $\boldsymbol{n}$ the unit normal vector. Here $s(\boldsymbol{x}, t)$ is the number of particles that need to be added to (if s is positive) or removed (if s is negative) the top bin of the overlap domain. Similarly, the momentum flux continuity is enforced by the equation

$$ms(\boldsymbol{x}, t)\langle\mathbf{u}'\rangle + \sum_i \mathbf{F}_i = A\boldsymbol{\Pi}\cdot\boldsymbol{n},$$

where $\mathbf{u}'$ is the velocity of the added or removed particles, and $\mathbf{F}_i$ is an external force acting on particle i in the region of flux-exchange from continuum to particles $(P \leftarrow C)$. Flekkoy *et al.* employed the *compressible* Navier–Stokes equations in the coupling, and thus the momentum flux tensor $\boldsymbol{\Pi}$ has the form

$$\boldsymbol{\Pi} = \rho\boldsymbol{v}\boldsymbol{v} + p - \mu(\nabla\boldsymbol{v} + \nabla\boldsymbol{v}^T - \nabla\cdot\boldsymbol{v}) - (\mu/3)\nabla\cdot\boldsymbol{v}.$$

Combining the two equations, we observe that the momentum equation is satisfied if the mass equation is satisfied, but in addition we need to enforce

$$\langle\mathbf{u}'\rangle = \boldsymbol{v} \text{ and } \sum_i \mathbf{F}_i = A(\boldsymbol{\Pi} - \rho\boldsymbol{v}\boldsymbol{v})\cdot\boldsymbol{n}.$$

In order to avoid drifting of particles, a weight function $g(x)$ was introduced. In particular, this function obeys $g(x) = g'(x) = 0$ for $x \leq 0$ and diverges as

$$g(x \to L/2) \propto \frac{1}{L/2 - x}$$

at the edge of the region $(P \leftarrow C)$. The coordinate x runs parallel to $\boldsymbol{n}$, and $x = 0$ is in the middle of the region $(P \to C)$ where flux-exchange from continuum to particles take place. Also, L is the size of the bins in the $\boldsymbol{n}$ direction. In addition, in order to maintain thermal equilibrium it was found necessary to thermalize the particles in the subdomain $P \leftarrow C$ using Langevin dynamics. Specifically, a force of the form

$$\mathbf{F}_{Li} = -\gamma(\mathbf{u}_i - \boldsymbol{v}) + \tilde{\mathbf{F}}, \left\langle\tilde{\mathbf{F}}(t)\tilde{\mathbf{F}}(t')\right\rangle = 2k_B T_\gamma \delta(t - t')$$

was added to the force $\mathbf{F}_i$; here γ is a measure of dissipation.

This scheme was tested for steady Couette and Poiseuille flow using a shifted Lennard–Jones potential, with time steps $\Delta t_{\mathrm{MD}} = 0.0017\tau$ on the MD side and $\Delta t = 100\Delta_{\mathrm{MD}}$ on the Navier–Stokes side.

In addition to MD simulations coupled to Navier–Stokes solutions it is also useful and computationally advantageous to develop coupling procedures between

MD and *mesoscopic* descriptions or between *mesoscopic* and *continuum* descriptions. For example, in Dupuis *et al.* (2007), LBM is coupled to MD models using the alternating Schwarz algorithm. A state-based algorithm is followed whereby velocity but also velocity gradients are exchanged at the interface. The hybrid MD-LBM method was used in simulations of liquid argon past and through nanotubes. A similar hybrid LBM-MD method was presented in Fyta *et al.* (2006) for simulating DNA translocation through a nanopore. It was shown that the hybrid method scales linearly with both solute size and solvent volume, making it an efficient tool for simulating the dynamics and energetics of *long* biopolymers. The LBM-MD approach models hydrodynamics by local interactions so there is no need to resort to long-range representations such as the Oseen tensor used in Brownian dynamics, see Section 3.2.2.

It is also possible to interface a mesoscopic method like DPD to Navier–Stokes equations (Fedosov, 2010) using an overlap region. In order to eliminate density fluctuations due to the pressure imbalance at the boundaries, we need to impose a pressure force which is computed by integrating the repulsive force multiplied by the radial distribution function over a spherical cap, the size of which is a function of the distance of the particle from the interface. In addition, to avoid erroneous slip velocities at the interface we need to impose adaptively a shear force whose value depends on the distance of the particle from the interface (Pivkin and Karniadakis 2006).

Since quantum effects can be significant for nanoflows, a multiscale method combining quantum, MD, and continuum theories based on the Poisson–Nernst–Planck and Stokes equations have also been developed. This method is described below.

3.4.1 *Hierarchical multiscale methods*

In this section, we describe a hierarchical multiscale approach for a full-scale characterization of water and ion transport in silica nanochannels. First, using the density functional theory (DFT), the molecular electrostatic potential which accounts for the quantum effects at the silanol surface is calculated. Partial atomic charges on the wall atoms are computed using a molecular electrostatic potential fit which are then used as inputs for molecular dynamics. Transport coefficients such as mobilities and diffusion coefficients are then computed from molecular dynamics simulations. Finally, the transport coefficients are used in the PNP equations to calculate the experimentally observable current-voltage characteristics.

In order to understand the electrokinetic transport in channels with a net zero surface charge, and to characterize the quantum-mechanical effects, we investigate transport in two types of channels (i) with wall partial charges from quantum calculations, and (ii) without any partial charges. We report electrokinetic transport in α-quartz channels of four different widths (4.1, 2.75, 2.1, and 1.1 nm) with 1 M KCl. Figure 3.22 shows the three levels of simulation hierarchy and the construction of the channel. In the following sections, we discuss in detail the

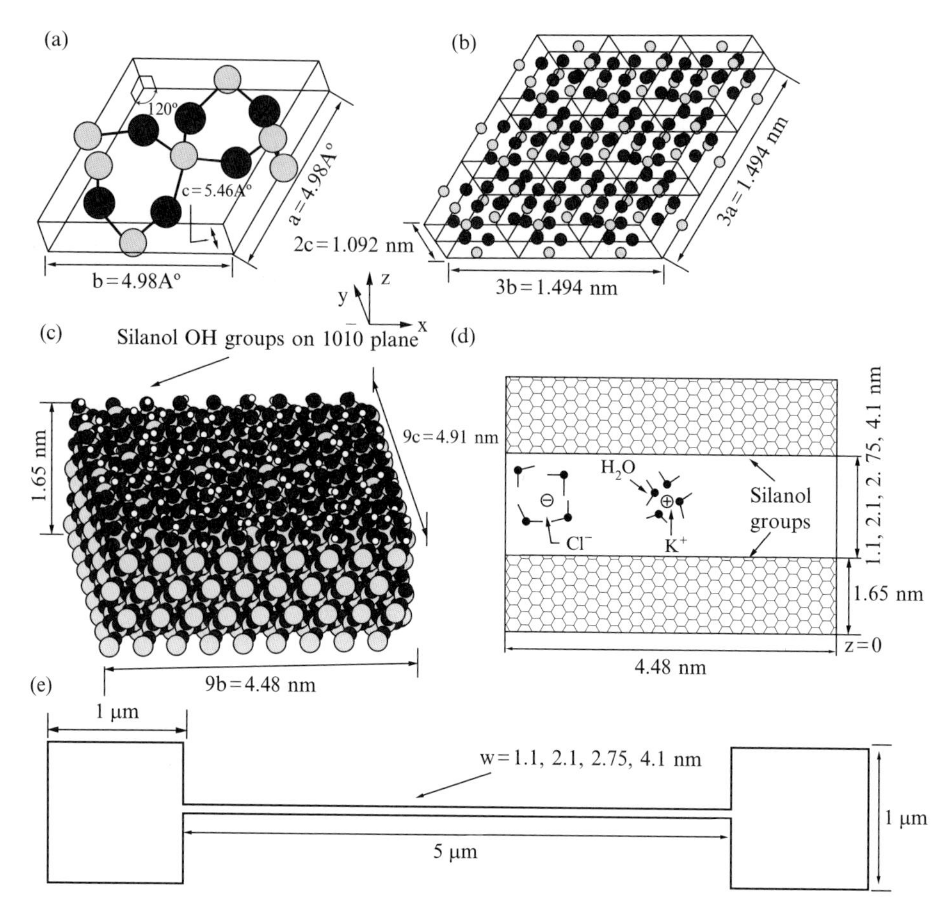

Fig. 3.22: Simulation system setup. (a) quartz unit cell. Circles with darker shade represent oxygen atoms and the circles with lighter shade represent silicon atoms. The unit cell has 3 Si and 6 O atoms. In the figure, silicon atoms at the unit cell edges and surfaces have also been included. (b) The DFT simulation domain comprising of 33 $\times$ 2 unit cells. (c) Channel wall for MD simulations made by periodic replication of the unit cells. In the x and y directions, 9 cells were replicated. In the z direction, a transformation was made from the parallelepiped to the rectangular structure. Silanol terminations on the surface after removing a layer of surface silicon atoms is also shown. (d) Schematic of the MD simulation. The axial length is 4.48 nm, and the width of each channel wall is 1.65 nm. (e) The continuum simulation domain.

multiscale framework used. We first describe the calculation of partial charges. Then, using molecular dynamics, we calculate the transport properties as a function of the channel width. Finally, the transport properties from molecular dynamics simulations are used to compute macroscopic I-V curves in the channels using continuum theory.

3.4.1.1 *Computation of quantum partial charges*

As far as silica–electrolyte interactions are concerned, among all the interaction parameters in the force field, the partial atomic charges are the most important since (i) they are the only parameters to capture bond polarization arising from quantum effects and (ii) their effects are long range and they can significantly influence the static and dynamic properties of water and ions. When the partial charges are greater than $\pm$ 1.0e they cause the coulombic interaction energy to be greater than that of bond, angle, or torsion energies by approximately two orders of magnitude (Heinz and Suter 2004). In force fields commonly cited in literature, the partial charges of bulk silicon span a wide range from $+0.5$e to $+4.0$e. Most of the existing calculations were based on geometry optimized silica clusters consisting of a few atoms rather than on silica slabs. Moreover, in many cases, surface silicon and oxygen atoms have been assumed to have the same charges as bulk atoms. So we performed density functional quantum calculations using Gaussian 03 (*et al.*, 2004) with the B3LYP model (Vosko *et al.* 1980; Lee *et al.* 1988) as the exchange correlation and a 6-31G* split valence polarization basis set to obtain the partial charges both for the surface and bulk silica and compared them with existing computations and experimental results.

The bulk quartz structure is a trigonal system characterized by 4-fold coordinated silicon atoms and 2-fold coordinated oxygen atoms. The unit cell has three Si and six O atoms and the primitive vectors are $a = b = 4.98$ Å, $c = 5.46$ Å with $\alpha = \beta = 90^\circ$ and $\gamma = 120^\circ$ (Fig. 3.22 (a)). To form the channel wall surface, the unit cells were periodically replicated along b and c directions with 9 unit cells and with 4 unit cells along the a direction. The surface structure was obtained by removing silicon atoms along the 1010 plane and attaching hydrogen atoms to the oxygens with a bond length of 1 Å to form a fully coordinated structure with silanol group terminations (Fig. 3.22 (c)).

To obtain the partial charges of the bulk and surface atoms, three types of systems were considered: (i) A quartz unit cell consisting of 6 oxygen and 3 silicon atoms which are periodic in all three directions representing bulk silica. (ii) A supercell which is a composite of $3 \times 1 \times 1$ unit cells that is periodic in b and c directions with 1 unit cell each but with 3 unit cells in the a direction the oxygens being terminated with silanol groups representing a slab geometry. (iii) A $3 \times 3 \times 2$ supercell structure (Fig. 3.22 (b)) which is non-periodic with silanol terminations in all three directions. Both Mulliken (Mulliken, 1955) partial charges based on partioning of the electronic charge density within different orbitals and CHelpG partial charges (Chirlian and Francl 1987; Francl *et al.* 1996) obtained by fitting point charges to the molecular electrostatic potential

TABLE 3.5: Partial charge computations on different structures.

Atom type	1 unit cell	3 × 1 × 1 supercell	3 × 3 × 2 supercell	3 × 3 × 2 supercell
Charge Type	Mulliken	Mulliken	Mulliken	CHelpG
Periodicity	3D	2D	Non-periodic	Non-periodic
Bulk Si	1.302	1.302	1.30	1.36
Bulk O	−0.651	−0.651	−0.65	−0.68
Surface Si	–	1.200	1.197	1.38
Surface O	–	−0.625	−0.621	−0.71
Silanol O	–	−0.70	−0.69	−0.77
Silanol H	–	0.45	0.45	0.44

were computed. Mulliken charges being based on charge partioning correlate well with experimental electron density observations. Since CHelpG charges are based on a molecular electrostatic potential fit, they can capture, though in an approximate way, higher order effects arising from dipoles and multipoles and are better suited for molecular dynamics simulations where partial charges are the only parameters used to include the effects of polarization.

Table 3.5 shows the results from the calculations. It was found that for the $3 \times 3 \times 2$ supercell, the interior atoms had the same Mulliken partial charge as that of a 3-D-periodic single unit cell system. Comparisons of the Mulliken charges of the surface atoms on the non periodic $3 \times 3 \times 2$ supercell and that of the surface atoms on the non periodic surface of the 2-D-periodic $3 \times 1 \times 1$ supercell gave very similar values. A non periodic supercell, if it is sufficiently large enough can capture the local electron density distribution as that of a periodic system.

3.4.1.2 *Molecular dynamics simulations* Molecular dynamics simulations were performed with the computed partial charges to obtain the structure and dynamics of water and ions for various channel widths. The computational domain consisting of the surface atoms and the confined electrolyte is shown in Fig. 3.22 (d). The structure of each interface was created by periodic replication of the unit cells. The system has two identical 1010 α-quartz interfaces which are mirror images of each other with respect to the channel center plane. Water and ions are confined in the region between the interfaces. The channel dimensions were $L_x = 9b = 4.48$ nm and $L_y = 9c$=4.91 nm and the thickness of the wall is 1.65 nm in the z-direction. Periodic replication gives a parallelepiped structure, which was transformed into a rectangular structure with silanol terminations as shown in Fig. 3.22 (c). Four different channel geometries with separation distance between the walls of 1.1 nm, 2.1 nm, 2.75 nm, and 4.1 nm were considered. The largest distance of 4.1 nm is sufficiently large enough to establish bulk water in the center of the channel.

We consider two types of potentials—Lennard-Jones and Coulomb. The Lennard-Jones parameters for Si and water were taken from the GROMACS force field (Lindahl *et al.* 2001), the ions from Koneshan *et al.* (1998) and the ion–Si and O–Si pairs were obtained by using the linear combination rule and the Si–Si parameters from Lindahl *et al.* (2001). The ions are modeled as Lennard–Jones atoms with point charges. Since the interface is dominated by the silanol groups, the wall-electrolyte van der Waals forces would be dominated by the contribution from oxygen atoms rather than silicon atoms. Most of the past force fields have been primarily developed to characterize the structural properties of silica and they have complex potentials with many body terms to accurately capture the conformational changes in silica. Since we are primarily interested in the electrokinetic transport and not in the structural properties of silica, flexible bonds are used to model only the surface silanol groups while the interior atoms are held fixed to reduce the computational cost. The flexible bonds in the silanol groups allow for some exchange of momentum and energy of the water and ions with the channel walls. For the silica surface, a 1010 α-quartz structure was chosen because recent experimental results in literature could be used to compare the interfacial water structure (Schlegel *et al.* 2002).

For water, we used the rigid, nonpolarizable SPC/E model (Berendsen *et al.* 1987) with partial charges for oxygen, $q_o = -0.8476$e, and two hydrogens, $q_H = +\ 0.4238$ and Lennard–Jones potential for oxygen with constraints using the SHAKE algorithm. The SPC/E model has been known to reproduce accurately the bulk properties of water especially relevant to electrokinetic transport such as diffusivity and dielectric properties (Sorenson *et al.* 2000). The number of water molecules is chosen such that the water concentration in the center of the channel is within 1.0 percent of 55 M. The wall silicon and oxygen atoms were fixed in space. The bonded interactions of the surface silanol groups were modeled using harmonic potentials with the force constants being the same as that in the SPC/E flexible model of water (Lindahl *et al.* 2001).

The simulation box dimensions along the periodic axes were $L_x = 4.48$ nm and $L_y = 4.91$ nm. Along the non-periodic z direction, in addition to the wall–electrolyte–wall system, a vacuum of $\approx$ 2.5–3 times the width of the wall electrolyte–wall system was added to obtain L_z = 15–18 nm depending on the channel width. Such a vacuum is more than enough for accurate computation of electrostatic interactions using a 2-D periodic PME. Berendsen (Berendsen *et al.* 1984) thermostat was used to maintain the temperature at 300 K with a time constant of 0.1 ps. The simulation was started from a random initial configuration and equilibrated for 2 ns before taking the statistics. For transport properties such as diffusion and mobility, a simulation time of 7 ns was used for generating enough statistics for calculating the properties as a function of the wall normal distance. For electroosmotic velocities, 50–70 ns simulation time was required to obtain reliable velocity profiles.

Figure 3.23 shows the comparison of the interfacial electron density calculated from atomic density distribution from molecular dynamics with that obtained from experimental X-ray diffraction methods on a water α-quartz interface (Schlegel *et al.* 2002). Results from simulations of 4.1 nm wide channels with both uncharged and partially charged wall atoms are presented. For the partially charged case, the positions of the peak of interfacial water layer and the silanol OH

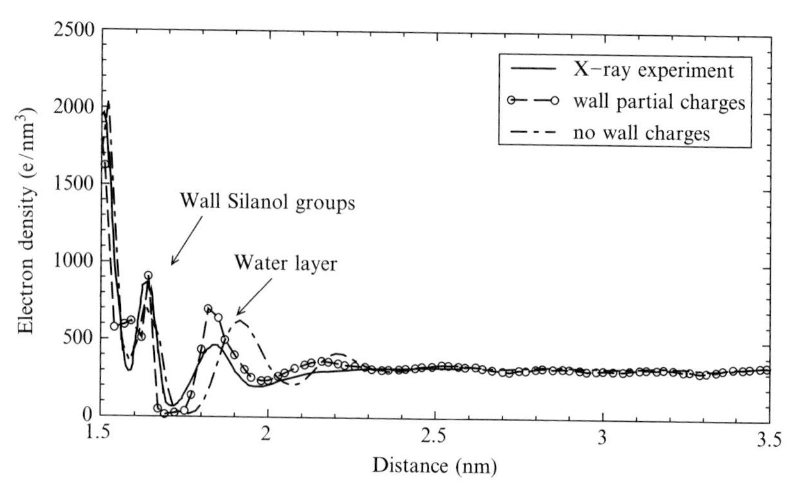

FIG. 3.23: Electron density plots at the surface of the 4.1 nm channel with partial charges compared to X-ray reflectivity data for water at a 1010 quartz surface. The experimental data is from Schlegel *et al.* (2002).

groups match reasonably well with that of experiments. Clearly, when there are no partial charges, interfacial atomic densities do not match well with experimental observations.

Figure 3.24 shows ion concentrations and water velocity profiles for uncharged and partially charged channels with widths of 4.1, 2.75, 2.1, and 1.1 nm. The partial charges affect the concentrations till about 1 nm from the wall. One of the most interesting phenomena arising from these concentration distributions is the existence of an electrical double layer near the surface when the net surface charge is zero. In the traditional double layer, there is a higher concentration of the counter ions than the co-ions near a charged surface. Here, though the net surface charge is zero, the cations and anions have different propensities to penetrate the surface water layer. This leads to differences in ionic concentration within the interfacial layer which in turn builds a finite charge and a diffuse layer near the interface. In order to ascertain whether there are any immobile layers of ions and water near the surface, we calculated the mean square displacement of water and ions and obtained the diffusion coefficients. In the case of the 4.1 nm channel, both for partially charged and uncharged cases, the diffusion coefficient of water is around 38 percent of the bulk value very close to the surface in the first water layer. But within about 1.5 nm from the surface, the bulk value is reached. This shows that the interfacial layer, though highly ordered, is not static but dynamic.

The oscillations in the ion concentration profiles give rise to electroosmotic flow under an applied electric field of 0.35 V/nm even though the solution in the channel is electroneutral (see Fig. 3.25 (left)). The direction of the flow is different in both partially charged and uncharged cases. The ionic mobilites obtained by computing the mean velocities of the ions are shown in Fig. 3.25 (right).

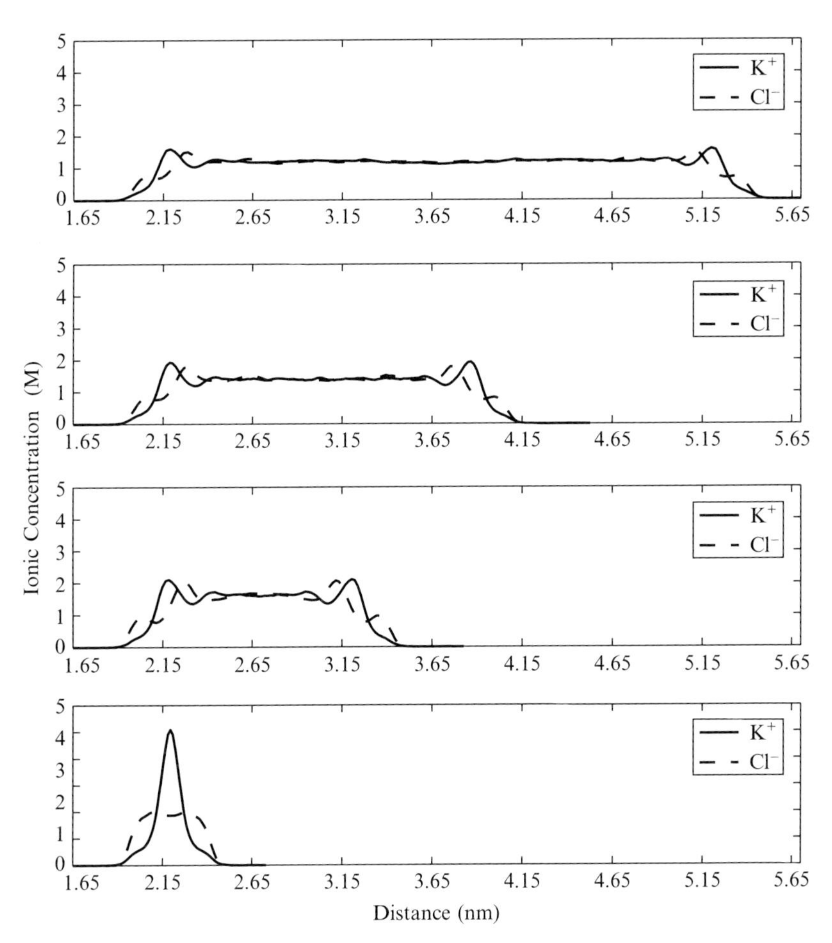

FIG. 3.24: Ionic concentrations (left) uncharged and partially charged channels (right).

3.4.1.3 *Continuum calculations*

The macroscopic ionic transport in the silica channel is described by the coupled Poisson–Nernst–Planck (PNP) and the Stokes equations (Oldham and Myland 1994; White 1994; Karniadakis *et al.* 2005) in continuum simulations which are solved self-consistently to obtain the concentration of the ions, electrical potential, and the bulk velocity. Using transport properties from atomistic simulations, we obtained the (current(I) versus Voltage(V)) I-V curves in channel bath systems using continuum theory. First, to ascertain whether the continuum models can adequately reproduce conductivities from MD simulations, we computed ionic conductances from continuum theory for the same channel dimensions as used in MD simulations with the same external electric field. The only parameter used

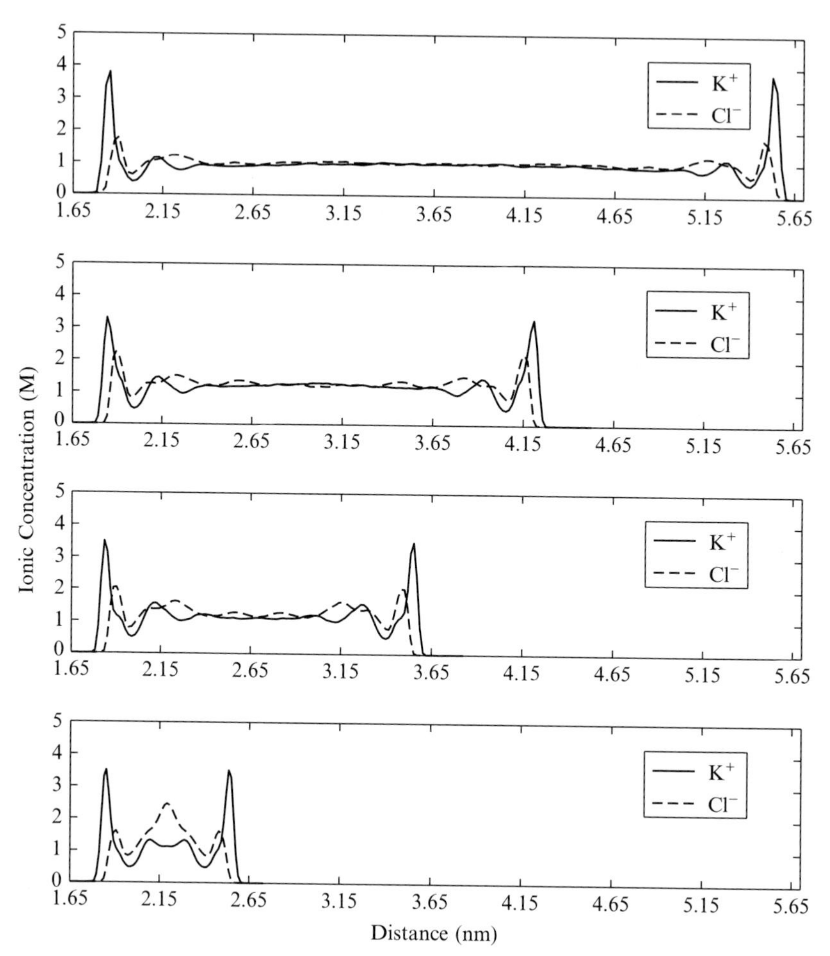

Fig. 3.24: (*continued*).

from MD was the mobilities at discrete bins. The MD and PNP calculations show very similar conductances. Once the continuum model has been shown to produce the same conductances as MD, we performed continuum calculations on channel–bath systems. Channels of the same widths as in MD simulations but with lengths of 5 μm were used to connect baths of micrometer dimensions (see Fig. 3.22(e)). Figure 3.26 shows the macroscopic I-V curves from continuum theory for different channel widths. At strengths of 1 M KCl, for channel widths of the order of 1–2 nm, the quantum partial charges reduce the magnitude of the current significantly but when the channel is 4.1 nm in width, the effect of the quantum partial charges is not so pronounced. As the channel width increases

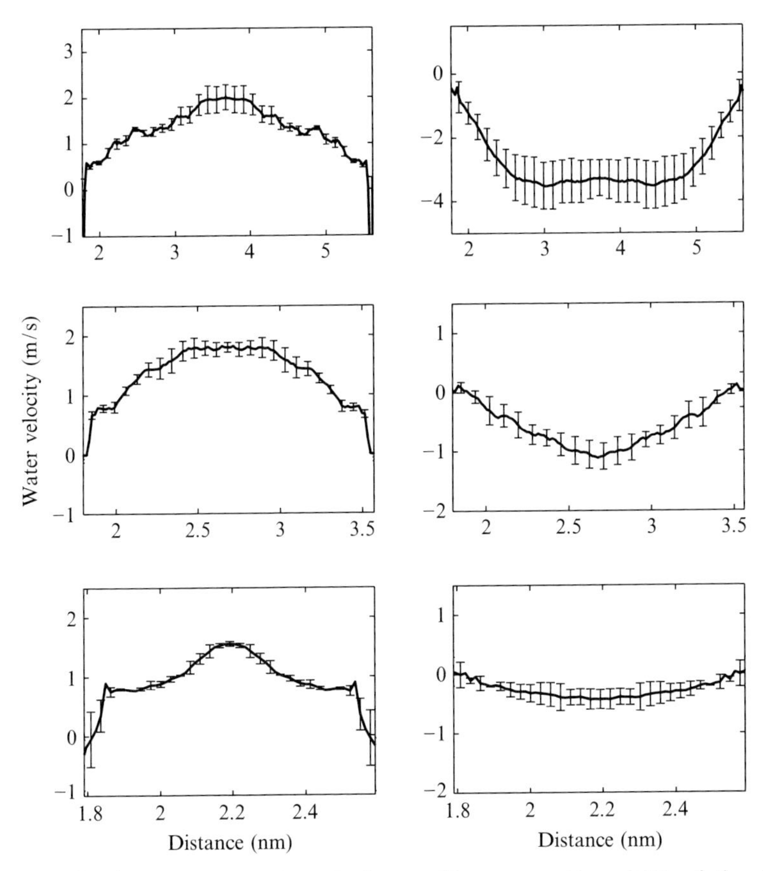

Fig. 3.25: (left) Electroosmotic velocity profiles across the width of the 4.1 nm (top), 2.1 nm (middle), and 1.1 nm (bottom) channel widths under an external electric field of 0.35 V/nm. Uncharged channels are on the left column and the partially charged channels are on the right column. (right) Ion mobilities for uncharged and charged channels. 4.1 nm (top), 2.75 nm (second from top), 2.1 nm (second from bottom), and 1.1 nm (bottom) channel widths. Uncharged channels are on the left column and the partially charged channels on the right column. The error bars denote the uncertainty in the computed values of the mobility.

the contribution of the surface current is smaller, so at 4.1 nm channel width, the net conductivity of partially charged and uncharged channels is the same.

3.5 Summary and outlook

Numerical simulation of biological fluids at micro and nano scale poses many challenges and in this chapter we have presented various methods that can be

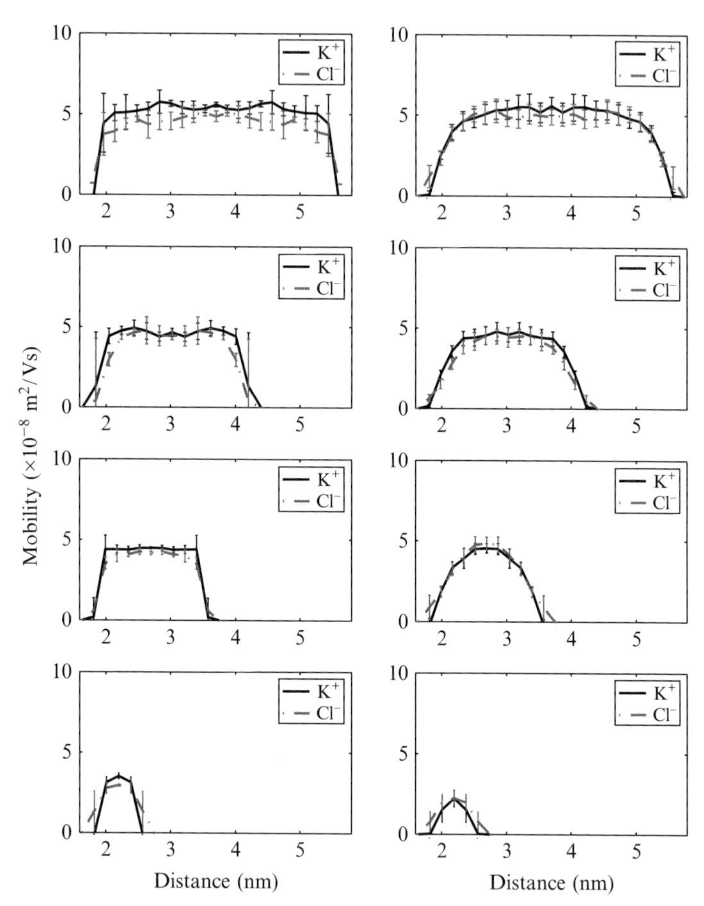

Fig. 3.25: (*continued*).

used to model fluids at various length and time scales. For microfluidics modeling, we have presented Lattice Boltzmann, Brownian dynamics, stochastic rotational dynamics, smooth particle hydrodynamics, and dissipative particle dynamics methods. These methods are effective in modeling fluid physics at large time and length scales. For nanoscale modeling of charged flows, we have presented self-consistent and grand canonical non-equilibrium molecular dynamics methods. The non-equilibrium molecular dynamics methods are typically limited to small length and short time scales. We have also presented multiscale approaches to seamlessly couple quantum, atomistic, and continuum theories.

While progress has been made in understanding and quantifying new physical mechanisms governing micro and nanoflows, a lot of questions remain open. Advances in experimental techniques combined with computational approaches

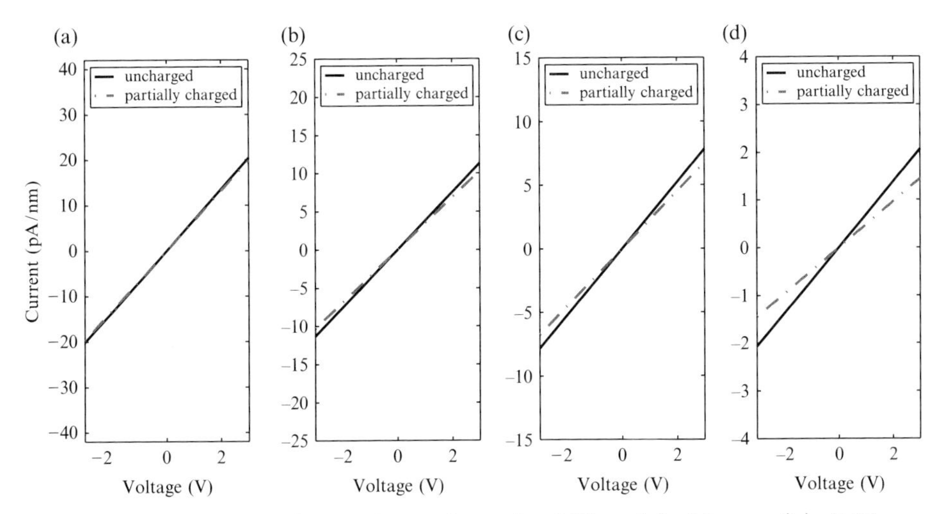

Fig. 3.26: IV curves for various channel widths: (a) 4.1 nm, (b) 2.75 nm, (c) 2.1 nm, and (d) 1.1 nm.

to probe micro and nano scale physics can reveal important information towards our understanding of fluids at small scales. In microflows, transport of biological fluids remains a great challenge. In nanoflows, with much progress being made in the discovery of new nanomaterials, understanding fluid flow and interaction of fluids with nanomaterials remains a great challenge. Over the next decade, we can expect significant advances in computational modeling of microflows and nanoflows—especially in the development of multiscale methods which can seamlessly combine physics at various length and time scales.

Acknowledgments

The authors would like to acknowledge support of their work by NSF. The computational results presented here were obtained using resources at the NCSA Supercomputing center.

References

Agarwal, U. S., Dutta, A., and Mashelkar, R. A., 1994. 'Migration of macromolecules under flow: The physical origin and engineering implications.' *Chemical Engineering Science*, **49**, p. 1693.

Allen, R., Melchionna, S., and Hansen, J. P., 2002. 'Intermittent permeation of cylindrical nanopore by water.' *Physical Review Letters*, **89**, p. 175502.

Allen, T. W., Kuyucak, S., and Chung, S. H., 1999. 'The effect of hydrophobic and hydrophilic channel walls on the structure and diffusion of water and ions.' *Journal of Chemical Physics*, **111**, pp. 7985–99.

Ansumali, S., and Karlin, I. V., 2002. 'Kinetic boundary conditions in the lattice Boltzmann equation.' *Physical Review E*, **66**, p. 026311.

Attard, P., 2004. 'Statistical mechanical theory for the structure of steady state systems: Application to a Lennard-Jones fluid with applied temperature gradient.' *Journal of Chemical Physics*, **121**, p. 7076.

Ballenegger, V., and Hansen, J. P., 2005. 'Dielectric permittivity profiles of confined polar fluids.' *Journal of Chemical Physics*, **122**, p. 114711.

Beckstein, O., and Sansom, M. S. P., 2003. 'Liquid-vapor oscillations of water in hydrophobic nanopores.' *Proceedings of the National Academy of Science*, **100**, p. 7063–7068.

Berendsen, H. J. C., Postma, J. M., DiNola, A., and Haak, J. R., 1984. 'Molecular-dynamics with coupling to an external bath.' *Journal of Chemical Physics*, **81**, p. 3684.

Berendsen, H. J. C., Grigera, J. R., and Straatsma, T. P., 1987. 'The missing term in effective pair potentials.' *Journal of Physical Chemistry*, **91**, p. 6269.

Boek, E. S., Coveney, P. V., Lekkererker, H. N. W., and van der Schoot, P., 1997. 'Simulating the rheology of dense colloidal suspensions using dissipative particle dynamics.' *Physical Review E*, **55**, pp. 3124–3133.

Boghosian, B. M., Love, P. J., Coveney, P. V., Karlin, I. V., Succi, S., and Yepez, J., 2003. 'Galilean-invariant lattice- Boltzmann models with H-theorem.' *Physical Review E Rapid Communications*, **68** (2), p. 025103.

Boinepalli, S., and Attard, P., 2003. 'Grand canonical molecular dynamics.' *Journal of Chemical Physics*, **119**, p. 12769.

Borgnia, M. J., and Agre, P. J., 2001. 'Reconstitution and functional comparison of purified GLPF and AQPZ., the glycerol and water channels from escherichia coli.' *Proceedings of the National Academy of Science U.S.A.*, **98**, p. 2888.

Brovchenko, I. V., Geiger, A., and Paschek, D., 2001. 'Simulation of confined water in equilibrium with a bulk reservoir.' *Fluid Phase Equilibria*, **183**, pp. 331–339.

Chang, C. W., Han, W. Q., and Zettl, A., 2005. 'Thermal conductivity of b-c-n and bn nanotubes.' *Journal of Vacuum Science & Technology B*, **23**, p. 1883.

Chen, S., and Doolen, G. D., 1998. 'Lattice Boltzmann method for fluid flows.' *Annual Review of Fluid Mechanics*, **30**, p. 329.

Chen, S., Martinez, D., and Mei, R., 1996. 'On boundary conditions in lattice Boltzmann methods.' *Physics of Fluids*, **8**, pp. 2527–2536.

Chen, Y. L., Graham, M. D., de Pablo, J. J., Jo, K., and Schwartz, D. C., 2005. 'DNA molecules in microfluidic oscillatory flow.' *Macromolecules*, **38** (15), pp. 6680–6687.

Cheng, L., Fenter, P., Nagy, K. L., Schlegel, M. L., and Sturchio, N. C., 2001. 'Molecular-scale density oscillations in water adjacent to a mica surface.' *Physical Review Letters*, **87** (15), pp. 156103.

Chirlian, L. E., and Francl, M. M., 1987. 'Atomic charges derived from electrostatic potentials – a detailed study.' *Journal of Computational Chemistry*, **8**, p. 894.

Chopra, N. G., and Zettl, A., 1998. 'Measurement of the elastic modulus of a multi-wall boron nitride nanotube.' *Solid State Communications*, **105**, p. 297.

Crozier, P. S., Henderson, D., Rowley, R. L., and Busath, D. D., 2001. 'Model channel ion currents in nacl-extended simple point charge water solution with applied-field molecular dynamics.' *Biophysical Journal*, **81**, p. 3077.

Darden, T., York, D., and Pedersen, L., 1993. 'Particle mesh Ewald - An N.log(N) method for Ewald sums in large systems.' *Journal of Chemical Physics*, **98**, p. 10089.

de Gennes, P. G., 1979. *Scaling Concepts in Polymer Physics.* Cornell University Press, Ithaca, NY.

de Groot, B. L., and Grubmller, H., 2005. 'The dynamics and energetics of water permeation and proton exclusion in aquaporins.' *Current Opinion in Structural Biology*, **15**, p. 176.

Ding, E. J., and Aidun, C. K., 2006. 'Cluster size distribution and scaling for spherical particles and red blood cells in pressure-driven flows at small Reynolds number.' *Physical Review Letters*, **96**, p. 204502.

Ding, K. J., and Valleau, J. P., 1994. 'Umbrella-sampling realization of widom chemical-potential estimation.' *Journal of Chemical Physics*, **98**, p. 3306.

Discher, D. E., Boal, D. H., and Boey, S. K., 1998. 'Simulations of the erythrocyte cytoskeleton at large deformation. II. Micropipette aspiration.' *Biophysical Journal*, **75**, pp. 1584–1597.

Doi, M., and Edwards, S. F., 1986. *The Theory of Polymer Dynamics.* Clarendon, Oxford.

Doyle, P. S., Ladoux, B., and Viory, J. L., 2000. 'Dynamics of a tethered polymer in shear flow.' *Physical Review Letters*, **84**, p. 4769.

Dupuis, A., Kotsalis, E. M., and Koumoutsakos, P., 2007. 'Coupling lattice Boltzmann and molecular dynamics models for dense fluids.' *Physical Review E*, **75**, p. 046704.

Egorov, A., and Brodskaya, E., 2003. 'The effect of ions on solid-liquid phase transition in small water clusters. A molecular dynamics study.' *Journal of Chemical Physics*, **118**, pp. 6380–6386.

Eisenriegler, E., and Maassen, R., 2002. 'Center-of-mass distribution of a polymer near a repulsive wall.' *Journal of Chemical Physics*, **116** (1), pp. 449–450.

Espanol, P., 1995. 'Hydrodynamics for dissipative particle dynamics.' *Physical Review E*, **52**, p. 1734.

Espanol, P., and Warren, P., 1995. 'Statistical mechanics of dissipative particle dynamics.' *Europhysics Letters*, **30** (4), p. 191–196.

Frisch, M. J. *et al.*, 2004. *Gaussian 03, Revision C.02.* Gaussian, Inc., Wallingford, CT.

Fan, X., Phan-Thien, N., Yong, N. T., Wu, X., and Xu, D., 2003. 'Microchannel flow of a macromolecular suspension.' *Physics of Fluids*, **15** (1), pp. 11–21.

Fang, L., Hu, H., and Larson, R. G., 2005. 'DNA configurations and concentration in shearing flow near a glass surface in a microchannel.' *Journal of Rheology*, **49**, pp. 127–138.

Fedosov, D., 2010. *Multiscale Modeling of Blood Flow and Soft Matter*. Ph.D. thesis, Brown University, Providence, RI.

Fixman, M., 1986. 'Construction of *langevin* forces in the simulation of hydrodynamic interaction.' *Macromolecules*, **19**, p. 1204.

Flekkoy, E. G., Wagner, G., and Feder, J., 2000. 'Hybrid model for combined particle and continuum dynamics.' *Europhysics Letters*, **52** (3), pp. 271–276.

Francl, M. M., Carey, C., Chirlian, L. E., and Gange, D. M., 1996. 'Charges fit to electrostatic potentials. 2. Can atomic charges be unambiguously fit to electrostatic potentials?' *Journal of Computational Chemistry*, **17**, p. 367.

Freund, J., 2002. 'Electro-osmosis in a nanometer-scale channel studied by atomistic simulation.' *Journal of Chemical Physics*, **116** (5), pp. 2194–2200.

Fyta, M. G., Melchionna, M., Kaxiras, E., and Succi, S., 2006. 'Multiscale coupling of molecular dynamics and hydrodynamics: Application to DNA translocation through a nanopore.' *Multiscale Modeling and Simulation*, **5**, pp. 1156–1173.

Galle, J., and Vortler, H., 1999. 'Monte Carlo simulation of primitive water in slit-like pores: Networks and clusters.' *Surface Science*, **421**, pp. 33–43.

Gallo, P., Rovere, M., Ricc, M. R., Harting, C., and Sphor, E., 1999. 'Evidence of glassy behavior of water molecules in confined states.' *Philosophical Magazine*, **79**, pp. 1923–1930.

Gallo, P., Rovere, M., and Spohr, E., 2000. 'Glass transition and layering effects in confined water: A computer simulation study.' *Journal of Chemical Physics*, **113**, pp. 11324–11334.

Gallo, P., Rapinesi, M., and Rovere, M., 2002*a*. 'Confined water in low hydration regime.' *Journal of Chemical Physics*, **117**, pp. 369–375.

Gallo, P., Ricci, M. A., and Rovere, M., 2002*b*. 'Layer analysis of the structure of water confined in vycor glass.' *Journal of Chemical Physics*, **116**, pp. 342–346.

Gingold, R. A., and Monaghan, J. J., 1977. 'Smoothed particle hydrodynamics: Theory and application to non-spherical stars.' *Monthly Notices of Royal Astronomical Society*, **181**, p. 375.

Gordillo, M., and Marti, J., 2003. 'Water on the outside of carbon nanotube bundles.' *Physical Review E*, **67**, p. 205425.

Green, M., and Lu, J., 1997. 'Simulation of water in small pore: Effect of electric field and density.' *Journal of Physical Chemistry B*, **101**, pp. 6512–6524.

Greengard, L., and Rokhlin, V., 1987. 'A fast algorithm for particle simulations.' *Journal of Computational Physics*, **73**, p. 325.

Groot, R. D., and Warren, P. B., 1997. 'Dissipative particle dynamics: Bridging the gap between atomistic and mesoscopic simulation.' *Journal of Chemical Physics*, **107** (11), pp. 4423–4435.

Hadjiconstantinou, N. G., and Patera, A. T., 1997. 'Heterogeneous atomistic-continuum representations for dense fluid systems.' *International Journal of Modern Physics C*, **8**, pp. 967–976.

Hammersley, J. M., and Handscomb, D. C., 1964. *Monte Carlo Methods.* John Wiley & Sons Inc., New York.
Hecht, M., Harting, J., Ihle, T., and Herrmann, H. J., 2003. 'Simulation of claylike colloids.' *Physical Review E*, **73**, pp. 011408.
Heffelfinger, G. S., and vanSwol, F., 1994. 'Diffusion in lennard-jones fluids using dual control-volume grand-canonical molecular-dynamics simulation (DEVGCMD).' *Journal of Chemical Physics*, **100**, p. 7548.
Heinz, H., and Suter, U. W., 2004. 'Atomic charges for classical simulations of polar systems.' *Journal of Physical Chemistry B*, **108**, p. 18341.
Henderson, D., Busath, D. D., and Rowley, R., 2001. 'Fluids near surfaces and in pores and membrane channels.' *Progress in Surface Science*, **65**, pp. 279–295.
Henderson, J. R., 1983. 'Statistical-mechanics of fluids at spherical structureless walls.' *Molecular Physics*, **50**, p. 741.
Hille, B., 2001. *Ion Channels of Excitable Membranes.* Sinauer Associates, Inc, Sunderland, MA.
Holdych, D. J., Noble, D. R., Georgiadis, J. G., and Buckius, R. O., 2004. 'Truncation error analysis of lattice Boltzmann methods.' *Journal of Computational Physics*, **193**, pp. 595–619.
Hoogerburgge, P. J., and Koelman, J. M., 1992. 'Simulating microscopic hydrodynamic phenomena with dissipative particle dynamics.' *Europhysics Letters*, **19** (3), pp. 155–160.
Hu, J. G., Goldman, S., Gray, C. G., and Guy, H. R., 2000. 'Calculation of the conductance and selectivity of an ion-selective potassium channel (irk1) from simulation of atomic scale models.' *Molecular Physics*, **98**, p. 535.
Hummer, G., Rasaiah, J., and Noworya, J. P., 2001*a*. 'Water conduction through the hydrophobic channel of a carbon nanotube.' *Nature*, **414**, p. 188.
Hummer, G., Rasaiah, J. C., and Noworya, J. P., 2001*b*. 'Water conduction through the hydrophobic channel of a carbon nanotube.' *Nature*, **414**, pp. 188–190.
Hur, J. S., Shaqfeh, E. S G., and Larson, R. G., 2000. 'Brownian dynamics simulations of single DNA molecules in shear flow.' *Journal of Rheology*, **44** (4), pp. 713–742.
Ihle, T., and Kroll, D. M., 2001. 'Stochastic rotation dynamics: A. Galilean-invariant mesoscopic model for fluid flow.' *Physical Review E*, **63**, p. 020201.
Ihle, T., and Kroll, D. M., 2003. 'Stochastic rotation dynamics. I. Formalism, Galilean invariance, and Green-Kubo relations.' *Physical Review E*, **67**, p. 066705.
Jenderjack, M., Schwartz, D. C., de Pablo, J., and Graham, M. D., 2004. 'Shear-induced migration in flowing polymer solutions: Simulation of long-chain DNA in microchannels.' *Journal of Chemical Physics*, **120**, pp. 2513–2529.
Jendrejack, R. M., de Pablo, J. J., and Graham, M. D., 2002. 'Stochastic simulations of DNA in flow: Dynamics and the effects of hydrodynamic interactions.' *Journal of Chemical Physics*, **116**, pp. 7752–7759.

Karlin, I. V., Ferrante, A., and Ottinger, H. C., 1999. 'Perfect entropy functions of the lattice Boltzmann method.' *Europhysics Letters*, **47**, pp. 182–188.

Karniadakis, G. E., Beskok, A., and Aluru, N. R., 2005. *Microflows and Nanoflows*. Springer, New York.

Khare, R., Graham, M. D., and de Pablo, J. J., 2006. 'Cross-stream migration of flexible molecules in a nanochannel.' *Physical Review Letters*, **96**, p. 224505.

Kjenllander, R., and Greberg, H., 1998. 'Mechanisms behind concentration profiles illustrated by charge and concentration distributions around ions in double layers.' *Journal of Electoanalytical Chemistry*, **450**, p. 233.

Koneshan, S., Rasaiah, J. C., Lynden-Bell, R. M., and Lee, S. H., 1998. 'Solvent structure, dynamics, and ion mobility in aqueous solutions at 25 degrees c.' *Journal of Physical Chemistry B*, **102**, p. 4193.

Koplik, J., and Banavar, J. R., 1995. 'Continuum deductions from molecular hydrodynamics.' *Annual Review of Fluid Mechanics*, **27**, pp. 257–292.

Koumoutsakos, P., 2005. 'Multiscale flow simulations using particles.' *Annual Review of Fluid Mechanics*, **37**, pp. 457–487.

Kratky, O., and Porod, G., 1949. 'Rontgenuntersuchung geloster fadenmolekule.' *Recueil des Travaux Chimiques Des Pays-Bas*, **68**, pp. 1106–1115.

Lamura, A., Gompper, G., Ihle, T., and Kroll, D. M., 2001. 'Multi-particle collision dynamics: Flow around a circular and a square cylinder.' *Europhysics Letters*, **56**, p. 319.

Larson, R. G., Perkins, T. T., Smith, D. E., and Chu, S., 1997. 'Hydrodynamics of a DNA molecule in a flow field.' *Physical Review E*, **55** (2), pp. 1794–1797.

Lee, C. T., Yang, W. T., and Parr, R. G., 1988. 'Development of the Colle-Salvetti correlation-energy formula into a functional of the electron-density.' *Physical Review B*, **37**, p. 785.

Lees, A. W., and Edwards, S. F., 1972. 'The computer study of transport processes under extreme conditions.' *Journal of Physics C*, **5**, p. 1921.

Leo, J., and Maranon, J., 2003. 'Confined water in nanotube.' *Journal of Molecular Structure (Theochem)*, **623**, pp. 159–166.

Levinger, N. E., 2002. 'Water in confinement.' *Science*, **29**, pp. 1722–1723.

Li, B., and Kwok, D. Y., 2004. 'Electrokinetic microfluidic phenomena by a lattice Boltzmann model using a modified Poisson-Boltzmann equation with an excluded volume effect.' *Journal of Chemical Physics*, **120**, pp. 947–953.

Li, S. C., Hoyles, M., Kuyucak, S., and Chung, S. H., 1998. 'Brownian dynamics study of ion transport in the vestibule of membrane channels.' *Biophysical Journal*, **74**, p. 37.

Lindahl, E., Hess, B., and van der Spoel, D., 2001. 'Gromacs 3.0: A package for molecular simulation and trajectory analysis.' *Journal of Molecular Modeling*, **7**, p. 306.

Lísal, M., Brennan, J. K., Smith, W. R., and Siperstein, F. R., 2004. 'Dual control cell reaction ensemble molecular dynamics: A method for simulations of reactions and adsorption in porous materials.' *Journal of Chemical Physics*, **121**, p. 4901.

Liu, M., Meakin, P., and Huang, H., 2006. 'Dissipative particle dynamics of multiphase flow in microchannels and microchannel networks.' *Physics of Fluids*, **18**, p. 027103.

Lynden-Bell, R. M., and Rasaiah, J. C., 1996. 'Mobility and solvation of ions in channels.' *Journal of Chemical Physics*, **105**, pp. 9266–9280.

Majumder, M., Chopra, N., Andrews, R., and Hinds, B. J., 2005. 'Nanoscale hydrodynamics – enhanced flow in carbon nanotubes.' *Nature*, **438**, p. 44.

Malevanets, A., and Karpal, R., 1999. 'Mesoscopic model for solvent dynamics.' *Journal of Chemical Physics*, **110**, p. 8605.

Malevanets, A., and Karpal, R., 2000. 'Solute molecular dynamics in a mesoscale solvent.' *Journal of Chemical Physics*, **112**, p. 7260.

Marko, J. F., and Siggia, E. D., 1995. 'Stretching DNA.' *Macromolecules*, **28**, pp. 8759–8770.

Marry, V., Dufreche, J. F., Jardat, M., and Turq, P., 2003. 'Equilibrium and electrokinetic phenomena in charged porous media from microscopic and mesoscopic models: Electro-osmosis in montmorillonite.' *Molecular Physics*, **101** (20), pp. 3111–3119.

Mashl, R. J., Joseph, S., Aluru, N. R., and Jakobsson, E., 2003. 'Anomalously immobilized water: A new water phase induced by confinement in nanotubes.' *Nano Letters*, **3**, pp. 589–592.

Mei, R., and Shyy, W., 1998. 'On the Finite Difference-based Lattice Boltzmann method in curvilinear coordinates.' *Journal of Computational Physics*, **143**, pp. 426–448.

Millan, J. A., Jiang, W., Laradji, M., and Wang, Y., 2007. 'Pressure driven flow of polymer solutions in nanoscale slit pores.' *Journal of Chemical Physics*, **126**, p. 124905.

Monaco, E., Luo, K. H., and Qin, R. S., 2007. 'Lattice Boltzmann simulations for microfluidics and mesoscale phenomena.' In *New Trends in Fluid Mechanics Research, Proceedings of the Fifth International Conference on Fluid Mechanics*, Shanghai, China, August 15–19.

Monaghan, J. J., 1992. 'Smoothed particle hydrodynamics.' *Annual Review of Astronomy and Astrophysics*, **30**, p. 543.

Morris, J. P., Fox, P. J., and Zhu, Y., 1997. 'Modeling low Reynolds number incompressible flows using SPH.' *Journal of Computational Physics*, **136**, pp. 214–226.

Muller, E. A., and Bubbins, K. E., 1998. 'Molecular simulation study of hydrophilic and hydrophobic behavior of activated carbon surfaces.' *Carbon*, **36**, pp. 1433–1438.

Mulliken, R. S., 1955. 'Electronic population analysis on LCAO[single bond]mo molecular wave functions. I.' *Journal of Chemical Physics*, **23**, p. 1833.

Nie, X. B., Chen, S. Y., E., W. N., and Robbins, M. O., 2004. 'A continuum and molecular hybrid method for micro- and nano-fluid flow.' *Journal of Fluid Mechanics*, **500**, pp. 55–64.

Noble, D. R., Chen, S., Georgiadis, J. G., and Buckius, R. O., 1995. 'A consistent hydrodynamic boundary condition for the lattice Boltzmann method.' *Physics of Fluids*, **7**, pp. 203–209.

Noguchi, H., and Gompper, G., 2005. 'Shape transitions of fluid vesicles and red blood cells in capillary flows.' *PNAS*, **102** (40), pp. 14159–14164.

O'Connell, S. T., and Thompson, P. A., 1995. 'Molecular dynamics-continuum hybrid computations: A tool for studying complex fluid flows.' *Physical Review E*, **52**, p. R5792.

Oldham, K. B., and Myland, J. C., 1994. *Fundamentals of Electrochemical Science*. Academic Press, San Diego, CA.

Ottinger, H. C., 1996. *Stochastic Processes in Polymeric Fluids*. Springer-Verlag, New York.

Palma, P. De Valentini, P., and Napolitano, M., 2006. 'Dissipative particle dynamics simulation of a colloidal micropump.' *Physics of Fluids*, **18**, p. 027103.

Paul, S., and Chandra, A., 2003. 'Dynamics of water molecules at liquid-vapour interfaces of aqueous ionic solutions: Effects of ion concentration.' *Chemical Physics Letters*, **373**, pp. 87–93.

Pivkin, I., 2005. *Continuum and Atomistic Methods for Biological Flows*. Ph.D. thesis, Brown University, Providence, RI.

Pivkin, I., and Karniadakis, G. E., 2005. 'A new method to impose no-slip boundary conditions in dissipative particle dynamics.' *Journal of Computational Physics*, **207**, pp. 114–128.

Pivkin, I., and Karniadakis, G. E., 2006. 'Controlling density fluctuations in wall-bounded dissipative particle dynamics systems.' *Physical Review Letters*, **96**, p. 206001.

Puibasset, J., and Pellenq, R., 2003. 'Water adsorption on hydrophilic mesoporous and plane silica substrates: A grand canonical Montte-Carlo simulation study.' *Journal of Chemical Physics*, **118**, pp. 5613–5622.

Qiao, R., and Aluru, N. R., 2003. 'Ion concentrations and velocity profiles in nanochannel electroosmotic flows.' *Journal of Chemical Physics*, **118** (10), pp. 4692–4701.

Qiao, R., and Aluru, N. R., 2004. 'Charge inversion and flow reversal in a nanochannel electro-osmotic flow.' *Physical Review Letters*, **92** (19).

Raghunathan, A. V., and Aluru, N. R., 2007. 'Self-consistent molecular dynamics formulation for electric-field-mediated electrolyte transport through nanochannels.' *Physical Review E*, **76**, p. 011202.

Ramakrishnan, V., Henderson, D., and Busath, D. D., 2004. 'Applied field nonequilibrium molecular dynamics simulations of ion exit from a beta-barrel model of the l-type calcium channel.' *Biochimica et Biophysica Acta*, **1664**, p. 1.

Revenga, M., Zuniga, I., and Espanol, P., 1998. 'Boundary model in DPD.' *International Journal of Modern Physics C*, **9**, p. 1319.

Revenga, M., Zuniga, I., and Espanol, P., 1999. 'Boundary conditions in dissipative particle dynamics.' *Computer Physics Communications*, **121–122**, pp. 309–311.

Ripoli, M., Mussawisade, K., Winkler, R. G., and Gompper, G., 2004. 'Low-Reynolds-number hydrodynamics of complex fluids by multi-particle-collision dynamics.' *Europhysics Letters*, **68** (1), pp. 106–112.

Rojnuckarin, A., Kim, S., and Subramaniam, S., 1998. 'Brownian dynamics simulations of protein folding: Access to milliseconds time scale and beyond.' *PNAS*, **95**, pp. 4288–4292.

Rovere, M., and Gallo, P., 2003. 'Strong layering effects and anomalous dynamical behaviour in confined water at low hydration.' *Journal of Physics: Condensed Matter*, **15**, pp. S145–S150.

Schlegel, M. L., Nagy, K. L., Fenter, P., and Sturchio, N. C., 2002. 'Structures of quartz (10(1)over-bar-0)- and (10(1)over-bar-1)-water interfaces determined by x-ray reflectivity and atomic force microscopy of natural growth surfaces.' *Geochimica et Cosmochimica Acta*, **66**, pp. 3037–3054.

Shan, X., and Chen, S., 1994. 'Lattice Boltzmann method for fluid flows.' *Physical Review E*, **49**, pp. 2941–2948.

Shanahan, M. E. R., 1988. 'Statics and dynamics of wetting on thin solids.' *Revue de Physique Appliquee*, **33**, pp. 1031–1037.

Shrivastava, I. H., and Sansom, M. S. P., 2000. 'Simulations of ion permeation through a potassium channel: Molecular dynamics of kcsa in a phospholipid bilayer.' *Biophysical Journal*, **78**, p. 557.

Sigaloti, L., Klapp, J., Sira, E., Melean, Y., and Hasmy, A., 2003. 'SPH simulations of time-dependent Poiseuille flow at low Reynolds numbers.' *Journal of Computational Physics*, **191**, pp. 622–638.

Sinnott, S. B., and Aluru, N. R., 2006. *Carbon Nanotechnology* (Edited by L. Dai). Elsevier.

Skordos, P. A., 1993. 'Initial and boundary conditions for the lattice Boltzmann method.' *Physical Review E*, **48**, pp. 4823–4842.

Smith, D. E., Babcock, H. P., and Chu, S., 1999. 'Single polymer dynamics in steady shear flow.' *Science*, **283**, p. 1724.

Sorenson, J. M., Hura, G., Glaeser, R. M., and Head-Gordon, T., 2000. 'What can x-ray scattering tell us about the radial distribution functions of water?' *Journal of Chemical Physics*, **113**, p. 9149.

Spohr, E., Hartnig, C., Gallo, P., and Rovere, M., 1999. 'Water in porous glasses: A computer simulation study.' *Journal of Molecular Liquids*, **80**, pp. 165–178.

Spohr, E., Trokhymchuk, A., and Henderson, D., 1998. 'Adsorption of water molecules in slit pore.' *Journal of Electroanalytical Chemistry*, **450**, pp. 281–287.

Stern, H. A., and Feller, S. E., 2003. 'Calculation of the dielectric permittivity profile for a nonuniform system: Application to a lipid bilayer simulation.' *Journal of Chemical Physics*, **118**, p. 3401.

Succi, S., 2001. *The Lattice Boltzmann Equation For Fluid Dynamics and Beyond.* Oxford University Press, Oxford.

Sui, H., Han, B., Lee, J. K., Walian, P., and Jap, B. K., 2001. 'Structural basis of water-specific transport through the AQP1 water channel.' *Nature*, **414**, pp. 872–878.

Sun, C., and Munn, L. L., 2005. 'Particulate nature of blood determines macroscopic rheology: A 2-D. Lattice Boltzmann analysis.' *Biophysical Journal*, **88**, pp. 1635–1645.

Sun, S. F., 1994. *Physical Chemistry of Macromolecules.* John Wiley & Sons, New York.

Symeonidis, V., and Karniadakis, G. E., 2006. 'A family of time-staggered schemes for integrating hybrid DPD models for polymers: Algorithms and applications.' *Journal of Computational Physics*, **218**, pp. 82–101.

Symeonidis, V., Karniadakis, G. E., and Caswell, B., 2006. 'Schmidt number effects in dissipative particle dynamics simulation of polymers.' *Journal of Chemical Physics*, **125**, p. 184902.

Takeda, H., Miyama, S. M., and Sekiya, M., 1994. 'Numerical simulation of viscous flows by smoothed particle hydrodynamics.' *Progress of Theoretical Physics*, **92**, p. 939.

Tartakovsky, A., and Meakin, P., 2005. 'Modeling of surface tension and contact angles with smoothed particle hydrodynamics.' *Physical Review E*, **72**, p. 026301.

Tartakovsky, A., Meakin, P., Scheibe, T. D., and West, R. E., 2007. 'Simulations of reactive transport and precipitation with smoothed particle hydrodynamics.' *Journal of Computational Physics*, **222**, pp. 654–672.

Tikhonov, A. N., and Samarskii, A. A., 1990. *Equations of Mathematical Physics.* Dover Publications, Inc., New York.

Tomaiuolo, G., Preziosi, V., Simeone, M., Guido, S., Ciancia, R., Martinelli, V. *et al.*, 2007. 'A methodology to study the deformability of red blood cells flowing in microcapillaries in vitro.' *Annali dell'Istituto Superiore di Sanita*, **43**, pp. 186–192.

Toney, M. F., Howard, J. N., Richer, J., Borges, G. L., Gordon, J. G., Melroy, O. R. *et al.*, 1994. 'Voltage-dependent ordering of water molecules at an electrode/electrolyte interface.' *Nature*, **368**, pp. 444–446.

Trofimov, S. Y., Nies, E. L., and Michels, M. A., 2002. 'Thermodynamic consistency in dissipative particle dynamics simulations of strongly nonideal liquids and liquid mixtures.' *Journal of Computational Physics*, **117**, p. 9383.

Tsang, T., 1997. *Classical Electrodynamics.* World Scientific, New Jersey.

Tuckerman, M. E., and Martyna, G. J., 2000. 'Understanding modern molecular dynamics: Techniques and applications.' *Journal of Physical Chemistry*, **104**, p. 159.

Underhill, P. T., and Doyle, P. S., 2004. 'On the coarse-graining of polymers into bead-spring chains.' *Journal of Non-Newtonian Fluid Mechanics*, **122** (1), p. 3–31.

Usta, O. B., Butler, J. E., and Ladd, A. J. C., 2006. 'Flow-induced migration of polymers in dilute solution.' *Physics of Fluids*, **18** (3), p. 031703.

Usta, O. B., Butler, J. E., and Ladd, A. J. C., 2007. 'Transverse migration of a confined polymer driven by an external force.' *Physical Review Letters*, **98**, p. 098301.

Vosko, S. H., Wilk, L., and Nusair, M., 1980. 'Accurate spin-dependent electron liquid correlation energies for local spin-density calculations – a critical analysis.' *Canadian Journal of Physics*, **58**, p. 1200.

Wagner, A. J., and Pagonabarraga, I., 2002. 'Lees-Edwards boundary conditions for Lattice Boltzmann.' *Journal of Statistical Physics*, **107**, pp. 521–537.

Walther, J. H., Jaffe, R., Halicioglu, T., and Koumoutsakos, P., 2001. 'Carbon nanotubes in water: Structure characteristics and energies.' *Journal of Physical Chemistry B*, **105**, pp. 9980–9987.

Wang, Y., and He, G., 2007. 'A dynamic coupling model for hybrid atomistic-continuum computations.' *Chemical Engineering Science*, **62**, pp. 3574–3579.

Warren, P. B., 2003. 'Vapor-liquid coexistence in many-body dissipative particle dynamics.' *Physical Review E*, **68**, p. 066702.

Werder, T., Walther, J. H., and Koumoutsakos, P., 2005. 'Hybrid atomistic-continuum method for the simulation of dense fluid flows.' *Journal of Computational Physics*, **205**, pp. 373–390.

White, F. M., 1994. *Fluid Mechanics*. McGraw Hill, New York.

Willemsen, S. M., Hoefsloot, H. C., and Iedema, P. D., 2000. 'No-slip boundary condition in dissipative particle dynamics.' *International Journal of Modern Physics*, **11** (5), pp. 881–890.

Won, C. Y., and Aluru, N. R., 2007. 'Water permeation through a subnanometer boron nitride nanotube.' *Journal of the American Chemical Society*, **129**, p. 2748.

Won, C. Y., Joseph, S., and Aluru, N. R., 2006. 'Effect of quantum partial charges on the structure and dynamics of water in single-walled carbon nanotubes.' *Journal of Chemical Physics*, **125**, p. 114701.

Yamakawa, H., 1971. *Modern Theory of Polymer Solutions*. Harper & Row, New York.

Yeh, I. C., and Berkowitz, M. L., 1999. 'Ewald summation for systems with slab geometry.' *Journal of Chemical Physics*, **111**, p. 3155.

Yeh, I., and Berkowitz, M. L., 2000. 'Effects of the polarizability and water density constraints on the structure of water near charged surfaces: Molecular dynamics simulations.' *Journal of Chemical Physics*, **112**, pp. 10491–10495.

Yong, W. A., and Luo, L. S., 2003. 'Nonexistence of H theorems for the athermal lattice Boltzmann models with polynomial equilibria.' *Physical Review E*, **67**, p. 051105.

Zhang, Q., Zheng, J., Shevade, A., Zhang, L., Gehrke, S., Heffelfinger, G. S. *et al.*, 2002*a*. 'Transport diffusion of liquid water and methanol through membranes.' *Journal of Chemical Physics*, **117**, pp. 808–817.

Zhang, Q. X., Zheng, J., Shevade, A., Zhang, L. Z., Gehrke, S. H., Heffelfinger, G. S. *et al.*, 2002*b*. 'Transport diffusion of liquid water and methanol through membranes.' *Journal of Chemical Physics*, **117**, p. 808.

Zheng, J., Lennon, E. M., Tsao, H. K., Sheng, Y. J., and Jiang, S., 2005. 'Transport of a liquid water and methanol mixture through carbon nanotubes under a chemical potential gradient.' *Journal of Chemical Physics*, **122**, p. 214702.

Zhou, J. D., Cui, S. T., and Cochran, H. D., 2003. 'Molecular simulation of aqueous electrolytes in model silica nanochannels.' *Molecular Physics*, **101**, pp. 1089–1094.

Zhu, G., Alexeev, A., Kumacheva, E., and Balazs, A. C., 2007. 'Modeling the interactions between compliant microcapsules and pillars in microchannels.' *Journal of Chemical Physics*, **127**, p. 034703.

Zhu, W., Singer, S. J., Zheng, Z., and Conlisk, A. T., 2005. 'Electro-osmotic flow of a model electrolyte.' *Physical Review E*, **71** (4).

4

PRESSURE-DRIVEN MICROFLUIDICS

Scott Miserendino and Yu-Chong Tai

Pressure force driven microfluidics is the foundation for volumetric scaling into the micro regime and, hence, for unique functionalities unachievable by macro-sized devices. Interestingly, the physics that govern the operation of pressure force driven microfluidics are just a subset of principles governing macro scale fluidics and low Reynold's number fluid mechanics is essential for the design and operation of pressure force driven microfluidic devices. This is because even fluidic theories cover a wide range of flow regimes, only regimes with mostly slow velocity and/or small sizes are encountered in microfluidic devices (Ho and Tai 1998). Section 4.1 provides a brief introduction to the necessary microfluidics basics required for understanding pressure force driven microfluidics, primarily focusing on laminar, viscous flow, and the importance of force-scaling relative to device size. More extensive coverage of fluidic theory is widely available in the literature (Bird *et al.* 2002; Whitaker 1981; White 1986, 1991) and in Chapters 2 and 3 of this book. Section 4.2 starts with pressure-driven flows in microchannels, the simplest but most important geometry. Section 4.3 covers devices designed to control fluid flow on-chip including passive and active valves and mixers. Section 4.4 reviews the variety of pressure sources used in microfluidic devices including those located on-chip and off-chip. Section 4.5 gives a review of flow sensors. Section 4.6 discusses microfluidic packaging issues. Section 4.6 illustrates some current and future applications for pressure force driven microfluidic devices and some of the remaining challenges facing microfluidics designers.

4.1 Microfluidics basics

The continuum postulate stipulates that the quantum nature of matter, namely that matter is composed of discrete particles, can be approximated by a continuous medium in which properties of a particle such as mass, velocity, pressure, and viscosity exist in and are continuously distributed over the entire space that the quantity of matter occupies. Here, most pressure driven microfluidic devices do satisfy the continuum postulate. Hence, we speak of space and time varying density, ρ, velocity, $\mathbf{v}$, and stress, τ. Of course, at some device length scale the continuum postulate will break down. A general rule of thumb is that the continuum postulate holds for length scales greater than 1 μm for gases and 10 nm for liquids (Ho and Tai 1998; Nguyen and Wereley 2002).

Besides the continuum assumption, two other major assumptions on the nature of fluid flow are often made and generally true for most microfluidic devices. First, the stress in the fluid is directly proportional to the strain, i.e. the fluid is Newtonian. Second, any heat flow is proportional to the temperature gradient hence satisfying Fourier's law of conduction. In the case of the liquid based devices, we usually assume incompressibility (constant density) and constant viscosity. However, all of these assumptions must be reconsidered when addressing extreme cases and complex fluid composition.

4.1.1 *Navier–Stokes equation*

The behavior of liquid flows under these five assumptions can be summarized by the equation of motion (Newton's Second Law), the continuity equation (conservation of mass), and the Newtonian definition of viscous stress. In vector notation, the equation of motion is

$$\rho \frac{d\mathbf{v}}{dt} = \rho \mathbf{g} + \nabla \bullet \mathbf{T}, \tag{4.1}$$

where $\boldsymbol{g}$ is the force due to gravity and $\mathbf{T}$ is the stress tensor defined as

$$\mathbf{T} = -p\mathbf{I} + \boldsymbol{\tau}, \tag{4.2}$$

and p, the pressure, is the stress experienced by the fluid at rest which always acts in the outwardly directed unit normal, $\mathbf{I}$ is an identity matrix, and $\boldsymbol{\tau}$ is the viscous stress tensor due to the motion of the fluid. Substituting equation (4.2) into (4.1-1) and applying the definition of $\frac{d}{dt}$* we have

$$\rho\left(\frac{\partial \mathbf{v}}{\partial t} + \mathbf{v}\bullet\nabla\mathbf{v}\right) = -\nabla p + \rho\mathbf{g} + \nabla\bullet\boldsymbol{\tau}. \tag{4.3}$$

For a Newtonian fluid, the viscous stress tensor can be represented as a combination of the rate of strain tensor, $\boldsymbol{d}$, fluid velocity vector, shear coefficient of viscosity, μ, and the bulk coefficient of viscosity, κ, as

$$\boldsymbol{\tau} = 2\mu\mathbf{d} + \left[\left(\kappa - \frac{2}{3}\mu\right)\nabla\bullet\mathbf{v}\right]\mathbf{I}. \tag{4.4}$$

The rate of strain tensor can be calculated from the velocity vector as

$$d_{ij} = \frac{1}{2}\left(\frac{\partial v_i}{\partial x_j} + \frac{\partial v_j}{\partial x_i}\right). \tag{4.5}$$

* $\frac{D}{Dt} = \frac{\partial}{\partial t} + v_x\frac{\partial}{\partial x} + v_y\frac{\partial}{\partial y} + v_z\frac{\partial}{\partial z} = \frac{\partial}{\partial t} + \mathbf{v}\bullet\nabla$ is called the substantial derivative and represents the time rate of change in the moving reference frame of the substance.

The derivation of the strain tensor is not trivial and a detailed derivation is available in Whitaker (1981). The continuity equation, a statement of conservation of mass, states

$$\frac{\partial \rho}{\partial t} + \nabla \bullet (\rho \mathbf{v}) = 0, \tag{4.6}$$

which for an incompressible fluid reduces to

$$\nabla \bullet \mathbf{v} = 0 \Rightarrow \frac{\partial v_i}{\partial x_i} = 0. \tag{4.7}$$

Substituting (4.7) and (4.4) into (4.3), one has

$$\rho\left(\frac{\partial \mathbf{v}}{\partial t} + \mathbf{v}\bullet\nabla\mathbf{v}\right) = -\nabla p + \rho\mathbf{g} + \nabla\bullet(2\mu\mathbf{d}). \tag{4.8}$$

Assuming a constant viscosity solution, substituting in equation (4.5) and once again applying (4.7) reveals the Navier–Stokes equation

$$\boxed{\rho\left(\frac{\partial \mathbf{v}}{\partial t} + \mathbf{v}\bullet\nabla\mathbf{v}\right) = -\nabla p + \rho\mathbf{g} + \mu\nabla^2\mathbf{v},} \tag{4.9}$$

which simply states that the time rate of change in linear momentum (left-hand side) must equal the applied body and surface forces (right-hand side). The Navier–Stokes equation is sufficient for isothermal systems; however, if a fluid flow experiences changes in temperature than another equation must be introduced and the assumption of constant viscosity should be reconsidered. For the vast majority of liquid fluid flows not involving chemical reactions in microfluidics, the restricted form of the equation of change for temperature is sufficient and states

$$\rho C_p \frac{\partial T}{\partial t} = k\nabla^2 T, \tag{4.10}$$

where T is temperature, C_p is heat capacity at constant pressure, and k is thermal conductivity (often assumed to be constant).

It is convenient to derive a dimensionless form of the Navier–Stokes equation. Three scale factors must be identified. A characteristic length (such as channel diameter), l_0, velocity (such as average flow velocity), v_0, and modified pressure (such as the drive pressure at the end of a channel), $\mathbf{P}_0 = p_0 + \rho g h_0$. Using the following definitions, the Navier–Stokes equation can be recast

$$\tilde{x} = \frac{x}{l_0} \qquad \tilde{y} = \frac{y}{l_0} \qquad \tilde{z} = \frac{z}{l_0} \qquad \tilde{t} = \frac{t}{l_0} \tag{4.11}$$

$$\tilde{\mathbf{v}} = \frac{\mathbf{v}}{v_0} \qquad \tilde{\mathbf{P}} = \frac{(\mathbf{P} - \mathbf{P}_0)l_0}{\mu v_0} \tag{4.12}$$

$$\tilde{\nabla} = l_0 \nabla \qquad \tilde{\nabla}^2 = \nabla^2 \qquad \frac{d}{d\tilde{t}} = \left(\frac{l_0}{v_0}\right)\frac{d}{dt} \tag{4.13}$$

$$\mathrm{Re} = \frac{l_0 v_0 \rho}{\mu} \tag{4.14}$$

$$\mathrm{Fr} = \frac{v_0^2}{l_0 g} \tag{4.15}$$

$$\mathrm{We} = \frac{\sigma}{l_0 v_0^2 \rho}, \tag{4.16}$$

where eqn (4.14) defines the Reynold's number, eqn (4.15) defines the Froude number, and eqn (4.16) defines the Weber number in which σ is the interface tension which could appear in the boundary conditions used to solve the partial differential equations. The dimensionless Navier–Stokes equation is

$$\mathrm{Re}\left(\frac{d}{d\tilde{t}}\tilde{\mathbf{v}}\right) = -\tilde{\nabla}\tilde{\mathbf{P}} + \tilde{\nabla}^2\tilde{\mathbf{v}}. \tag{4.17}$$

Equation (4.17) shows the importance of the Reynold's number as a metric for determining whether the viscous forces (right-hand side) or inertial forces (left-hand side) dominate. For this reason, the Reynold's number is used to characterize different flow regimes. The first regime is referred to as laminar or creep flow and is characterized by a low Reynold's number($\mathrm{Re} < 1500$) (Ho and Tai 1998; Nguyen and wereley 2002). In the laminar flow regime, the fluid dynamics are dominated by viscous forces and the inertial components of the Navier–Stokes equation can be ignored without introducing large amounts of error into the analysis.

4.1.2 *Low Reynold's Number Flow—Stokes Flow*

The reduced form of the Navier–Stokes equation in the laminar flow regime is the Stokes flow or creeping flow equation:

$$0 = -\nabla p + \rho \mathbf{g} + \mu \nabla^2 \mathbf{v}. \tag{4.18}$$

The Stokes equation is critical for understanding fluid dynamics in microfluidic pressure driven devices since almost all microfluidic devices satisfy the assumptions and conditions for Stokes flow. Figure 4.1 shows a plot of the Reynold's number for a pressure driven room temperature water flow through a cylindrical tube. As is evident from this example, microfluidic devices with typical channel diameters in the 10 to 1000 μm range and volumetric flow rates between 0.1 and 10 μL/min are low Reynold's number flows. The second regime, 'inviscid' or Euler flow, is characterized by a high Reynold's number ($\mathrm{Re} > 2100$) (Whitaker 1981).

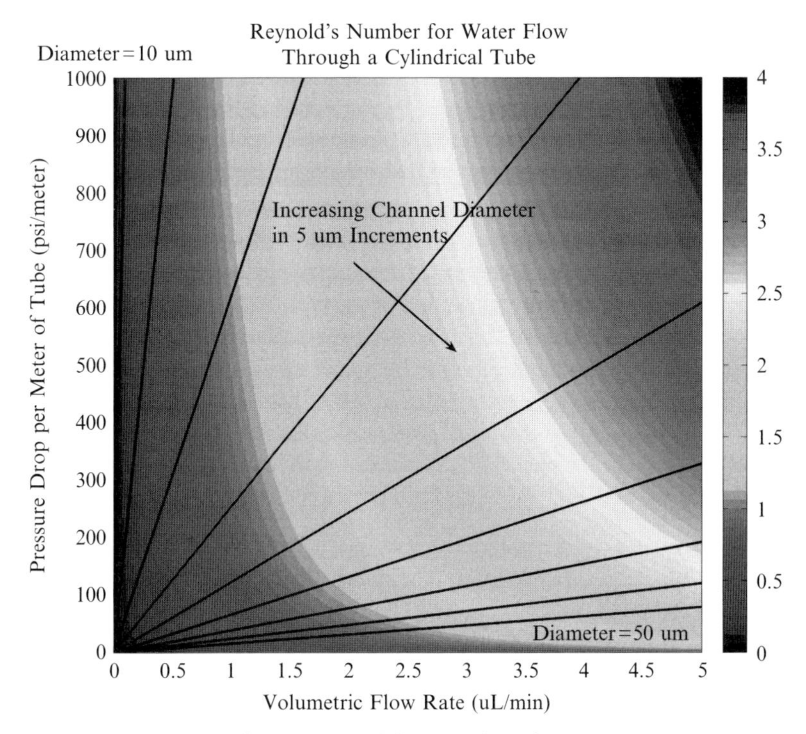

FIG. 4.1: Shaded plot of the Reynold's number for room temperature water flowing in a cylindrical tube. Lines indicating various tube diameters are given for reference.

This type of flow is almost never encountered in microfluidic systems, but is common in larger-scale flows such as those around airplane wings. In inviscid flow, the viscous terms of the Navier–Stokes equation can be ignored resulting in the Euler equation:

$$\rho \frac{d\mathbf{v}}{dt} = -\nabla p + \rho \mathbf{g}. \tag{4.19}$$

The third regime is the transitory regime in which neither the viscous or inertial components of the flow can be safely ignored.

To solve the Navier–Stokes equations, a set of boundary conditions must be determined. The standard boundary conditions to apply are no-slip and continuous-temperature

$$\begin{aligned} \mathbf{v}_{wall}\left.\right|_{liquid} &= \mathbf{v}_{liquid}\left.\right|_{wall} \\ T_{wall}\left.\right|_{liquid} &= T_{liquid}\left.\right|_{wall}. \end{aligned} \tag{4.20}$$

These conditions state that the velocity and temperature of the liquid at the point of contact with a wall are equivalent to the velocity and temperature of the wall at the point of contact with the liquid. For most applications in microfluidics, these conditions are excellent models of fluid behavior; however, there are two cases that can result in a violation of the no-slip condition. The first case is fairly uncommon and has been studied by Thompson and Troian (1997). Under cases of extremely high shear stress along the wall, the fluid will no longer behave as a Newtonian fluid resulting in a velocity slip at the fluid-wall interface. A more common example of a fluid violating the no-slip condition is that of capillary action in which a fluid creeps along a stationary wall.

4.1.2 *Capillary action and laplace pressure*

Capillary action can occur at an interface of a solid, liquid, and gas. At such a three-phase interface, several forces will act on the molecules at the liquid's surface. First consider the gas–liquid interface. The molecules of liquid at the interface experience two opposing attractive intermolecular forces, commonly referred to as *van der Waals forces.* The interface molecules are weakly attracted by the relatively low density of gas molecules on one side of the interface, and strongly attracted to the tightly packed bulk liquid molecules on the other side. This force differential over the surface area of the interface induces both a surface tension, σ, and a pressure drop across the interface. In the case of a freely suspended liquid droplet of radius, r_{drop}, the pressure drop is

$$\Delta p = \frac{2\sigma}{r_{drop}}. \tag{4.21}$$

At the liquid–solid interface, however, the interface liquid molecules have strong attractions to both the bulk liquid and the solid. If the intermolecular force balance favors the solid, the liquid will seek to minimize the total surface energy by spreading out along the solid's surface, a process called *wetting* (Diez *et al.* 1994). A solid surface that water will wet is *hydrophilic.* If the intermolecular force balance favors the bulk liquid, the liquid will not wet the surface but instead form a bead or spherical shape in an effort to minimize the surface energy. A solid surface that induces this behavior from water is *hydrophobic.* The angle the bead of liquid makes with the solid surface is called the contact angle, θ_c, and can be related to the surface tension of the various interfaces in the following equation:

$$\sigma_{sl} + \sigma_{lg} \cos \theta_c = \sigma_{sg}. \tag{4.22}$$

This relationship was first suggested by Thomas Young (1805) and later proven from first principles by Collins and Cooke (1959). A quantitative definition of hydrophobicity is based on the contact angle and states that a hydrophilic surface is one with a contact angle less than 90 °, while a hydrophobic surface has a contact angle greater than 90 °.

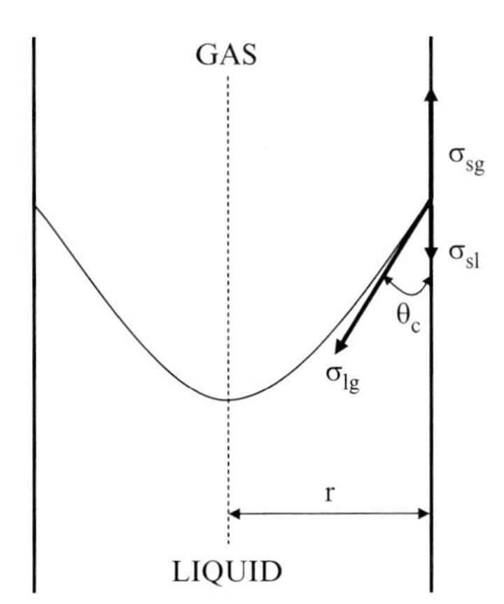

Fig. 4.2: Diagram of a meniscus in a capillary tube. Arrows show surface tension effects due to the gas–liquid–solid interface.

As with the liquid–gas interface, there is a pressure drop across a three-phase interface. By combining eqns (4.21) and (4.22), the pressure drop across a meniscus in a capillary tube of radius, r, can be determined (Nguyen and Wereley 2002; Wapner and Hoffman 2000; Yang *et al.* 2004) to be:

$$\Delta p = \frac{2\sigma_{lg} \cos \theta_c}{r}. \tag{4.23}$$

This surface tension induced pressure, referred to as the *Laplace pressure*, can play a significant role in microfluidics. For example, a glass capillary with a 5 μm radius placed in contact with room temperature water ($\sigma_{lg} = 0.073$ N/m and $\theta_c = 0°$) will have a Laplace pressure of 4.24 psi. Laplace pressures can affect device design, fabrication and even performance.

4.2 Microflows in microchannels

The flow through a straight channel is one of the simplest but most common configurations in microfluidic systems (Ho and Tai 1998). As a matter of fact, most pressure-driven microflows are flows in microchannels so it is also the most important fundamental case to be discussed here.

4.2.1 *Laminar flow in a cylindrical microchannel*

The microchannel flows can be studied by solving the Stokes flow equation (4.18) using the no-slip boundary conditions for liquids (4.20). The most general geometry to consider is that of a long, horizontal, cylindrical tube of radius, r_0, and of length, L, that extends along the z-axis. We can make several intuitive arguments about the flow in this geometry. First, the pressure drop across the tube will only occur in the z-direction and hence p will not be a function of r or θ. Second, because the fluid is assumed to have a constant density it must also have a constant volumetric flow rate through any cross-section of the tube, hence, the velocity can not be a function of the θ or z. Third, the velocity of the fluid must be finite. Using these three arguments the Stokes equation can be reduced to

$$\frac{\partial p}{\partial z} = \frac{p_{z=0} - p_{z=L}}{L} = -\frac{\Delta p}{L} = \mu \frac{1}{r} \frac{\partial}{\partial r}\left(r \frac{\partial v_z}{\partial r}\right), \tag{4.24}$$

where v_z is the z-component of the fluid velocity. The no-slip boundary condition becomes

$$v_z(r = r_0) = 0. \tag{4.25}$$

This partial differential equation is easily solvable with a double integration over r as

$$r\frac{dv_z}{dr} = -\left(\frac{\Delta p}{L}\right)\frac{r^2}{2\mu} + C_1 \tag{4.26}$$

$$v_z = -\left(\frac{\Delta p}{L}\right)\frac{r^2}{4\mu} + C_1 \ln(r) + C_2. \tag{4.27}$$

Since r can reach 0 and v_z must be finite, it is clear that C_1=0. Now applying the boundary condition of (4.25), we find

$$C_2 = \left(\frac{\Delta p}{L}\right)\frac{r_0^2}{4\mu} \tag{4.28}$$

The velocity profile is now fully determined:

$$v_z = \left(\frac{\Delta p}{L}\right)\frac{r_0^2}{4\mu}\left[1 - \left(\frac{r}{r_0}\right)^2\right]. \tag{4.29}$$

The functional form of the velocity profile indicates that the fluid front is a parabolic shape. A parabolic-shaped fluid flow profile is characteristic of almost all pressure driven flows in microfluidic devices. From this velocity profile, several important parameters can be calculated. The maximum linear flow velocity, $v_{z,max}$, which occurs at r=0 is

$$v_{z,\,\max} = \left(\frac{\Delta p}{L}\right)\frac{r_0^2}{4\mu}. \tag{4.30}$$

The volumetric flow rate, Q, can be calculated as the integral of the linear velocity profile over the cross-section

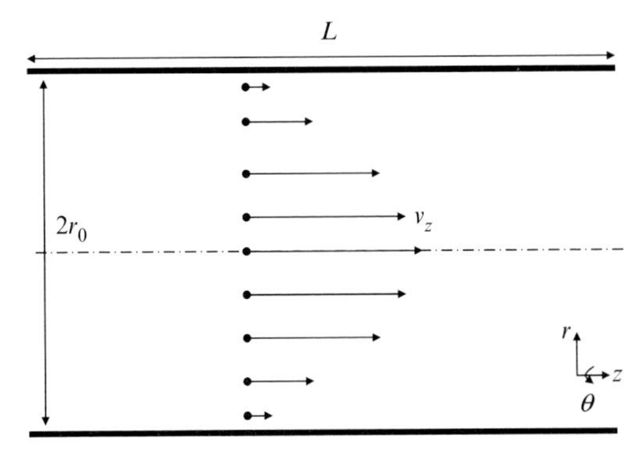

FIG. 4.3: Cross-sectional diagram of laminar flow in a cylindrical tube.

$$Q = \int_0^{2\pi} \int_0^{r_0} v_z r \; dr d\theta = \frac{\pi r_0^4}{8\mu} \left(\frac{\Delta p}{L} \right). \tag{4.31}$$

Equation (4.31) is known as the Hagen–Poiseuille Law. The average linear flow velocity, $\bar{v}_z$, which is defined by the ratio of the volumetric flow rate to the cross-sectional area, A,

$$\bar{v}_z = \frac{Q}{A} = \frac{Q}{\pi r_0^2} = \frac{r_0^2}{8\mu} \left(\frac{\Delta p}{L} \right). \tag{4.32}$$

4.2.2 *Laminar flow with non-circular cross-sections*

The Stokes equation with no-slip boundary conditions can be applied to a variety of cross-section geometries. In many cases, analytical solutions to the equations exist describing the velocity profile and the volumetric flow rate in the laminar flow regime. Analytical solutions by Berker (1963) and Shah and London (1978) for rectangular and isosceles-triangular are given here for reference:

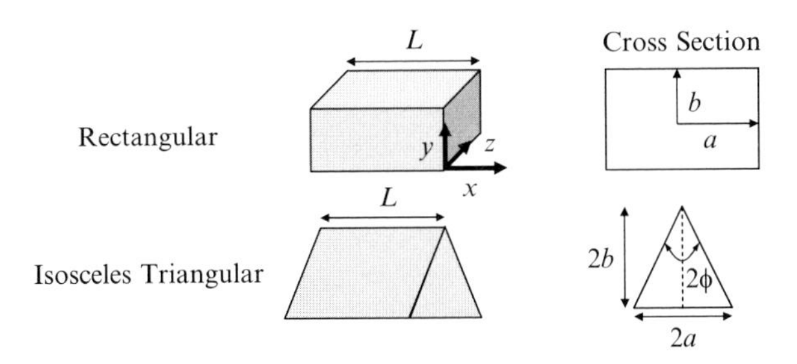

FIG. 4.4: Diagrams of some channel geometries with analytical solutions.

Rectangular:

$$v_x(y,z) = -\frac{16a^2\Delta p}{\mu\pi^3 L}\sum_{i=1,3,5,\ldots}^{\infty}(-1)^{(i-1)/2}\left[1-\frac{\cosh(i\pi z/2a)}{\cosh(i\pi b/2a)}\right]\frac{\cos(i\pi y/2a)}{i^3} \tag{4.33}$$

$$Q = -\frac{4ba^3\Delta p}{3\mu L}\left[1-\frac{192a}{\pi^5 b}\sum_{i=1,3,5,\ldots}^{\infty}\frac{\tanh(i\pi b/2a)}{i^5}\right] \tag{4.34}$$

Isosceles Triangular:

$$v_x(y,z) = -\frac{\Delta p}{\mu L}\frac{y^2 - z^2\tan^2\phi}{1-\tan^2\phi}\left[\left(\frac{z}{2b}\right)^{B-2} - 1\right] \tag{4.35}$$

$$Q = -\frac{4ab^3\Delta p}{3\mu L}\frac{(B-2)\tan^2\phi}{(B+2)(1-\tan^2\phi)} \tag{4.36}$$

where

$$B = \sqrt{4+\frac{5}{2}\left(\frac{1}{\tan^2\phi}-1\right)}. \tag{4.37}$$

For arbitrary cross-sections, a couple of general methods have been proposed. Bahrami *et al.* (2006) propose using the square root of the cross-sectional area, $\sqrt{A}$, as the defining length scale and drive approximations to the pressure drop across the channel. Bahrami's method achieves an accuracy of 8 per cent compared to known analytical solutions. The conventional choice, however, for the defining length scale is the hydraulic diameter, D_h, which is defined as

$$D_h = \frac{4 \times \text{Cross Sectional Area}}{\text{Wetted Perimeter}} = \frac{4A}{P_{wet}}. \tag{4.38}$$

For a channel that is entirely filled with liquid, the wetted perimeter is equivalent to the geometric perimeter. The concept behind the hydraulic diameter is to model any arbitrary channel cross-section as an equivalent circular cross-section channel with a diameter of D_h. Once an appropriate diameter has been calculated, an approximation to the flow rate and pressure drop across a channel can be made by using the analytical solutions for a channel with a circular cross-section. One must be very careful in using this empirical approach. Its accuracy improves at higher Reynold's numbers and for channel cross-sections more closely resembling a circle meaning its application to microfluidic channels may result in relatively large estimation error (White 1991). For laminar, low Reynold's number flows in

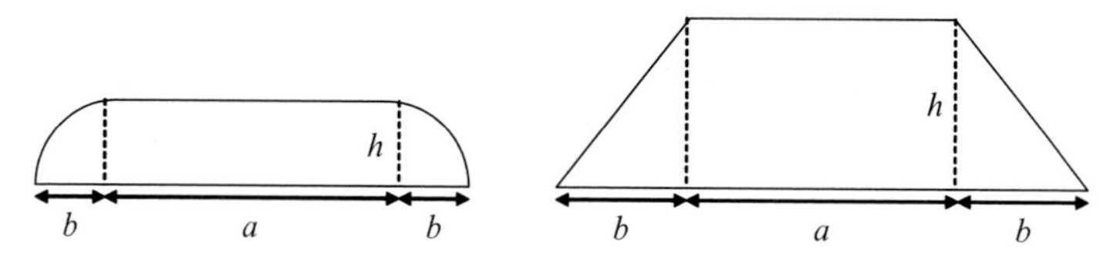

Fig. 4.5: Common channel cross-sections in microfluidic devices.

high-aspect ratio channels the hydraulic diameter estimation technique can result in errors as high as 50 per cent (White 1991; Papautsky *et al.* 1999).

There are two common channel cross-sections in microfluidics (rounded rectangle and trapezoid). The hydraulic diameters for these geometries are:

Rounded rectangle:

$$D_h = \frac{4\left(ah + \frac{\pi}{2} bh\right)}{2(a+b) + 2b \int_0^{\pi/2} \sqrt{1 - \frac{b^2-h^2}{h^2} \sin^2 \theta d\theta}} \approx \frac{2ah + \pi bh}{a + b + \frac{\pi}{2}\sqrt{\frac{(b^2+h^2)}{2}}} \tag{4.39}$$

Trapezoid:

$$D_h = \frac{4\left(\frac{h}{2}(a+b)\right)}{a + b + 2h\sqrt{1 + \left(\frac{a-b}{2h}\right)^2}} = \frac{2h}{1 + \frac{2h}{a+b}\sqrt{1 + \left(\frac{a-b}{2h}\right)^2}} \tag{4.40}$$

The hydraulic diameter is a useful tool in estimating flow rates and pressure drops across microfluidic channels. The most common estimation for the pressure drop across a channel of arbitrary cross-section and length L is taken from eqn. (4.32) and is expressed in terms of the Reynold's number

$$\Delta p = \mathrm{Re}\, f \frac{\mu L}{2D_h^2} \bar{v} \tag{4.41}$$

where f is the Fanning friction factor* and $\bar{v}$ is the average linear flow velocity. Most microfluidic devices, however, do not consist just of long, constant cross-section channels with fully developed flows. We must gain some understanding of when a flow can be considered fully developed.

4.2.3 *Entrance length effect*

A laminar flow will develop a steady-state flow profile some distance from the inlet of the channel. This distance is referred to as the entrance length, L_e. The entrance length is defined quantitatively by Shah and London (1978) as the point at which the centerline velocity in the channel reached 99 per cent of its fully developed value. The entrance length has been analytically and experimentally related to the channel diameter and the Reynold's number (Whitaker 1981; White 1991; Nguyen and Wereley 2002; Shah and London 1978) of the flow according to

* The Fanning friction factor is generally a function of the flow regime (Reynold's number), roughness of the channel sidewalls, level of development of the flow, channel cross-section and possibly other factors. For circular cross-section, channels in a well-developed laminar flow regime Re $f = 64$. For rectangular cross-section channels Re f is typically between 50 and 60.

$$L_e \approx D_h \left(\frac{0.6}{1 + 0.035\text{Re}} + 0.056\text{Re} \right). \tag{4.42}$$

This expression shows that the entrance length for low Reynold's number flows is about 60 per cent of the hydraulic diameter. Hence, any analysis of the velocity profile assuming a fully developed flow is only valid at distances of at least 0.6 D_h away from any entrance or exit. For example, consider a 100 μm wide by 25 μm tall rectangular cross-section microchannel operating at a constant volumetric flow rate of 1 μL/min (estimated linear flow velocity of 0.027 m/sec and Re = 1.064). Using this assumption, the entrance length would be approximately $0.634 D_h = 25.38\,\mu\text{m}$. Further discussion of entrance effects, non-Newtonian flow behavior, viscosity changes, and other second order considerations in analysis and visualization of fluid flows in microfluidic systems can be found in articles by Koo and Kleinstreuer (2003) and Thompson *et al.* (2005).

4.2.4 *Stream function and velocity potential*

Hydraulic diameter is a useful tool in determining the input and output behavior of a microfluidic system. To gain some understanding of the internal flow velocity profile in a channel of complicated geometry in which analytically solving the Navier–Stokes equations is difficult or impossible, it is often beneficial to consider a two-dimensional model of the flow through the device. This simplification is reasonable for most microfluidic devices. When using this simplification, one must remember that the analysis will only be valid in regions far-from any changes in geometry involving the ignored dimension. A good rule of thumb is that the analysis is valid for areas in which the flow is fully developed (i.e. at least an entrance length away from any geometrical variations). The most powerful tools in two-dimensional analyses of flows are the *stream function* and *velocity potential.* While these tools are used typically in two-dimensional analysis the concepts can be extended to three dimensions (Hsiao *et al.* 2001; Greywall 1988, 1993; Knight and Mallison 1996).

The stream function, $\psi(x, y)$, and velocity potential, $\phi(x, y)$, are mathematical entities constructed such that they satisfy the Cauchy–Riemann equations,

$$\frac{\partial \phi}{\partial x} = \frac{\partial \psi}{\partial y} \tag{4.43}$$

and

$$\frac{\partial \phi}{\partial y} = -\frac{\partial \psi}{\partial x}. \tag{4.44}$$

The velocity field of a two-dimensional incompressible, irrotational flow ($\nabla \times \mathbf{v} = 0$) must satisfy

$$\text{(irrotational)} \qquad \nabla \times \mathbf{v} = \frac{\partial v_x}{\partial y} - \frac{\partial v_y}{\partial x} = 0 \tag{4.45}$$

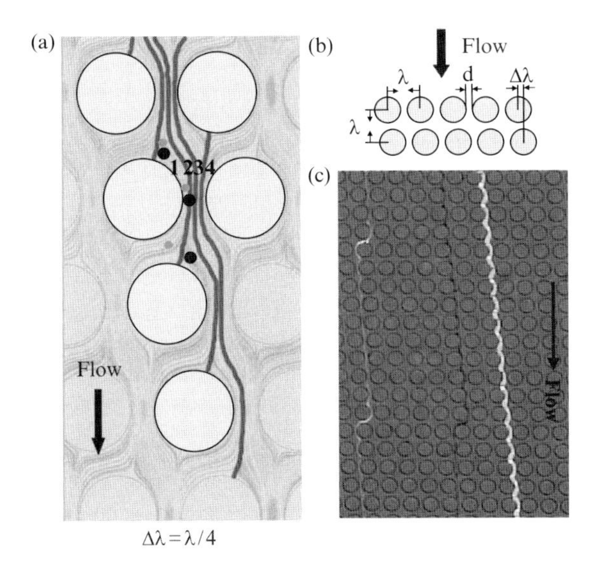

Fig. 4.6: Stream lines used to analyze flow behavior of 5μm, 8μm, and 10μm fluorescent beads in a deterministic lateral displacement filter. (a) Finite-element analysis of periodic column geometry with flow lines between stagnation points highlighted. (b) Diagram of device geometry. (c) Experimentally visualized separation of beads in continuous flow. (Images adopted, with permission, from Zheng *et al.* (2005))

$$\text{(continuity equation)} \quad \nabla \bullet \mathbf{v} = \frac{\partial v_x}{\partial x} + \frac{\partial v_y}{\partial y} = 0. \tag{4.46}$$

By combining the Cauchy–Riemann equations with eqs. (4.45) and (4.46) the stream function and velocity potential can be differentially defined in terms of the flow velocity field

$$\frac{\partial \psi}{\partial x} = v_y \text{ and } \frac{\partial \psi}{\partial y} = -v_x \tag{4.47}$$

$$\frac{\partial \phi}{\partial x} = -v_x \text{ and } \frac{\partial \phi}{\partial y} = -v_y. \tag{4.48}$$

Once a suitable stream function is found, it is possible to construct a family of stream lines by setting the function equal to a constant, thus generating a different stream line for each constant chosen. A similar approach can be used with the velocity potential to produce a family of equipotential lines. Stream lines and equipotential lines are useful in the study of microfluidic devices because of several convenient analytical and experimental properties. A few of these properties are:

a. Stream lines and equipotential lines never intersect themselves.
b. Stream lines are everywhere perpendicular to equipotential lines.
c. Stream lines are everywhere tangent to the fluid velocity.
d. A solid surface placed along a stream line will not affect the flow velocity.
e. Changes in distance between stream lines are indications of changes in fluid flow velocity with denser stream lines indicating higher flow velocity.
f. For steady flows the stream lines represent the path an infinitesimal particle will take through the channel when that particle is placed on the stream line.

The last property given about stream lines is the most commonly used in the study of microfluidics. By injecting small particles or dye molecules into the fluid flow stream, lines can be traced or visualized experimentally giving an idea of the behavior of larger particles in the flow (Huang *et al.* 2004; Inglis *et al.* 2006; Zheng *et al* 2005). Stream lines are also often used in analytical or finite element analysis of fluid flows to better understand device function.

4.3 Microvalves for microflow control

Pressure driven microfluidic devices can be organized in a variety of ways. Microvalves are the most important devices for microfluidics control. Microvalves can incorporate various different ways such as mechanical, thermal or electrical transducer to convey energy to the fluid and hence cause it to flow in a preferential manner. To be considered 'on-chip' the transducer must be fabricated using

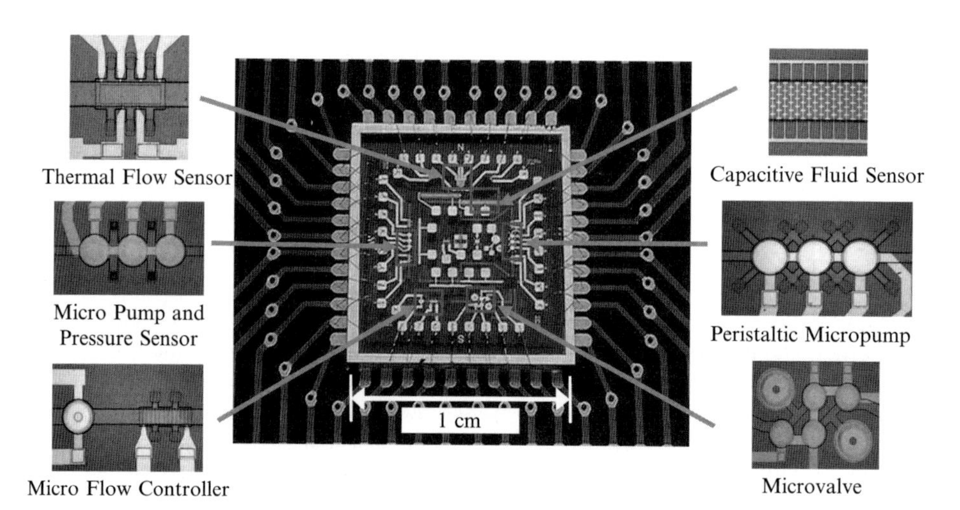

Fig. 4.7: A fully integrated pressure-driven microfluidic circuitry enabled by a thin-film parylene MEMS technology (Tai 2003). Multiple devices like valves, pumps, flow sensors, and micro flow controllers can all be made on a single substrate.

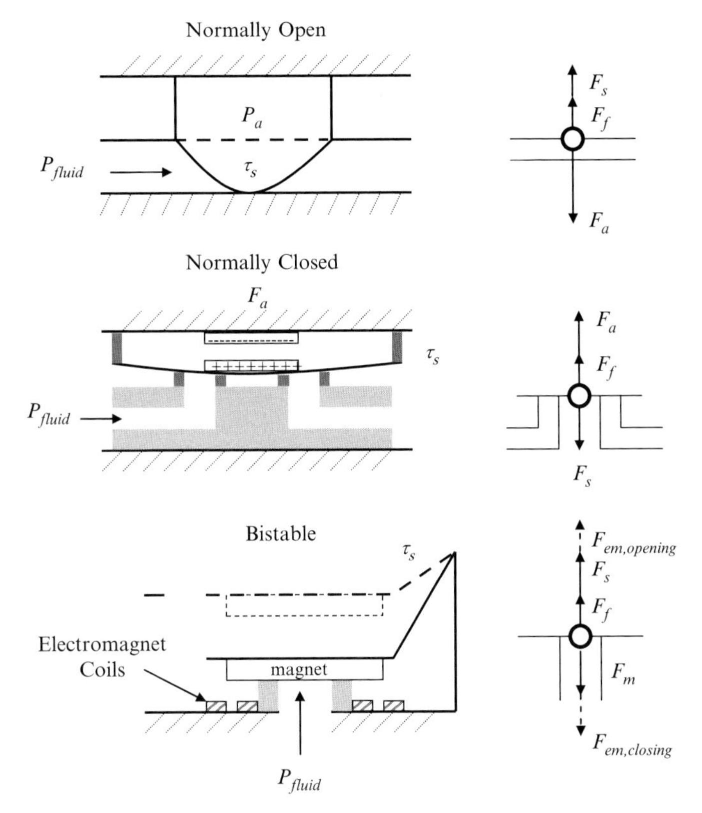

Fig. 4.8: Diagrams of a normally open pneumatic valve (Unger *et al.* 2000), normally closed electrostatic valve (Oh *et al.* 2002), and a bistable electromagnetic valve (Sutanto *et al.* 2006) in their closed state. Force body diagrams are given to the right. Dashed portions of the diagrams indicate valve configuration in the open state.

microfabrication or MEMS processes such as bulk micromachining, surface micromachining, or soft lithography. Other microfluidic devices do not incorporate transducers but instead rely on external ('off-chip') ones and then incorporate a coupling mechanism to allow flowing fluid into the device. Both types of devices have their benefits and drawbacks.

On-chip flow control (microvalves) and pressure sources (micropumps) have the advantages of lower dead volume and integrated fabrication (Tie 2003; Xie *et al.* 2004a). The major disadvantages to on-chip flow control and pressure sources are increases in fabrication complexity for the system, incompatibility with some microfluidic fabrication technologies, and in many cases custom design based on the ultimate desired function. Off-chip sources and valves (Unger *et al.* 2000) provide for simplicity in operating conditions, easy replacement or exchange of components, and can be used with almost every microfluidic device regardless of the microfabrication technology used to make the device. Off-chip components however require some sort of coupling mechanism to deliver fluid to the micro-device and often result in vastly increased dead volume. Packaging of microfluidic devices is a challenging problem for both on-and off-chip pressure sources and flow control and will be dealt with in a later section.

4.3.1 *Microvalve function*

Microvalves are a fundamental part of most fluidic systems. Valves provide the fundamental capability of selectively starting, stopping, and redirecting fluid flow while a device is operating. Review articles by Oh and Ahn (2006), Shoji and Esashi (1994), and several books (Nguyen and Wereley 2002; Koch *et al.* 2000) explore the wide variety of microvalve designs published in the literature. Two methods of categorizing microvalves are by their operating methodology and by their actuation mechanism. Microvalves exist with nearly every combination of operating methodology and actuation mechanism. The two most common operating methodologies are *normally open* and *normally closed* valves. The term *normally* is used to indicate the natural or unactuated state of the valve. Hence a normally open valve will allow fluid to flow through the valve when no energy is applied to the actuation mechanism but will stop fluid from flowing through the valve when energy is applied to the actuator. A normally closed valve acts in the opposite manner by preventing fluid flow with no energy consumption but requiring energy to open the valve. A valve that requires energy only during the transition from the open to closed state but requires no energy to remain in either state is said to be *bistable* or *latching*.

Active microvalves, valves requiring energy to function, all operate by the same governing principle. A *valve seat* is fabricated that can be moved by a *closing force* with the remainder of the valve body composing a *valve spring* that supplies a restoring force which acts against the closing force. *Passive* microvalves are valves that function without an actuating mechanism and hence require no energy to operate. Passive valves may or may not have a valve seat or valve spring. A valve seat is the part of the valve that physically prevents the fluid from moving through the valve. It is responsible for forming a seal capable of withstanding the operating pressure of the fluid. The closing force is supplied by the actuation mechanism and varies according to valve type. The valve spring is a function of the material and geometry of the valve and is designed such that the valve will require no energy to remain fully in either the normally open or normally closed position under standard operating conditions for that valve.

4.3.2 *Microvalve figures of merit*

Several figures of merit are important in valve design. The most important figures of merit in most applications are the maximum sustainable operating pressure, called the *maximum* or *blowout pressure*, an open valve can withstand without breaking and the maximum sustainable pressure, called the *breakdown pressure*, a closed valve can withstand without leaking. There exists both a forward and a reverse breakdown pressure because a closed valve can fail either by lifting off the valve seat (forward breakdown) or by collapsing under high back pressure (reverse breakdown). Since valves are not ideal many will exhibit some leakage even when fully closed. The leakage ratio, L_r, is defined as the ratio of the volumetric flow rate in closed state to the volumetric flow rate in the open state and is a function of fluid operating pressure

$$L_r = \frac{Q_{closed}}{Q_{open}}. \tag{4.49}$$

Two other major figures of merit are power consumption and response time. Power consumption is measured in the actuated state or in the case of a bistable valve it is the average power over the transition between states. Response time is a measure of how quickly the valve can transition between states. For passive valves, the pressure at which the valve is considered to be open, called the *cracking pressure*, and the minimum back pressure (pressure directed opposite to the primary fluid flow direction) needed to reseal the valve, called the *reseat pressure*, are additional important figures of merit.

For any actuated valve it is useful to define a figure of merit that describes the valves efficient use of space. A convenient metric that accomplishes this is the *energy density*, $E_a^{'}$, of the actuator. The energy density of an actuator is the work

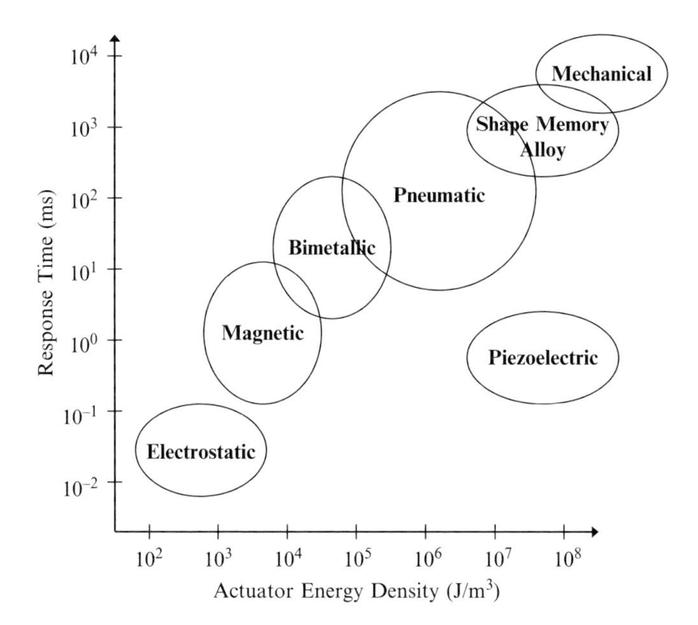

FIG. 4.9: Response time verse energy density for a variety of microvalve actuators.

the actuator is capable of performing divided by the volume of the actuator, V_a. The work an actuator is capable of is the product of the force of the actuation, F_a, and the maximum displacement of the valve seat, s_a,

$$E_a^{'} = \frac{W_a}{V_a} = \frac{F_a s_a}{V_a}. \tag{4.50}$$

A trade-off between response time and energy density is typical of microvalve designs. Figure 4.9 illustrates this trade-off over the many possible actuator types (Shoji and Esashi 1994; Weibel *et al.* 2005).

In microsystems valve size is also of critical importance. Other than the footprint of the valve in the microsystem there are three other considerations relating to the valve size that can affect the performance of the larger microfluidic system of which the valve is a part. The volume of fluid displaced by the valve when transitioning from an open to closed state, called the *displacement volume*, is important because it determines how precisely a user can control the flow of fluid in the system. Displacement volume can be critical in microfluidic system using off-chip valves which often displace volumes up to 10 per cent of the entire system volume. Hence low displacement volume is a critical advantage of microvalves relative to large-scale macrovalves. Another consideration is the *sweep volume* of the valve. This is the total volume of fluid that is in the valve structure when the valve is in the open state. Low sweep volume valves allow for minute sample sizes to be used without having to dilute them. Again low sweep volume is a major advantage of microvalves over macrovalves which can have volumes easily in the tens of microliters. This can make a large difference in microfluidic systems working with sample volumes in the tens of nanoliters.

Finally, a valve's *dead volume* is important. Dead volume is an often used yet poorly defined (and in some cases undefined) term. Because of inconsistence in definition one should be careful when making comparisons in the literature. It has been used by authors to refer to what is here called displacement and sweep volumes as well as the volume of fluid in the valve that achieves a flow velocity significantly less than that of the fluid at the inlet of the valve. We will adapt a quantitative definition of dead volume as that volume of fluid in the valve with a flow velocity below one per cent of the maximum inlet flow velocity (Miserendino and Tai 2007; Puntambekar and Ahn 2002). Dead volume can be detrimental to many microfluidic systems because analytes trapped in these regions can take a long time to diffuse back into the main fluid path and, in the case of liquid chromatography, can lead to accelerated band broadening and peak asymmetry. In other microfluidic devices regions of dead volume are often gathering centers for bubbles which can become difficult to flush out because of the relatively low fluid velocity.

4.3.3 *Passive microvalves*

Passive valves come in two varieties: check valves and capillary valves. *Check valves* are open or closed by the operating pressure of the fluid and hence require no

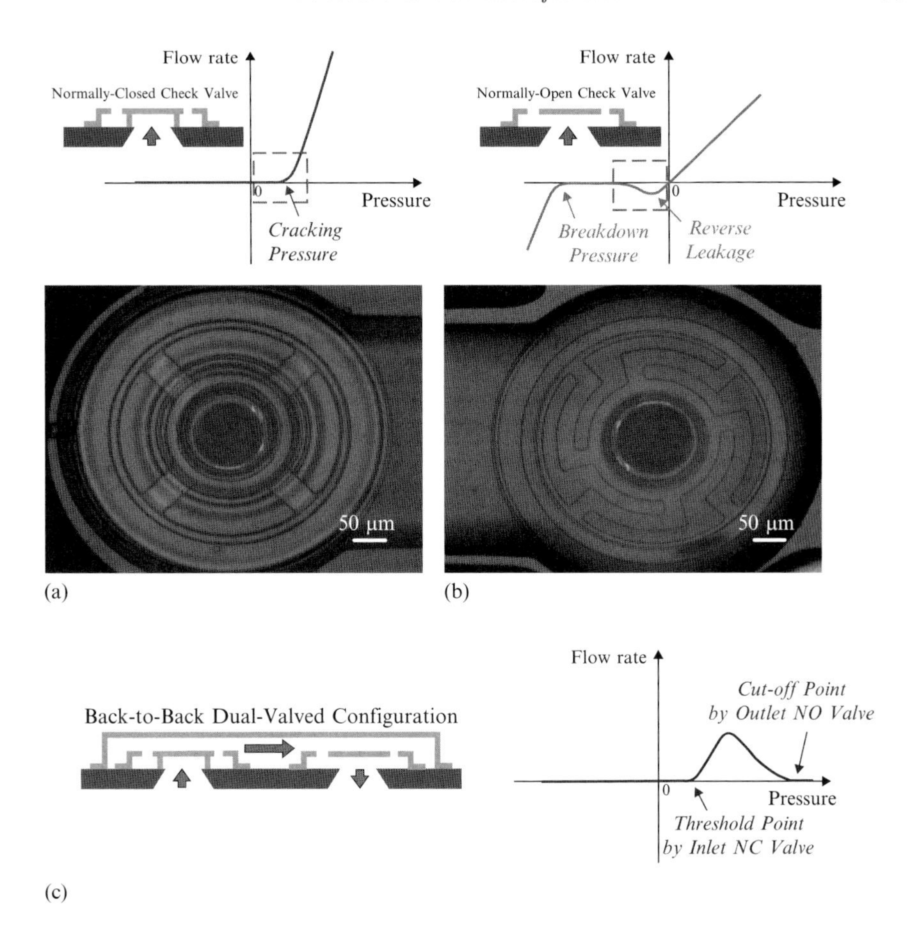

Fig. 4.10: Diagram of (a) normally closed check valve, (b) normally open check valve, and (c) dual valves made using back-to-back normally open and normally closed check valves. (Images adopted, with permission, from Chen *et al.* (2006).

outside energy to actuate. They can be either normally open (Chen *et al.* 2006; Wang *et al.* 1999) or normally closed (Chen *et al.* 2006; Wang and Tai 2000; Xie *et al.* 2001). A pressure operating window can be defined using back-to-back normally open and normally closed check valves called a dual valve (Chen *et al.* 2006). Because the operating pressure only acts in one direction at a time check valves are directional, meaning that ideally they only allow fluid to flow in one direction similar in concept to a diode in electronic circuits. Figure 4.10 illustrates the concepts behind check and dual valve operation. Check valves are commonly incorporated into micropumps to rectify the displacement behavior of the fluid due to the oscillating membrane. See the next section for more details on micropumps.

Capillary valves rely on the change in Laplace pressure between two adjacent channels of different hydraulic diameter. Capillary valves are based solely on

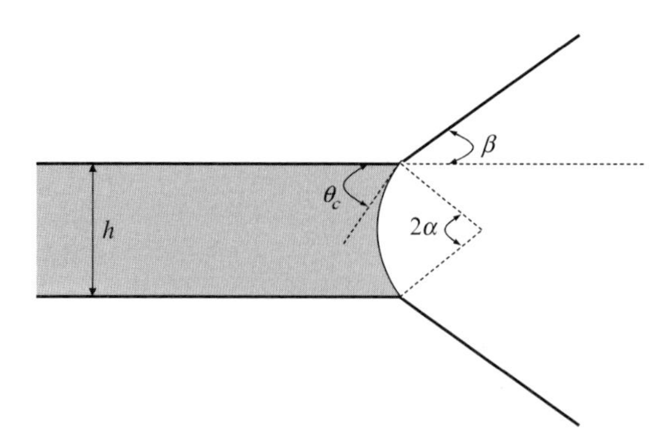

Fig. 4.11: Capillary valve geometry definitions.

channel geometry and material and have no moving parts. Liquid will freely enter a smaller diameter channel from a larger channel if the smaller channel is hydrophilic. However, a force is needed to eject a fluid from a capillary into a larger channel. The pressure needed to push a fluid out of a capillary is known as the *barrier pressure* and can be considered the equivalent of a cracking pressure. Man *et al.* developed a 2-D model for predicting the barrier pressure at a capillary expansion (Man *et al.* 1998). They found that for a capillary whose width is much greater than its height, h, the barrier pressure, $P_{barrier}$, achieved is

$$P_{barrier} = \frac{2\sigma_{lg}}{h}\left(\frac{\cos\theta_c - \frac{\alpha}{\sin\alpha}\sin\beta}{\cos\beta + \frac{\sin\beta}{\sin\alpha}\left(\frac{\alpha}{\sin\alpha} - \cos\alpha\right)}\right), \tag{4.49}$$

where σ_{lg} is the surface tension of the liquid–gas interface, θ_c is the contact angle, β is the wedge angle, and α is half of the angle subtended by the arc of the meniscus (see Fig. 4.11). Capillary valves can be made unidirectional by making asymmetric expansion regions. For an extension of this model into 3-D see Leu and Chang (2004).

4.3.4 *Mechanically actuated microvalves*

The simplest type of an active microvalve is one that relies on mechanical actuation. This mechanical actuation can be an external piston (Shoji and Esashi 1994) or screw (Weibel *et al.* 2005) that is manipulated by the user to deflect the top of the channel causing the channel to collapse. For the case of the screw, or torque, actuated mechanical microvalve the energy density can be estimated as

$$E_a' = \frac{T}{2\pi K r_s^3} \tag{4.50}$$

where T is the applied torque, K is the friction factor of the screw (typically 0.20 for dry screws and 0.16 for lubricated screws (Czernik 1996)), and r_s is the radius of the screw. Mechanical actuators typically deliver very high energy efficiency but have a very slow response rate because the valve is manually operated. While manual operation reduces response time it does increase valve reliability. These valves also offer no dead volume and extremely small sweep volume. Mechanical microvalves have seen very limited study but may provide solutions for microfluidic systems needing high valve reliability and low complexity and requiring rare state changes.

4.3.5 *Pneumatically actuated microvalves*

Pneumatic microvalves are the simplest type of microvalve after mechanical microvalves and unlike mechanical microvalves have seen a tremendous amount of study and use in microfluidic systems (Oh and Ahn 2006). An externally actuated pneumatic valve requires external equipment, usually some source of compressed gas (typically nitrogen or air), pressure regulator, and electronic valve or electronically controlled gas manifold. These external equipment requirements make pneumatic valve systems much larger than is indicated by the on-chip valve structure but allow for a wide range of operability with different microfabrication technologies and materials. The external equipment also limits response time. Fully integrated pneumatic valves will be covered later.

Pneumatically actuated valves function by applying a pressure to the backside of the valve seat causing it to close, if the valve is normally open, or open, if the valve is normally closed. The valve seat is typically a membrane made of either silicon (Takao and Ishida 2003; Huff and Schmidt), polydimethylsiloxane (PDMS) (Unger *et al.* 2000) and/or silicone rubber (Yang *et al.* 1999), or Parylene (Yang *et al.* 1998a; Hua *et al.* 2004). The pressure necessary to actuate such a valve is given by

$$p_{act} = p_{in} + \frac{k\delta}{A_m}, \tag{4.51}$$

where p_{in} is the pressure of the fluid at the inlet of the valve, k is the spring constant of the membrane, δ is the maximum distance the membrane must deflect, and A_m is the cross-sectional area of the membrane. Typical membrane geometries are rectangular and circular. For circular membranes a closed form equation for the spring constant is

$$k_{circular} = \frac{16\pi Et^3}{3r^2(1-v^2)}, \tag{4.52}$$

where E is the Young's modulus, ν is the Poisson ratio, t is the membrane thickness, and r is the membrane radius (Nguyen and Wereley 2002). The energy density for a circular membrane pneumatic actuator can be expressed as

$$E_a' = \frac{16Et^3}{3r^4(1-v^2)}. \tag{4.53}$$

The spring constant for a rectangular membrane is significantly more complicated than a circular membrane and in general is a function of the membrane displace-

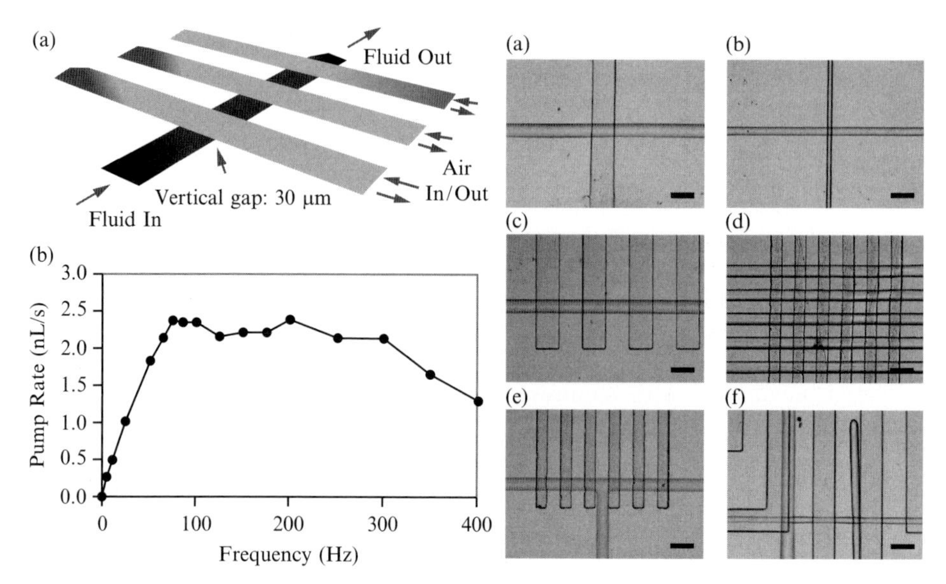

FIG. 4.12: Examples of pneumatic valves made of PDMS. Note that three pneumatic valves in series can be designed into a peristaltic pump. (Images adopted, with permission, from Unger *et al.* (2000)).

ment. The pressure/displacement relationship derived by Tabata *et al.* (1989) can be used to calculate the spring constant at the point of sealing (maximum displacement, δ)

$$k_{rect} = \frac{4}{n}\left(C_1 \sigma t + \frac{C_2 E t \delta^2}{a^2}\right) \tag{4.54}$$

$$C_1 = \frac{\pi^4(1+n^2)}{64} \tag{4.55}$$

$$C_2 = \left[\frac{\pi^6}{32(1-\upsilon^2)}\right] \times \left[\frac{9+2n^2+9n^4}{256} - \frac{(4+n+n^2+4n^3-3n\upsilon(1+n))^2}{256n+256\upsilon+18\pi^2(1+n^2)(9-\upsilon)}\right] \tag{4.3--10}$$

$$n = \frac{a}{b}, \tag{4.56}$$

where $2a$ is the length of the shorter side of the rectangle, $2b$ is the length of the longer side, and σ is the residual stress in the membrane. For a square ($n = 1$) silicone rubber membrane ($\nu = 0.5$) C_1 and C_2 are 3.04 and 2.55, respectfully. Pneumatic valves, whether circular or rectangular, typically operate in the 10 to

600 kPa range of actuating pressures with typically membrane thicknesses between 10 and 250 μm (Oh and Ahn 2006). These valves have author reported dead volumes between 0 and 160 nL although most are under 10 nL (Oh and Ahn 2006).

4.3.6 *Electrically actuated microvalves*

Electrically actuated valves come in four main varieties: electrostatic, piezoelectric, electromagnetic, and electrochemical. Electrostatic, piezoelectric, and electromagnetic actuated valves are typically microcantilevers or several microcantilevers suspending a membrane that will seal the orifice. Electrochemical actuated valves are fundamentally a passive capillary valve in geometry. The geometry of the capillary valve is used to prevent fluid flow after a gas bubble is generated by electrolysis of the working solution. Electrically actuated microvalves produce the fastest response times of any microvalve class and their operation can be easily integrated into a computer controlled system. Hence electrically actuated microvalves are well suited for microfluidic applications requiring high switching rates and automation. Electrically actuated valves' drawbacks can include limited valve seat displacement, difficult microfabrication, and relatively low power efficiency.

Electrostatic valves rely on the attraction of two oppositely charged plates to move the valve seat. The closing force, or *pull-in force*, can be approximated using a parallel plate capacitor model

$$F_{pullin} = \frac{1}{2}\varepsilon_r\varepsilon_0 A\left(\frac{V}{d}\right)^2, \tag{4.57}$$

where ε_r is the relative permittivity of the dielectric, ε_0 is the permittivity of free space ($8.85418 \times 10^{-12}\,\mathrm{F/m}$), A is the area of the electrodes, V is the applied potential difference, and d is the separation between the electrodes. In most cases a dielectric insultation layer must be added to the top of one electrode to prevent short circuits when the valve is closed. The effect of this insulation layer (with relative permittivity, $\varepsilon_{r,i}$, and thickness, d_i) is to change the closing force to

$$F_{pullin} = \frac{1}{2}\varepsilon_r\varepsilon_0 A\left(\frac{V}{d}\right)^2\left(\frac{\varepsilon_{r,i}d}{\varepsilon_r d_i + \varepsilon_{r,i}d}\right)^2. \tag{4.58}$$

Using eqn. (4.57) the energy density of an electrostatic actuator can be estimated as

$$E_a' = \frac{1}{2}\varepsilon E_{el}^2 \approx \frac{1}{2}\varepsilon\left(\frac{V}{d}\right)^2, \tag{4.59}$$

where $\varepsilon = \varepsilon_r\varepsilon_0$, and E_{el} is the electric field. There is only exact equality if the actuator is actually a parallel plate capacitor. The voltage needed to actuate the valve can be calculated using a force balance between the electrostatic pull-in force

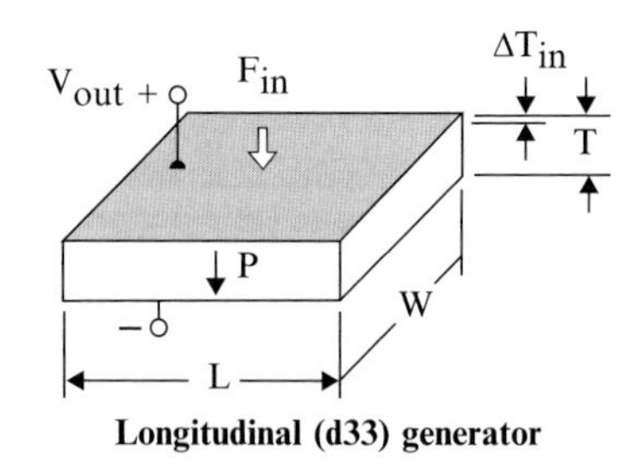

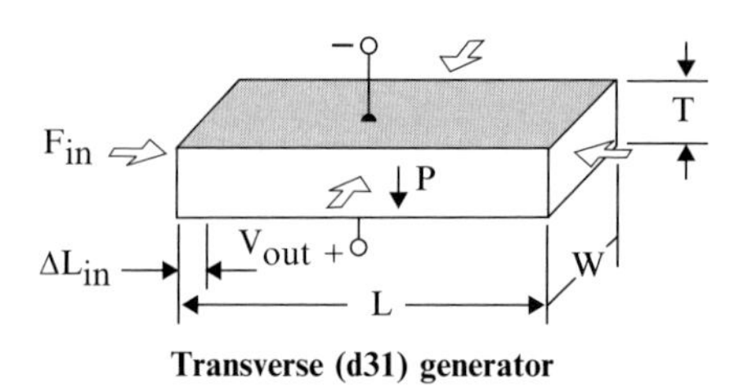

FIG. 4.13: Diagram of piezoelectric strain with electrodes on the top and bottom.

and the valve spring force. Since both the electrostatic and valve spring forces are dependent on the gap distance between the valve seat and the opposing electrode a critical gap distance, d_{cr}, exists where an unstable equilibrium (zero net force) exists. A critical voltage, V_{cr}, can then be defined as that voltage needed to achieve the critical gap distance. Voltages above the critical voltage will move the valve seat past the unstable equilibrium point causing the valve to fully actuate. For an electrostatic valve using air as the primary dielectric ($\varepsilon_r = 1$) the critical gap distance and voltage are

$$d_{cr} = \frac{d_0}{3} - \frac{2d_i}{3\varepsilon_{r,i}}, \tag{4.60}$$

$$V_{cr} = \frac{2\sqrt{k}\left(d_{cr} + \frac{d_i}{\varepsilon_i}\right)^{3/2}}{\sqrt{\varepsilon_0 A}}, \tag{4.61}$$

where k is the valve spring constant and d_0 is the initial gap thickness. For the valve to release the voltage must be decreased below the pull up voltage which is defined by the force balance of a fully closed valve

$$\frac{\varepsilon_{r,i}\varepsilon_0 A V_{pullup}^2}{2d_i^2} = kd_0 \rightarrow V_{pullup} = \sqrt{\frac{2kd_0 d_i^2}{\varepsilon_{r,i}\varepsilon_0 A}}. \tag{4.62}$$

Electrostatic valves typically operate at back pressures between 0 and 13.5 kPa for pull in voltages between 150 and 250 V and can have response times with 0.1–1 msec (Nguyen and Wereley 2002; Oh and Ahn 2006).

Piezoelectric valves consist of piezostack valves which offer high closing force but limited valve seat displacement and bimorph piezocantilevers which offer large displacement but relatively low closing force. Piezoelectric valves typically consist of an off-chip piezoelectric element and a microfabricated valve structure. The microfabricated valve seat is typically a rectangular or circular membrane or a suspended plate with cantilever beam supports. The piezoelectric effect is characterized by a change in the crystalline structure of a material (usually a ceramic) in the presence of an electric field. The most common piezoelectric materials used in microvalves are polyvinylidene fluoride (PVDF), lead zirconate titanate (PZT), and zinc oxide (ZnO). The strain, ε, induced by the electric field within the piezoelectric material is characterized by a set of coefficients, d_{ij}. The strain in the j axis due to an electric field, E, can be expressed as

$$\varepsilon_j = \sum_{i=1}^{3} d_{ij} E_i \tag{4.63}$$

The most critical coefficients are typically d_{33} and d_{31} because they represent the ratio of strain to electric field strength in the axial and transverse directions respectfully for a flat piece of piezoelectric with electrodes on its top and bottom (i.e. electric field only in the z or number 3 axis).

The closing force of a piezoelectric actuator at a fixed operating voltage can be assumed to be nearly linear from zero to some maximum displacement, d_{max} (Nguyen and Wereley 2002). Hence,

$$F(d) = F_{\max}\left(1 - \frac{d}{d_{\max}}\right), \tag{4.64}$$

where d is the distance the valve seat has travel from its rest position and F_{max} is the force delivered by the piezoelectric actuator at zero displacement. The maximum inlet pressure a piezoelectric actuator can withstand is

$$p_{\max} = \frac{F_{\max}\left(1 - \frac{d_0}{d_{\max}}\right) - kd_0}{A}, \tag{4.65}$$

where k is the valve spring constant, d_0 is the valve gap between the open and closed states, and A is the area of the valve seat. The energy density of a piezoelectric actuator can be estimated as

$$E_a' = \frac{E_{el}}{2(Ed_{33})^2}, \tag{4.66}$$

where E is the Young's modulus and E_{el} is the electric field strength.

Electromagnetic actuators consist of two magnetic field generating elements. These elements can either be an off-chip solenoid, on-chip microcoil, or an integrated permanent magnet. Integrated permanent magnets are typically fabricated either by electroplating or sputtering nickel, iron, or some nickel or iron based alloy. The magnetic force developed by the actuator in the case of an integrated permanent magnet separated from a microcoil by an air gap is

$$F_{closing} = \frac{\mu_0 A(NI + M_{sat})^2}{4d_0^2} \tag{4.67}$$

$$M_{sat} = \sqrt{\frac{4(BH)_{\max}}{\mu_0}} \tag{4.68}$$

where m_0 is the magnetic permeability of free space, A is the cross-sectional area of the permanent magnet, N is the number of turns of the microcoil, I is the current in the microcoil, M_{sat} is the saturation magnetization of the permanent magnet, d_0 is the gap length, $(BH)_{max}$ is the maximum energy product of the permanent magnet (Chowdhury *et al.* 2000). The energy density of the actuator can be estimated by the energy in the magnetic field

$$E_a' = \frac{B^2}{2\mu} \tag{4.69}$$

where B is the magnetic flux density and $\mu = \mu_r \mu_0$ is the permeability of the operating medium with a relative permeability, μ_r. Since both microcoils and off-chip solenoids require a relatively high current due to their high resistances and need to maintain a strong magnetic field, electromagnetic actuators tend to consume a lot of power relative to other actuator technologies. Development of lower power bistable or latching electromagnet actuators is critical to the competitiveness of this technology (Sutanto *et al* 2006).

Electrochemical actuators work by the production of gas through the electrolysis of water. The electrolysis typically occurs at a platinum electrode that can also catalyze the reverse reaction:

$$2H_2O \underset{Pt}{\rightleftarrows} 2H_2 + O_2 \tag{4.70}$$

Since the forward electrolysis happens at a faster rate than the reverse, recombination, reaction a net volume of gas can be maintained in a closed system at any fixed current. Electrochemically actuated valves can either be of the gate-variety or the capillary-variety. In gated electrochemical valves the generated gas bubble acts as a pneumatic pressure source deflecting a membrane and hence actuating the valve. Capillary electrochemical valves work by using the gas bubble and a constriction in the geometry of the channel to produce a passive capillary valve. An operating pressure above the Laplace pressure (see eqn. (4.68)) is then needed to move the bubble through the capillary. Water electrolysis occurs at voltages close to 1 V. The low operating voltage and small currents necessary to produce the gas bubble mean electrochemical valves (especially of the capillary type) are very efficient. The actual energy efficiency of a capillary electrochemical valve is

$$E_a^{'} = \frac{h_f^0\left(\rho_{O_2} + 2\rho_{H_2}\right)}{3M_{H_2O}} \tag{4.71}$$

Where h_f^0 is the enthalpy of formation of water, ρ is the density of the gas at the operating pressure and temperature, and M_{H_2O} is the molar mass of water (18 g/mol). Capillary electrochemical valves can operate with power consumption below 5 μW while gated electrochemical actuators can withstand over 200 kPa of back pressure and induce membrane deflections over 50 μm consuming just 80 μW making them ideal for low power applications (Nguyen and Wereley 2002; Neagu *et al.* 1997, 2000).

4.3.7 *Thermally actuated microvalves*

The last class of active microvalves contains those actuated using thermally energy. (Thermal energy can be used to either heat a fluid causing it to expand and in a closed system cause a pressure increase in which case the actuation mechanism is thermo-pneumatic.) In a thermomechanical actuator, the difference in thermal expansion coefficients between two dissimilar materials can be used to create mechanical bending in a bimorph (a cantilever composed of two layers). Finally, some materials known as shape memory alloys (SMA) undergo a solid–solid phase transformation when heated causing the metal to deform and hence providing a third method of using thermal energy to make a valve. The energy density for thermopneumatic and thermomechanical actuator can be estimated as

$$E_a^{'} = \frac{E(\gamma\Delta T)^2}{2} \tag{4.72}$$

where E is the Young's modulus, γ is the coefficient of thermal expansion (in the case of a bimorph the difference in γ should be used instead), and ΔT is the change in temperature. Estimating the energy density of a SMA actuator is difficult because the temperature verse strain/strain relationship of a SMA is highly non-linear.

Thermopneumatic valves work with both liquid and gas operating fluids. In many cases, the actuator makes use of the liquid–vapor phase change to induce higher sealing pressures. Henning has proposed a comprehensive model for the operation of thermopneumatic valves (Henning 2006). Henning's model holds that for a square membrane valve seat of side length, a, gap thickness, d_0, and membrane thickness, t, Young's modulus, E, and with an all liquid working fluid

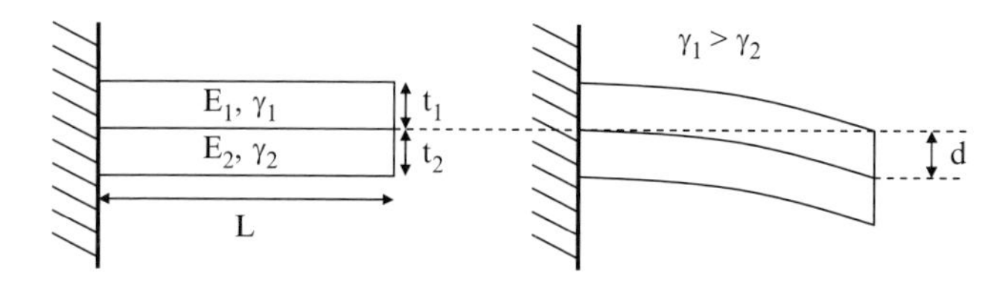

FIG. 4.14: Cross-sectional diagram of a bimetallic cantilever. Diagram on the left shows the cantilever in an unbent state which naturally occurs only for $\gamma_1 = \gamma_2$.

$$\Delta V = V_0 \gamma \Delta T, \tag{4.73}$$

$$s = \frac{V_0 \gamma \Delta T}{\eta v a^2} = \frac{d_0 \gamma \Delta T}{\eta v}, \tag{4.74}$$

$$\frac{P_{trans} a^2}{E t^2} = A_s \left(\frac{s}{t}\right) + B_s \left(\frac{s}{t}\right)^3, \tag{4.75}$$

Where ΔV is the volume change of the pneumatic chamber, ΔT is the temperature change of the liquid, γ is the liquid's coefficient of thermal expansion, s is the deflection at the center of the membrane, ηv is the volume expansion factor ($0 < \eta v \leq 1$ and is typically between 0.25 and 0.38 with an average value of 0.345 (Henning 2006)), P_{trans} is the transmembrane pressure, A_s and B_s are constants with some material dependence. For an all vapor working fluid the actuator chamber pressure can be estimated from the ideal gas law for a gas in a sealed, fixed-volume chamber

$$\frac{T_0}{T_f} = \frac{p_0}{p_f}. \tag{4.76}$$

Thermopneumatic valves can be made from a wide range of materials and can have maximum pressure ratings between 1 and 700 kPa and typically consume between 50 and 200 mW (Nguyen and Wereley 2002).

Thermomechanical valves are based on the deflection of a bimorph under thermal stress. The deflection of cantilevers under various conditions of stress has been well studied (Chu *et al.* 1993; Gehring *et al.* 2000; Huang *et al.* 2004). The maximum deflection of a bimorph cantilever under thermal stress, also called a bimetallic cantilever, is

$$d_{\max} = \frac{L^2}{2R}, \tag{4.77}$$

$$R = \frac{(b_1 E_1 t_1)^2 + (b_2 E_2 t_2)^2 + 2 b_1 b_2 E_1 E_2 t_1 t_2 \left(2t_1^2 + 3 t_1 t_2 + 2 t_2^2\right)}{6(\gamma_2 - \gamma_1) b_1 b_2 E_1 E_2 t_1 t_2 (t_1 + t_2) \Delta T}, \tag{4.78}$$

where L is the length of the cantilever, b is the width of the material, E is the material's Young's modulus, t is the materials thickness, γ is the materials coefficient of thermal expansion, and ΔT is the change in temperature (Chu *et al.* 1993). The intrinsic force at the tip of the cantilever is

$$F_{int} = \frac{3(EI)_{beam} d_{\max}}{L^3}, \tag{4.79}$$

$$(EI)_{beam} = \frac{(b_1 E_1 t_1)^2 + (b_2 E_2 t_2)^2 + 2b_1 b_2 E_1 E_2 t_1 t_2 \left(2t_1^2 + 3t_1 t_2 + 2t_2^2\right)}{12(b_1 E_1 t_1 + b_2 E_2 t_2)}, \tag{4.80}$$

where $(EI)_{beam}$ is the flexural rigidity of the beam. For beams with equal width materials ($b_1 = b_2 = b$) the tip force can be reduced to (Nguyen and Wereley 2002):

$$F_{int} = \frac{3bt_1 t_2 E_1 E_2 (t_1 + t_2)(\gamma_1 - \gamma_2)\Delta T}{4L(t_1 E_1 + t_2 E_2)}. \tag{4.81}$$

Typically thermomechanical valves are made by depositing a metal layer at a high temperature using either electron beam or thermal deposition on top of a silicon microheater. The valve seat is suspended using several bimetallic actuators. Because the metal layer was deposited at high temperature (usually in excess of 400 °C) and then cooled a thermomechanical force is present on the valve seat according to eqn. (4.79). Since the valve seat is rigid the differential equation and boundary conditions describing the deflection of the cantilever is

$$\begin{gathered} y''' = \frac{-F}{(EI)_{beam}} \\ \text{BC:}\ \ y'(0) = 0 \quad y'(L) = 0 \quad y(0) = 0 \end{gathered} \tag{4.82}$$

Solving (4.82) for F at a seat valve a distance, $d_0 < \frac{F_{int} L^3}{12(EI)_{beam}}$, from the valve opening we find the minimum force per actuator required to keep a valve open as

$$F_{\min} = \frac{12 d_0 (EI)_{beam}}{L^3}. \tag{4.83}$$

The total closing force of a valve with N identical bimetallic actuators can be calculated from the difference between the intrinsic tip force and the minimum tip force per actuator

$$F_{close} = N(F_{int} - F_{min}) = \frac{3N(EI)_{beam}}{L}\left(\frac{1}{2R} - \frac{4d_0}{L^2}\right). \tag{4.84}$$

Thermally actuated valves, whether thermopneumatic or thermomechanical, provide high energy density, high maximum back pressures and reasonable power efficiency. Their major drawbacks are the long response times and high power consumption due to the heat capacitance of the materials.

4.4 Source of on-chip pressure—micropumps

Microfluidic devices designed to move fluid from an inlet to an outlet are micropumps. Micropumps can use several transduction mechanisms to impart a directional flow to fluid. The most common mechanism used to achieve this goal is the mechanical displacement of a diaphragm, which in micropumps is identical to that in microvalves. Almost all of the mechanical actuation mechanisms used in microvalves can be incorporated into micropumps. A method for displacement of fluid alone, however, is not sufficient to form a mechanical micropump. The displacement caused by the forward stroke of the actuator results in a decrease in the stroke volume, the volume of the pump chamber. Similarly, the backstroke of the actuator also causes a change in volume. The flow due to the membrane displacement on the forward stroke is equal but opposite to that of the backward stroke. To achieve the desired unidirectional net flow, the pumping action must be rectified. There are three rectification schemes in common use: 1) check valves, 2) peristaltic pumping, and 3) diffuser/nozzle structures. Mechanical micropumps that do not have a backstroke, such as rotary pumps and ultrasonic pumps, do not require rectification. There are also several non-mechanical micropumping techniques, such as electrochemical and ferrofluidic, that build a pressure head capable of moving a fluid. Non-mechanical pumping techniques that do not require pressure gradients, such as electrokinetic, electrooptic, and surface tension force driven, are discussed in other chapters. Detailed information on demonstrated micropump performance is available in several review articles (Laser and Santiago 2004; Nguyen *et al.* 2002; Woias 2005).

As in the case of microvalves, micropumps are also characterized by several figures of merit. The *maximum volumetric flow rate*, or pump capacity, is the amount of fluid delivered per unit time at zero backpressure. Backpressure is the pressure developed at the outlet of the pump due to the increase in fluid velocity developed by the pump and the total flow resistance of the down stream components (channels, valves, etc.). The *maximum backpressure* is the backpressure at which the flow rate of the pump is zero. The *pump head* is the net work by the pump per unit weight of liquid between the inlet and outlet and is given by Nguyen and Wereley (2002):

$$h = \left(\frac{p}{\rho} + \frac{\bar{v}^2}{2g} + z\right)_{out} - \left(\frac{p}{\rho} + \frac{\bar{v}^2}{2g} + z\right)_{in}, \tag{4.85}$$

where p is the static pressure, ρ is the density of the fluid, $\bar{v}$ is the average linear fluid velocity, g is the acceleration due to gravity, and z is the elevation of the inlet/outlet. The maximum pump power, P_{pump}, is defined in terms of the maximum volumetric flow rate, Q_{max}, and the maximum back pressure, p_{max}, as:

$$P_{pump} = \frac{p_{\max} Q_{\max}}{2}. \tag{4.86}$$

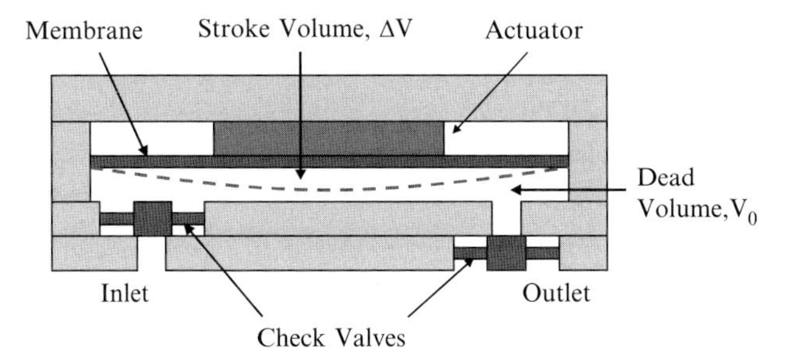

Fig. 4.15: Illustration of a check-valved displacement micropump. The dotted line indicates the position of the bottom of the membrane at the end of the forward stroke. The stroke volume is the volume above the dotted line and the dead volume is the volume below the dotted line and above the check valves.

The *pump efficiency*, η, is then:

$$\eta = \frac{P_{pump}}{P_{actuator}}, \tag{4.87}$$

where $P_{actuator}$ is the input power to the actuator. In the case of mechanical displacement micropumps, an additional figure of merit is the *compression ratio*, ψ, which is the ratio of the stroke volume, ΔV, to the pump dead volume, V_0, as

$$\psi = \frac{\Delta V}{V_0}. \tag{4.88}$$

4.4.1 *Check valve displacement micropumps*

Displacement mechanical micropumps consist of an actuator that moves a membrane or diaphragm in a periodic motion. The fluid flow through the pump must be rectified to achieve a net unidirectional flow. One common method is to attach a check valve to the inlet and outlet of the pump. The inlet check valve is oriented to allow flow into the pump during the backward stroke and is closed during the forward stroke. The outlet check valve is open during the forward stroke, allowing fluid to exit the pump, but is closed during the backward stroke. Figure 4.15 illustrates a typical design for a check-valve, displacement pump. The geometry and material of the check valves determine the pressure necessary to open them: the critical pressure. Use of small Young's modulus materials is ideal for the check valve and membrane since these materials allow for the largest stroke volumes and smallest critical pressures (Nguyen *et al.* 2002). Softer materials such as silicone rubber (Meng *et al.* 2000; Pan *et al.* 2005), polyimide (Mercanzini *et al.* 2005; Bustgens *et al.* 1994), and Parylene (Xie *et al.* 2004b), have been successfully integrated as pump membranes and valves. In addition to maximizing stroke volume and minimizing critical pressure, ideal displacement micropumps will also exhibit high compression ratios most often achieved by minimizing pump dead volume.

Displacement micropumps are capable of moving both gas and liquids. If properly designed, the displacement pump can remove the air in the pump chamber and fill

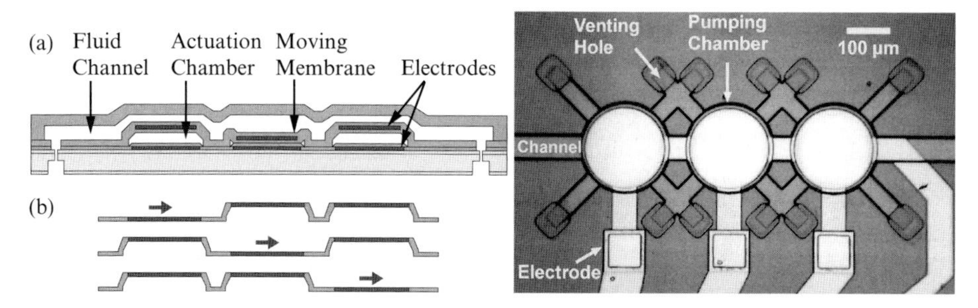

FIG. 4.16: Diagrams of a three-phase, electrostatically-driven peristaltic micropump. (a) Device schematic. (b) Three-phase actuation sequence. (Images adopted, with permission, from Xie *et al* (2004b)).

itself with liquid, a process referred to as self-priming. To self-prime, the work done by the pump must over come the surface energy of the wetted inlet orifice. In cases where the surface energy is dominated by capillary force on the check valve, the following is true for a self-priming pump with inlet orifice of diameter, d:

$$W_{\min} = \Delta p_{crit} \frac{\pi d^2}{4} z = \sigma \pi d z = U_{surface}, \tag{4.89}$$

$$\Delta p_{crit} = \frac{4\sigma}{d}, \tag{4.90}$$

Where W_{min} is the minimum amount of work necessary to open the check valve, Δp_{crit} is the cracking pressure of the check valve, z is the small gap between the check valve and the inlet orifice, and σ is the surface tension of the working liquid (Nguyen and Wereley 2002). The minimum compression ratio needed to achieve the necessary minimum work is given by Nguyen and Wereley (2002):

$$\psi_{\min} = \frac{1}{k} \frac{|\Delta p_{crit}|}{p_0}, \tag{4.91}$$

where k is the specific heat ratio of air ($k = 1.4$) and p_0 is the ambient air pressure in the pump chamber (typically taken as 1 atm). Equation (4.91) puts a lower limit on the stroke volume an actuator must displace for a pump to be self-priming.

4.4.2 *Peristaltic micropumps*

Peristaltic micropumps are two or more membranes actuated in sequence to produce a rectified fluid flow (Fig. 4.16). While two stage peristaltic pumps are possible, three stage pumps are more common because the additional stage increases maximum flow rates and pressure heads while simultaneously reducing back flow (Berg *et al.* 2003). The actuation mechanisms described in Section 4.3 can all be used to operate peristaltic pumps. Peristaltic micropumps have the advantage of not requiring a check valve or other recertification scheme and the phased actuation sequence alone provides for the necessary flow rectification. The fluid channel in the pump is never completely sealed and hence peristaltic micropumps are susceptible to back flows when they are not actively pumping. In applications where back flows are not tolerable, a check valve should be integrated at the outlet of the peristaltic pump. The volumetric flow rate through a peristaltic micropump can be estimated as the product of the stroke volume and the actuation frequency, f, by

$$Q = \Delta V f. \tag{4.92}$$

Peristaltic micropumps have been reported with maximum volumetric flow rates between 0.14 and 100 μL/min with water as the working fluid and maximum pressure heads between 0.17 and 34.5 kPa (Laser and Santiago 2004). A variety of membrane materials including glass, silicon, silicone rubbers, Parylene, PDMS, and titanium have all been used to fabricate peristaltic micropumps (Laser and Santiago 2004). Most peristaltic pumps reported in the literature operate at actuation frequencies between 1 and 100 Hz with electrostatically actuated pumps achieving the highest actuation frequencies (Nguyen and Wereley 2002).

4.4.3 *Valveless displacement micropumps*

Two designs exist for the construction of single chamber displacement micropumps that do not require the use of check valves to rectify the fluid flow. One design makes use of preferential flow through a nozzle/diffuser (Figure 4.17). A nozzle is a structure of constant cross-sectional shape but constantly decreasing cross-sectional area in the direction of the fluid flow. A diffuser is a structure of constant cross-sectional shape as well but constantly increasing cross-sectional area in the

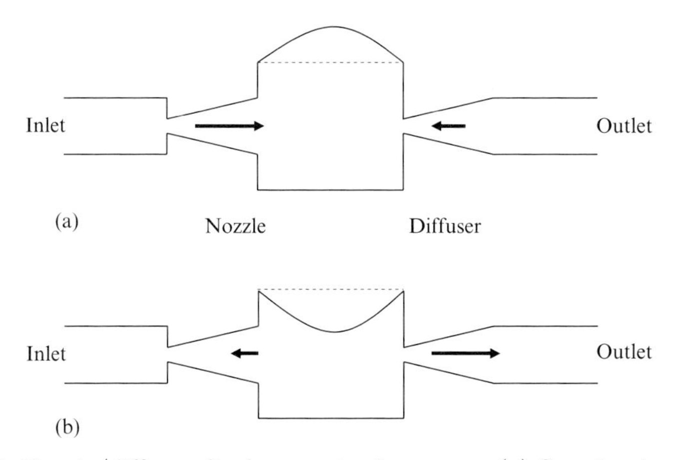

FIG. 4.17: Nozzle/diffuser displacement micropump. (a) Supply phase showing preferential flow from the inlet. (b) Pump phase showing preferential flow to the outlet.

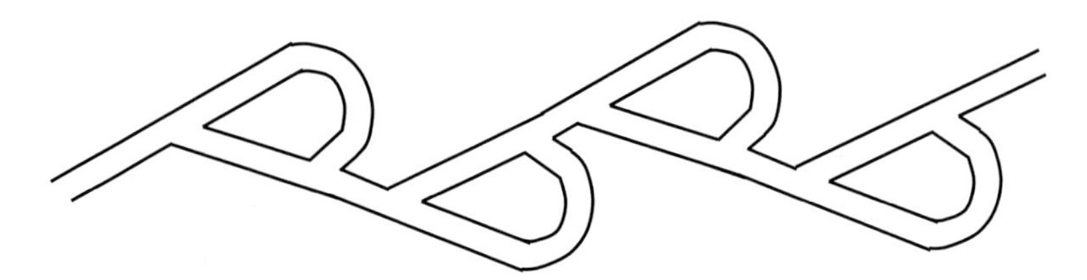

FIG. 4.18: Valvular conduit geometry for flow rectification.

direction of the fluid flow. Due to the change in cross-sectional area, diffusers convert the kinetic energy present in the flow to potential energy in the form of pressure. The pressure change in a nozzle/diffuser can be expressed as:

$$\Delta p = \xi \frac{\rho \bar{v}^2}{2}, \tag{4.93}$$

where ξ is the pressure loss coefficient, ρ is the fluid density, and $\bar{v}$ is the average linear velocity of the fluid in the rectifier. The pressure loss coefficient will be higher for a nozzle than for a diffuser and for a nozzle/diffuser rectifier the fluidic diodicity, η, is defined as the ratio of the these pressure loss coefficients:

$$\eta = \frac{\xi_{nozzle}}{\xi_{diffuser}}. \tag{4.94}$$

Maximal fluidic diodicity is achieved with a rounded diffuser entrance, 2.5 diffuser expansion angle (diffuser walls are slanted at an angle of 2.5 relative to the centerline flow direction), and the exit of the diffuser has sharp corners (White 1986). Using this maximal geometry, diodicity of 4.25 can be achieved (White 1986).

The volumetric flow rate through the nozzle/diffuser displacement micropump is given by Stemme and Stemme (1993).

$$Q = 2f\Delta V \frac{\sqrt{\eta} - 1}{\sqrt{\eta} + 1}, \tag{4.95}$$

where f is the actuation frequency and ΔV is the stroke volume. A dimensionless parameter known as the flow rectification efficiency, γ, is defined from equation (4.95) as (Ahmadian and Mehrabian 2006):

$$\gamma = \frac{\sqrt{\eta} - 1}{\sqrt{\eta} + 1}. \tag{4.96}$$

Nozzle/diffuser micropumps made of standard microfluidic materials with pressure heads between 0.38 and 74 kPa and maximum volumetric flow rates between 1.4 and 1100 μL/min have been reported in the literature (Laser and Santiago 2004). Stemme and Stemme (1993) reported the first nozzle/diffuser micropump

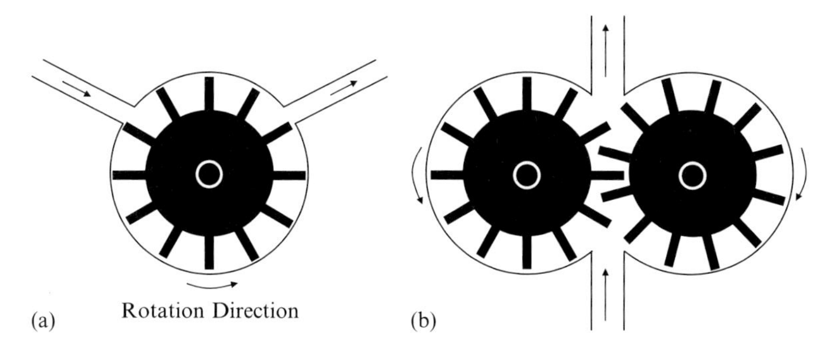

FIG. 4.19: Rotary micropump designs. (a) Single gear. (b) Two-gear.

but the structure was made from brass and was relatively large. It was capable of flow rates as high as 16 mL/min and pressure heads of 21 kPa.

A second static rectification geometry, invented by Nikola Tesla, is based on a valvular conduit (Figure 4.18) (Tesla 1920). Rectification is performed by providing a different fluid flow path for the forward and reverse direction through the valve. The forward direction has a relatively straight and hence lower resistance path, while the reverse path is more convoluted and hence has higher resistance. Valvular micropumps were implemented by Froster *et al.* (1995) in silicon and found to have higher diodicity than nozzle/diffuser rectifiers fabricated in a similar manner. Several authors have observed an increase in diodicity of the valvular conduit with increasing flow rate, a behavior not observed in nozzle/diffuser rectifiers and not expected from eqn. (4.94) (Froster *et al.* 1995; Feldt and Chew 2002; Bardell 2000). Diodicity up to 1.74 is possible with valvular conduits (Bardell 2000).

4.4.4 *Rotary micropumps*

Rotary pumps operate by displacing a constant volume of liquid during a single revolution of the pumping elements, typically gear teeth, around the pump shaft. Rotary pumps are particularly well suited for pumping highly viscous fluids, but due to high viscous forces encountered during pumping, powerful actuators capable of delivering high torque are required. Rotary pumps have been miniaturized with single and multiple gear designs as well as on-chip and off-chip actuators (Fig. 4.19) (Ahn and Allen 1995; Doppler *et al.* 1997). For a two-gear rotary pump, the flow rate is given by Doppler *et al.* (1997)

$$Q = h\pi n\left(\frac{d_g^2}{2} - \frac{s^2}{2} - \frac{m_g^2\pi^2}{6}\cos^2 a\right), \tag{4.97}$$

$$m_g = \frac{d_g}{N}, \tag{4.98}$$

where h is the height of the microfluidic channel, n is the spin speed in number of revolutions per minute, d_g is the pitch circle diameter of the gears, s is the center-to-center shaft distance, α is the engagement angle, m_g is the gear module, and N is the number of teeth per gear.

4.4.5 *Ultrasonic micropumps*

Ultrasonic micropumps move liquid in a gentle manner by inducing a flexural plate wave (FPW). A FPW is a standing mechanical wave typically induced by electrical excitation of an interdigitated electrode pattern on a piezoelectric film which has been added to the back of a thin membrane (White and Wenzel 1998). Other FPW excitation methods include electrostatic (Giesler and Meyer 1994) and magnetic (Martin *et al.* 1998) deflection of the thin membrane. The energy in a high intensity acoustic wave propagating in a medium will be absorbed and scattered by the medium. This attenuation of the high-intensity wave is non-linear and causes the medium itself to move, a process known as acoustic streaming (Nguyen *et al.* 1998). The fluid velocity away from the membrane is given by Nguyen and Wereley (2002):

$$v = v_{\max} \exp\left(-\frac{z}{\delta_a}\right), \tag{4.99}$$

$$\delta_a = \frac{\lambda}{2\pi\sqrt{1-(c_{ph}/c_s)^2}} \approx \frac{\lambda}{2\pi}, \tag{4.100}$$

where z is the distance away from the membrane in the direction normal to the membrane surface, δ_a is the acoustic evanescent length, λ is the FPW wavelength, c_{pm} is the phase velocity of the FPW, and c_s is the speed of sound in the fluid. The spacing between the interdigitated electrodes can be used to control the FPW wavelength. Ultrasonic micropumps have achieved linear flow velocities of 1.5 mm/sec (Meng *et al.* 2000).

4.4.6 *Electrochemical micropumps*

Electrochemical micropumps use the electrolysis of water to create pressure gradients that produce fluid flow. By electrolyzing water in a sealed chamber, the internal pressure in the chamber increases as the volume of gas increases. The electrolysis process produces both hydrogen and oxygen. The hydrogen is reduced at the cathode and oxygen is oxidized at the anode:

$$\text{Cathode: } 2H_2O + 2e^- \underset{recombination}{\overset{electrolysis}{\rightleftarrows}} H_2 + 2OH^- \tag{4.101}$$

$$\text{Anode: } 2H_2O \underset{recombination}{\overset{electrolysis}{\rightleftarrows}} O_2 + 4H^+ + 4e^- \tag{4.102}$$

$$\text{Net: } 2H_2O \underset{recombination}{\overset{electrolysis}{\rightleftarrows}} O_2 + 2H_2. \tag{4.103}$$

The two electrodes can be quasi-separated to reduce recombination effects, but there must be an ion flow path between the two chambers such that a current can be supported. A supporting electrolyte, typically 25 per cent potassium hydroxide, is added to the water to increase its current capacity. A major concern for large scale electrolysis and for power sensitive microfluidic applications is the conversion efficiency of the electrolysis reaction. A portion of the power introduced into the system will be dissipated as heat and, as a consequence, is not used to electrolyze water. Conversion efficiency is a strong function of electrode material and solution activation (Kaninsky *et al.* 2006). Nickel electrodes in an activated solution produce reasonable efficiencies relative to cost while a platinum-molybdenum alloy (Pt_2Mo) has been shown to have the most catalytic effect on electrolytic hydrogen evolution (Kaninsky *et al.* 2006).

The volumetric pumping rate is equivalent to the volume rate of change of the gas bubbles (Bohm *et al.* 1999):

$$Q = \frac{dV}{dt} = \frac{3}{4}\frac{i}{F}V_m, \tag{4.104}$$

where i is the applied current, F is Faraday's constant (96.49 C/mol), and V_m is the molar gas volume at 25°C and atmospheric pressure (24.7×10^{-3} m^3/mol). Electrochemical actuation can achieve high displacement force but is relatively slow (Pang *et al.* 2006). Electrochemical pumps are relatively easy to integrate into

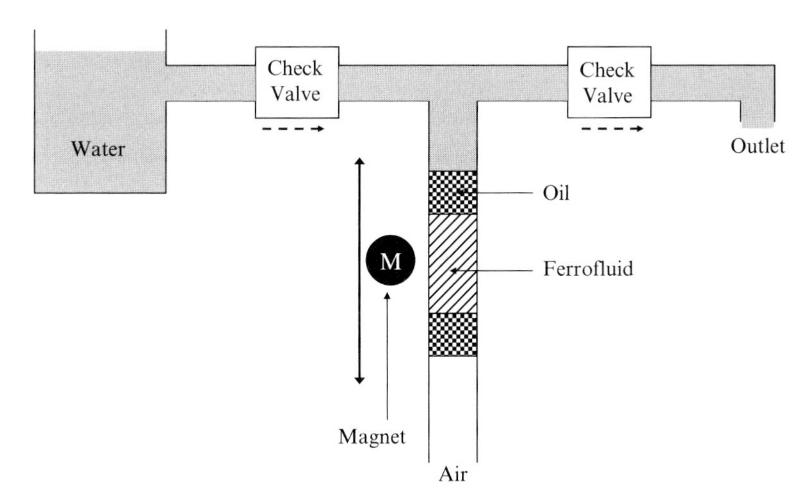

FIG. 4.20: Example of ferrofluidic micropump using check valve flow rectification (Yamahata *et al.* 2005). The oil is used to separate the ferrofluid from the water. As the permanent magnet is oscillated back and forth, the fluid flows into and out of the chamber between the two check valves. The check valve on the supply side remains closed during the pumping stroke and the valve on the outlet side closed during the suction stroke.

microfluidic devices and have been incorporated into dosing systems (Bohm *et al*, 1999), rare blood cell sorting systems (Furdui *et al*, 2003), and on-chip gradient liquid chromatography systems (Xie *et al.* 2004a).

4.4.7 *Ferrofluidic micropump*

Ferrofluids are stable colloidal suspensions of nano size magnetic particles in an aqueous or non-polar solvent (Mao and Koser 2006). The magnetic nanoparticles must be coated with stabilizing dispersion agent to prevent agglomeration (Mao and koser 2006). Magnetic field gradients and viscous fluid forces allow for continuous actuation and positioning of ferrofluids in microchannels. Several ferrofluidic pumping principles have been developed and reported in the literature and the pumping techniques fall into two classes: 1) use of immiscible ferrofluidic plugs to pump a separate working fluid, and (2) pumping ferrofluids themselves. Ferrofluids are easily controlled by external magnetic fields but can also be moved through integrated electromagnets. To pump a ferrofluid in a desired direction, a technique similar to flexural wave generation is used. Instead of creating a directional mechanical wave, a traveling electromagnetic wave is used. The ferrohydrodynamic equations describing the fluid's motion must be solved numerically, but

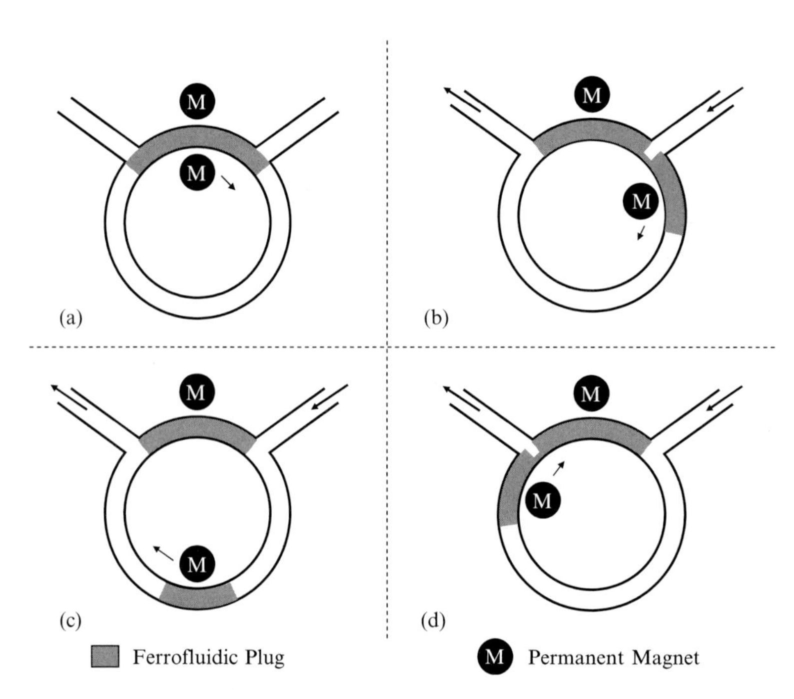

FIG. 4.21: Rotary ferrofluidic micropump. Diagrams show stationary and moving permanent magnets that, respectively, hold a ferrofluidic plug in place to act as a closed valve to the working fluid and use a second plug to produce the pumping action on the working fluid (Hatch *et al.* 2001).

experimental results show pumping pressures of up to 2 Pa can be attained achieving average linear flow velocities on the order of 5 mm/sec (Mao and Koser 2006).

One ferrofluidic pump for use on separate working fluid uses integrated check valves to rectify the flow generated by an oscillating ferrofluidic plug (Yamahata *et al.* 2005) (Figure 4.20). The force experienced by the working fluid resulting from the one dimensional motion of the ferrofluidic plug can be estimated by (Yamahata *et al.* 2005):

$$F_x = \frac{m}{\mu_0} \frac{\partial B_x}{\partial x}, \tag{4.105}$$

$$m = V \chi_r \bar{B}_x, \tag{4.106}$$

where m is the magnetic moment of the ferrofluidic plug, μ_0 is the magnetic permeability of free space, V is the volume of the plug, χ_r is the relative susceptibility of the ferrofluid, and B_x (and $\bar{B}_x$ indicates the average of B_x over the distance the plug moves) is the magnetic field in the x-direction.

Another ferrofluidic micropump uses a ferrofluidic plug for both a pumping action and for flow rectification (Hatch *et al.* 2001). The device consists of two external permanent magnets one of which is stationary and the other rotates. Figure 4.21 shows this rotary ferrofluidic micropump in various stages of operation. When the two magnets are aligned there is a single large ferrofluidic plug that blocks both the inlet and outlet. As the rotating magnet moves away from the stationary magnet a portion of the ferrofluidic plug is dragged along and eventually separates from the remaining ferrofluid. This moving plug pushes the working fluid out the outlet as it rotates while simultaneously drawing working fluid in through the inlet. The pressure difference is maintained by the stationary ferrofluid which acts as a closed valve. As the rotating magnet returns to its initial position the moving and stationary plugs recombine and the cycle resumes. This rotary ferrofluidic micropump could achieve flow rates above 45 μL/min and pressure gradients of 0.192 psi (Hatch *et al.* 2001). Ferrofluidic devices, however, typically suffer from short life spans due to derogation of the ferrofluid and fouling of the microfluidic channel walls (Hatch *et al.* 2001).

4.5 Microflow sensors

Microflow sensors will come to play a large role in advanced microfluidic systems. Microflow sensors use a variety of transduction mechanisms to measure the volumetric flow rate through a microchannel (Nguyen and Wereley 2002). No matter which transduction mechanism is used, all microflow sensors have five important characteristics: 1) dynamic range, 2) sensitivity, 3) response time, 4) power consumption, and 5) compatibility. Dynamic range is the difference between the lower detection limit of the sensor and the sensor's output saturation level. It is generally desirable to have a linear relationship between the quantity being sensed and the sensor output. Often, the reported dynamic range of a microflow sensor is restricted to the flow rates that produce this linear response. Sensitivity is the ratio of the change in sensor signal to the change in flow rate (or average linear flow velocity) and, hence, within the linear dynamic range is a constant

$$S = \frac{dV}{dQ} \text{ or } S = \frac{dV}{d\bar{v}}. \tag{4.107}$$

Response time is the time interval required for the sensor output to stabilize after a stepped change in flow rate. Power consumption is the amount of power required to operate the sensor. Compatibility can refer to a sensor's biological or chemical compatibility depending on its application.

One common technique is to use the relationship developed in Section 4.1 expressed in eqn. (4.31) that relates the volumetric flow rate to the pressure drop across a length of channel. These differential pressure flow sensors function by using two pressure sensors separated by a known length of channel to calculate the average linear flow velocity (Oosterbroek *et al.* 1999; Pedersen *et al.* 2005). The pressure sensors operate using a number of mechanisms, most of which involve a pressure sensitive membrane deflection similar to those described in Section 4.2. The two most common pressure sensors are capacitive and piezoresistive. Capacitive sensors rely on a change in the gap distance between two parallel plates due to a movable membrane Chang and Allen (2004). Piezoresistive pressure sensors change electrical resistance in response to a strain as

$$\frac{\Delta R}{R} = \alpha\varepsilon \tag{4.108}$$

where R is the resistance of the piezoresistive material, α is the gauge factor, and ε is the strain (Pedersen *et al.* 2005; Wu *et al.* 2006). Metals have gauge factors on the order of two while polycrystalline silicon has a gauge factor between 10 and 40 and single-crystalline silicon between 50 and 150 (Nguyen and Wereley 2002). For additional information on micromachined pressure sensors, see the review published by Eaton and Smith (1997) or other texts covering this technology (Fraden 1996; Kovacs 1998; Sze 1994).

Mechanical flow rate sensors measure the volumetric flow rate or linear flow velocity by producing a signal that is correlated to the mechanical bending of a cantilever or plane. This mechanical force can be due either to the drag (Grosse *et al.* 2006; Gass *et al.* 1993) or lift force (Czaplewski *et al.* 2004; Svedin *et al.* 1998, 2003) on the sensing element. The drag force acts parallel to the flow direction. The total force on a cantilever of characteristic length, L, in a fluid flow of volumetric flow rate, Q, and can be expressed as (Nguyen and Wereley 2002):

$$F_{total} = F_{drag} + F_{pressure\_drop} = \frac{C_1 L\mu}{A_g} Q + \frac{C_2 \rho}{A_0 A_g^2} Q^2, \tag{4.109}$$

where C_1 and C_2 are constants depending on the form of the cantilever, μ is the viscosity, ρ is the fluid density, A_g is the cross-sectional area of the gap around the cantilever, and A_0 is the surface area of the cantilever parallel to the flow. Lift force, which acts perpendicular to the flow direction, can be expressed as (Svedin *et al.* 1998):

$$F_{lift} = \frac{C_L A_0 \rho \bar{v}^2}{2}, \tag{4.110}$$

Where C_L is the lift force coefficient, A_0 is the surface air of the airfoil, and $\bar{v}$ is the linear flow velocity of the fluid. Since both fluid density and viscosity are strongly dependent on temperature, mechanical flow rate sensors typically require temperature compensation.

The most common integrated flow rate sensors use variations in thermal energy to detect flow rate. Thermal sensors are often easy to integrate into microfluidic devices and do not usually require special packaging. There are two primary types of thermal flow sensors: 1) hot wire/film and 2) time of flight. Hot wire sensors work by monitoring the power needed to maintain a constant temperature of the wire. As fluid flows over the wire at various rates, the wire requires either more or less power. The amount of power needed to heat a hot wire in a fluid is (Nguyen *et al.* 1996)

$$P = \frac{0.664 \kappa A_H \Delta T \sqrt{\mathrm{Re}^3 \, \mathrm{Pr}}}{L} \tag{4.111}$$

$$\mathrm{Pr} = \frac{c_p \mu}{\kappa} \tag{4.112}$$

where Re is the Reynold's number, Pr is the Prandtl number, L is the characteristic length of the device, κ is thermal conductivity of the fluid, μ is the viscosity, c_p is the specific heat, A_H is the heater area, ΔT is the temperature difference between the heater and the fluid. Increases in the amount of power needed to maintain a constant temperature are the result of an increase in Reynold's number corresponding to a higher flow rate.

Time of flight sensors work by measuring the time it takes for a temperature pulse produced at one location in the microchannel to travel to a thermal sensor located some distance, L, away. A thermal pulse can be described by an evolving two-dimensional Gaussian according to the following equation (van Kuijk *et al.* 1995):

$$T(x, y, t) = \frac{q_0}{4\pi\kappa t} \exp\left[-\frac{(x - \bar{v}t)^2 + y^2}{4at} \right], \tag{4.113}$$

where q_0 is the initial thermal energy (W/m), $\bar{v}$ is the average linear flow velocity, and a is the thermal diffusivity of the fluid. The time of flight can be found from (4.113) (van Kuijk *et al.* 1995):

$$\tau = \frac{-2a + \sqrt{4a^2 + \bar{v}^2 L^2}}{\bar{v}^2} \tag{4.114}$$

and for heater to sensor lengths much greater than $4at$ eqn. (4.114) can be reduced to

$$\tau = \frac{L}{t}. \tag{4.115}$$

A variety of thermal flow rate sensors have been described in the literature (van Kuijk *et al.* 1995; Ashauer *et al.* 1999; Meng and Tai 2003; Rasmussen *et al.* 2001). The time of flight sensing principle can even be extended to different actuation mechanisms such as electrochemical (Wu and Sansen 2002). A wide variety of mechanisms exist for measuring absolute and differential temperature in microfluidic devices. See any of the text books covering thermal sensor technology for additional information (Nguyen and Wereley 2002; Fraden 1996; Kovacs 1998; Sze 1994).

4.6 Packaging of pressure-driven microfluidic devices and systems

Packaging describes any method, material, or technology used to interface a micro scale device to the macro scale world or to link multiple devices together into a complete system. Packaging is one of the most challenging and often overlooked aspects of microfluidic system design. It plays a major role in overall system performance and can have an enormous influence on the extent a technology is used in the wider world. When done properly good packaging will hardly be noticed, but when done poorly packaging and the problems arising from it can easily dominate a user's time and adversely affect the device performance. Perhaps the reason that packaging is such a difficult problem is that so much is required of it. Ideally, microfluidic packaging will exhibit low or no leak rates, high maximum sealing pressures, no clogging, excellent thermal stability, chemical inertness, high mechanical strength, low dead volume, reusability, be easily integrated into the fabrication process, be scalable with device feature size, and result in a user interface that has both ease of use and will fail gracefully if misused. In extreme circumstances, such as those involved in space flight, biomedical implants, and military applications, packaging is also needed that is highly durable, bio-compatible, and unaffected by long exposure times to extreme heat, cold, and radiation. The ultimate goal for packaging is that it is never the limiting factor in device performance.

In the case of microfluidic devices, an aspect of packaging that is of critical importance is the fluidic interconnect. Similar to electrical interconnects in many other MEMS applications, fluidic interconnects provide the physical path for fluid to enter and leave the microsystem. Fluidic interconnects are the center of microfluidic packaging. A variety of interconnect technologies have been developed since the advent of microfluidics, yet, the vast majority of reported microfluidic devices still use very rudimentary interconnect technology. A review article by Fredrickson and Fan covers many of the current and historical approaches to the problem Fredrickson and Fan (2004). Interconnect technologies can be categorized into four types: reservoirs, glues and epoxies, press-fit, and compression. Several interconnect schemes consists of varied combinations of these four basic technologies.

Reservoirs are by far the simplest interconnect technology. Reservoirs are simply large volume containers, either machined into the device or attached to it, that store the working fluid. Reservoir systems typically rely on capillary action to fill the system with fluid. This interconnect technology is particularly useful in electroosmotic systems where fluid flow is induced between reservoirs through the application of large electric fields. Reservoir systems are sensitive to evaporation of the working fluid and hydrostatic pressure flows. *Hydrostatic pressure* arises from the collective weight of the fluid. In a reservoir, the pressure at the bottom of the reservoir containing a static ($\mathbf{v} = 0$) fluid column of height, h, can be easily derived from (4.1-18)

$$p_{static} = p_{atm} + \rho g h, \tag{4.116}$$

where p_{atm} is the atmospheric pressure above the liquid column, ρ is the density of liquid, and g is the acceleration due to gravity. In a system of two or more connected reservoirs the hydrostatic pressure in each reservoir must be equal otherwise a pressure driven flow will result from the reservoir of higher pressure to that of lower pressure until an equilibrium is reached.

The maximum unrestricted evaporation rate of a liquid can be estimated using the Hertz–Knudsen–Schrage relationship

$$E_{max} = \eta_{ev} A P_v \sqrt{\frac{M}{2\pi RT}} \tag{4.117}$$

where η_{ev} is Schrage's correction factor (traditionally taken as 2 but more recently found to be 1.667 (Koffman *et al.* 1984)), A is the area of the fluid/gas interface, M is the molar mass of the liquid, R is the ideal gas law constant, P_v is the equilibrium vapor pressure at the liquid surface, and T is the absolute temperature (Koffman *et al.* 1984; Beverley *et al.* 1999). This maximum is hardly ever achieved in real devices because of the presence of a relatively stagnant vapor layer between the liquid interface and top of the reservoir. Because hydrostatic pressure is not a function of reservoir cross-sectional area, the fluid level in all reservoirs must be at

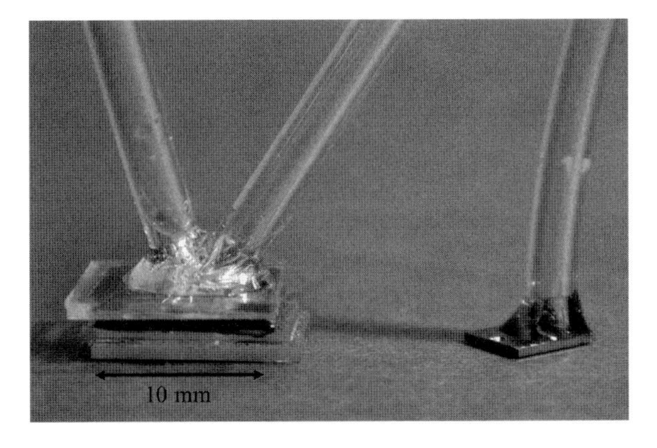

FIG. 4.22: Photograph of microfluidic device with glued tube interconnect.

TABLE 4.1: Operation ranges of various interconnect technologies. (Meng 2003)

Interconnect Technology	Operation Range	Reference
Polyethylene Coupler/Tubing and Epoxy Reinforcement	> 5 kPa (0.7 psi)	van der Wingaart *et al.* (2000)
Mylar Sealant and Epoxy Reinforcement with Capillary	∼ 190 kPa (28 psi)	Tsai and Lin (2001)
Polyimide/Parylene Ribbon Cable Style Interconnect	∼ 200 kPa (29 psi)	Man *et al.* (1997)
Silicon Finger Microjoint with Silicone Gasket & Tygon Tubing	> 210 kPa (30 psi)	Gonzalez *et al.* (1998)
Thermoplastic Retaining Flange with PEEK Tubing	> 210 kPa (30 psi)	Puntambekar and Ahn (2002)
PDMS Press-fit	∼300-700 kPa (40 -100 psi)	Christensen *et al.* (2005)
Silicon/Plastic Coupler with Silicone Gasket & Capillary	∼ 410 kPa (60 psi)	Gray *et al.* (1999)
Silicone Gasket Sealed Silicon Coupler with Capillary	∼ 550 kPa (80 psi)	Yao *et al.* (2000)
Photopatternable Silicone O-rings	> 1.7 × 10^3 kPa (250 psi)	Miserendino and Tai (2007)
Silicon Sleeve Coupler with Capillary	∼ 3.6 × 10^3 kPa (500 psi)	Gray *et al.* (1999)
Polymer Coupler with Fused Silica Capillary	∼ 6.2 × 10^3 kPa (900 psi)	Meng (2003)
Silicon Bulk Coupler with Cryogenically inserted PEEK Tubing	> 9 × 10^3 kPa (1300 psi)	Meng *et al.* (2001)
Silicon Post Coupler with Fused Silica Capillary or PEEK Tubing	> 9 × 10^3 kPa (1300 psi)	Meng *et al.* (2001)
Silicon Flanged Coupler with PEEK Tubing	> 1 × 10^3 kPa (1500 psi)	Meng (2003)
Mismatched Silicon Coupler with Capillary	> 1.2 × 10^3 kPa (1740 psi)	Spiering *et al.* (1997)

the same height for a system in equilibrium. Evaporation rate, however, is a function of reservoir geometry and hence if reservoirs are not carefully designed and filled a near constant non-equilibrium state of the system could result. Furthermore, because the equilibration process is driven by relatively weak hydrostatic pressure differences through high resistance fluidic channels systems, using reservoir interconnects can take a long time to equilibrate often exasperating the evaporation problem. Although minimization of evaporations effects is typically desirable, they can be used as an actuating mechanism for reservoir-based pressure driven devices (Zimmermann *et al.* 2005).

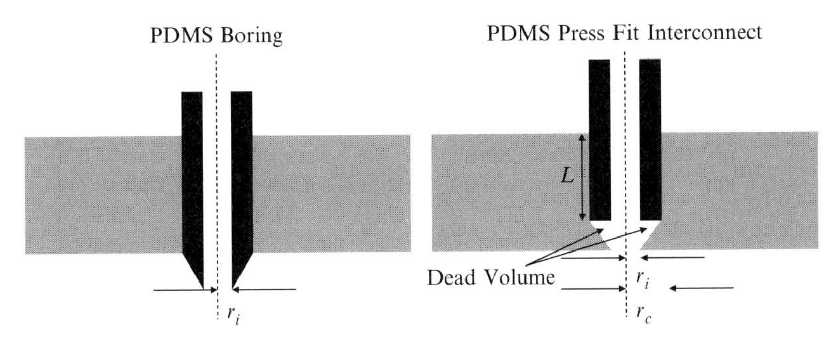

FIG. 4.23: (left) PDMS interconnect being bored with sharp coring tool. (right) PDMS press fit interconnect showing possible regions of dead volume.

4.6.1 *Glues and epoxies*

The simplest pressure-tolerant interconnect technology used in microfluidics are tubes manually glued to the surface of the microfluidic device (Fig. 4.22). This technique has many obvious disadvantages including possible clogging, manual alignment, and incompatibility with some surface machined devices. The process of attaching the tubes can also be time consuming and is prone to a low success rate. Glued or epoxied interconnects are permanent and tend to have relatively low maximum operating pressures. The major advantage of this technique is its simplicity and the lack of any special microfabrication process. More recently, glues and epoxies have been used to augment other interconnect technologies to improve mechanical strength and pressure tolerance.

While glues and epoxies might be the simplest interconnect technology, the most popular is probably the press-fit interconnect. There are several version of press-fit interconnects but all more or less work by sliding a capillary tube into a specially made via or coupler. A variety of press-fit interconnect schemes have been developed for use in silicon based microfluidics Meng *et al.* (2001). For silicon and surface micromachined based microfluidics press-fit connectors are more difficult to use because silicon is not elastic enough to provide sidewall compression and surface machined microfluidic devices are too thin to provide the mechanical stability necessary for interconnects. The silicon based press-fit interconnects require a coupler that attaches to capillary and can then be either glued or snapped into a specially machined receiving port on the device. The capillary is inserted into the coupler after being frozen in liquid nitrogen and then allowed to warm to room temperature. The warming process causes the capillary to expand and press up against the walls of the coupler producing a locking force.

4.6.2 *Press fit interconnects*

The reason press fit, also referred to as interference fit, interconnects are so popular is that they are the ideal type of connector for a soft, compressible material such as

PDMS. Since use in PDMS microfluidics dominate the use of press fit interconnects it is valuable to examine this case specifically. Typically the interconnection is made by first boring a via with a needle or other sharp coring tool. The boring procedure can have a large effect on the pressure tolerance of the final interconnect. The best interconnects are form in at least 3 mm of PDMS using a sharpened flat tip needle and a slight downward twisting motion (Christensen *et al.* 2005). A capillary, usually made of stainless steel, is then inserted into the bored hole. The capillary outer diameter must be larger than the boring tools inner diameter, ideally several hundred microns larger. A typical boring tool is a 20 gauge (610 μm inner diameter) flat bottom needle. An unmodified 20 gauge needle flat bottom adapter (953 μm outer diameter) is a convenient capillary since it allows for easy connections to other fluidic components off-chip. Any spiral or vertical cracks in the bored hole will result in leaking even at low operating pressures.

Press fit locking force, the force required to remove a press fit connector, can be estimated by first assuming that the elasticity of the PDMS is much greater than the elasticity of the capillary and hence no deformation of the capillary will occur. It should be noted that this assumption is not generally true of all press fit connections since most involve fitting of two metals, but in the case of PDMS and either a metal or glass capillary this assumption is reasonable. We also assume that the frictional force on the sidewall of the capillary can be approximated by a Coulomb friction (linear) relationship with coefficient of friction, μ_f, and we find:

$$\varepsilon_r = \frac{r_c - r_i}{r_i} \tag{4.116}$$

$$\sigma_r = E_r \varepsilon_r \tag{4.117}$$

$$F_{lock} = 2\pi r_c \mu_f \int_0^L \sigma_r dz \tag{4.118}$$

$$F_{lock} = 2\pi \mu_f E_r L (r_c - r_i) \frac{r_c}{r_i} \tag{4.119}$$

where ε_r is the radial strain, σ_r is the radial stress, E_r is the Young's modulus in the radial direction (for PDMS typically between 360 and 870 kPa and usually close to 600 kPa (Bhushan and Burton 2005)), L is the length of capillary inserted into the PDMS, r_c is the outer radius of the capillary, and r_i is the initial radius of the via before capillary insertion. Equation (4.119) makes the further assumption that the radial stress is constant along the length of the capillary. Many factors can effect the locking force of a press fit interconnect and several more complex estimation methods have been proposed (Friedrich *et al.* 2005; Goel 1978). A compilation of coefficients of friction for PDMS is given in Table 4.2. Researchers have also noted that the locking force tends to reduce over time due to material relaxation and repeated insertions of capillary tubes.

TABLE 4.2: Table of PDMS coefficients of friction.

Materials	Coefficient of Friction, μ_f	Reference
PDMS Bulk-Silicon	0.11 ± 10%	Bhushan and Burton (2005)
PDMS Film-PDMS Film	0.43 ± 10%	Bhushan and Burton (2005)
PDMS Bulk-PDMS Bulk	0.51 ± 10%	Bhushan and Burton (2005)
PDMS Bulk-Stainless Steel	0.58 ± 10%	Friedrich *et al.* (2005)

To prevent the interconnection from blowing out, the locking force must exceed the maximum hydrostatic end force, $H_{G,max}$, applied to the capillary by the working pressure of the fluid, p, in the system;

$$H_{G,\,max} = \pi r_c^2 p, \tag{4.120}$$

$$p \leq 2\mu_f E_r L \left(\frac{1}{r_i} - \frac{1}{r_c} \right). \tag{4.121}$$

The locking force for the PDMS press fit connection described above is approximately 1.76 N and is theoretically capable of withstanding a maximum working pressure of 2.46 MPa (357 psi). Practically, these types of interconnects are rarely operated over 700 kPa (~100 psi) because the PDMS/substrate adhesion is not sufficient to withstand such high-pressures (Christensen *et al.* 2005). PDMS press fit interconnects can contain some dead volume as depicted in Fig. 4.23 when the insertion length is less than the thickness of the PDMS. This interconnect dead volume is usually quite small and tends not to be a factor in device performance. There is no doubt that the availability of an easy-to-use, unrestrictive, reliable, and versatile interconnect methodology is a key factor in the wide spread use of PDMS in modern microfluidics.

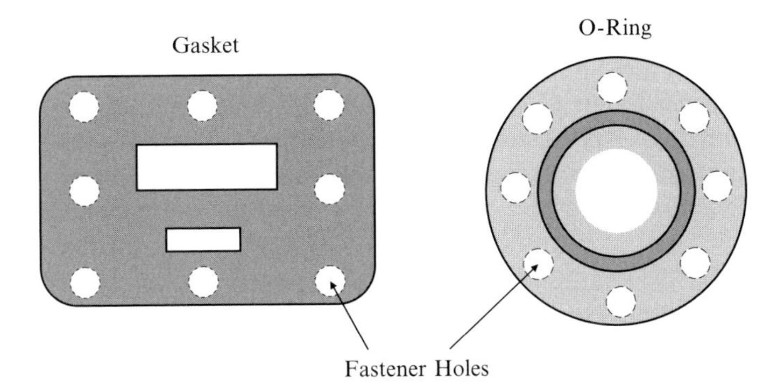

FIG. 4.24: Diagrams of a gasket and O-ring connection. The gasket is shown with two fluid flow paths and cut-outs for fasteners. The O-ring is shown with a single fluid flow path and fastener holes surrounding the O-ring.

4.6.3 *Compression interconnects*

While the press fit interconnect has gained wide acceptance in PDMS microfluidics, its use has been very limited amongst other microfluidic fabrication technologies. Microfluidic devices made using fabrication technologies such as bulk micromachining and surface micromachining tend to rely on compression interconnects when a reusable, unrestrictive, and reliable interconnect is required. Compression interconnects in microfluidic systems are directly related to the interconnections found in macroscopic fluidic systems. Specifically the fundamental sealing technology is that of gaskets and O-rings. Gaskets are compressible membranes of any shape that typically seal multiple fluidic paths simultaneouscly and have holes in them through which fasteners are placed in order to apply the compressive force between the bottom flange and the top flange. O-rings are continuous rings of constant cross-sectional area that are placed between the fluid flow path and an arrangement of fasteners that are used to compress the O-ring and seal the connection. O-rings are typically compressed individually and seal only one fluidic path at a time. Both gaskets and O-rings require a compressive force first to seat them and then to maintain enough pressure to prevent the working fluid from leaking. The compressive force developed by a set of N fasteners is

$$F_{compress} = \sum_{i=1}^{N} \frac{T_i}{KD_{f,i}}, \tag{4.122}$$

where T_i is the torque on the ith bolt, K is the fastener friction factor (typically 0.20 for a dry fastener and 0.16 for a lubricated fastener), and $D_{f,i}$ is the nominal fastener diameter of the ith bolt (Czernik 1996). For a set of identical and equally torqued bolts eqn. (4.122) reduces to

$$F_{compress} = \frac{NT}{KD_f}. \tag{4.123}$$

Gaskets require a minimum seating stress to initially seat the gasket during joint assembly (Czernik 1996). This minimum seating stress, σ_y, is a function of gasket material and flange geometry and must be experimentally determined. Once σ_y is known a minimum fastener torque must be applied to seat the gasket such that

$$T \geq \frac{A_s KD_f \sigma_y}{N} \tag{4.124}$$

where A_s is the seated area of the gasket. After seating the gasket a constant compressive force must be maintained such that the gasket stay seated and that it overcome any hydrostatic end force that develops.

In the case of O-rings, it is unclear exactly at what point in the O-ring the hydrostatic load reaction occurs causing difficulty in defining the seated area. The diameter associated with the point of load reaction is termed G. According to the American Society of Mechanical Engineering (ASME) Code for Pressure Vessels Section VIII Division 1 (ASME Boiler and Pressure Vessel Code 2001) for typical O-ring sizes used in microfluidic systems, G can be taken as the mean gasket diameter

$$G = D_g - W_g \tag{4.125}$$

where D_g is the outer O-ring diameter and W_g is the width of the O-ring. For other types of flange and O-ring geometries G is defined differently often moving it closer to the edge of the O-ring for O-rings wider than 0.5 inches (ASME Boiler and Pressure Vessel Code 2001). The ASME Code, however, does not offer an analytical definition of G and has changed how it is calculated over time (ASME Boiler and Pressure Vessel Code 2001; Rossheim and Markl 1943). Some authors have expressed discontent at the ASME Code and use the inner O-ring diameter instead of G. Bouzid *et al.* (2002) define $G/2$ analytically as the radial location at which the total torque on the gasket due to compressive stress, $S_g(r)$, is equivalent to the torque developed from a force applied at a distance $G/2$ equal to the total geometric gasket area, A_g, times the average compressive stress per gasket, $\bar{S}_g$. This analytical definition closely conforms to the ASME Code for thin gaskets and can be used for nonlinear gasket behavior induced by flange rotation. Furthermore, it can be expressed mathematically as

$$\int_{D_g/2-W_g}^{D_g/2} 2\pi r^2 S_g(r)dr = \frac{G}{2} A_g \bar{S}_g. \tag{4.126}$$

Under uniform radial stress

$$G = D_g - W_g + \frac{W_g^2}{3(D_g - W_g)}. \tag{4.127}$$

Hence, to seat an O-ring the minimum seating torque per fastener, using the AMSE definition of G, is

$$T \geq \frac{\pi K D_f W_g (D_g - W_g)\sigma_y}{2N}. \tag{4.128}$$

The maximum fluid working pressure, p, with an O-ring interconnect formed with N identical and equally torqued fasteners is

$$p \leq \frac{4NT}{\pi K D_f (D_g - W_g)^2}. \tag{4.129}$$

Microfluidic systems have a major advantage in using compression interconnects over macro-scale fluidic systems due to the inverse square relationship between maximum sustainable pressure and O-ring diameter. While compression interconnections are ideal for high-pressure microfluidic applications their major drawback is the need for a more complex packaging scheme. The only way to apply compression interconnects is with the use of a jig. The jig is used to hold the microfluidic device, external fluidic tubing, and help alignment of the O-rings or gasket with the vias on the device (Fig. 4.25). Jigs tend to be custom-made for each microfluidic system meaning substantial increases in the system cost, size, and development time. Jigs, however, offer excellent mechanical stability, the opportunity to readily interface with the wide array of commercially available products for handing fluids, and protection for the microfluidic device. Most jig designs also attempt to incorporate a method for providing electrical interconnects to the microfluidic system. By solving both the electrical and fluidic interconnect problems, jigs can prove to be well worth the investment. In any commercial application, jigs will likely be a requirement because most microfluidic devices are too delicate to be used outside of a laboratory setting in an unpackaged state.

4.6.4 *Microgaskets, MEMS O-rings, and modular microfluidics*

Commercial O-rings are the most common sealing mechanism used in compression interconnects. Commercial O-rings, however, must be manually aligned and ultimately are limited in minimum size. The desire for arbitrary shaped fluidic vias, high density interconnections, and extremely high accuracy alignment has led to

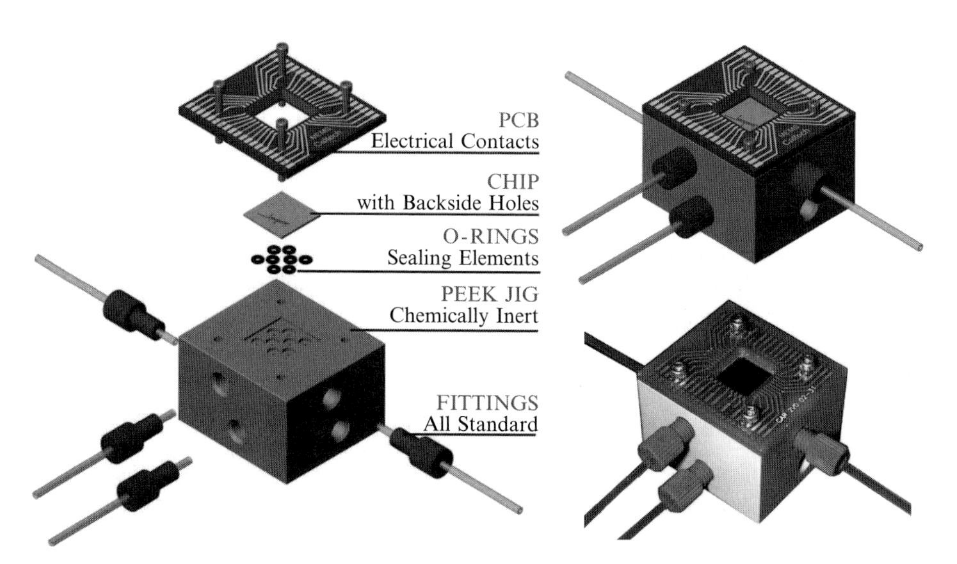

Fig. 4.25: Example of high-pressure microfluidic system using compression interconnects. Left and top-right images are rendered computer models. Bottom-right image is a photograph of the assembled system. (Images adopted, with permission, from He (2006).)

the need for fully integrated MEMS O-rings. Silicone rubbers have proven successful as a material for commercial O-rings and gaskets. Silicone rubber, of which PDMS is a specific type, is the most likely candidate for MEMS O-rings because of its relative thermal stability, chemical inertness, and great compressibility. In addition, silicone rubber is a good candidate because it has weak adhesion to many materials, which is especially important because the O-ring should not stick or bond to the flange (Grosjean 2001). The major difficulty with silicone MEMS has been finding a way to pattern the silicone. Silicone has been successfully incorporated into microfluidic devices through bulk micromachining, surface micromachining, molding, and replica casting (Grosjean 2001). Ideally, however, the silicone should be UV photopatternable. Dow Corning has developed a photodefinable silicone for MEMS application that is ideal for use as a microgasket or MEMS O-ring.

Microgaskets and MEMS O-rings function similarly to their commercial counterparts but because of their extremely small thickness (usually less than 25 μm), the topology of the flanges is a much greater consideration. The critical feature of the flange surface is the height variation across the gasket seating area. Generally, for a gasketed system to be capable of forming a seal, it must satisfy:

$$\max_{A_s}(y) - \min_{A_s}(y) \leq (C_{\max} - C_{seat})t \tag{4.130}$$

where A_s is the seated area of the gasket, y is the surface profile of the flange, C_{max} is the maximum compression tolerance, C_{seat} is the per cent compression necessary to seat the gasket, and t is the initial thickness of the gasket (Miserendino and Tai 2007). Rubber O-rings typically work best when compressed between 15 and 30 per cent of their initial thickness (Czernik 1996). Compression of less than 15 per cent is usually insufficient to insure a complete seating of the O-ring. Compression above 30 per cent increases the likelihood that the O-ring will be permanently deformed or possibly even crack. For MEMS, O-rings are based on a 25 μm thick material which means the surface height variation they can seal across is only about 7.5 μm. Actual integrated silicone microgaskets have demonstrated seating compression at approximately 12 per cent and a maximum compression of approximately 25 per cent (Miserendino and Tai 2007). Bulk micromachined microfluidics' surface height variation is usually well within this limit. Surface micromachined microfluidics, however, tend to have channel heights in excess of 10 μm hence they will require some type of surface planarization if MEMS O-rings are to be used. Another major advantage of fully integrated microgaskets is extremely small interconnect dead volume.

With the use of integrated microgaskets, it is possible to consider the construction of microfluidic systems using multiple, separately fabricated devices. The use of multiple independent microfluidic devices to form a microfluidic system is known as modular microfluidics. Integrated microgaskets are an excellent solution to bridge what could be called the micro/micro divide inherent in modular microfluidic architectures. The advantages of modular microfluidics are that it allows separate fabrication technologies to be used on the various components, individual replacement of devices when they fail or when improved versions have been

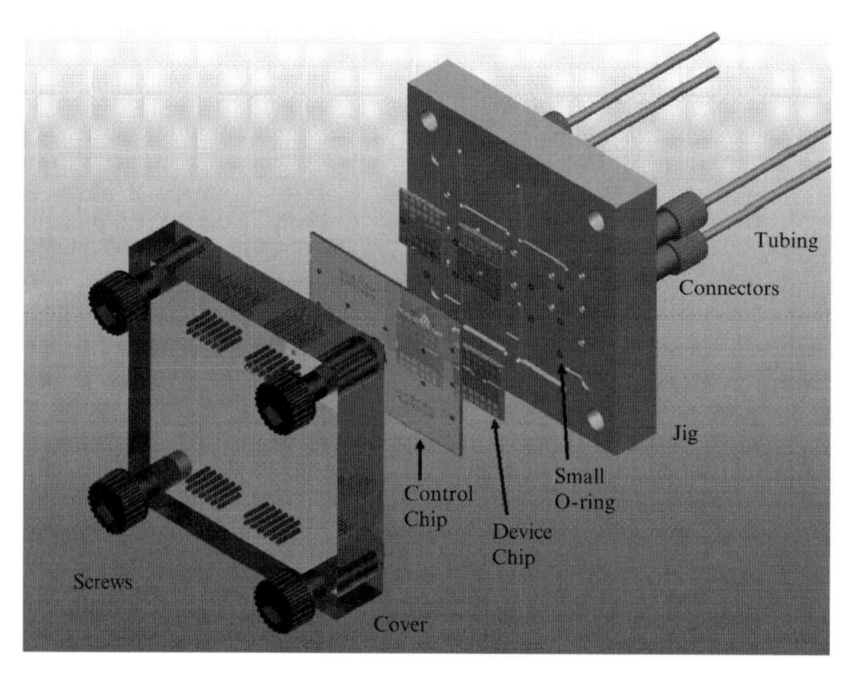

Fig. 4.26: Computer rendering of a modular microfluidic system. The control chip acts as a microfluidic breadboard connecting the device chips to one another and also interconnecting with the jig through small commercial O-rings. The entire system is sealed by tightening the four screws.

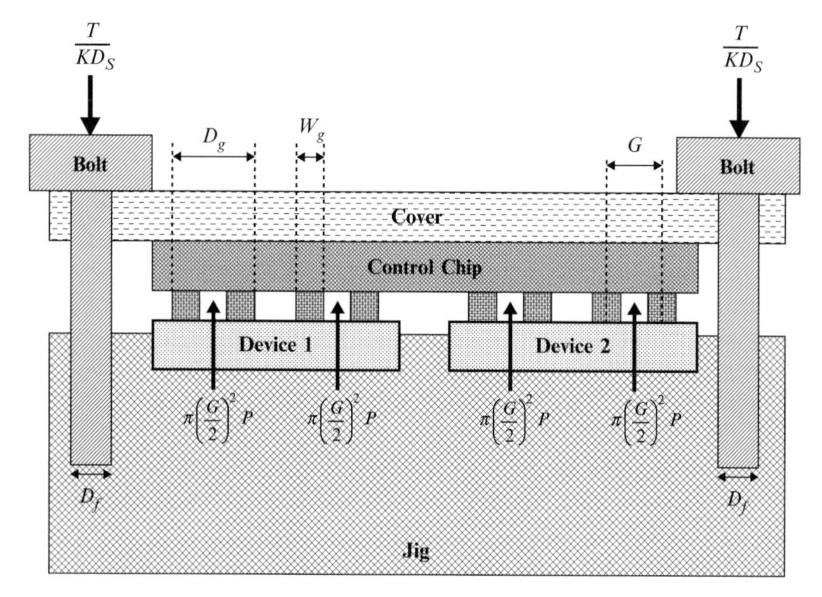

Fig. 4.27: A modular microfluidic system sealed using compression interconnects made from photodefinable silicone MEMS O-rings (Miserendino and Tai 2007).

developed, and it allows devices initially designed for one system to be used in a different system. These advantages can help reduce overall system design and testing time, improve system life span, and encourage parallel development and improvement of system components. The major disadvantages to modular microfluidic systems are the need for a method of connecting the individual components (called a microfluidic breadboard) and the precision alignment required in mating the components with the breadboard. The use of modular versus monolithic microfluidic systems is a trade-off between the additional costs associated with a modular system and the additional benefit of increased flexibility.

Figure shows a modular microfluidic system using MEMS O-rings to seal between two device chips and a control chip (the fluidic breadboard) (Miserendino and Tai 2007). For a modular system using many MEMS O-rings that are simultaneously sealed, the maximum working pressure formula given in equation (4.129) must be modified to account for the cumulative hydrostatic end force. For a system with M total vias the working pressure must satisfy:

$$p \leq \frac{4NT}{\pi MKD_f(D_g - W_g)^2}. \qquad (4.131)$$

This modular microfluidic system can operate at pressures above 250 psi. Each via contributes less than 1 nL of dead volume to the system (Miserendino and Tai 2007).

Modular microfluidic systems have been designed using both surface micromachined fluidics and PDMS fluidics. Shaikh *et al.* (2005) developed a highly complex PDMS based modular microfluidic architecture that requires through-wafer deep reactive ion etching to form the fluidic vias and multilayer PDMS processing to make the devices. This system was used to detect ultra low levels of prostate specific antigen and then was reconfigured to detect the presence of lead. Suk *et al.* (2005) also used PDMS to make a modular microfluidic system that incorporated electrical, as well as fluidic, interconnects. In Suk's system, PDMS was used to form both interconnects between the device chips and microchannels on the device chips, which were fabricated on PC board. In both cases, PDMS bonding was used to seal the multi-device system. A complete high-pressure modular microfluidic system was developed by Miserendino (2007) using surface micromachined, all-Parylene microchannels.

4.7 Select applications of pressure-driven microfluidics

Pressure-force driven microfluidics is becoming more diversified in terms of design and application, and improvements in this technology continue to be rapidly made. New application areas have been found by looking at improvements in performance of common technologies when devices could be made on smaller scales. Even in cases in which a device does not produce better results on the micro scale, they may be more attractive since microfluidic fabrication technologies result in a reduction in cost, size, weight, and power compared to their macro counterparts. Finally, some microfluidic devices are dedicated to improving our understanding of physical, chemical, and biological phenomenon at the micro and nano length scale. Although it is beyond the scope of this chapter to discuss all the possible applications of pressure driven microfluidics in full detail, some select applications will be briefly highlighted and references offered where additional information on these applications can be readily found.

TABLE 4.3: Diffusion coefficients in water (Lide 1996).

Solute	T (°C)	D ($\times 10^{-5}$ cm^2/s)	Solute	T (°C)	D ($\times 10^{-5}$ cm^2/s)
Acetone	25	1.28	HCO_3^-	25	1.185
Acetonitrile	15	1.26	Isopropyl alcohol	15	0.87
Cl^-	25	2.032	K^+	25	1.957
Ethanol	25	1.24	Na^+	25	1.334
F^-	25	1.475	NH_4^+	25	1.957
Glucose	25	0.67	NO_2^-	25	1.912
Glycine	25	1.05	NO_3^-	25	1.912
H^+	25	9.311	OH^-	25	5.273

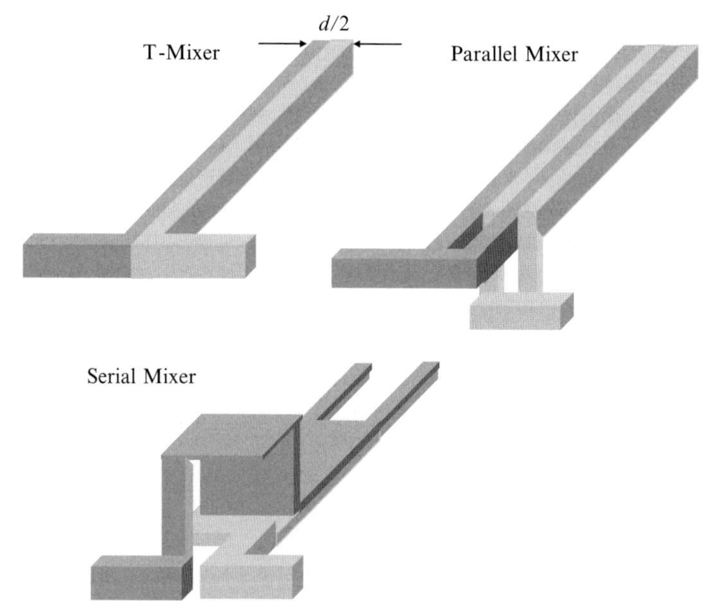

FIG. 4.28: T-mixer, parallel mixer (n=2), and serial mixers (m=2). Lamination aids mixing by decreasing diffusion distance.

4.7.1 *Micromixers*

As shown in Section 4.1, almost all pressure driven microfluidic devices operate in the low Reynold's number flow regime. One consequence of this is the almost total lack of turbulence in the fluid flow which makes efficient mixing in microfluidic systems difficult. It is therefore desirable for many applications to be able to mix two or more reagent flow streams. Since there is little to no turbulent flow in most microfluidic systems, mixing is strictly limited by the diffusion of reagents. Diffusion processes are described by Fick's First Law:

$$J = -D\Delta C \tag{4.132}$$

where J is the flux of reagent, D is the diffusion coefficient, and C is the concentration of the reagent. For particles undergoing strict Brownian motion, Einstein derived the functional form of the diffusion coefficient Einstein (1956):

$$D = \frac{RT}{N_A}\frac{1}{6\pi\mu r} \tag{4.133}$$

where N_A is Avogadro's number, R is the ideal gas law constant, T is the absolute temperature, μ is the solvent viscosity, and r is the radius of the particle. Using the diffusion coefficient, a useful dimensionless quantity known as Peclet number is defined as:

$$Pe = \frac{\bar{\upsilon}L}{D}, \tag{4.134}$$

where L is the characteristic length and v̄ is the characteristic linear flow velocity. The Peclet number represents the relative effects of convective and diffusion on mass transport. Higher Peclet numbers indicate convective dominated transport while low Peclet numbers indicate diffusion dominated transport. The ratio of the Peclet number to the Reynold's number is around 1000 for most liquid flows and around 1 for gas flows Nguyen and Wu (2005).

$$\frac{Pe}{Re} = \frac{\mu}{D}. \tag{4.135}$$

Table 4.3 is a list of diffusion coefficients of common reagents in water.

In pure diffusion mixing, the mixing time is the most important figure of merit. The mixing time, τ, is determined by the diffusion distance, d, and the coefficient of diffusion, D, according to Einstein (1956),

$$\tau = \frac{d^2}{2D}. \tag{4.136}$$

The major class of static micromixers is based on the concept of flow lamination. Lamination micromixers work by dividing the two input fluid flows into multiple streams and then interlacing the streams to form a single output flow. The simplest mixer design is shown in Fig. 4.28 and is the T-mixer. In the T-mixer, the input flows are not separated, but directly interlaced. The resulting output flow has a diffusion distance defined by the width of the microchannel. A common variant of the T-mixer is the Y-mixer which works the same way as a T-mixer except the input flow streams form an angle less than 180°. The lamination concept can be achieved in either a parallel or serial fashion. The parallel lamination micromixer can be easily designed in a planer manner, which makes its fabrication easier than the sequential lamination micromixer which requires more complicated three dimensional structures. Using n divisions of the inlet flow streams, a parallel mixer can achieve a reduced mixing time relative to the T-mixer, which can be considered a single stage parallel mixer, as

$$\frac{\tau_{parallel}}{\tau} = \frac{\left(\frac{d_{in}}{n}\right)^2}{2D} \times \frac{2D}{d_{in}^2} = \frac{1}{n^2}. \tag{4.137}$$

A serial lamination micromixer uses multiple stages of splitting and laminating to achieve the final mixing. After the first stage, the output of the previous stage is split either horizontally or vertically and then laminated in the opposite manner to that which it was split to form the input into the next stage. The diffusion length at the end of each stage is

$$\begin{aligned} d_0 &= d_{in} \\ d_1 &= \frac{d_{in}}{2} \\ d_2 &= \frac{d_{in}}{4} \\ d_3 &= \frac{d_{in}}{8} \\ &\vdots \\ d_n &= \frac{d_{n-1}}{2} = \frac{d_{in}}{2^n} \end{aligned} \tag{4.138}$$

The savings in diffusion time in a serial lamination micromixer over a T-mixer (a T-mixer can be thought of as a single stage serial micromixer) is (Branebjerg *et al.* 1996)

$$\frac{\tau_{serial}}{\tau} = \frac{\left(\frac{d_{in}}{2^n}\right)^2}{2D} \times \frac{2D}{\left(\frac{d_{in}}{2}\right)^2} = \frac{1}{4^{n-1}}. \tag{4.139}$$

A mixing method similar to parallel lamination mixing is injection mixing. An injection, or microplume, mixer splits only one input stream into many substreams (typically several hundred) and subsequently injects those substreams simultaneously at different points along the other input stream's flow (Miyake *et al.* 1996;

Voldman *et al.* 2000). This results in a large increase in contact area between the two steams, and hence a reduced diffusion distance.

Another method used to decrease mixing times in microchannels is to induce *chaotic advection.* Advection describes how a particle moves in a fluid flow (the mathematical representation of the flow field as a set of differential equations is often referred to as the advection equations). Chaotic advection is a type of advection characterized by a strong sensitivity to initial conditions (Aref 1990). Therefore, two particles initially located very close to one another will follow drastically different paths through the chaotic system. This behavior can be exploited to improve mixing efficiency since the goal of mixing is to homogenize the particle distribution. Chaotic advection can arise in microflows due to rips or grooves on the channel walls (Stroock *et al.* 2002). The chaotic advection causes the fluid flow to fold back on itself and, as with the lamination approach, this folding causes a decrease in diffusion path length. The change in mixing time due to chaotic advection is given by Stroock *et al.* (2002) as

$$\tau_{chaotic} = \frac{d^2}{D} \exp\left(-\frac{2\Delta y}{\lambda}\right) \tag{4.140}$$

where d is the characteristic distance of the channel (usually taken as the channel height or width), D is the coefficient of diffusivity, Δy is the length traversed along the channel, and λ is a characteristic length determined by the trajectories of the chaotic flow. The mixing length, Δy_m, for large Peclet number flows which are typically of liquid micromixers is (Stroock *et al.* 2002)

$$\Delta y_m \approx \lambda \ln(Pe). \tag{4.141}$$

Active micromixing techniques include the use of pressure disturbances either along the channel Glasgow and Aubry (2003) or at the flowinlet (Deshmukh *et al.* 2001), ultrasonic waves (Moroney *et al.* 1991; Rife *et al.* 2000), magneto hydrodynamic disturbances (Bau *et al.* 2001), thermal disturbances (Mao *et al.* 2002), and several others. Since active micromixers are limited in development to only a few device designs per actuation mechanism, it is difficult to focus on developing trends at this time. Two literature review articles are available for further investigation of both passive and active micromixer technology (Nguyen and Wu 2005; Kakuta *et al.* 2001). While micromixer technology is relatively mature, its fundamental importance to most applications guarantee continued interest.

4.7.2 *Separation on-a-chip*

In contrast to micromixers, separation devices seek to take a nearly homogeneous solution and recover the constituent components of that solution. There are two major separation technologies that have been miniaturized involving pressure driven liquid flows: 1) microfilters and 2) nano high-performance liquid chromatography (nHPLC). Other miniaturized separation technologies, such as capillary and gel electrophoresis systems, do not require pressure driven flows and will not be discussed. Micro gas chromatographic systems have been developed as well (Bhushan *et al.* 2007; Noh *et al.* 2002; Whiting *et al.* 2001), but this section will focus on liquid based microfluidic systems.

Microfilters are typically used to either clean a solution of particulate contaminants or to separate particles of interest from a buffer solution. In both cases, the fundamental separation methodology is size exclusion. Size exclusion is separation based on the size of a particle relative to the pore size of a filter. Depending on the filters fabrication, pore sizes can be in the micron to nanometer range for thin membrane filters and down to the angstrom range for gap filters. Thin membrane filters have pores that are defined using photolithography and hence are ultimately restricted in pore size by the resolution of the mask and photolithographic technology. Gap filters can use either a carefully timed etch or a sacrificial layer of metal or silicon dioxide to define the pore size. If the final pore size is defined using a controlled etching process, the filter is termed a subtractive gap filter, and if a sacrificial layer is used, it is called an additive gap filter. Additive gap filters are typically more reproducible and have lower pressure losses than subtractive gap filters. The three major figures of merit for microfilters are the filter pore size, the maximum sustainable pressure, and the pressure loss across the filter. The pressure loss across a membrane filter can be expressed as

$$\Delta p_{filter} = \frac{1}{2} K \rho \bar{v}^2 = f(\mathrm{Re}, \beta, t/D_h), \tag{4.142}$$

where K is the pressure coefficient, ρ is the fluid density, $\bar{v}$ is the average linear flow velocity, β is the opening factor, t is the membrane thickness, and D_h is the hydraulic diameter of the pore. The opening factor is defined by the ratio of surface area of the pores to the total surface area of the filter as

$$\beta = \frac{A_{pores}}{A_{total}}. \tag{4.143}$$

Authors have offered several empirical models for the functional relationship of K relative to the device geometry and flow characterization. Hasegawa *et al.* (1997) studied low Reynold's number flow of liquids through micron size pores and found

$$K = K_1 + \frac{K_2}{\mathrm{Re}} \tag{4.144}$$

where K_1 and K_2 are constants with K_2 between 27 and 37.7. Yang *et al.* (1998b) studied low Reynold's number flow of gases through micron size pores and found a similar result using numerical simulations and associated the relationship to the device geometry as

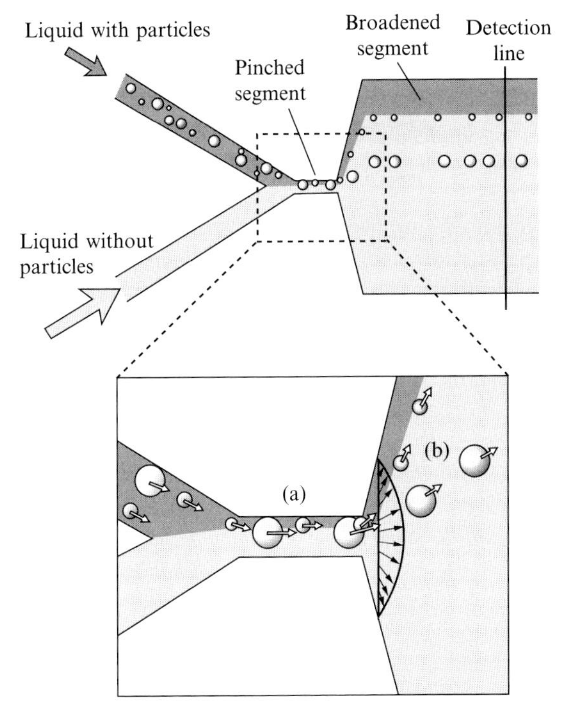

FIG. 4.29: Illustration of a pinched flow fractionation device from Yamada *et al.* (2004). (a) Shows the pinched flow region. (b) Shows the outlet region and particle separation. The grey area indicates the width of the sample flow.

$$K = \beta^{-2}\left(\frac{t}{D_h}\right)^{0.28}\left[\frac{73.5}{\mathrm{Re}} + 1.7\right]. \qquad (4.145)$$

However, empirical experiments indicate that (4.145) consistently over-estimates the pressure loss across the membrane. While exact values for the pressure coefficient may be difficult to determine analytically, a strong dependence on the opening factor and Reynold's number has been proven.

Some microfilters do not depend on pores to perform size exclusion separation. One such type is a lateral displacement filter. Lateral displacement filters make use of the microflow around obstacles in the buffer flow path to separate different sized particles. Lateral displacement filters have an array of posts separated by a center-to-center distance of λ and an edge-to-edge distance, or gap distance, of d. The rows of the array are horizontally shifted a fractional amount, ε, of λ from the previous row. Figure 4.6 illustrates a lateral displacement filter in which $\Delta\lambda$ is ε. Separation in a lateral displacement filter occurs because as particles flow through the array, they either 'bump' into the posts or 'zig-zag' through them. Whether a particle travels in 'bump' mode or 'zig-zag'

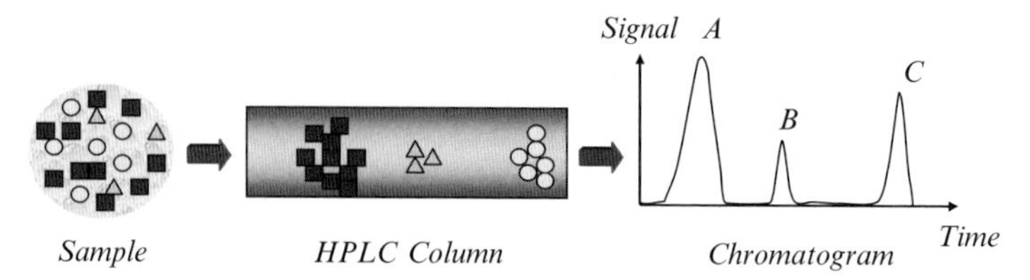

FIG. 4.30: Illustration of analytical liquid chromatography. As a mixed sample traverses the separation column different analytes within the sample will interact more or less strongly with the stationary phase and buffer solution. Differences in interaction strength produce differences in migration velocity of individual analytes through the column, resulting in analyte separation. As analytes elute off the column they are detected, and the detection signal as a function of time is known as a chromatogram.

mode is determined by the critical particle diameter, D_c, of the filter. The critical particle diameter can be estimated from the device geometry by assuming a parabolic flow profile between the posts (Inglis *et al.* 2006)

$$\frac{1}{4}\left(\frac{D_c}{d}\right)^3 - \frac{3}{4}\left(\frac{D_c}{d}\right)^2 + \varepsilon = 0. \tag{4.146}$$

Particles with diameters below D_c will travel in 'zig-zag' mode and separate from particles above D_c which will travel in 'bump' mode. Variations on this concept include multiple channel separations through variation of the gap size along the array (Huang *et al.* 2004) and using side channels instead of posts Zheng and Tai (2006). This technique has been applied to the separation of bacterial chromosomes, blood cells, and DNA.

A second type of continuous flow microfiltration technique that is useful for micron size particle separation is pinched flow fractionation (Fig. 4.29). Pinched flow fractionation is achieved by using hydrodynamic focusing in a microchannel to push the sample fluid against the edge of a channel. The channel is designed with a rapidly expanding outlet. When the pinched flow reaches the outlet, it will expand causing the particles to separate according to their size. Yamada *et al.* (Yamada *et al.* 2004) estimated the horizontal location of the particle relative to the bottom of the channel in the outlet region as:

$$y = \frac{w_o}{w_p}\left(w_p - \frac{D_p}{2}\right), \tag{4.147}$$

where w_p is the width of the pinched flow just prior to the expansion, w_o is the width of the flow in the outlet, and D_p is the particle diameter.

Unlike microfilters, nHPLC systems are designed to separate multiple analyte mixtures. The analytes can range in size from small inorganic ions to large proteins and other macromolecules such as DNA. Chromatography describes any separation technique based on the differences in relative migration velocities of analytes and typically involves a stationary and a mobile phase. Typically, during liquid chromatography, the mobile phase is a liquid and the stationary phase is a column of tightly packed silica beads with diameters of 10 m or less. As the mobile phase passes through the stationary phase, the analytes in the mobile phase interact with the stationary phase. Each analyte interacts differently which causes each to have a different migration velocity, thus allowing each analyte to either elute or exit the stationary phase at different times (Fig. 4.30). High performance chromatography refers to the high pressures used to reduce separation times. When HPLC is used for analytical studies, the eluent must be passed through a detector. Many different types of detectors are used in HPLC but some of the most common are mass spectrometry, ultraviolet absorption, laser induced florescence (LIF), electrochemical, and conductivity.

As the analytes travel through the separation column, they are sorted by their migration velocities and simultaneously brought closer together by diffusion. The longer they remain in the column, the more time they have to diffuse toward one another. The effect of diffusion can be seen most easily on a chromatogram as an increase in the peak width. Chromatography works because the band broadening effects of diffusion scale at a slower rate relative to the band separating effects of differential migration. More specifically, the band width scales as the square root of the length, L, traveled along the column while the band separation scales linearly with the length traveled. The resolution of the separation, R_s, can then be said to scale as follows

$$R_s \propto \frac{\Delta L}{w} \approx \frac{L}{\sqrt{L}} = \sqrt{L} \tag{4.148}$$

Equation (4.148) indicates that separations are possible even with very short columns. Although this follows theoretically, the initial band width caused by finite injection volume places limits on minimum column length.

The separation efficiency of a chromatography column has been historically characterized by the number of theoretical plates,

$$N = \frac{L^2}{\sigma^2} = 16\frac{t_r^2}{w_p^2}, \tag{4.149}$$

where σ^2 is the variance of the peak, typically determined by measuring the peak width, w_p, at 13.4 per cent of the maximum peak height, and t_r is the peak retention time Neue (1997). The use of the theoretical plate count as a measure of column efficiency is not ideal. The theoretical plate count is not a property of the column itself but of the ratio of migration to diffusion of a particular analyte in the column. Hence comparisons of plate counts amongst various columns must be considered carefully. Unfortunately, this is a detail often overlooked by HPLC column manufacturers. Typical plate counts for modern macro HPLC systems are around 20,000.

Traditional, or normal-phase, chromatography was performed with a polar stationary phase and a non-polar mobile phase, but this has been replaced in 70 to 80 per cent of chromatography applications with reverse-phase chromatography, which has a non-polar stationary phase and polar mobile phase Neue (1997). Other chromatography variants include hydrodynamic chromatography, hydrophobic interaction chromatography, ion-exchange chromatography, and size-exclusion chromatography. On-chip HPLC systems have been developed for hydrodynamic chromatography (Blom *et al.* 2002; 2003), ion-exchange chromatography (He 2006; Miserendino 2007; He *et al.* 2004; Murrihy *et al.* 2001), reverse-phase chromatography (Tai *et al.* 2002; Xie *et al.* 2005), and temperature gradient chromatography (Shih *et al.* 2006a, 2006b).

A major difficulty in designing nHPLC systems is fabricating devices that can withstand the relatively high packing and operating pressures. To avoid the high packing pressures, monolithic HPLC separation columns based on either organic polymers or silica are being developed (Lee *et al.* 2004; Ro *et al.* 2006; Ericson *et al.* 2000). On-chip monolithic separation columns have reported porosity of around 80 per cent, separation times of only a few minutes, and separation efficiency exceeding typical macro HPLC columns (Ro *et al.* 2006). Other nHPLC systems cope with the high-pressures by using high strength materials. Another problem encountered in isocratic elution systems, such as ion-exchange chromatography, is achieving repeatable on-chip injection of nanoliter sample volumes. Of course, one of the major advantages of nHPLC is its ability to work with extremely small sample volumes. Other advantages of miniaturizing HPLC systems are reduced separation times and improved separation efficiency.

4.7.3 *Microbioassays*

Microbioassays are microfluidic devices concerned primarily with the analysis, recognition, or amplification of biological material. The biological material is typically macromolecules of deoxyribonucleic acid (DNA), ribonucleic acid (RNA), or proteins. Microbioassays base their function on the high affinity binding of biomolecules. This high affinity binding, called biorecognition, is either due to some form of hybridization, as in the case of DNA or, in the case of proteins, antigen/antibody binding. The role of the microfluidic device is to provide a controlled environment in which the biochemical reactions can occur. Pressure-driven flows of biochemical reagents can be used to alter the chemical environment or to move a reaction to a different physical environment. On-chip temperature controllers, micromixers, and biosensors play a key role in pressure-driven microbioassays. This section will focus solely on pressure-driven microbioassays. For additional information on microbioassays where pressure-driven microfluidics does not play a significant role, the reader is referred to any of the several review articles (Selvaganapathy *et al.* 2003; Sims and Allbritton 2007; Sia and Whitesides 2003; Polla *et al.* 2000).

One of the most important examples of a pressure-driven microbioassay is the polymerase chain reaction (PCR) device developed by Kopp *et al.* (1998). PCR is a

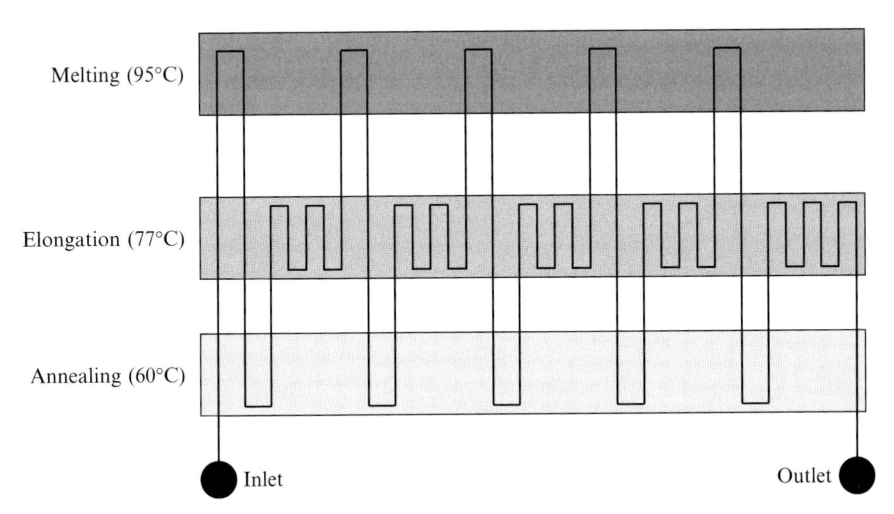

Fig. 4.31: Continuous flow PCR microbioassay developed by Kopp *et al.* (1998). Note the three different regions, each for a different stage in the cycle; 1) annealing, 2) elongation, and 3) melting, each kept at the optimal temperature for that particular phase based on the DNA fragment's length and composition. The chip shown above is shown with five PCR cycles ($n = 5$), each represented by five different loops through the chip.

classical biology technique that allows for the exponential amplification of DNA through a series of DNA copying reactions, called cycles. The amplification factor, Γ, of a PCR system of n cycles is given by

$$\Gamma = [1 + E_{PCR}(n)]^n, \tag{4.150}$$

where E_{PCR} is the efficiency factor of the system and a function of the number of cycles but is approximately 1 for 20 or fewer cycles (Nguyen and Wereley 2002). Each PCR cycle consists of three steps: (1) melting of the double stranded DNA (denaturation), (2) binding of the specific primer to its target site (annealing), (3) extending the primers using thermostable DNA polymerase (elongation). Each step in a PCR cycle must occur in a particular temperature range. Denaturation typically occurs between 94 and 96°C, annealing between 50 and 60°C, and elongation between 72 and 77°C depending on the specific length and composition of the DNA fragment being amplified (Kopp *et al.* 1998). Classically the reactions occur in a static solution while an external heater cycles the temperature. Kopp *et al.* developed a microchip that consisted of three temperature zones with microchannels of various lengths in each zone (Fig. 4.31). The reaction solution is pumped through the zone at a constant flow rate with the length of the channels determining the residence time in each zone. This continuous PCR method allows for substantial reductions in processing time and power consumption (Selvaganapathy *et al.* 2003). The major challenge with on-chip PCR system is the high surface area to volume ratio which increases the amount of DNA adsorption on the channel walls and reduces overall system efficiency. To combat the DNA adsorption problem, passivated silicon, fused silica glass, and chemically modified plastics have been used to fabricate PCR microbioassays (Selvaganapathy *et al.* 2003).

In addition to PCR, other bioassays have been successfully miniaturized resulting in decreases in process time. One such assay is the enzyme-linked immunosorbent assay (ELISA). ELISA is a type of heterogeneous immunoassay. Heterogeneous immunoassays require the antibody/antigen complex to be bound to a substrate, as opposed to homogeneous immunoassays which do not require a substrate. Because protocols require both a substrate and flushing of unbound antibodies, heterogeneous immunoassays stand to benefit greatly from microfluidics. On-chip ELISA devices have demonstrated a 12-fold decrease in processing time compared to standard microtiter plate processing (Eteshola and Balberg 2004).

Pressure-driven microfluidic devices have also seen use in the study and manipulation of cells. On-chip pressure driven flow cytometry has been developed to sort cells based on differences in fluorescence (Fu *et al.* 2002; Huh *et al.* 2002). The on-chip flow cytometers suffer from significantly lower throughput, lower efficiency in capturing viable cells, and analysis has been limited to much smaller parameter spaces when compared to standard macro scale systems. With additional work and integration, however, on-chip flow cytometry may produce systems with simpler sample preparation methods and smaller sample sizes (Sims and Allbritton 2007). Other microfluidic devices focus on precise control and sensing of the microenvironment around living cells. Application of temporal and spatially discrete chemical stimuli, not possible without microfluidic systems, have been applied to the study of cellular signal transduction mechanisms, blood cell homing, and tumor metastasis (Sims and Allbritton 2007).

References

Ahmadian, M. T., and Mehrabian, A., 2006. 'Design optimization by numerical characterization of fluid flow through the valveless diffuser micropumps.' *Journal of Physics: Conference Series*, **34**, pp. 379–384.

Ahn, C., and Allen, M., 1995. *Fluid Micropumps Based on Rotary Magnetic Actuators, in MEMS'95.* Amsterdam, The Netherlands.

Aref, H., 1990. 'Chaotic advection of fluid particles.' *Philosophical Transactions: Physical Sciences and Engineering*, **333** (1631), pp. 273–288.

Ashauer, M., Glosch, H., Hedrich, F., Hey, N., Sandmaier, H., and Lang, W., 1999. 'Thermal flow sensor for liquids and gases based on combinations of two principles.' *Sensors and Actuators a-Physical*, **73** (1–2), pp. 7–13.

ASME Boiler and Pressure Vessel Code, 2001. ASME., New York.

Bahrami, M., Yovanovich, M. M., and Culham, J. R., 2006. 'Pressure drop of fully developed, laminar flow in rough microtubes.' *Journal of Fluids Engineering-Transactions of the Asme*, **128** (3), pp. 632–637.

Bardell, R., 2000. *The Diodicity Mechanism of Tesla-Type No-Moving-Parts Valves, in Mechanical Engineering.* University of Washington, Seattle, WA.

Bau, H. H., Zhong, J. H., and Yi, M. Q., 2001. 'A minute magneto hydro dynamic (MHD) mixer.' *Sensors and Actuators B-Chemical*, **79** (2–3), pp. 207–215.

Berg, J. M., Anderson, R., Anaya, M., Lahlouh, B., Holtz, M., and Dallas, T., 2003. 'A two-stage discrete peristaltic micropump.' *Sensors and Actuators a-Physical*, **104** (1), pp. 6–10.

Berker, A. R., 1963. 'Integration des equations du movement d'un fluide visqueux incompressible.' *Handbuch der Physik*. In: Flugge S., Editor. Springer, Berlin, p. 384.

Beverley, K. J., Clint, J. H., and Fletcher, P. D. I., 1999. 'Evaporation rates of pure liquids measured using a gravimetric technique.' *Physical Chemistry Chemical Physics*, **1** (1), pp. 149–153.

Bhushan, A., Yemane, D., Trudell, D., Overton, E. B., and Goettert, J., 2007. 'Fabrication of micro-gas chromatograph columns for fast chromatography.' *Microsystem Technologies-Micro-and Nanosystems-Information Storage and Processing Systems*, **13** (3–4), pp. 361–368.

Bhushan, B., and Burton, Z., 2005. 'Adhesion and friction properties of polymers in microfluidic devices.' *Nanotechnology*, **16** (4), pp. 467–478.

Bird, R. B., Stewart, W. E., and Lightfoot, E. N., 2002. *Transport Phenomena*, 2nd international ed. John Wiley & Sons, New York, xii, 895.

Blom, M. T., Chmela, E., Gardeniers, J. G. E., Tijssen, R., Elwenspoek, M., and van den Berg, A., 2002. 'Design and fabrication of a hydrodynamic chromatography chip.' *Sensors and Actuators B-Chemical*, **82** (1), pp. 111–116.

Blom, M. T., Chmela, E., Oosterbroek, R. E., Tijssen, R., and van den Berg, A., 2003. 'On-chip hydrodynamic chromatography separation and detection of nanoparticles and biomolecules.' *Analytical Chemistry*, **75** (24), pp. 6761–6768.

Bohm, S., Olthuis, W., and Bergveld, P., 1999. 'An integrated micromachined electrochemical pump and dosing system.' *Journal of Biomedical Microdevices*, **1** (2), pp. 121–130.

Bouzid, A. H., and Derenne, M., 2002. 'Analytical modeling of the contact stress with nonlinear gaskets.' *Journal of Pressure Vessel Technology-Transactions of the ASME*, **124** (1), pp. 47–53.

Branebjerg, J., Gravesen, P., Krog, J. P., and Nielsen, C. R., 1996. *Fast Mixing by Lamination, in MEMS'96*. San Diego, CA.

Bustgens, B., Bacher, W., Menz, W., and Schomburg, W. K., 1994. *Micropump Manufactured by Thermoplastic Molding, in MEMS'94*. Oiso, Japan.

Chang, S. P., and Allen, M. G., 2004. 'Capacitive pressure sensors with stainless steel diaphragm and substrate.' *Journal of Micromechanics and Microengineering*, **14** (4), pp. 612–618.

Chen, P. J., Rodger, D. C., Meng, E., Humayun, M. S., and Tai, Y. C., 2006. *Surface-Micromachined In-Channel Parylene Dual Valves for Unpowered Microflow Regulation, in Hilton Head 2006 Workshop*. Hilton Head, SC.

Chowdhury, S., Ahmadi, M., Jullien, G. A., and Miller, W. C., 2000. *A Modular MEMS Electromagnetic Actuator for Use in A Hearing Instrument, in 43rd IEEE Midwest Symposium on Circuits and Systems*.

Christensen, A. M., Chang-Yen, D. A., and Gale, B. K., 2005. 'Characterization of interconnects used in PDMS microfluidic systems.' *Journal of Micromechanics and Microengineering*, **15** (5), pp. 928–934.

Chu, W. H., Mehregany, M., and Mullen, R., 1993. 'Analysis of tip deflection and force of a bimetallic cantilever microactuator.' *Journal of Micromechanics and Microengineering*, **3**, pp. 4–7.

Collins, R. E., and Cooke, C. E., 1959. 'Fundamental basis for the contact angle and capillary pressure.' *Transactions of the Faraday Society*, **55** (9), pp. 1602–1606.

Czaplewski, D. A., Ilic, B. R., Zalalutdinov, M., Olbricht, W. L., Zehnder, A. T., Craighead, H. G., and Michalske, T. A., 2004. 'A micromechanical flow sensor for microfluidic applications.' *Journal of Microelectromechanical Systems*, **13** (4), pp. 576–585.

Czernik, D. E., 1996. *Gaskets: Design, Selection, and Testing.* McGraw-Hill, New York, xiii, 335.

Deshmukh, A., Liepmann, D., and Pisano, A. P., 2001. *Characterization of a Micro-Mixing, Pumping, and Valving System, in Transducers'01.* Munich, Germany.

Diez, J. A., Gratton, R., Thomas, L. P., and Marino, B., 1994. 'Laplace pressure-driven drop spreading.' *Physics of Fluids*, **6** (1), pp. 24–33.

Dopper, J., Clemens, M., Ehrfeld, W., Jung, S., Kamper, K. P., and Lehr, H., 1997. 'Micro gear pumps for dosing of viscous fluids.' *Journal of Micromechanics and Microengineering*, **7** (3), pp. 230–232.

Eaton, W. P., and Smith, J. H., 1997. 'Micromachined pressure sensors: Review and recent developments.' *Smart Materials & Structures*, **6** (5), pp. 530–539.

Einstein, A., 1956. *Investigations on the Theory of the Brownian Movement.* Furth R., Editor. Dover, New York.

Ericson, C., Holm, J., Ericson, T., and Hjerten, S., 2000. 'Electroosmosis- and pressure-driven chromatography in chips using continuous beds.' *Analytical Chemistry*, **72** (1), pp. 81–87.

Eteshola, E., and Balberg, M., 2004. 'Microfluidic ELISA: On-chip fluorescence imaging.' *Biomedical Microdevices*, **6** (1), pp. 7–9.

Feldt, C., and Chew, L., 2002. 'Geometry-based macro-tool evaluation of non-moving-part valvular microchannels.' *Journal of Micromechanics and Microengineering*, **12** (5), pp. 662–669.

Fraden, J., 1996. *Handbook of Modern Sensors: Physics, Designs, and Applications*, 2nd ed. American Institute of Physics, Woodbury, NY., xiv, 556.

Fredrickson, C. K., and Fan, Z. H., 2004. 'Macro-to-micro interfaces for microfluidic devices.' *Lab on a Chip*, **4** (6), pp. 526–533.

Friedrich, C. R., Avula, R. R. K., and Gugale, S., 2005. 'A fluid microconnector seal for packaging applications.' *Journal of Micromechanics and Microengineering*, **15** (6), pp. 1115–1124.

Froster, F., Bardell, R., Afromowitz, M., Sharma, N., and Blanchard, A., 1995. *Design, Fabrication, and Testing of Fixed-Valve Micro-Pumps, in ASME Fluids Engineering Division, IMECE 1995.* San Francisco, CA.

Fu, A. Y., Chou, H. P., Spence, C., Arnold, F. H., and Quake, S. R., 2002. 'An integrated microfabricated cell sorter.' *Analytical Chemistry*, **74** (11), pp. 2451–2457.

Furdui, V. I., Kariuki, J. K., and Harrison, D. J., 2003. 'Microfabricated electrolysis pump system for isolating rare cells in blood.' *Journal of Micromechanics and Microengineering*, **13** (4), pp. S164–S170.

Gass, V., van der Schoot, B. H., and de Rooij, N. F., 1993. *Nanofluid Handling by Micro-Flow-Sensor Based on Drag Force Measurements, in Transducers '93*. Yokohama, Japan.

Gehring, G. A., Cooke, M. D., Gregory, I. S., Karl, W. J., and Watts, R., 2000. 'Cantilever unified theory and optimization for sensors and actuators.' *Smart Materials & Structures*, **9** (6), pp. 918–931.

Giesler, T., and Meyer, J. U., 1994. 'Electrostatically excited and capacitively detected flexural plate waves on thin silicon-nitride membranes with chemical sensor applications.' *Sensors and Actuators B-Chemical*, **18** (1–3), pp. 103–106.

Glasgow, I., and Aubry, N., 2003. 'Enhancement of microfluidic mixing using time pulsing.' *Lab on a Chip*, **3** (2), pp. 114–120.

Goel, R. P., 1978. 'Analysis of an interference-fit pin connection.' *IEEE Transactions on Components Hybrids and Manufacturing Technology*, **1** (3), pp. 248–251.

Gonzalez, C., Collins, S. D., and Smith, R. L., 1998. 'Fluidic interconnects for modular assembly of chemical microsystems.' *Sensors and Actuators B-Chemical*, **49** (1–2), pp. 40–45.

Gray, B. L., Jaeggi, D., Mourlas, N. J., van Drieenhuizen, B. P., Williams, K. R., and Maluf, N. I., 1999. 'Novel interconnection technologies for integrated microfluidic systems.' *Sensors & Actuators A*, **77** (1), pp. 57–65.

Greywall, M. S., 1988. 'Streamwise computation of 3-dimensional incompressible potential flows.' *Journal of Computational Physics*, **78** (1), pp. 178–193.

Greywall, M. S., 1993. 'Streamwise computation of 3-dimensional flows using 2 stream functions.' *Journal of Fluids Engineering-Transactions of the Asme*, **115** (2), pp. 233–238.

Grosjean, C., 2001. *Silicone MEMS for Fluidics, in Electrical Engineering*. California Institute of Technology, Pasadena, CA.

Grosse, S., Schroder, W., and Brucker, C., 2006. 'Nano-Newton drag sensor based on flexible micro-pillars.' *Measurement Science & Technology*, **17** (10), pp. 2689–2697.

Hasegawa, T., Suganuma, M., and Watanabe, H., 1997. 'Anomaly of excess pressure drops of the flow through very small orifices.' *Physics of Fluids*, **9** (1), pp. 1–3.

Hatch, A., Kamholz, A. E., Holman, G., Yager, P., and Bohringer, K. F., 2001. 'A ferrofluidic magnetic micropump.' *Journal of Microelectromechanical Systems*, **10** (2), pp. 215–221.

He, Q., 2006. *Integrated Nano Liquid Chromatography System On-a-Chip, in Electrical Engineering*. California Institute of Technology, Pasadena, CA.

He, Q., Pang, C., Tai, Y. C., and Lee, T. D., 2004. 'Ion liquid chromatography on-a-chip with beads-packed parylene column.' In *MEMS 2004*. Maastricht, the Netherlands.

Henning, A. K., 2006. 'Comprehensive model for thermopneumatic actuators and microvalves.' *Journal of Microelectromechanical Systems*, **15** (5), pp. 1308–1318.

Ho, C. M., and Tai, Y. C., 1998. 'Micro electro mechanical systems (MEMS) and fluid flows.' *Annual Review of Fluid Mechanics*, **30**, pp. 579–612.

Hsiao, H. S. S., Hamrock, B. J., and Tripp, J. H., 2001. 'Stream functions and streamlines for visualizing and quantifying side flows in EHL of elliptical contacts.' *Journal of Tribology-Transactions of the Asme*, **123** (3), pp. 603–607.

Hua, Z., Srivannavit, O., Xia, Y., and Gulari, G., 2004. *A Compact Chemical-Resistant Microvalve Array Using Parylene Membrane and Pneumatic Actuation. in International Conference on MEMS., NANO and Smart Systems (ICMENS).* Banff, Canada.

Huang, C., Lin, Y. Y., and Tang, T. A., 2004. 'Study on the tip-deflection of a piezoelectric bimorph cantilever in the static state.' *Journal of Micromechanics and Microengineering*, **14** (4), pp. 530–534.

Huang, L. R., Cox, E. C., Austin, R. H., and Sturm, J. C., 2004. 'Continuous particle separation through deterministic lateral displacement.' *Science*, **304** (5673), pp. 987–990.

Huff, M. A., and Schmidt, M. A., 1992. *Fabrication, Packaging, and Testing of a Wafer-Bonded Microvalve, in Solid-State Sensor and Actuator Workshop, 1992. 5th Technical Digest, IEEE.*

Huh, D., Tung, Y. C., Wei, H. H., Grotberg, J. B., Skerlos, S. J., Kurabayashi, K., and Takayama, S., 2002. 'Use of air-liquid two-phase flow in hydrophobic microfluidic channels for disposable flow cytometers.' *Biomedical Microdevices*, **4** (2), pp. 141–149.

Inglis, D. W., Davis, J. A., Austin, R. H., and Sturm, J. C., 2006. 'Critical particle size for fractionation by deterministic lateral displacement.' *Lab on a Chip*, **6** (5), pp. 655–658.

Kakuta, M., Bessoth, F. G., and Manz, A., 2001. 'Microfabricated devices for fluid mixing and their application for chemical synthesis.' *Chemical Record*, **1** (5), pp. 395–405.

Kaninski, M. P. M., Stojic, D. L., Saponjic, D. P., Potkonjak, N. I., and Miljanic, S. S., 2006. Comparison of different electrode materials – Energy requirements in the electrolytic hydrogen evolution process. *Journal of Power Sources*, **157** (2), pp. 758–764.

Knight, D., and Mallinson, G., 1996. 'Visualizing unstructured flow data using dual stream functions.' *IEEE Transactions on Visualization and Computer Graphics*, **2** (4), pp. 355–363.

Koch, M., Evans, A., and Brunnschweiler, A., 2000. *Microfluidic Technology and Applications. Microtechnologies and Microsystems Series.* Research Studies Press, Philadelphia, PA., xxi, 321.

Koffman, L. D., Plesset, M. S., and Lees, L., 1984. 'Theory of evaporation and condensation.' *Physics of Fluids*, **27** (4), pp. 876–880.

Koo, J. M., and Kleinstreuer, C., 2003. 'Liquid flow in microchannels: Experimental observations and computational analyses of microfluidics effects.' *Journal of Micromechanics and Microengineering*, **13** (5), pp. 568–579.

Kopp, M. U., de Mello, A. J., and Manz, A., 1998. 'Chemical amplification: Continuous-flow PCR on a chip.' *Science*, **280** (5366), pp. 1046–1048.

Kovacs, G. T. A., 1998. *Micromachined Transducers Sourcebook*. WCB., Boston, MA., xx, 911, 3.

Laser, D. J., and Santiago, J. G., 2004. 'A review of micropumps.' *Journal of Micromechanics and Microengineering*, **14** (6), pp. R35–R64.

Lee, D., Svec, F., and Frechet, J. M. J., 2004. 'Photopolymerized monolithic capillary columns for rapid micro high-performance liquid chromatographic separation of proteins.' *Journal of Chromatography A*, **1051** (1–2), pp. 53–60.

Leu, T. S., and Chang, P. Y., 2004. 'Pressure barrier of capillary stop valves in micro sample separators: Sensors and actuators.' *Physical A*, A115 (2–3), pp. 508–515.

Lide, D. R., ed. 1996. *CRC Handbook of Chemistry and Physics*, 77 ed. CRC Press, Inc., Boca Raton, FL.

Man, P. F., Jones, D. K., and Mastrangelo, C. H., 1997. *Microfluidic Plastic Interconnects For Multi-bioanalysis Chip Modules, in SPIE – International Society for Optical Engineering. Proceedings of SPIE – Micromachined Devices and Components III.* Austin, TX.

Man, P. F., Mastrangelo, C. H., Burns, M. A., and Burke, D. T., 1998. 'Microfabricated capillarity-driven stop valve and sample injector.' In *MEMS '98*. Heidelberg, Germany.

Mao, H. B., Yang, T. L., and Cremer, P. S., 2002. 'A microfluidic device with a linear temperature gradient for parallel and combinatorial measurements.' *Journal of American Chemical Society*, **124** (16), pp. 4432–4435.

Mao, L. D., and Koser, H., 2006. 'Towards ferrofluidics for mu-TAS and lab on-a-chip applications.' *Nanotechnology*, **17** (4), pp. S34–S47.

Martin, S. J., Butler, M. A., Spates, J. J., Schubert, W. K., and Mitchell, M. A., 1998. 'Magnetically-excited flexural plate wave resonator.' *IEEE Transactions on Ultrasonics Ferroelectrics and Frequency Control*, **45** (5), pp. 1381–1387.

Meng, A. H., Nguyen, N. T., and White, R. M., 2000. 'Focused flow micropump using ultrasonic flexural plate waves.' *Biomedical Microdevices*, **2** (3), pp. 169–174.

Meng, E., 2003. *MEMS Technology and Devices for a Micro Fluid Dosing System, in Electrical Engineering*. California Institute of Technology, Pasadena, CA.

Meng, E., and Tai, Y. C., 2003. *A Parylene MEMS Flow Sensing Array, in Transducers '03*. Boston, MA.

Meng, E., Wang, X. Q., Mak, H., and Tai, Y. C., 2000. *A Check-Valved Silicone Diaphragm Pump, in MEMS 2000*. Miyazaki, Japan.

Meng, E., Wu, S. Y., and Tai, Y. C., 2001. 'Silicon couplers for microfluidic applications.' *Fresenius Journal of Analytical Chemistry*, **371** (2), pp. 270–275.

Mercanzini, A., Bachmann, M., Jordan, A., Amstutz, Y., De Rooij, N., and Stergiopulos, N., 2005. *A Low Power, Polyimide Valved Micropump for Precision Drug Delivery, in Microtechnology in Medicine and Biology (EMBS) 2005.* Oahu, HI.

Miserendino, S., 2007. *A Modular Microfluidic Approach to Nano High Performance Liquid Chromatography with Electrochemical Detection, in Electrical Engineering.* California Institute of Technology, Pasadena, CA.

Miserendino, S., and Tai, Y. C., 2007. 'Photodefinable silicone MEMS gaskets and O-rings for microfluidics packaging.' In *MEMS 2007.* Kobe, Japan.

Miyake, R., Lammerink, T. S. J., Elwenspoek, M., and Fluitman, J. H. J., 1993. *Micro Mixer with Fast Diffusion, in MEMS'93.* Fort Lauderdale, FL.

Moroney, R. M., White, R. M., and Howe, R. T., 1991. *Ultrasonically Induced Microtransport, in MEMS'91.* Nara, Japan.

Murrihy, J. P., Breadmore, M. C., Tan, A. M., McEnery, M., Alderman, J., O'Mathuna, C., O'Neill, A. P., O'Brien, P., Advoldvic, N., Haddad, P. R., and Glennon, J. D., 2001. 'Ion chromatography on-chip.' *Journal of Chromatography A*, **924** (1–2), pp. 233–238.

Neagu, C. R., Gardeniers, J. G. E., Elwenspoek, M., and Kelly, J. J., 1997. 'An electrochemical active valve.' *Electrochimica Acta*, **42** (20–22), pp. 3367–3373.

Neagu, C., Jansen, H., Gardeniers, H., and Elwenspoek, M., 2000. 'The electrolysis of water: An actuation principle for MEMS with a big opportunity.' *Mechatronics*, **10** (4–5), pp. 571–581.

Neue, U. D., 1997. *HPLC Columns: Theory, Technology, and Practice.* Wiley-VCH., New York, xvi, 393.

Nguyen, N T and Wereley, S. T., 2002. *Fundamentals and Applications of Microfluidics.* Microelectromechanical systems series. Artech House, Boston, MA., xiii, 471.

Nguyen, N. T., and Wu, Z. G., 2005. 'Micromixers – a review.' *Journal of Micromechanics and Microengineering*, **15** (2), pp. R1–R16.

Nguyen, N. T., Bochnia, D., Kiehnscherf, R., and Dotzel, W., 1996. 'Investigation of forced convection in microfluid systems.' *Sensors and Actuators A-Physical*, **55** (1), pp. 49–55.

Nguyen, N. T., Doering, R. W., Lal, A., and White, R. M., 1998. *Computational Fluid Dynamics Modeling of Flexural Plate Wave Pumps, in IEEE Ultrasonics Symposium.* Sendai, Japan.

Nguyen, N. T., Huang, X. Y., and Chuan, T. K., 2002. 'MEMS-micropumps: A review.' *Journal of Fluids Engineering-Transactions of the Asme*, **124** (2), pp. 384–392.

Noh, H. S., Hesketh, P. J., and Frye-Mason, G. C., 2002. 'Parylene gas chromatographic column for rapid thermal cycling.' *Journal of Microelectromechanical Systems*, **11** (6), pp. 718–725.

Oh, K. W., and Ahn, C. H., 2006. 'A review of microvalves.' *Journal of Micromechanics and Microengineering*, **16** (5), pp. R13–R39.

Oh, K. W., Han, A., Bhansali, S., and Ahn, C. H., 2002. 'A low-temperature bonding technique using spin-on fluorocarbon polymers to assemble microsystems.' *Journal of Micromechanics and Microengineering*, **12** (2), pp. 187–191.

Oosterbroek, R. E., Lammerink, T. S. J., Berenschot, J. W., Krijnen, G. J. M., Elwenspoek, M. C., and van den Berg, A., 1999. 'A micromachined pressure/flow-sensor.' *Sensors and Actuators A-Physical*, **77** (3), pp. 167–177.

Pan, T. R., McDonald, S. J., Kai, E. M., and Ziaie, B., 2005. 'A magnetically driven PDMS micropump with ball check-valves.' *Journal of Micromechanics and Microengineering*, **15** (5), pp. 1021–1026.

Pang, C., Tai, Y. C., Burdick, J. W., and Andersen, R. A., 2006. 'Electrolysis-based diaphragm actuators.' *Nanotechnology*, **17** (4), pp. S64–S68.

Papautsky, I., Brazzle, J., Ameel, T., and Frazier, A. B., 1999. 'Laminar fluid behavior in microchannels using micropolar fluid theory.' *Sensors and Actuators A-Physical*, **73** (1–2), pp. 101–108.

Pedersen, C., Jespersen, S. T., Krog, J. P., Christensen, C., and Thomsen, E. V., 2005. 'Combined differential and static pressure sensor based on a double-bridged structure.' *IEEE Sensors Journal*, **5** (3), pp. 446–454.

Polla, D. L., Erdman, A. G., Robbins, W. P., Markus, D. T., Diaz-Diaz, J., Rizq, R., Nam, Y., Brickner, H. T., Wang, A., and Krulevitch, P., 2000. 'Microdevices in medicine.' *Annual Review of Biomedical Engineering*, **2**, pp. 551–576.

Puntambekar, A., and Ahn, C. H., 2002. 'Self-aligning microfluidic interconnects for glass- and plastic-based microfluidic systems.' *Journal of Micromechanics and Microengineering*, **12** (1), pp. 35–40.

Rasmussen, A., Mavriplis, C., Zaghloul, M. E., Mikulchenko, O., and Mayaram, K., 2001. 'Simulation and optimization of a microfluidic flow sensor.' *Sensors and Actuators a-Physical*, **88** (2), pp. 121–132.

Rife, J. C., Bell, M. I., Horwitz, J. S., Kabler, M. N., Auyeung, R. C. Y., and Kim, W. J., 2000. 'Miniature valveless ultrasonic pumps and mixers.' *Sensors and Actuators a-Physical*, **86** (1–2), pp. 135–140.

Ro, K. W., Nayalk, R., and Knapp, D. R., 2006. 'Monolithic media in microfluidic devices for proteomics.' *Electrophoresis*, **27** (18), pp. 3547–3558.

Rossheim, D. B., and Markl, R. C., 1943. 'Gasket-loading constants.' *Mechanical Engineering*, **65** (9), pp. 647–648.

Selvaganapathy, P. R., Carlen, E. T., and Mastrangelo, C. H., 2003. 'Recent progress in microfluidic devices for nucleic acid and antibody assays.' *Proceedings of the IEEE*, **91** (6), pp. 954–975.

Shah, R. K., and London, A. L., 1978. 'Laminar flow forced convection in ducts: A source book for compact heat exchanger analytical data.' *Advances in Heat Transfer: Supplement.* Academic Press, New York, xiv, 477.

Shaikh, K. A., Ryu, K. S., Goluch, E. D., Nam, J. M., Liu, J. W., Thaxton, S., Chiesl, T. N., Barron, A. E., Lu, Y., Mirkin, C. A., and Liu, C., 2005. 'A modular microfluidic architecture for integrated biochemical analysis.' *Proceedings of the National Academy of Science U S A*, **102** (28), pp. 9745–9750.

Shih, C. Y., Chen, Y., Li, W., Xie, J., He, Q., and Tai, Y. C., 2006a. 'An integrated system for on-chip temperature gradient interaction chromatography.' *Sensors and Actuators A-Physical*, **127** (2), pp. 207–215.

Shih, C. Y., Chen, Y., Xie, J., He, Q., and Tai, Y. C., 2006b. 'On-chip temperature gradient interaction chromatography.' *Journal of Chromatography A*, **1111** (2), pp. 272–278.

Shoji, S., and Esashi, M., 1994. 'Microflow devices and systems.' *Journal of Micromechanics and Microengineering*, **4** (4), pp. 157–171.

Sia, S. K., and Whitesides, G. M., 2003. 'Microfluidic devices fabricated in poly (dimethylsiloxane) for biological studies.' *Electrophoresis*, **24** (21), pp. 3563–3576.

Sims, C. E., and Allbritton, N. L., 2007. 'Analysis of single mammalian cells on-chip.' *Lab on a Chip*, **7** (4), pp. 423–40.

Spiering, V. L., van der Moolen, J. N., Burger, G. J., and van den Berg, A., 1997. *Novel Microstructures and Technologies Applied in Chemical Analysis Techniques, in Transducers '97.* Chicago, IL.

Stemme, E., and Stemme, G., 1993. 'A valveless diffuser/nozzle-based fluid pump.' *Sensors and Actuators A-Physical*, **39** (2), pp. 159–167.

Stroock, A. D., Dertinger, S. K. W., Ajdari, A., Mezic, I., Stone, H. A., and Whitesides, G. M., 2002. 'Chaotic mixer for microchannels.' *Science*, **295** (5555), pp. 647–651.

Suk, S. D., Chang, S., and Cho, Y. H., 2005. *Design, Fabrication, and Characterization of Electrical and Fluidic Interconnections for a Multi-chip Microelectrofluidic Bench, in Proceedings of MEMS 2005.* Miami, FL.

Sutanto, J., Hesketh, P. J., and Berthelot, Y. H., 2006. 'Design, microfabrication and testing of a CMOS compatible bistable electromagnetic microvalve with latching/unlatching mechanism on a single wafer.' *Journal of Micromechanics and Microengineering*, **16** (2), pp. 266–275.

Svedin, N., Kalvesten, E., Stemme, E., and Stemme, G., 1998. 'A new silicon gas-flow sensor based on lift force.' *Journal of Microelectromechanical Systems*, **7** (3), pp. 303–308.

Svedin, N., Kalvesten, E., and Stemme, G., 2003. 'A new edge-detected lift force flow sensor.' *Journal of Microelectromechanical Systems*, **12** (3), pp. 344–354.

Sze, S. M., 1994. *Semiconductor Sensors.* John Wiley & Sons, New York, xii, 550.

Tabata, O., Kawahata, K., Sugiyama, S., and Igarashi, I., 1989. 'Mechanical property measurements of thin-films using load deflection of composite rectangular membranes.' *Sensors and Actuators*, **20** (1–2), pp. 135–141.

Tai, Y. C., 2003. 'Parylene MEMS: Material, technology and applications.' In *Japanese Sensor Symposium.* Tokyo, Japan.

Tai, Y. C., Xie, J., He, Q., Liu, J., and Lee, T., 2002. 'Integrated micro/nano fluidics for mass-spectrometry protein analysis.' *International Journal of Nonlinear Sciences and Numerical Simulation*, **3** (3–4), pp. 739–741.

Takao, H., and Ishida, M., 2003. 'Microfluidic integrated circuits for signal processing using analogous relationship between pneumatic microvalve and MOSFET.' *Journal of Microelectromechanical Systems*, **12** (4), pp. 497–505.

Tesla, N., 1920. *Valvular Conduit.* United States.

Thompson, B. R., Maynes, D., and Webb, B. W., 2005. 'Characterization of the hydrodynamically developing flow in a microtube using MTV.' *Journal of Fluids Engineering-Transactions of the Asme*, **127** (5), pp. 1003–1012.

Thompson, P. A., and Troian, S. M., 1997. 'A general boundary condition for liquid flow at solid surfaces.' *Nature*, **389** (6649), pp. 360–362.

Tsai, J. H., and Lin, L., 2001. 'Micro-to-macro fluidic interconnects with an integrated polymer sealant.' *Journal of Micromechanics and Microengineering*, **11** (5), pp. 577–581.

Unger, M. A., Chou, H. P., Thorsen, T., Scherer, A., and Quake, S. R., 2000. 'Monolithic microfabricated valves and pumps by multilayer soft lithography.' *Science*, **288** (5463), pp. 113–116.

van der Wijngaart, W., Andersson, H., Enoksson, P., Noren, K., and Stemme, G., 2000. *The First Self-Priming and Bi-Directional Valve-less Diffuser Micropump for Both Liquid and Gas, in MEMS 2000.* Miyazaki, Japan.

van Kuijk, J., Lammerink, T. S. J., de Bree, H. E., Elwenspoek, M., and Fluitman, J. H J., 1995. 'Multiparameter detection in fluid-flows.' *Sensors and Actuators A-Physical*, **47** (1–3), pp. 369–372.

Voldman, J., Gray, M. L., and Schmidt, M. A., 2000. 'An integrated liquid mixer/valve.' *Journal of Microelectromechanical Systems*, **9** (3), pp. 295–302.

Wang, X. Q., and Tai, Y. C., 2000. 'A normally-closed in-channel micro check valve.' In *MEMS 2000.* Miyazakli, Japan.

Wang, X. Q., Lin, Q., and Tai, Y. C., 1999. *A Parylene Micro Check Valve, in MEMS 1999.* Orlando, FL.

Wapner, P. G. and Hoffman, W. P., 2000. 'Utilization of surface tension and wettability in the design and operation of microsensors.' *Sensors and Actuators B-Chemical*, **71** (1–2), pp. 60–67.

Weibel, D. B., Kruithof, M., Potenta, S., Sia, S. K., Lee, A., and Whitesides, G. M., 2005. 'Torque-actuated valves for microfluidics.' *Analytical Chemistry*, **77** (15), pp. 4726–4733.

Whitaker, S., 1981. *Introduction to Fluid Mechanics*, reprint ed. Krieger, Malabar, FL., xiii, 457.

White, F. M., 1986. *Fluid Mechanics*, 2nd ed. McGraw-Hill, New York, xii, 732.

White, F. M., 1991. *Viscous Fluid Flow*, 2nd ed. McGraw-Hill series in mechanical engineering. McGraw-Hill, New York, xxi, 614.

White, R. M., and Wenzel, S. W., 1988. 'Fluid loading of a lamb-wave sensor.' *Applied Physics Letters*, **52** (20), pp. 1653–1655.

Whiting, J. J., Lu, C. J., Zellers, E. T., and R. D., 2001. sacks. 'A portable, high-speed, vacuum-outlet GC vapor analyzer employing air as carrier gas and surface acoustic wave detection.' *Analytical Chemistry,* **73** (19), pp. 4668–4675.

Woias, P., 2005. 'Micropumps – past, progress and future prospects.' *Sensors and Actuators B-Chemical,* **105** (1), pp. 28–38.

Wu, C. H., Zorman, C. A., and Mehregany, M., 2006. 'Fabrication and testing of bulk micromachined silicon carbide piezoresistive pressure sensors for high temperature applications.' *IEEE Sensors Journal,* **6** (2), pp. 316–324.

Wu, J. A., and Sansen, W., 2002. 'Electrochemical time of flight flow sensor.' *Sensors and Actuators A-Physical,* **97–98**, pp. 68–74.

Xie, J., Yang, X., Wang, X. Q., and Tai, Y. C., 2001. 'Surface micromachined leakage proof parylene check valve.' In *MEMS 2001.* Interlaken, Switzerland.

Xie, J., Shih, J., Lin, Q. A., Yang, B. Z., and Tai, Y. C., 2004a. 'Surface micromachined electrostatically actuated micro peristaltic pump.' *Lab on a Chip,* **4** (5), pp. 495–501.

Xie, J., Miao, Y. N., Shih, J., He, Q., Liu, J., Tai, Y. C., and Lee, T. D., 2004b. An electrochemical pumping system for on-chip gradient generation. *Analytical Chemistry,* **76** (13), pp. 3756–3763.

Xie, J., Miao, Y. N., Shih, J., Tai, Y. C., and Lee, T. D., 2005. 'Microfluidic platform for liquid chromatography-tandem mass spectrometry analyses of complex peptide mixtures.' *Analytical Chemistry,* **77** (21), pp. 6947–6953.

Yamada, M., Nakashima, M., and Seki, M., 2004. 'Pinched flow fractionation: Continuous size separation of particles utilizing a laminar flow profile in a pinched microchannel.' *Analytical Chemistry,* **76** (18), pp. 5465–5471.

Yamahata, C., Chastellain, M., Parashar, V. K., Petri, A., Hofmann, H., and Gijs, M A. M., 2005. 'Plastic micropurnp with ferrofluidic actuation.' *Journal of Microelectromechanical Systems,* **14** (1), pp. 96–102.

Yang, L. J., Yao, T. J., and Tai, Y. C., 2004. 'The marching velocity of the capillary meniscus in a microchannel.' *Journal of Micromechanics and Microengineering,* **14** (2), pp. 220–225.

Yang, X., Grosjean, C., and Tai, Y. C., 1998a. *A Low Power MEMS Silicone/ Parylene Valve, in The Sensors and Actuators Workshop.* Hilton Head, SC., pp. 316–319.

Yang, X., Yang, J. M., Wang, X. Q., Meng, E., Tai, Y. C., and Ho, C. M., 1998b. *Micromachined Membrane Particle Filters, in MEMS' 98.* Heidelberg, Germany.

Yang, X., Grosjean, C., and Tai, Y. C., 1999. 'Design, fabrication, and testing of micromachined silicone rubber membrane valves.' *Journal of Microelectromechanical Systems,* **8** (4), pp. 393–402.

Yao, T. J., Lee, S. W., Fang, W., and Tai, Y. C., 2000. 'A micromachined rubber o-ring microfluidic coupler.' In *MEMS 2000.* Miyazaki, Japan, pp. 624–627.

Young, T., 1805. 'An essay on the cohesion of fluids.' *Philosophical Transactions of the Royal Society of London,* **95**, pp. 65–87.

Zheng, S., and Tai, Y. C., 2006. *Streamline Based Design of a MEMS Device for Continuous Blood Cell Separation, in Hilton Head 2006*. Hilton Head Island, SC.

Zheng, S., Tai, Y. C., and Kasdan, H., 2005. *A Micro Device for Separation of Erythrocytes and Leukocytes in Human Blood, in 27th Annual International Conference of the IEEE Engineering in Medicine and Biology Society (EMBS)*. Shanghai, China.

Zimmermann, M., Bentley, S., Schmid, H., Hunziker, P., and Delamarche, E., 2005. 'Continuous flow in open microfluidics using controlled evaporation.' *Lab on a Chip*, **5** (12), pp. 1355–1359.

5

ELECTROKINETICS OF PARTICLES AND FLUIDS

Hywel Morgan, Nicolas G. Green and Tao Sun

5.1 Introduction

Microfluidic systems integrate multiple functions on a single platform. Automated or remote manipulation and analysis of particles and fluids is a key element in microtechnologies and Lab on a Chip. In most Lab on a Chip systems, samples are suspended in an aqueous electrolyte: a conducting fluid medium. The reduction in size of these systems leads to a number of changes in system behavior, for example the behavior of fluid is dominated by viscosity when the scale is small. Also it is reasonable easy to generate very large electric fields and field gradients in microsystems using quite low voltages. External pumps are often used to move fluid, but there is a growing interest in using electrokinetics to move liquids and particles within microchips using integrated electrodes. Electrokinetics is particularly attractive on the scale of microfluidic systems. The forces are easy to control by designing optimum electrode structures and the choice of field and frequency. In this chapter we review the theory of electrokinetics in DC and AC fields, and its application to the manipulation of particles, including dielectrophoresis. We then discuss the theories and application of electrohydrodynamic effects in microsystems, focusing on electroosmosis (in DC and AC fields) and electrothermal effects in AC field.

5.2 DC electrokinetics—fundamentals

5.2.1 *Quasi-electrostatic systems and conduction*

The quasi-electrostatic approximation is used, meaning that the charges are stationary or have small velocities and accelerations. The movement of the charge therefore does not change the electric field and magnetic fields can be assumed to be negligible.

5.2.1.1 *Basic equations* In electrokinetic systems, we generally use Gauss' Law and Poisson's equation, which are solved to obtain analytical or numerical representations of the electric potential and then field in a system. Gauss' Law relates the vector electric field **E** to the volume electric charge density ρ.

$$\nabla \cdot \boldsymbol{E} = \frac{\rho}{\varepsilon_0} \tag{5.1}$$

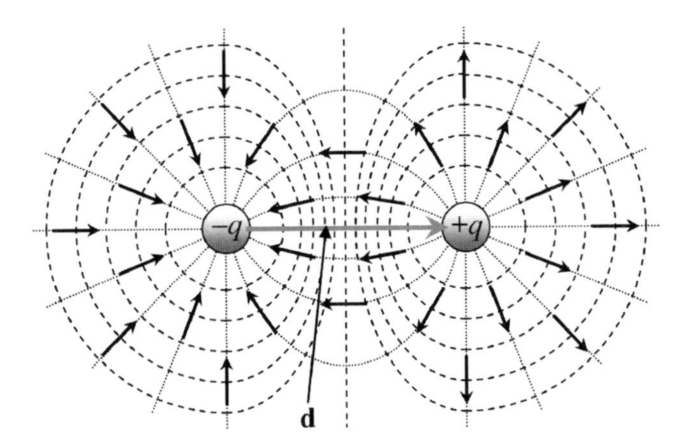

Fig. 5.1: Equipotentials (dashed), electric field lines (dotted), and electric field vectors (black arrows) around a dipole consisting of a negative and a positive charge. The grey arrow between the two charges indicates the vector **d** and the direction of the dipole moment.

and substituting for the potential gives

$$\nabla^2\phi = -\frac{\rho}{\varepsilon_o} \tag{5.2}$$

This is one form of Poisson's equation. For regions where the charge density is zero, this becomes Laplace's equation

$$\nabla^2\phi = 0 \tag{5.3}$$

The charge conservation equation relates the rate of change with respect to time of the volume charge density ρ to the current density $\mathbf{J}$

$$\nabla \cdot \mathbf{J} = -\frac{\partial\rho}{\partial t} \tag{5.4}$$

where in the steady state, $\frac{\partial\rho}{\partial t} = 0$ and therefore $\nabla \cdot \mathbf{J} = 0$.

An electrical dipole is formed from a distribution of charges and is fundamental to many aspects of electrokinetics. A dipole consists of two charges of the same magnitude q and opposite sign, separated by a distance d. The dipole moment is the vector $\mathbf{p} = q\mathbf{d}$ directed from the negative to the positive charge, as shown in Fig. 5.1. The vector $\mathbf{d}$ is shown in the Fig. 5.1. The dipole moment has units of Coulomb-meter or Debye (where 1 Debye $= 3.33 \times 10^{-30}$ Cm).

5.2.1.2 *Polarization, complex permittivity, interfacial charging* A dielectric material is a material that contains charges which polarize under the influence of an applied electric field. These charges are bound in some manner within the material and can only move short distances when the field is applied, the negative and positive charges moving in opposite directions to form induced dipoles. Dielectric materials can be classified into polar and non-polar. There are three basic molecular polarization mechanisms that can occur when an electric field is applied to a dielectric: electronic, atomic, and orientational (or dipolar) (Pethig 1979). In addition, there is a long-range polarization which is due to accumulation of charge carriers at interfaces in the dielectric, called interfacial polarization. Assuming that the polarizability mechanisms act independently, then the total polarizability of a dielectric is the sum of the polarizabilities.

Physically, the permittivity, ε, is a measure of the energy storage or charge accumulation in a system, through polarization by an applied electric field. Conductivity, σ, is a measure of the ease with which charge can move through a material under the influence of the applied field. A non-ideal dielectric has both permittivity and conductivity. In AC electric fields, the concept of complex permittivity $\tilde{\varepsilon}$, describes the frequency dependent response of the dielectric to the field:

$$\tilde{\varepsilon} = \varepsilon_0 \varepsilon_r - i\frac{\sigma}{\omega} \tag{5.5}$$

where ε_0 is the permittivity of free space (8.854×10^{-12} F/m) and ε_r is the relative permittivity of the material; i is the imaginary unit, $i^2 = -1$.

Driven by the applied electric field, the movement of charges inside the dielectric creates dipoles, leading to polarization. At low frequencies, the dipoles have sufficient time to align with the electric field. But as the frequency increases, the dipoles are no longer able to orient fast enough to align with the field and the polarizability from this polarization mechanism diminishes, leading to a decrease in the total polarizability. This decrease leads to the reduction of the energy storage (permittivity), referred to as a dielectric relaxation or dispersion. Assuming electronic and atomic polarization to be constant and including the contribution from the DC conductivity of the material, the complex permittivity of a dielectric material can be written as:

$$\tilde{\varepsilon} = \varepsilon_0 \left(\varepsilon_\infty + \frac{\varepsilon_s - \varepsilon_\infty}{1 + i\omega\tau} \right) - i\frac{\sigma}{\omega} \tag{5.6}$$

where ε_∞ is the relative permittivity at infinite frequency; ε_s is the relative permittivity measured in a static driving field, and τ is the relaxation time constant of the orientational polarization.

Equation (5.6) characterizes an ideal Debye behavior in which there is a single relaxation time. Separating the real and imaginary parts of the complex permittivity, ε' and ε'' are:

$$\varepsilon' = \varepsilon_0 \left[\varepsilon_\infty + \frac{\varepsilon_s - \varepsilon_\infty}{1 + (\omega\tau)^2} \right] \tag{5.7 a}$$

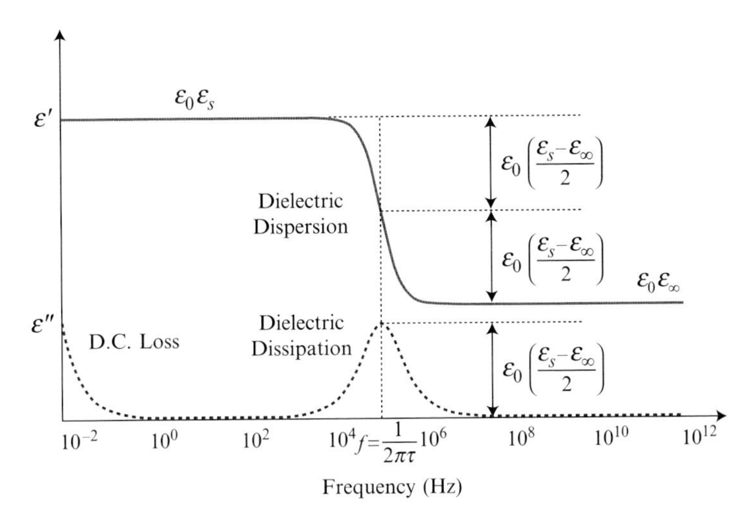

Fig. 5.2: Plot of the real and imaginary part of the complex permittivity versus frequency, showing an ideal Debye relaxation.

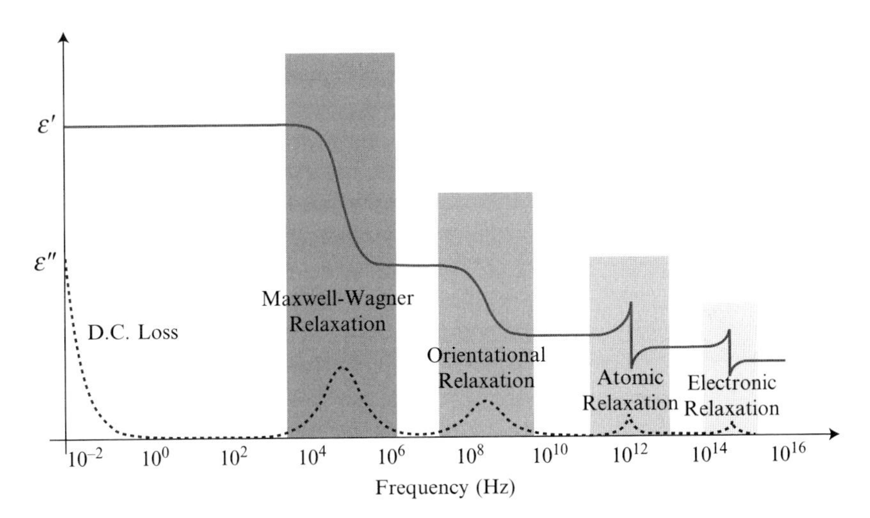

Fig. 5.3: Diagram showing typical relaxations found in the dielectric spectroscopy.

$$\varepsilon'' = \varepsilon_0 \left[\frac{(\varepsilon_s - \varepsilon_\infty)\omega\tau}{1 + (\omega\tau)^2} \right] + \frac{\sigma}{\omega} \tag{5.7 b}$$

Figure 5.2 shows the variation of ε' and ε'' with frequency, showing how the energy storage, ε' reduces with increasing frequency. The energy dissipation, ε'' is dominate by DC conductivity at low frequencies, and reaches a maximum at each characteristic relaxation frequency.

In most cases, real systems are heterogeneous rather than homogeneous. When an electric field is applied to a heterogeneous system, charge accumulates at the interfaces between any boundaries between different dielectrics. This mechanism is referred to as interfacial–or Maxwell–Wagner–polarization. It typically occurs at frequencies much lower than orientational atomic and electronic polarizations, as shown in Fig. 5.3. The magnitude of the Maxwell–Wagner polarization can often be extremely large and dominate other processes in biological systems.

5.2.2 *Ions and the double layer*

5.2.2.1 *Ions in solution* In an ionic solution, the current is carried by different types of ion, each with a different mobility. For an ideal system the conductivity of the solution is given by the sum of the contributions of each ion

$$\sigma = \sum_j \lambda_j c_j \tag{5.8}$$

At low ionic concentrations, the conductivity is directly proportional to the concentration. For higher concentrations, the ions interact and the molar conductivity is influenced by higher order effects.

When an ion is placed in water, ion–water interactions produce a change in the properties of the medium close to the ions, minimizing the energy of the system—a process known as solvation or hydration (Hasted 1973). The charge carried by the ion produces a local field which polarizes the water molecules immediately around it. This region of polarized water is the ionic atmosphere, a region which screens the ion from the other ions in the medium, ensuring electroneutrality. Locally, the electrical potential resulting from the ion falls off exponentially with distance; this length scale is given by the Debye length (the distance at which the potential falls to $1/e$ of its maximum value), written as λ_D or κ^{-1} which for a monovalent ion is

$$\kappa = \sqrt{\frac{q^2 n_o}{\varepsilon k_B T}} \equiv \sqrt{\frac{1}{D}\frac{\sigma}{\varepsilon}} \tag{5.9}$$

In this expression $\varepsilon = \varepsilon_o \varepsilon_r$, n_o is the number density of ions in the bulk (m^{-3}). k_B is the Boltzmann constant, T the temperature and D the diffusion constant.

Figure 5.4 is a diagram showing the water molecules orienting around an ion. The nearest neighbor water molecules are irrotationally bound to the ion and in this region the dielectric constant falls rapidly from the bulk value, decreasing with radial distance. The creation of the polarized ionic atmosphere produces a region around the ion that has very different properties to the bulk fluid. The water molecules are effectively structured so that as the ion moves, its water atmosphere moves with it, and the moving object is bigger than the naked ion. The hydration radius used to determine viscous drag on the ion represents this increased size.

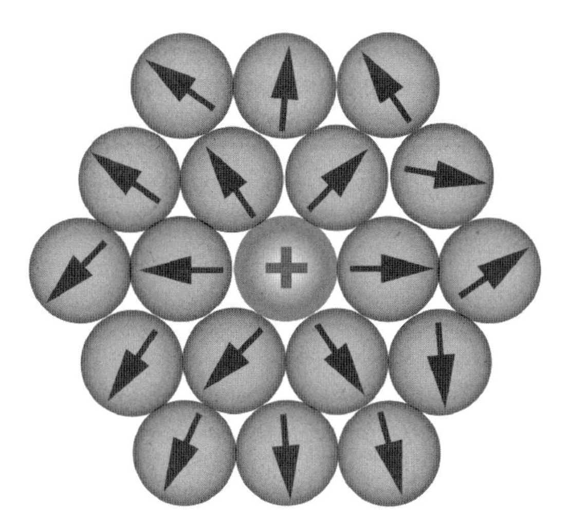

Fig. 5.4: Diagram showing the polarization of water molecules around a positive ion. The arrows indicate the direction of the dipole moments of the water molecules.

An important characteristic of an electrolyte is the charge relaxation time. It is a measure of the time required for an ion to move a distance of the order of the Debye length, κ^{-1} by diffusion (Bockris and Reddy 1973). This is given by:

$$\tau_q = \frac{1}{D\kappa^2} \tag{5.10}$$

Substituting the Debye length (eqn. 5.9) into this expression, we see that the charge relaxation time is also given by

$$\tau_q = \frac{\varepsilon}{\sigma} \tag{5.11}$$

The angular frequency associated with this time $\omega_q = 1/\tau_q$, is referred to as the charge relaxation frequency.

5.2.2.2 *Electrical double layer and electrode polarization* The electrical double layer plays a fundamental role in defining the electrokinetic behavior of colloidal particles. The perturbation of the double layer charge by the applied electric field affects the polarization of the particle. When the surface of a solid object (e.g. an electrode or a particle) is immersed into an electrolyte, the electrostatic potential created by the net charge on the surface attracts ions of opposite charges (counterions) from the medium and repels ions with like charges (co-ions). The density of the counterions is much higher than the density of the co-ions in the

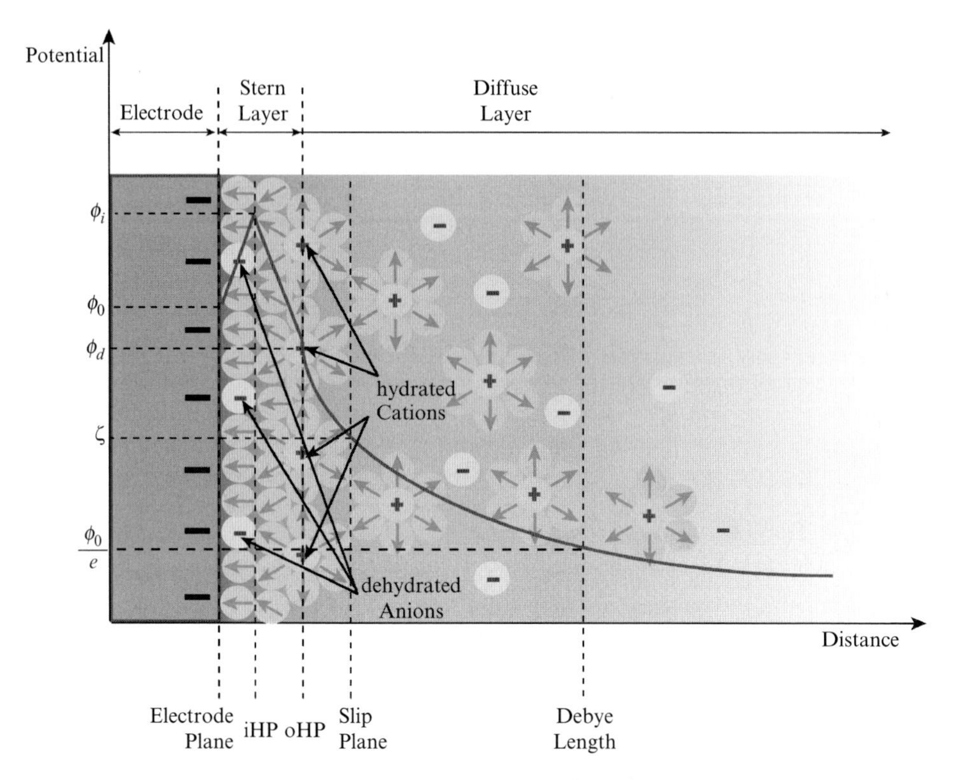

Fig. 5.5: Diagram of the electrical double layer (EDL), showing the distribution of the ions and the change in potential as a function of distance from the surface.

region close to the surface and decreases away from the surface. This accumulation of the counterions near the surface is called the Electrical Double Layer (EDL).

The general structure of the Double Layer is shown in Fig. 5.5. Close to the surface of the electrode, there is a compact layer of tightly associated counterions called the Stern layer (Stern 1924). This layer is of the order of one or two solvated ions thick and is often divided into two sub-layers (Bockris and Reddy 1973): ions adsorbed to the surface defined by the inner Helmholtz plane (iHP); and bound hydrated ions between the inner and outer Helmholtz planes. Within the Stern layer, the inner layer can contain both co- and counter ions, whilst the second outer 'diffuse' layer contains only counterions. The diffuse layer contains a higher density of counterions (compared to co-ions) and is delineated by the outer Helmholtz plane.

The variation of the electrical potential within the EDL region is also shown in Fig. 5.5. In this figure it first increases from the electrode surface potential to a value defined by ϕ_I, then drops linearly to ϕ_d, at the outer Helmholtz plane. In the diffuse layer, the potential decays exponentially with a characteristic distance given by the Debye length, at which point the potential is ϕ_d/e. For most practical applications, the potential at the beginning of the diffuse layer ϕ_d, is regarded as equivalent to the potential at the slip plane—the zeta potential ζ. Hydrodynamically, the slip plane is defined as the interface between a moving fluid and a stationary or immobile region, generally regarded as the Stern–diffuse layer interface.

The charge accumulation at the interface between an electrode and an electrolyte can be viewed as a capacitor with non-uniform charge density. The potential across the charge layer decays exponentially from a maximum at the electrode to nearly zero in the bulk. Application of a constant potential to the electrode changes the charge distribution, analogous to the charging of a capacitor—a phenomenon referred to as electrode polarization (Schwan 1968; Bard and Faulkener 1980). Because most of the applied potential is dropped across this capacitor, the potential in the bulk electrolyte may only be a fraction of that applied to the electrode (in the absence of electrochemical reactions).

Electrode polarization has practical consequences for the measurement of impedance at low frequencies. Dielectric measurements of aqueous suspensions show large reactance at low frequencies, particularly at high suspending medium conductivities, which is due to electrode polarization. AC electrokinetic measurements are similarly affected because the low frequency potentials in the suspending medium are reduced due to the EDL capacitance. This effect can be understood using the simple equivalent circuit model shown in Fig. 5.6.

At low frequencies, the reactive term from the EDL ($1/i\omega C_{DL}$) means that most of the excitation voltage is dropped across the interface and there is very little electric field in the sample. As shown by the circuit, the problem diminishes as the frequency increases, since the EDL capacitor becomes effectively short-circuited. A more general way of modeling electrode polarization is using a Constant Phase Angle (CPA) model, giving the complex impedance of the double layer as:

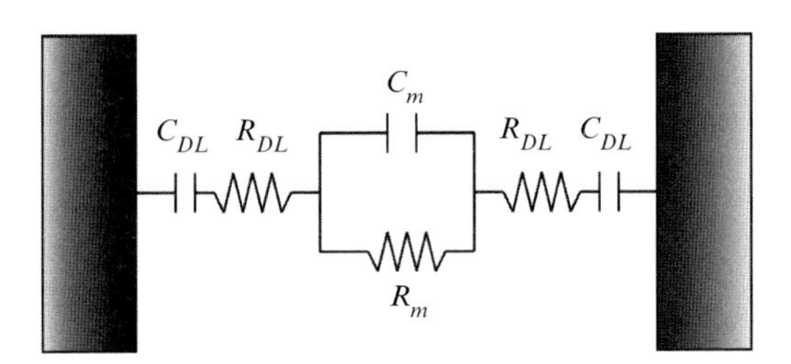

FIG. 5.6: Diagram showing an equivalent circuit model for an electrolyte between two electrodes. The impedance of the electrical double layer (EDL) is modeled as a capacitor C_{DL}, in series with a resistor R_{DL} and the impedance of the medium is modeled as a capacitor C_m in parallel with a resistor R_m.

$$\tilde{Z}_{DL} = \frac{R_{DL}}{(i\omega R_{DL} C_{DL})^{\beta}} \quad 0 \leq \beta \leq 1 \tag{5.12}$$

If $\beta = 0$, the EDL is a pure resistor, whilst if $\beta = 1$, EDL it is pure capacitance. Typical values are $\beta = 0.8-0.9$, but for most practical applications the EDL can be modeled as a simple capacitor, with a total capacitance given by the series sum of the Stern layer and diffuse layer capacitances.

Considering only the diffuse layer part of the double layer, the specific capacitance is approximated by $C_d = \varepsilon\kappa A$, and is proportional to the Debye length and the electrode area. This expression shows that in order to reduce the voltage drop across the interface, the capacitance must be maximized. Since the Debye length scales with the inverse of the square root of the ion concentration, the capacitance is maximum at low ion concentrations and the voltage in the medium will be maximum at lower conductivities (for low frequencies). Large electrodes are preferable, but this is not practical in microdevices. In these systems, the capacitance can be maximized by coating the electrode with a porous, conducting material such as platinum black (Schwan 1968); or using nanoporous materials such as iridium oxide (Blouin and Guay 1997).

5.2.3 *DC electroosmosis*

An understanding of how conducting electrolytes behave in microdevices is essential for analyzing experimental data. The basics of fluid dynamics can be found in numerous textbooks (Tritton 1988; Acheson 1990; Castellanos 1998). In this section we summarize the principal governing equations of fluid dynamics so that the behavior of fluids in microsystems can be analyzed.

5.2.3.1 *Laminar flow, Stokes force and DC electroosmosis* The Navier–Stokes equation is the equation of motion for the fluid and is derived from conservation of momentum arguments. For an incompressible Newtonian fluid, the Navier–Stokes equation is:

$$\rho_m \frac{\partial \mathbf{u}}{\partial t} + \rho_m (\mathbf{u} \cdot \nabla)\mathbf{u} = -\nabla p + \eta \nabla^2 \mathbf{u} + f \tag{5.13}$$

where ρ_m is the mass density, $\mathbf{u}$ is the velocity of the fluid, t is time, p is the pressure, η is the viscosity and $\boldsymbol{f}$ is the total applied body force (force per unit volume) acting on the fluid.

The ratio of the viscous term: $\eta \nabla^2 \mathbf{u}$ to the inertial term: $\rho_m (\mathbf{u} \cdot \nabla)\mathbf{u}$ is referred to as the Reynold's number, a parameter used to characterize microfluidic systems:

$$\mathrm{Re} = \frac{\rho_m u_o l_o}{\eta} \tag{5.14}$$

with l_o a length scale and u_o a typical velocity.

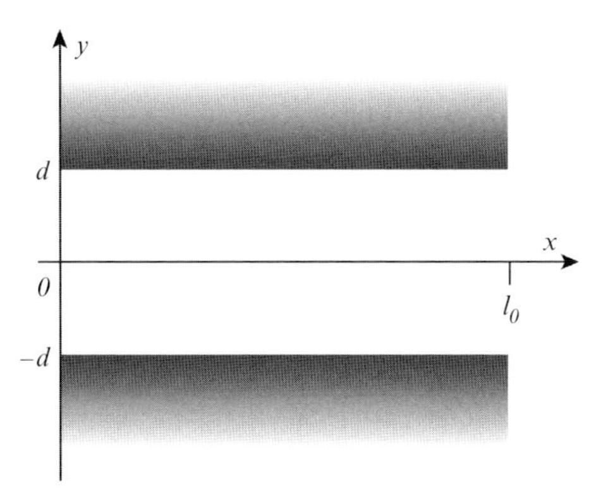

FIG. 5.7: Diagram of the simple channel, height $2d$ and length l_o, with the axes marked. The top and bottom of the channel are at $+d$ and $-d$.

For low values of Reynold's number (Re $\ll$ 1), the viscous term dominates, whilst for high values (Re $\gg$ 1) the inertial term dominates. In lab-on-a-chip systems, microfabricated channels and chambers are used to guide the fluid through the device. Typical dimensions for these channels vary between ~10 μm and 1 mm. The flow in such channels is generally laminar i.e. the fluid flow follows streamlines and is free of turbulence.

As an illustration, consider the problem of steady two-dimensional fluid flow through a chamber of height $2d$. Figure 5.7 shows the problem space, with the upper and lower parallel walls of the chamber at $y = \pm d$ respectively. The chamber has a width much greater than the height $2d$, so that the system can be considered two dimensional. The chamber has length l_o and a pressure p_o is applied at the start of the chamber ($x = 0$).

$$u_x = \frac{1}{2\eta}\frac{p_o}{l_o}(d^2 - y^2) \tag{5.15}$$

with maximum of

$$u_{\max} = \frac{p_o d^2}{2\eta l_o} \tag{5.16}$$

in the center of the chamber ($y = 0$) and an average value in the chamber of

$$u_{av} = \frac{d^2}{3\eta}\frac{p_o}{l_o} \tag{5.17}$$

The velocity is zero at the walls and maximum in the center of the channel. This type of steady state flow is referred to as Poisseuille flow.

For a particle moving in the fluid, the fluid exerts a drag force on the particle that affects the velocity of the particle. The force is known as the Stokes force

$$\mathbf{F}_{\eta} = -f\mathbf{u} \tag{5.18}$$

The constant f is the friction factor. It is a factor that depends on particle parameters such as size, shape, and surface characteristics. For a spherical particle, the Stokes force is given by:

$$\mathbf{F}_{\eta} = -6\pi\eta a\mathbf{u} \tag{5.19}$$

where a is the radius of the particle.

Electroosmosis in a DC field describes the movement of fluid across a charged surface due to an imposed electric field. Consider a surface with some charge density and an associated electrical double layer. If an electric field is applied tangential to the surface, the charges in the double layer experience a force which causes the double layer charges to move, pulling the fluid along due to the viscosity of the fluid. This generates a flow which is zero immediately at the surface but quickly rises to a maximum (and constant thereafter) at the slip plane. The electroosmotic flow velocity u_x is characterized by the Helmholtz–Smoluchowski equation:

$$u_x = -E_x \frac{\varepsilon\zeta}{\eta} \tag{5.20}$$

where E_x is the electric field along the direction in which the charge moves and ζ is the zeta potential. Equation (5.20) indicates that the DC electroosmotic velocity is linear with the applied electric field. Compared to the size of the channel, the double layer is very thin (1–100 nm), and the fluid appears to move at the surface of the channel with the maximum electroosmotic velocity.

(ii) Applications of DC electroosmosis for pumping and mixing.

Electroosmosis is a simple and highly effective method for manipulating fluids in low Reynold's number regimes, and has been widely used in lab-on-a-chip applications. Numerous review papers summarize recent progress in this field; recent examples include (Kakuta *et al.* 2001; Stone *et al.* 2004; Laser and Santiago 2004; Nguyen and Wu 2005; Hessel *et al.* 2005; Chang and Yang 2007).

Electroosmosis can be thought of as a dynamic pump that requires energy input to increase the fluid momentum. One of the first applications of electroosmosis was flow injection analysis (FIA) (Liu and Dasgupta 1992; Dasgupta and Liu 1994). Generally the magnitude of the electroosmotic flow (EOF) depends on the zeta potential of the surface, which is a function of the ion density and the pH. For a given fluid, the zeta potential reaches a maximum at the lowest ion density (Bard and Faulkener 1980), therefore maximum EOF is observed at very low ion density. The pH value of the fluid has a stronger influence on the zeta potential than that of the ion density (Hunter 1981). A relatively simple relationship for zeta potential as a function of ion density and pH value for silica surfaces has been discussed in detail by Yao *et al.* (2003). Recently, more complicated modeling of electroosmotic processes as a function of pH (Lemaire *et al.* 2007), ionic strength (Thormann *et al.* 2007), and surface heterogeneity (Adrover *et al.* 2007) have been reported.

Apart from ion density and pH, EOF also depends on the applied voltage and the geometry of the channel (eqn. 5.20). To quantitatively analyze the performance of an EOF pump two parameters are used (Zeng *et al.* 2001, 2002; Gan *et al.* 2003): (a) the maximum flow rate normalized to the applied voltage and flow cross-sectional area, $\Delta Q_{\mathrm{max,V,A}}$ (μl min^{-1} V^{-1} cm^{-2}) and (b) the maximum pressure normalized by the applied voltage, $\Delta p_{\mathrm{max,V}}$ (kPa V^{-1}). According to eqn. (5.20), an efficient strategy to increase the flow rate and pressure is to maximize the permittivity and minimize the viscosity of the fluid. Reichmuth *et al.* (2003) showed how additives such as trimethylammonium-propane sulfonate (TMAPS), which possesses a large inherent dipole moment, can be used to generate pressures up to 156 k Pa/V with an efficiency of up to 5.6 per cent. A major problem with conventional DC electroosmotic pumps is the requirement of high voltage (hundreds of volts) and the by-products caused by electrolysis of the electrolyte. Takamura *et al.* (2003) designed a low voltage electroosmotic pump using a narrow channel (<1 μm) and applying the voltage near the narrow channel using electrodes inserted into the flow path.

Efficient electroosmotic pumping requires a high surface-to-volume ratio, and this can be obtained using a single continuous piece of highly cross-linked porous polymer. Tripp *et al.* (2004) reported a monolithic pumping structure prepared from a porous polymer with grafted ionizable functionalities. The ease of preparation of the monolithic discs from liquid precursors makes the approach suitable for the fabrication of micropumps. Later Wang *et al.* (2006) fabricated a high-pressure electroosmotic pump using silica-based monoliths with high charge density and high porosity.

Rapid fluid mixing is difficult to achieve in microfluidics but is often required. Many bio-medical and bio-chemical applications such as DNA hybridization, PCR amplification, and immunoassays require fast mixing. Mixing occurs by diffusion across adjacent laminar streams, and a number of approaches have been pursued to enhance mixing by breaking up fluids into smaller stream lines, or using chaotic systems. Therefore, the development of efficient micromixers is crucial for microfluidic systems. Recent extensive reviews on micromixers have been presented by Nguyen and Wu (2005), Hessel *et al.* (2005) and Chang and Yang (2007). Generally micromixers can be classified as passive or active micromixers. Here we focus on the development of active electrokinetic micromixers that use electroosmosis. One way of reducing mixing time is to use grooved surfaces and/or heterogeneous surface charge patterning (Chang and Yang 2007; Wu and Liu 2005). Fu *et al.* (2005) presented an active double-T-form microfluidic mixer utilizing low-frequency switching of the electrosmotic flow. They have shown that a mixing efficiency of 97 per cent can be achieved with a 700 μm channel and a driving field (50 V/cm) switching at 1 Hz. Wu and Yang (2006) investigated electrokinetically driven flow mixing in a side channel type micromixer. A T-shaped micromixer featuring 45° parallelogram barriers within the channel was presented by Tai *et al.* (2006). The barriers are designed to generate a shedding flow to increase the mixing performance.

Electroosmosis is a general method for controling the fluids, but it can also be used for the manipulation of micro- and nanoparticles (Heeren *et al.* 2007) and for controling the interface position of two adjacent laminar flows (Wang *et al.* 2007).

5.2.4 *DC electrophoresis*

Electrophoresis is the movement of a charged particle in a fluid. Biological particles generally have a finite surface charge density (usually negative, due to the presence of acid groups on the surface) and observation of the movement of these particles in a uniform electric field is used both to characterize and to separate particles. The Coulomb force on the particle is given by the product of the electric field and the charge on the particle:

$$\mathbf{F}_{EP} = q\mathbf{E} = \int_S \sigma_q dS \ \mathbf{E} \tag{5.21}$$

where q is the total charge on the particle which, if the particle has a surface charge density σ_q, is given by the integral of this charge density over the closed surface of the particle S.

Consider a surface with free charge placed in an electric field. When charge moves along the surface under the influence of the field, a surface current flows which can be characterized by a surface conductance K_s. As previously discussed, the double layer can be separated into two distinct layers: the diffuse double layer and the Stern layer. It is reasonable to separate the surface conductance into two separated components (Lyklema 1995). The total surface conduction becomes the sum of conduction in the diffuse part of the double layer and conduction behind the slip plane in the Stern layer. The conductance in the Stern layer is given by the sum of the product of the surface charge density in this part of the double layer and the ion mobility for all ions. In the diffuse part of the double layer there is an added effect; the electroosmotic transport of charge carriers, which must be considered along with conduction. Therefore, surface current in this part of the double layer, $\Delta\mathbf{J}(y)$ is the sum of two components, one due to electroosmosis and the other due to conduction. For a symmetrical electrolyte the surface current is (Lyklema 1995):

$$\Delta\mathbf{J}(y) = \underbrace{q(z_- n_-(y) - z_+ n_+(y))\frac{\varepsilon}{\eta}(\zeta - \phi(y))\mathbf{E}}_{\text{Electroosmosis}} + \underbrace{q(z_- n_-(y) + z_+ n_+(y))\mathbf{E}}_{\text{Conduction}} \tag{5.22}$$

At low electrolyte conductivities (molarity) the diffuse layer conductance is small in comparison to the Stern layer conductance, which is typically of the order of 1 nS. However, at higher molarities the two are comparable and the diffuse layer conductance can make a significant contribution to the total conductance.

When a particle is placed in an aqueous suspending medium, the presence of the double layer means that to an observer the particle is electroneutral; the excess charge in the double layer is exactly equal and opposite to the charge on the particle. The particle therefore appears to have zero net charge, but when placed in an electric field, it moves. Clearly, if the ions in the double layer were fixed to the surface, then the particle would not move. However, in the diffuse part of the double layer the ions have the same mobility as the bulk solution; those in the Stern layer are bound but have non-zero mobility. Since the counterions have the opposite sign to the surface charge, they move in the direction opposite to that in which the particle would move if it were unscreened. This ion movement gives rise to fluid motion around the particle (electroosmosis). However, the particle is very small and the fluid is in the viscous limit for the fluid so that on the global scale the fluid is stationary. Because of this, the moving ions push the particle in the opposite direction, in other words in the same direction that the particle would move.

The situation is more complicated than this, since co-ions around the particle move ahead of the particle and counterions move in the direction opposite to the particle. As a result, a polarization field, arising from the displacement of the charge and counter-charge of the particle is established. This additional field can be with or against the applied field, either increasing or decreasing the electrophoretic velocity respectively.

Obviously, the electrophoretic mobility of a colloidal particle depends on the properties of the double layer. More particularly, it depends on the thickness of the double layer compared to the size of the particle. This is generally described by the ratio of the particle radius to the Debye length given by κa. There are two limiting cases to consider:

Thin double layer, $\kappa a \gg 1$

For a double layer which is thin compared with the particle radius $\kappa a \gg 1$ it has very low curvature and can be considered to be flat allowing the planar theory to be used. The derivation of the motion of the ions and the fluid is identical to the derivation of electroosmotic flow. However, in this case the surface moves relative to the stationary fluid. The electrophoretic mobility is

$$\mu_E = \frac{\varepsilon\zeta}{\eta} \tag{5.23}$$

This is referred to as the Helmholtz–Smoluchowski limit.

Thick double layer, $\kappa a \ll 1$

When a particle has a thick double layer (small particles or very dilute electrolytes), the forces acting on the double layer are not felt by the particle. The net force acting on the particle is simply the difference between the Coulombic force and the drag force given by Stokes law. This is: F $=qE=6\pi\eta av$, giving $\mu=v/E=q/6\pi\eta av$. The potential due to the point charge can be identified with the zeta potential, i.e. $\zeta = q/4\pi\varepsilon a$. Combining these expressions gives the electrophoretic mobility as

$$\mu_E = \frac{2\varepsilon\zeta}{3\eta} \tag{5.24}$$

This is referred to as the Hückel–Onsager limit.

Micro- and nanoparticles generally have some surface charges and move under an applied electric field. They exhibit different electrophoretic mobilities and can therefore be separated. Due to the automated, rapid, and high separation efficiency of capillary electrophoresis (CE), it is widely used to separate biological particles, including amino acids, viruses, bacteria, and cells. Excellent reviews on CE have been published by Schmitt-Kopplin and Frommberger (2003) and Kremser *et al.* (2004). CE is an important clinical tool for the analysis of proteins in physiological matrices, such as serum, urine, and cerebrospinal fluid (Oda *et al.* 1997). Michels *et al.* (2002) developed a system for automated protein analysis. In their system, the fluorescence labeled proteins were successively fractionized by submicellar CE at pH 7.5 and 11.1 in serial fashion. Laser-induced fluorescence was used as a sensitive detector of the separated proteins. Ye *et al.* (2004) developed a nanoliter volume enzyme analysis system for on-line CE peptide mapping of proteins, allowing picomole quantities of proteins to be digested. They enhanced the detection of peptide fragments in CE by post-derivatization and laser-induced fluorescence detection. Weissinger *et al.* (2004) used CE with mass spectrometry to identify and differentiate protein patterns in body fluids of healthy and diseased individuals. With this technology, polypeptide patterns from urine were established within 45 minutes. Okhonin *et al.* (2004) reported a sweeping capillary electrophoresis (SweepCE) to study protein-DNA interactions. The method is based on the sweeping of slowly migrating DNA by fast migrating protein using electrophoresis. It can directly measure the biomolecular rate constant of complex formation as a non-stopped-flow method. The separation of proteins using CE is generally monitored by UV detection. However, due to the small sample volume in CE and the short detection path-length in the optical setup, the detection sensitivity is normally low. Online sample pre-concentration techniques have been developed to overcome this problem, including sweeping (Quirino *et al.* 2002; Gong *et al.* 2007) and stacking (Chien 2003; Jung *et al.* 2003; Liu 2007).

5.3 AC electrokinetics

5.3.1 *Theory*

AC electrokinetics describes the translational motion of particles in AC electric fields and includes: dielectrophoresis (DEP), traveling wave dielectrophoresis (twDEP), and electrorotation (ROT). Generally, non-uniform electric fields are used in AC electrokinetics. The assumption that the uniform field solution for the dipole moment is valid is referred to as the dipole moment approximation. This approximation is valid if the size of the particle is small compared to the scale of the electric field non-uniformity, which is true for most cases

5.3.1.1 *Dielectrophoresis (DEP)* The DEP arises from the interaction between the induced dipole moment on a particle and the applied electric field. For an AC potential at a single frequency, the electric field can be expressed using complex phasor notation (Morgan and Green 2003) as: $\mathbf{E} = \mathrm{Re}[\tilde{\mathbf{E}}e^{i\omega t}]$, where $\mathbf{E}$ is the complex amplitude of the electric field. The time-averaged dielectrophoretic force on the dipole is:

$$\langle \mathbf{F}_{DEP} \rangle = \frac{1}{2}\mathrm{Re}[(\tilde{\mathbf{p}} \cdot \nabla)\tilde{\mathbf{E}}^*] = \frac{1}{2}v\,\mathrm{Re}[\tilde{\alpha}(\tilde{\mathbf{E}} \cdot \nabla)\tilde{\mathbf{E}}^*] \quad (5.25)$$

where $\tilde{\mathbf{p}}$ is the induced dipole moment phasor, v the volume of the particle, $\tilde{\alpha}$ the effective polarizability, and * indicates complex conjugate. If the non-uniform electric field has no spatialy dependent phase, the dielectrophoretic force simplifies to:

$$\langle \mathbf{F}_{DEP} \rangle = \frac{1}{4}v\,\mathrm{Re}[\tilde{\alpha}]\nabla|\tilde{\mathbf{E}}|^2 \quad (5.26)$$

For a spherical particle, eqn. (5.26) becomes:

$$\langle \mathbf{F}_{DEP} \rangle = \pi\varepsilon_m a^3\,\mathrm{Re}[\tilde{f}_{CM}]\nabla|\tilde{\mathbf{E}}|^2 \quad (5.27)$$

According to eqn. (5.27), if the electric field is uniform, the gradient of the magnitude of the field is zero, which means that there is no DEP force. The frequency dependence and the direction of the DEP force are governed by the real part of the Clausius–Mossotti factor. If the particle is more polarizable than the medium, the particle is attracted to high intensity electric field regions. This is termed as positive dielectrophoresis (pDEP). Conversely, if the particle is less polarizable than the medium, the particle is repelled from high intensity field regions and negative dielectrophoresis (nDEP) occurs. Therefore the real part of the Clausius–Mossotti factor characterizes the frequency dependence of the DEP force. DEP was first described in detail in Pohl's book (Pohl 1978).

5.3.1.2 *Travelling wave dielectrophoresis (twDEP)* In an electric field with spatialy varying phase, eqn. (5.25) becomes:

$$\langle \mathbf{F}_{DEP} \rangle = \frac{1}{4}v\,\mathrm{Re}[\tilde{\alpha}]\nabla|\tilde{\mathbf{E}}|^2 - \frac{1}{2}v\,\mathrm{Im}[\tilde{\alpha}](\nabla \times (\mathrm{Re}[\tilde{\mathbf{E}}] \times \mathrm{Im}[\tilde{\mathbf{E}}])) \quad (5.28)$$

For a spherical particle, this is:

$$\langle \mathbf{F}_{DEP} \rangle = \pi\varepsilon_m a^3\,\mathrm{Re}[\tilde{f}_{CM}]\nabla|\tilde{\mathbf{E}}|^2 - 2\pi\varepsilon_m a^3\,\mathrm{Im}[\tilde{f}_{CM}](\nabla \times (\mathrm{Re}[\tilde{\mathbf{E}}] \times \mathrm{Im}[\tilde{\mathbf{E}}])) \quad (5.29)$$

The first term on the right is the DEP force. The second term on the right is the additional traveling wave dielectrophoretic force, which propels a particle along an electrode array (e.g. interdigitated electrode arrays). If there is no spatially varying phase, the imaginary part of the electric field is zero, which means there is no twDEP. In order to produce twDEP, the frequency of the excitation voltage and the conductivity of the medium should be chosen to satisfy two conditions: (a) the particle experiences nDEP so that it levitates above the electrodes and (b) the imaginary part of the Clausius–Mossotti factor is non-zero. Masuda *et al.* (1987, 1988) were the first to demonstrate that traveling electric fields, generated by sequentially phase-shifted AC voltages, can be used to induce translational motion of particles.

5.3.1.3 *Electrorotation (ROT)* The interaction between the electric field and the effective dipole moment on the particle leads to a torque on the particle. For a lossy particle, there is a finite time between the applied electric field and the formation of the dipole. As the electric field vector changes direction, the dipole moment vector follows the changing field vector, resulting in rotation. The time-averaged torque is given by (Jones 1995):

$$\langle \Gamma_{ROT} \rangle = \frac{1}{2}\mathrm{Re}[\tilde{\mathbf{p}} \times \tilde{\mathbf{E}}^*] = -v\,\mathrm{Im}[\tilde{\alpha}](\mathrm{Re}[\tilde{\mathbf{E}}] \times \mathrm{Im}[\tilde{\mathbf{E}}]) \tag{5.30}$$

For a spherical particle, this becomes:

$$\langle \Gamma_{ROT} \rangle = -4\pi\varepsilon_m a^3\,\mathrm{Im}[\tilde{f}_{CM}]|\tilde{\mathbf{E}}|^2 \tag{5.31}$$

Equation (5.31) shows that the frequency-dependent property of the ROT torque depends on the imaginary part of the Clausius–Mossotti factor. The particle will rotate with or against the electric field, depending on whether the imaginary part of the Clausius–Mossotti factor is negative or positive. The phenomenon of ROT was explored in detail by Arnold and Zimmerman (1982, 1988) and is now widely used for single cell analysis such as viability assays (Zhou *et al.* 1995; Hodgson and Pethig 1998; Dalton *et al.* 2001), biocide testing (Zhou *et al.* 1996), or single cell characterization (Hölzel 1997; Zhou *et al.* 1998; Yang *et al.* 1999; Cristofanilli 2002; Dalton 2004).

5.3.2 *Applications*

AC electrokinetic techniques have been widely used for the manipulation, separation, focusing, trapping, and handling of latex spheres, viruses, bacteria, and cells (Morgan and Green 2003). A number of different geometries of electrodes have been designed to perform particle identification and separation in the electrokinetic area.

5.3.2.1 *Castellated electrode array* The castellated electrode design was first used by Pethig's group (Price *et al.* 1988; Burt *et al.* 1989) to dielectrophoretically collect particles close to the electrodes. Later, similar configurations (e.g. interleaved electrodes (Gascoyne *et al.* 1992) and the sawtooth electrode array (Green and Morgan 1997) were used for separating biological particles (Morgan *et al.* 1999; Hughes and Morgan 1999; Green *et al.* 2000a). Figure 5.8 (a) and (b) show photographs of pDEP and nDEP of latex spheres on castellated electrodes (Green *et al.* 2000a).

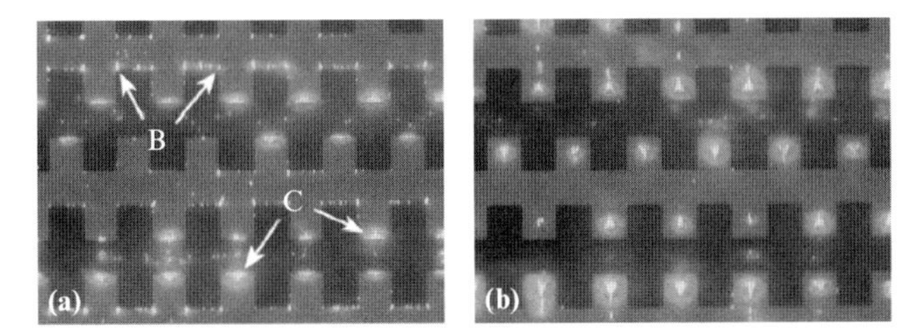

Fig. 5.8: (a) Image showing positive dielectrophoresis of 557 nm diameter latex spheres. The points B and C are high intensity field regions. (b) Image showing negative dielectrophoresis (Green *et al.* 2000a).

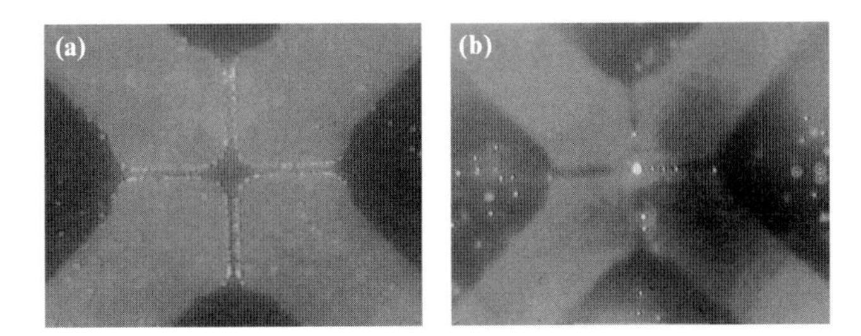

Fig. 5.9: (a) Image showing positive dielectrophoresis of 557 nm diameter latex spheres on polynomial electrodes. (b) Image showing negative dielectrophoresis (Green *et al.* 2000a).

5.3.2.2 *Polynomial electrodes* The polynomial electrode design has four electrodes with edges defined by a hyperbolic function in the center, and with parallel edges elsewhere to some arbitrary distance. The theoretical principle of this electrode design was described by Huang and Pethig (1991). It has been demonstrated that the polynomial electrode can be used for particle trapping and characterization (Green *et al.* 2000a; Huang and Pethig 1991; Huang *et al.* 1992; Markx *et al.* 1996; Hughes *et al.* 1998) using both pDEP and nDEP (Fig. 5.9 Green *et al.* 2000a).

5.3.2.3 *Interdigitated electrode arrays* The interdigitated electrode array (Fig. 5.10) is the most commonly used design for performing DEP and twDEP.

A number of papers have been published that derive analytical expressions for the electric fields, generated by the interdigitated electrode arrays. Methods include Green's theorem (Wang *et al.* 1996), Green's function (Garcia and Clague 2000; Liang *et al.* 2004), half-plane Green's function (Clague and Wheeler 2001), and Fourier series (Morgan *et al.* 2001; Chuang *et al.* 2003). However, these analytical solutions all involve approximations. Recently, completely analytical solutions without any approximation have been derived by Sun *et al.* (2007) using the Schwarz–Christoffel Mapping method.

5.3.2.4 *DEP-FFF* Field flow fraction (FFF) describes a method for separating particles based on combining some force field with hydrodynamic separation. In a typical configuration (Fig. 5.11) particles move to equilibrium positions in a flow,

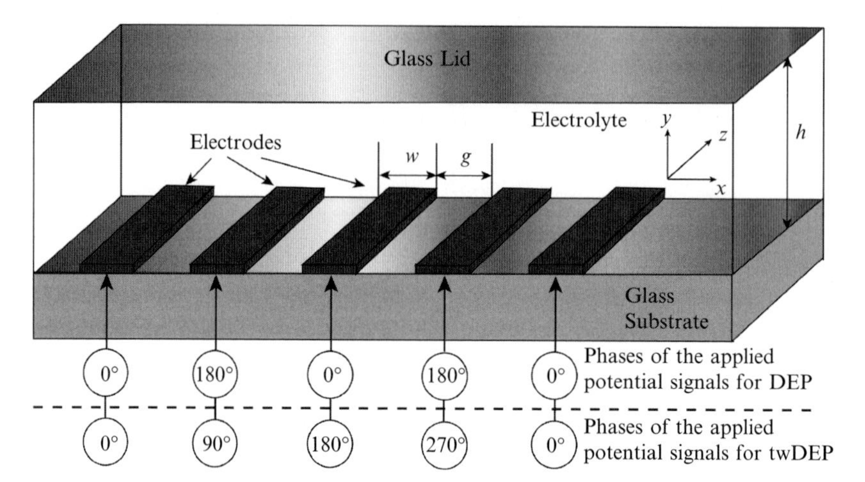

Fig. 5.10: Diagram showing a typical experimental system, consisting of an interdigitated electrode array fabricated on a solid substrate and used for dielectrophoresis and traveling wave dielectrophoresis. For DEP, the electrodes are connected to voltages with 180° phase shifts. For twDEP, the electrodes are connected to a frequency generator with a 90° phase shift. w is the electrode width, g is the electrode gap, and h is the height of the channel.

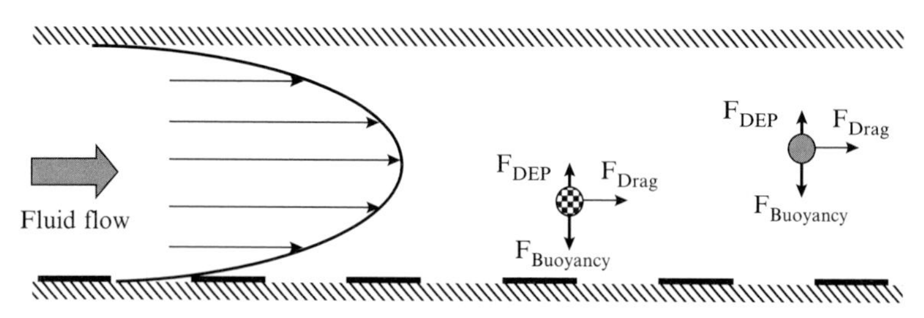

Fig. 5.11. Principle of dielectrophoretic field-flow fractionation (DEP-FFF).

according to the balance of different forces, such as dielectrophoresis (DEP), hydrodynamic lift (HDLF), and gravity (buoyancy). Different types of particles move to different equilibrium positions in the system. Therefore, the height-dependent hydrodynamic velocity profile of the fluid carries the particles at different velocities according to their positions. As a result, a sample can be separated and fractionated along the channel, since different types of the particles are transported at different rates (Markx *et al.* 1997a, 1997b; Rousselet *et al.* 1998; Gascoyne and Vykoukal 2004).

5.3.2.5 *Other electrode designs* A spiral electrode configuration that combined twDEP and ROT for particle analysis was proposed by Goater *et al.* (1997). This design was further developed by Gascoyne et al. (1997; 2002) where DEP was combined with twDEP to concentrate cells. DEP does not always have to be performed with conducting metal electrodes. For example, 3-D insulating-post arrays (Cummings and Singh 2003; Lapizco-Encinas *et al.* 2004a; 2004b) have been developed to trap and separate live and dead bacteria using DC voltages applied across the length of a microchannel. The insulating posts in the channel create obstructions in the pathways of the electric field producing a non-uniform electric field distribution in the channel, generating DEP. These devices have been used to separate bacteria or latex beads (Gascoyne and Vykoukal 2004; Goater *et al.* 1997; Hawkins *et al.* 2007). An excellent review on cell manipulation using electrical forces has recently been published by Voldman (2006). This work reviewed the quadrupole electrode, octopole electrode, nDEP microwell, points-and-lid geometry, and ring-dot geometry, for single cell positioning and trapping.

5.4 Electrohydrodynamics

In electrokinetic systems, the electric field acts on charges and dipoles in the fluid as well as the particles, causing an electrical force on the fluid which can lead to flow. In a DC field, the nature of the electroosmotic flow depends predominantly on the applied field and the charge density (zeta potential) of the surface. DC electroosmosis normally requires fairly high voltages; Faradaic reactions at the electrodes lead to bubble formation (and local changes in ion concentration) and to Joule heating.

An AC electric field applied to microelectrodes also generates flow. In 1998 and 1999, Ramos *et al.* (1998, 1999), showed that it was possible to generate a non-linear electroosmotic flow using low voltage AC signals, a phenomenon named AC electroosmosis. In AC fields, fluid is driven in motion either by electrothermal effects, or by induced charge on electrodes. The frequency window over which these two phenomena are found is summarized in Fig. 5.12. At low frequencies (<100 kHz), the dominant fluid flow is AC electroosmosis. At higher frequencies the magnitude of the AC electroosmotic flow decreases, and the dominant effect becomes electrothermally induced fluid flow. The behavior of the fluid and the resulting motion of particles are distinct and differentiable for each of the two mechanisms.

A general expression for the electrical force per unit volume on a liquid is (Stratton 1941):

$$\bar{f}_E = \rho_q \bar{E} - \frac{1}{2}\bar{E}^2 \nabla\varepsilon + \frac{1}{2}\nabla\left(\rho_m \frac{\partial\varepsilon}{\partial\rho_m}\bar{E}^2\right) \tag{5.32}$$

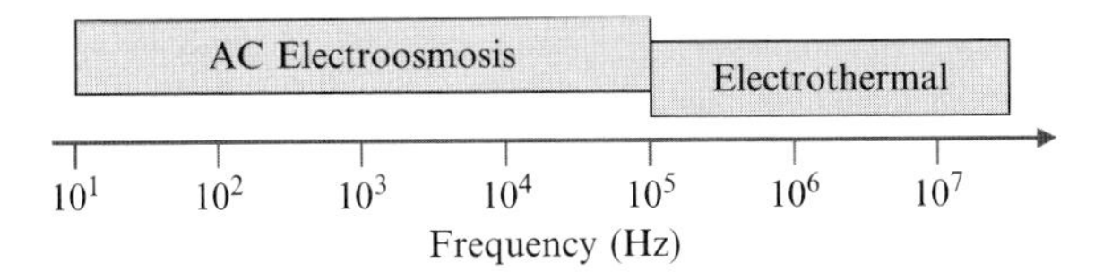

Fig. 5.12. An outline of the frequency range over which the different types of electric field induced fluid flow are observed in AC electrokinetic microsystems. AC electroosmosis dominates in the lower frequency range (<100 kHz) and electrothermal fluid flow at higher frequencies.

where ρ_q is the volume charge density and ρ_m is the mass density. The electric field can be written as the sum of two components, the applied field E_0, and the perturbation field E_1, where $\bar{E} = \bar{E}_0 + \bar{E}_1$ and $|\bar{E}_1| \ll |\bar{E}_0|$. For an incompressible fluid the last term in eqn. (5.32) has no effect on the dynamics (it is the gradient of a scalar) and therefore can be ignored (Stratton 1941). For an isothermal fluid there is no free charge and the permittivity gradient is zero so that the total force is zero. In addition to field driven flow, natural convection can also drive fluid if the conductivity is sufficiently high so that Joule heating produces significant temperature gradients and therefore gradients in the density of the fluid. However, in most practical cases, these effects are very small by comparison with other fluid effects and can be ignored.

5.4.1 *AC electroosmosis*

AC electroosmotic fluid flow occurs when the charge induced by the electric field at the electrode/electrolyte interface experiences a force due to the applied field. The simplest experimental setup (Green *et al.* 2000b, 2000c, 2002) that can be used to measure and analyze fluid flow is the two parallel co-planar microelectrode system shown in Fig. 5.13. The two electrodes are energized by a single phase AC supply of

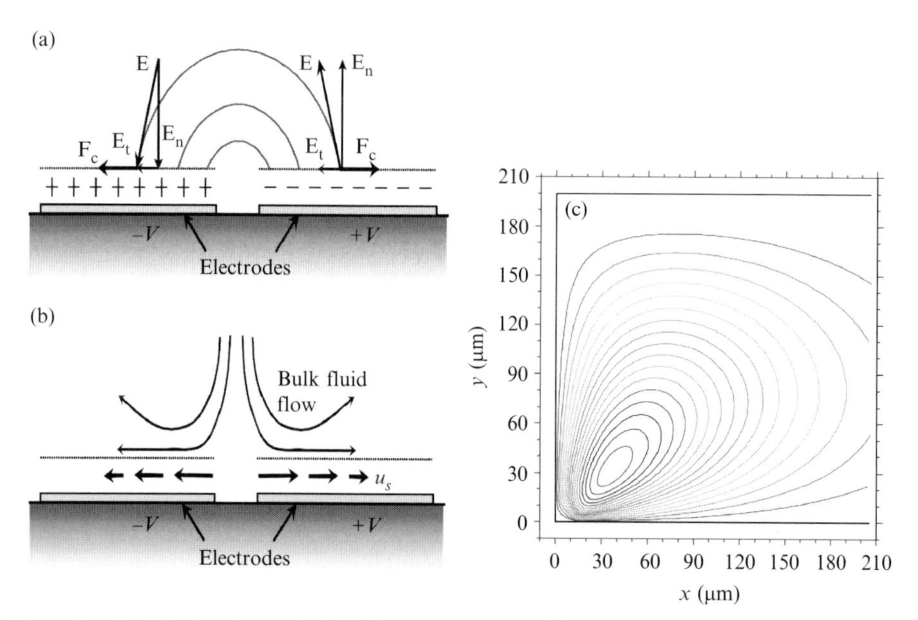

Fig. 5.13: Mechanism of AC electroosmosis. (a) The interaction between the induced surface charges and the tangential component of the applied electric field produces a horizontal force. (b) The force gives rise to the flow. (c) Numerically calculated streamlines of the fluid flow over one electrode.

variable voltage and frequency. As shown in Fig. 5.13(a), the normal electric field induces a charge on the electrodes, whilst the tangential component produces a force that moves the charges (and the fluid) along the surface of the electrodes, giving rise to fluid flow—Fig. 5.13(b). For any given frequency and voltage, the flow magnitude and direction is constant because when the field direction changes, the sign of the induced charge also changes and therefore the direction of the force remains unchanged. The fluid velocity depends on a number of factors: frequency, voltage, fluid conductivity, permittivity, viscosity, and importantly the scale of the system. A numerical simulation of the fluid streamlines over one electrode is shown in Fig. 5.13(c).

The phenomenon of AC electroosmosis is predominant at low frequencies (of the order of Hz to tens of kHz). In the low frequency limit, the magnitude of the flow tends to zero because the field in the bulk electrolyte disappears due to the screening effect of the electrical double layer. At high frequencies the AC electroosmotic flow again tends to zero because there is insufficient time for the induced charge to form in the double layer. Maximum flow occurs at a frequency proportional to the charge relaxation time of the fluid and the scale of the system (Ramos *et al.* 1999).

One attractive application of AC electroosmosis is the potential for producing pumps with no moving parts. Such a pump could operate at fairly low voltages and potentially overcome some of the issues of DC electroosmotic pumps, such as bubble formation and Joule heating. Ajdari (2000) first proposed the use of asymmetric electrode structures to pump fluids using AC electroosmosis. Brown *et al.* (2001) experimentally demonstrated an AC electroosmotic pump using an asymmetry created from pairs of planar electrodes of different widths and gaps. Ramos *et al.* (2003) presented a theoretical analysis of the pumping phenomena based on an electroosmotic model. The underlying principle of AC electroosmosis pumping using an asymmetric electrode array is shown in Fig. 5.14(a). As shown in Fig. 5.13, the tangential component of the electric field produces a force on the induced charge. However, the asymmetry in the system gives rise to two different

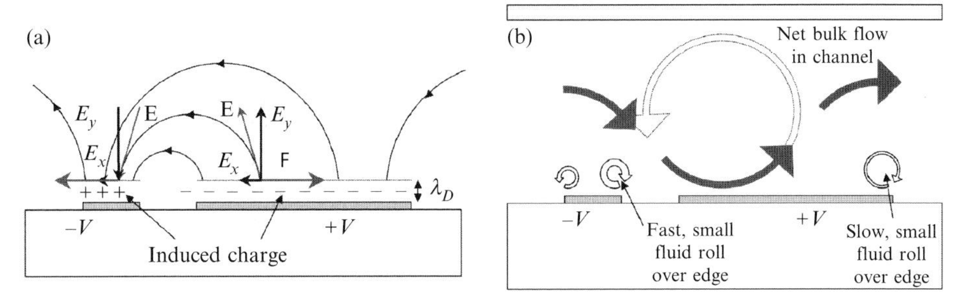

Fig. 5.14: Diagram showing the mechanism of AC electroosmosis in an asymmetric electrode array (Ramos *et al.* 2003). (a) Diagram showing the induced charge, the resulting electric field, and the force. (b) Diagram showing the resulting bulk flow.

fluid rolls above the two different sized electrodes. A small roll occurs over the edge of the small electrodes and a large roll over the inner edge of the large electrode. The net effect of these two rolls is to produce a unidirectional flow.

These pumps have been characterized by a number of researchers. Mpholo *et al.* (2003) used an array of asymmetric interdigitated electrodes, made of indium tin oxide (ITO) to produce fluid flow. They demonstrated that the velocity of the fluid increases with decreasing electrode dimensions. Debesset *et al.* (2004) have shown that circular pumps can pump electrolytes in a continuous loop—such a system has potential applications in separation and chromatography. Studer *et al.* (2004) fabricated a pump using interdigitated electrodes and reported velocities of up to 500 μms^{-1} in 20 μm deep and 100 μm wide channel. In 2004, Bazant and Squires (2004) generalized the concept of AC electroosmosis introducing the term 'induced-charge electroosmosis (ICEO)'. Bazant and Ben (2006) theoretically predicted how 3-D structures could be used to produce AC electroosmotic pumps with the potential to operate 20 times faster than planar AC electroosmotic pumps, for the same applied voltage and feature size. Bazant and coworkers (Urbanski *et al.* 2006) experimentally demonstrated a 3-D AC electroosmotic pump by electroplating steps on a symmetric electrode array. The 3-D pump had a faster flow rate than the conventional planar pump over the frequency range of 0.1–100 kHz (up to one order of magnitude).

An interesting development in AC electroosmotic pumping is the use of conventional arrays of interdigitated electrodes energized with traveling electric fields to produce fluid motion, (analogous to twDEP) as first demonstrated by Cahill *et al.* (2004). The principle is shown in Fig. 5.15; the electric field interacts with the electrical double layer charge to produce a fluid flow in the direction of the traveling wave. Due to the finite time for charging the double layar, a delay exists between the time of maximum induced charge and the maximum applied signal. The electric field acts on the induced charge, pulling the fluid in the

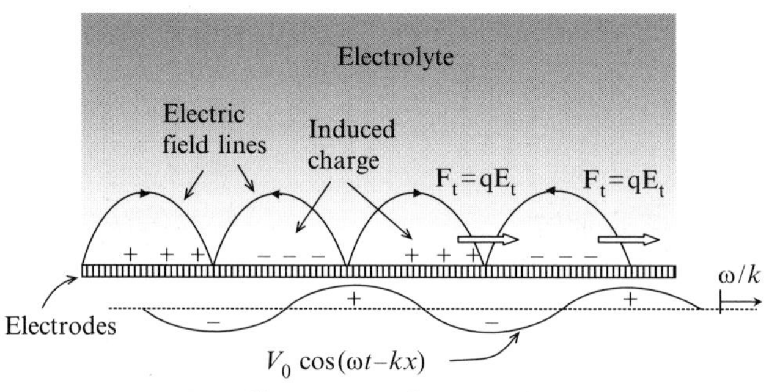

Fig. 5.15: Diagram showing the mechanism of traveling-wave electroosmosis (Ramos *et al.* 2005).

direction of the traveling wave. Ramos *et al.* (2005) reported experiments on traveling-wave electroosmosis using an interdigitated electrode array and observed two flow regimes: at small-voltage amplitudes the fluid flow follows the direction of the traveling wave, and at higher-voltage amplitudes the fluid flow is reversed. The experiments were analyzed using a linear electroosmotic model based upon the Debye–Huckel theory of the double layer and observations at low voltages were in qualitative agreement with the electroosmotic model. More recently, Ramos *et al.* (2007) developed a linear model to explain the effect of Faradaic currents on traveling-wave electroosmosis.

AC electroosmosis is a promising 'no moving parts' method of manipulating and transporting fluids. Recent work has shown that it can be used to enhance mixing in microfludic systems. For example, Sasaki *et al.* (2006) presented an AC electroosmotic micromixer using a coplanar meandering electrode configuration. Huang *et al.* (2007) presented an AC electroosmotic mixer which generated microvortex patterns in a microchannel. Kim *et al.* (2007) reported a numerical simulation of system for simultaneous mixing and pumping of electrolytes using microelectrodes.

Apart from micropumps and micromixers, EHD systems are now being used in a number of micro total analysis systems, including: particle separation (Green and Morgan 1998; Zhao *et al.* 2005), trapping (Hoettges *et al.* 2003; Wu *et al.* 2005a, 2005b; Grom 2006; Islam *et al.* 2007), and cell/molecule concentrating (Wong *et al.* 2004; Bhatt *et al.* 2005; Brown and Meinhart 2006). Green and Morgan (1998) firstly combined dielectrophoretic and electrohydrodynamic forces to separate nanoparticles. Hoettges *et al.* (2003) combined dielectrophoresis and EHD to trap particles from solution onto the top of electrode surfaces. Wu *et al.* (2005a, 2005b) combined AC electroosmosis with impedance measurement to improve the sensitivity of particle electrical detection.

Morgan *et al.* (2007) reported how AC electric fields can be used to deflect and manipulate two co-flowing streams of different electrolytes within a microfluidic channel, as shown in Fig. 5.16.

The electrode array on the bottom of the channel is connected to a single phase voltage supply. When a single phase AC potential was applied to the electrodes the boundary between the two fluids moves, as shown schematically in Fig. 5.16.

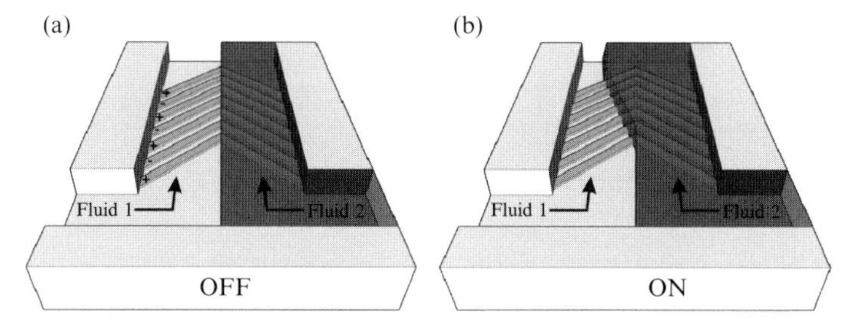

FIG. 5.16: Diagram of the experimental chip showing the two co-flowing fluid streams before and after application of the electric field.

5.4.2 *Electrothermal flow*

At high frequencies (>100 kHz) and/or high electrolyte conductivities, AC electroosmotic flow is negligible and the dominant fluid flow is electrothermal. This type of flow is driven by temperature gradients in the fluid, which can be generated by internal or external energy sources. The internal energy source is Joule heating, where the electric field causes power dissipation in the fluid, and the corresponding temperature rise diffuses through the system. This gives rise to gradients in conductivity and permittivity. The electric field acts on these gradients producing a body force on the fluid and consequently a flow. The velocity of the flow is proportional to the temperature rise in the fluid, which is in turn proportional to the conductivity of the electrolyte and the electric field. As a result, this type of fluid flow is observed mainly at high electrolyte conductivities.

The time-averaged force per unit volume on a fluid is (Ramos *et al.* 1998):

$$\langle \bar{f}_E \rangle = -\frac{1}{2}\left(\left(\frac{\nabla\sigma}{\sigma} - \frac{\nabla\varepsilon}{\varepsilon} \right) \cdot \bar{E}_o \frac{\varepsilon \bar{E}_o}{1 + (\omega\tau)^2} + \frac{1}{2}|E_o|^2 \nabla\varepsilon \right) \tag{5.33}$$

Where $\tau = \varepsilon/\sigma$, is the charge relaxation time of the liquid. The first term on the right side of the equation represents the Coulomb force and the second term the dielectric force. As shown by this equation, the temperature gradients produce gradients in σ or ε, and the magnitude of the flow depends on both the magnitude and frequency of the electric field. Temperature gradients can also be imposed externally using heat sources and heat sinks to control the geometry of the temperature field. For example, the fluid can be heated by the illumination source (Ramos *et al.* 1999).

Figure 5.17 shows an example of an electrothermal system where two parallel planar electrodes are on a solid substrate. An external temperature source establishes a temperature gradient as shown in the figure.

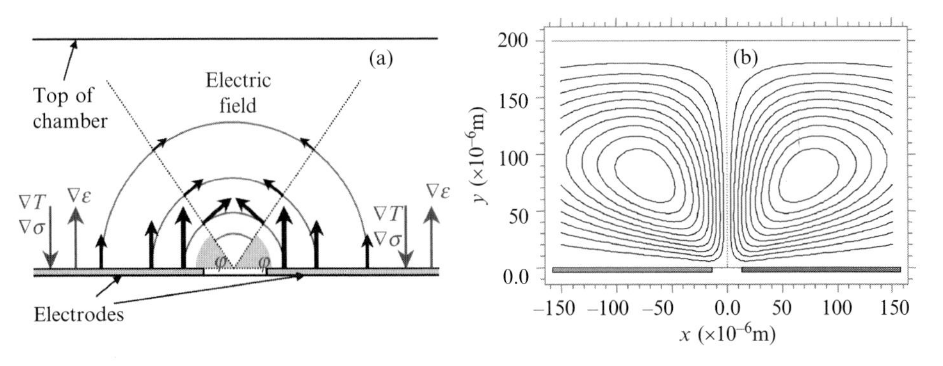

Fig. 5.17: (a) Diagram showing temperature gradients generated by a heated substrate beneath the electrode and the resulting gradients in conductivity and permittivity. (b) Numerical simulation of electrothermal flows above two parallel plane electrodes.

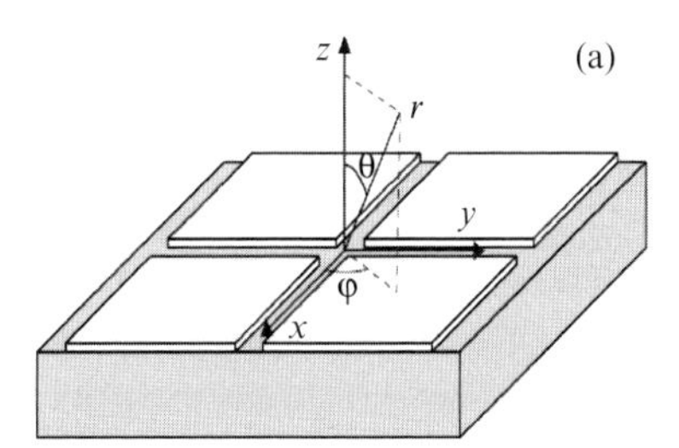

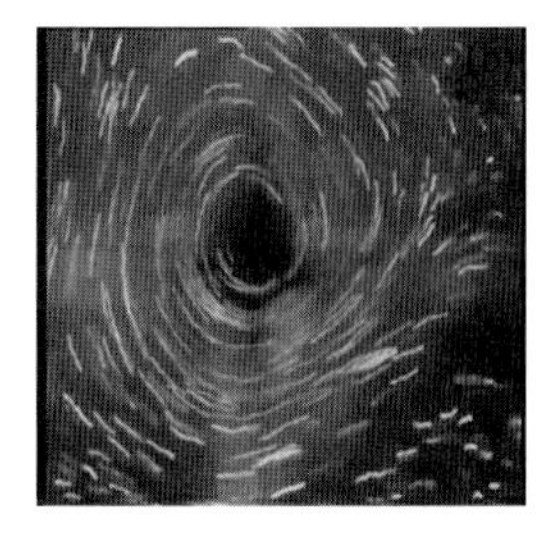

Fig. 5.18: (a) Diagram showing a microsystem with four coplanar electrodes and (b) the rolls generated using a 4-phase electric field (observed from the top) (González *et al.* 2006).

In this example, the heat radiates upwards through the electrolyte to produce gradients in conductivity and permittivity. This temperature gradient gives rise to two circular fluid flows located in the bulk fluid above each of the electrodes. The direction of the flow depends on frequency, at low frequencies the flow is dominated by the Coulomb force and moves across the electrodes into the gap. At higher frequencies the flow direction changes as permittivity effects dominate. A numerical simulation of a typical flow pattern, calculated at low frequencies, is shown in Fig. 5.17(b) (Green *et al.* 2001).

Different types of electrothermal flows can be generated by changing the electrode geometry. For example González *et al.* (2006) reported an electrothermal flow generated by a rotating electric field in a four-electrode system, as shown in Fig. 5.18.

Over the last 25 years, a number of groups have developed electrothermal pumps to move liquids. The principle of electrothermal pumping was first demonstrated by Melcher (1966) and Melcher and Firebaugh (1966, 1967) who showed that a traveling electric field could move induced charges in an insulating liquid giving unidirectional fluid motion. Maximum fluid flow occurred at a frequency corresponding to the charge relaxation time of the system. The earliest microfabricated EHD pumps were made by Bart *et al.* (1990), Müller *et al.* (1993), and Fuhr *et al.* (1992, 1994) in the early years of 1990s. EHD pumps have the advantages of simple fabrication and work at high frequencies, therefore avoiding electrochemical dissociation of the fluid and the accumulation of impurities on the surface of the electrodes (Wong *et al.* 2004). Electrothermal effects have also been used to induce stirring of fluid for lab-on-a-chip applications. Sigurdson *et al.* (2005) used electrothermally-generated micro-stirring to improve the binding rates in flow-through assays. Feng and Seyed-Yagoobi (2006) designed an EHD conduction pump for active thermal control of a parallel heat exchanger. The electrically driven flow generated by the pump successfully controlled the two-phase flow distribution between two channels. Felten *et al.* (2006) presented a 3-D electrothermal pump driven by traveling electric waves. The power of the pump was increased by using additional temperature fields. Wu *et al.* (2007) reported electrothermal flows for two different EHD pump designs (asymmetrical coplanar electrodes and T-shape electrodes).

Electrohydrodynamic effects cover a broad range of phenomena from Taylor cone formation to electrosprays. Many of these areas are now being investigated for lab-on-a-chip applications and reviews of these topics can be found in references (Saville 1997; Barrero and Loscertales 2007; and de la Mora 2007).

5.5 Conclusion and summary

The behavior of particles and fluids on the microscale is determined by a number of forces. Electrokinetic forces act on particles, either through forces on fixed charge (electrophoresis) or a second order force on the dipole (induced or fixed). Electrohydrodynamic forces on the liquid are predominantly due to double layer charge or body forces due to gradients in conductivity and/or permittivity created by thermal gradients. In separation systems, buoyancy force can be significant (as in FFF) but often the magnitude of this force is much lower than the other electrokinetic forces for micron sized particles. In electrophoresis the force depends only on the magnitude of the electric field, and is generally constant. By contrast, the DEP force depends on the gradient of the energy density, which changes on the length scale of the electrodes and is therefore a short range effect. The DEP force can also be modulated by changing the frequency and electrical properties of the suspending medium. The high strength electric fields used in DEP separation systems often give rise to fluid motion, which in turn results in a viscous drag on the particle. This fluid flow occurs because electric fields generate heat, leading to volume forces in the liquid. Gradients in conductivity and permittivity give rise to electrothermal forces; gradients in mass density to buoyancy. These forces scale in a complex manner with system dimensions, frequency, field etc. AC electroosmosis dominates fluid motion at low frequencies and small system sizes, electrothermal flow dominates at high frequencies and voltages. A full analysis of scaling in microsystems has been reviewed in detail (Castellanos *et al.* 2003).

References

Acheson, D. J., 1990. *Elementary Fluid Dynamics.* Oxford University Press, Oxford.

Adrover, A., Giona, M., Pagnanelli, F., and Toro, L., 2007. 'Influence of surface heterogeneity in electroosmotic flows – implications in chromatography, fluid mixing and chemical reactions in microdevices.' *Applied Surface Science*, **253**, pp. 5785–5790.

Ajdari, A., 2000. 'Pumping liquids using asymmetric electrode arrays.' *Physical Review E*, **61**, pp. R45–R48.

Arnold, W. M., and Zimmerman, U., 1982. 'Rotating-field-induced rotation and measurement of the membrane capacitance of single mesophyll cells of Avena sativa.' *Zeitschrift für Naturforschung*, **37c**, pp. 908–915.

Arnold, W. M., and Zimmerman, U., 1988. 'Electrorotation: Development of a technique for dielectric measurements on individual cells and particles.' *Journal of Electrostatistics*, **21**, pp. 151–191.

Bard, A. J., and Faulkener, L. R., 1980. *Electrochemical Methods, Fundamentals and Applications.* John Wiley and Sons, New York.

Barrero, A., and Loscertales, I. G., 2007. 'Micro- and nanoparticles via capillary flows.' *Annual Review of Fluid Mechanics*, **39**, pp. 89–106.

Bart, S. F., Tavrow, L. S., Mehregany, M., and Lang, J. H., 1990. 'Microfabricated electrohydrodynamic pumps.' *Sensors and Actuators*, **A21**–**A23**, pp. 193–197.

Bazant, M. Z., and Ben, Y., 2006. 'Theoretical prediction of fast 3D AC electro-osmotic pumps.' *Lab on a Chip*, **6**, pp. 1455–1461.

Bazant, M. Z., and Squires, T. M., 2004. 'Induced-charge electrokinetic phenomena: Theory and microfluidic applications.' *Physical Review Letters*, **92**, pp. 066101.

Bhatt, K. H., Grego, S., and Velev, O. D., 2005. 'An AC electrokinetic technique for collection and concentration of particles and cells on patterned electrodes.' *Langmuir*, **21**, pp. 6603–6612.

Blouin, M., and Guay, D., 1997. 'Activation of ruthenium oxide, iridium oxide and mixed ruxir1-x oxide electrodes during cathodic polarization and hydrogen evolution.' *Journal of Electrochemical Society*, **144**, pp. 573–581.

Bockris, J. O'M., and Reddy, A. K. N., 1973. *Modern Electrochemistry.* Plenum Press, New York.

Brown, A. B. D., Smith, C. G., and Rennie, A. R., 2001. 'Pumping of water with ac electric fields applied to asymmetric pairs of electrodes.' *Physical Review E*, **63**, pp. 016305.

Brown, M. R., and Meinhart, C. D., 2006. 'AC electroosmotic flow in a DNA concetrator.' *Microfluid Nanofluid*, **2**, pp. 513–523.

Burt, J. P. H., A-Ameen, T. A. K., and Pethig, R., 1989. 'An optical dielectrophoresis spectrometer for low-frequency measurements on colloidal suspensions.' *Journal of Physics E: Scientific Instruments*, **22**, pp. 952–957.

Cahill, B. P., Heyderman, L., Gobrecht, J., and Stemmer, A., 2004. 'Electroosmotic streaming on application of travelling-wave electric fields.' *Physical Review E*, **70**, pp. 036305.

Castellanos, A., (Ed.), 1998. *Electrohydrodynamics.* Springer, Wien/New York.

Castellanos, A., Ramos, A., Gonzalez, A., Green, N. G., and Morgan, H., 2003. 'Electrohydrodynamics and dielectrophoresis in microsystems: Scaling laws.' *Journal of Physics D: Applied Physics*, **36**, pp. 2584–2597.

Chang, C. C., and Yang, R. J., 2007. 'Electrokinetic mixing in microfluidic systems.' *Microfluid Nanofluid*, **3**, pp. 501–525.

Chien, R. L., 2003. 'Sample stacking revisited: A personal perspective.' *Electrophoresis*, **24**, pp. 486–497.

Chuang, D. E., Loire, S., and Mezić, I., 2003. 'Closed-form solutions in the electric field analysis for dielectrophoretic and travelling wave interdigitated electrode arrays.' *Journal of Physics D: Applied Physics*, **36**, pp. 3073–3078.

Clague, D. S., and Wheeler, E. K., 2001. 'Dielectrophoretic manipulation of macromolecules: The electric field.' *Physical Review E*, **64** (026605), pp. 1–8.

Cristofanilli, M., Gasperis, G. D., Zhang, L. S., Huang, M. C., Gasconye, P. R. C., and Horotobagyi, G. N., 2002. 'Automated electrorotation to reveal dielectric variations related to HER-2/neu Overexpression in MCF-7 sublines.' *Clinical Cancer Research*, **8**, pp. 615–619.

Cummings, E. B., and Singh, A. K., 2003. 'Dielectrophoresis in microchips containing arrays of insulating posts: Theoretical and experimental results.' *Analytical Chemistry*, **75**, pp. 4724–4731.

Dalton, C., Goater, A. D., Burt, J. P. H., and Smith, H. V., 2004. 'Analysis of parasites by electrorotation.' *Journal of Applied Microbiology*, **96**, pp. 24–32.

Dalton, C., Goater, A. D., Drysdale, J., and Pethig, R., 2001. 'Parasite viability by electrorotation.' *Colloids and Surfaces A: Physicochemical and Engineering Aspects*, **195**, pp. 263–268.

Dasgupta, P. K., and Liu, S., 1994. 'Electroosmosis: A reliable fluid propulsion system for flow injection analysis.' *Analytical Chemistry*, **66**, pp. 1792–1798.

de la Mora, J. F., 2007. 'The fluid dynamics of Taylor cones.' *Annual Review of Fluid Mechanics*, **39**, pp. 217–243.

Debesset, S., Hayden, C. J., Dalton, C., Eijkel, J. C. T., and Manz, A., 2004. 'An AC electroosmotic micropump for circular chromatographic applications.' *Lab in a Chip*, **4**, pp. 396–400.

Felten, M., Geggier, P., Jager, M., and Duschi, C., 2006. 'Controlling electrohydrodynamic pumping in microchannels through defined temperature fields.' *Physics of Fluids*, **18**, pp. 1707.

Feng, Y., and Seyed-Yagoobi, J., 2006. 'Control of adiabatic two-phase dielectric fluid flow distribution with EHD conduction pumping.' *Journal of Electrostatics*, **64**, pp. 621–627.

Fu, L. M., R.-J., Yang, C. H., Lin, and Chien, Y. S., 2005. 'A novel microfluidic mixer utilizing electrokinetic driving forces under low switching frequency.' *Electrophoresis*, **5**, pp. 1814–1824.

Fuhr, G., Hagedorn, R., Müller, T., Benecke, W., and Wagner, B., 1992. 'Microfabricated electrohydrodynamic (EHD) pumps for liquids of higher conductivity.' *Journal of Microelectromechanical Systems*, **1**, pp. 141–146.

Fuhr, G., Schnelle, T., and Wagner, B., 1994. 'Travelling wave-driven microfabriacated electrohydrodynamic pumps for liquids.' *Journal of Micromechanics and Microengineering*, **4**, pp. 217–226.

Gan, W. E., Yang, L., He, Y., Zeng, R., Cervera, M. L., and Guardia, M., 2003. 'Mechanism of porous core electroosmotic micropump flow injection system and its application to determination of chromium (VI) in waste-water.' *Talanta*, **51**, pp. 667–675.

Garcia, M., and Clague, D. S., 2000. 'The 2D electric field above planar sequence of independent strip electrodes.' *Journal of Physics D: Applied Physics*, **33**, pp. 1747–1755.

Gascoyne, P. R. C., and Vykoukal, J., 2004. 'Dielectrophoresis-based sample handling in general-purpose programmable diagnostic instruments.' *IEEE Proceedings*, **92**, pp. 22–42.

Gascoyne, P. R. C., Huang, Y., Pethig, R., Vykoukal, J., and Becker, F. F., 1992. 'Dielectrophoretic separation of mammalian cells studied by computerized image analysis.' *Measurement Science and Technology*, **3**, pp. 439–445.

Gasconye, P. R. C., Mahidol, C., Ruchirawat, M., Satayavivad, J., Watcha-rasit, P., and Becker, F. F., 2002. 'Microsample preparation by dielectrophoresis: Isolation of malaria.' *Lab in a Chip*, **2**, pp. 70–75.

Goater, A. D., Burt, J. P. H., and Pethig, R., 1997. 'A combined travelling wave dielectrophoresis and electrorotation device: Applied to the concentration and viability determination of Cryptosporidium.' *Journal of Physics D: Applied Physics*, **30**, pp. L65–L69.

Gong, M., Wehmeyer, K. R., Limbach, P. A., and Heineman, W. R., 2007. 'Flow manipulation for sweeping with a catonic surfactant in microchip capillary electrophoresis.' *Journal of Chromatography A*, **1167**, pp. 217–224.

González, A., Ramos, A., Morgan, H., Green, N. G., and Castellanos, A., 2006. 'Electrothermal flows generated by alternating and rotating electric fields in microsystems.' *Journal of Fluid Mechanics*, **564**, pp. 415–433.

Green, N. G., and Morgan, H., 1997. 'Manipulation and trapping of sub-micron bioparticles using dielectrophoresis.' *Journal of Biochemical and Biophysical Methods*, **35**, pp. 89–102.

Green, N. G., and Morgan, H., 1998. 'Separation of submicrometre particles using a combination of dielectrophoretic and electrohydrodynamic forces.' *Journal of Physics D: Applied Physics*, **31**, pp. L25–L30.

Green, N. G., Ramos, A., and Morgan, H., 2000a. 'AC electrokinetics: A survey of sub-micrometer particle dynamics.' *Journal of Physics D: Applied Physics*, **33**, pp. 632–641.

Green, N. G., Ramos, A., González, A., Castellanos, A., and Morgan, H., 2000b. 'Electric field induced fluid flow on microelectrodes: The effect of illumination.' *Journal of Physics D: Applied Physics*, **33**, pp. L13–L17.

Green, N. G., Ramos, A., González, A., Morgan, H., and Castellanos, A., 2000c. 'Fluid flow induced by non-uniform AC electric fields in electrolytic solutions on microelectrodes: Part I: Experimental measurements.' *Physical Review E*, **61**, pp. 4011–4018.

Green, N. G., Ramos, A., González, A., Castellanos, A., and Morgan, H., 2001. 'Electrothermally induced fluid flow on microelectrodes.' *Journal of Electrostatistics*, **53**, pp. 71–87.

Green, N. G., Ramos, A., González, A., Morgan, H., and Castellanos, A., 2002. 'Fluid flow induced by non-uniform AC electric fields in electrolytic solutions on microelectrodes: Part III: Observation of streamlines and numerical simulation.' *Physical Review E*, **66**, p. 026305.

Grom, F., Kentsch, J., Müller, T., Schnelle, T., and Stelzle, M., 2006. 'Accumulation and trapping of hepatitis A virus particles by electrohydrodynamic flow and dielectrophoresis.' *Electrophoresis*, **27**, pp. 1386–1393.

Hasted, J., 1973. *Aqueous Dielectrics*. Chapman & Hall, London.

Hawkins, B. G., Smith, A. E., Syed, Y. A., and Kirby, B. J., 2007. 'Continuous-flow particle separation by 3D insulative dielectrophoresis using coherently shaped, DC-biased, AC electric fields.' *Analytical Chemistry*, **79**, pp. 7291–7300.

Heeren, A., Luo, C. P., Henschel, W., Fleischer, M., Kern, D. P., 2007. 'Manipulation of micro- and nano-particles by electro-osmosis and dielectrophoresis.' *Microelectronic Engineering*, **84**, pp. 1706–1709.

Hessel, V., Löwe, H., Schönfeld, F., 2005. 'Micromixers – a review on passive and active mixing principles.' *Chemical Engineering Science*, **60**, pp. 2479–2501.

Hodgson, C. E., and Pethig, R., 1998. 'Determination of the viability of Escherichia coli at the single organism level by electrorotation.' *Clinical Chemistry*, **44**, pp. 2049–2051.

Hoettges, K. F., McDonnel, M. B., and Hughes, M. P., 2003. 'Use of combined dielectrophoretic/electrohydrodynamic forces for biosensor enhancement.' *Journal of Physics D: Applied Physics* **36**, pp. L101–L104.

Hölzel, R., 1997. 'Electrorotation of single yeast cells at frequencies between 100 Hz, and 1.6 ghz.' *Biophysical Journal*, **73**, pp. 1103–1109.

Huang, S. H., Wang, S. K., Khoo, H. S., and Tseng, F. G., 2007. 'AC electro-osmotic generated in-plane microvortices for stationary or continuous fluid mixing.' *Sensors and Actuators B*, **125**, pp. 326–336.

Huang, Y., and Pethig, R., 1991. 'Electrode design for negative dielectrophoresis.' *Measurement Science and Technology*, **2**, pp. 1142–1146.

Huang, Y., Hölzel, R., Pethig, R., and Wang, X. B., 1992. 'Differences in the AC electrodynamics of viable and non-viable yeast cells determined through combined dielectrophoresis and electrorotation studies.' *Physics in Medical and Biology*, **37**, pp. 1499–1517.

Hughes, M. P., Morgan, H., Rixon, F. J., Burt, J. P. H., and Pethig, R., 1998. 'Manipulation of herpes simplex virus I by dielectrophoresis.' *Biochemica et Biophysica Acta*, **1425**, pp. 119–126.

Hughes, M. P., and Morgan, H., 1999. 'Dielectrophoretic characterization and separation of antibody-coated submicrometer latex spheres.' *Analytical Chemistry*, **71**, pp. 3441–3445.

Hunter, R. J., 1981. *Zeta Potential in Colloid Science.* Academic, San Diego, CA.

Islam, N., Lian, M., and Wu, J., 2007. 'Enhancing microcantilever capability with integrated AC electroosmotic trapping.' *Microfluid Nanofluid*, **3**, pp. 369–375.

Jones, T. B., 1995. *Electromechanics of Particles.* Cambridge University Press, Cambridge, UK.

Jung, B., Bharadwaj, R., and Santiago, J. G., 2003. 'Thousandfold signal increase using field-amplified sample stacking for on-chip electrophoresis.' *Electrophoresis*, **24**, pp. 3476–3483.

Kakuta, M., Bessoth, F. G., and Manz, A., 2001. 'Microfabricated devices for fluid mixing and their application for chemical synthesis.' *The Chemical Record*, **1**, pp. 395–405.

Kim, B. J., Yoon, S. Y., Sung, H. J., and Smith, C. G., 2007. 'Simultaneous mixing and pumping using asymmetric microelectrodes.' *Journal of Applied Physics*, **102**, pp. 074513.

Kremser, L., Blaas, D., and Kenndler, E., 2004. 'Capillary electrophoresis of biological particles: Viruses, bacteria and eukaryotic cells.' *Electrophoresis*, **25**, pp. 2282–2291.

Lapizco-Encinas, B. H., Simmons, B. A., Cummings, E. B., and Fintschenko, Y., 2004a. 'Insulator-based dielectrophoresis for the selective concentration and separation of live bacteria in water.' *Electrophoresis*, **25**, pp. 1695–1704.

Lapizco-Encinas, B. H., Simmons, B. A., Cummings, E. B., and Fintschenko, Y., 2004b. 'Dielectrophoretic concentration and separation of live and dead bacteria in an Array of insulators.' *Analytical Chemistry*, **76**, pp. 1571–1579.

Laser, D. J., and Santiago, J. G., 2004. 'A review of micropumps.' *Journal of Micromechanics and Microengineering*, **14**, pp. R35–R64.

Lemaire, T., Moyne, C., and Stemmelen, D., 2007. 'Modelling of electro-osmosis in clayey materials including Ph effects.' *Physics and Chemistry of the Earth*, **32**, pp. 441–452.

Liang, E., Smith, R. L., and Clague, D. S., 2004. 'Dielectrophoretic manipulation of finite sized species and the importance of the quadrupolar contribution.' *Physical Review E*, **70** (066617), pp. 1–8.

Liu, F. K., 2007. 'A high-efficiency capillary electrophoresis-based method for characterizing the sizes of Au nanoparticles.' *Journal of Chromatography A*, **1167**, pp. 231–235.

Liu, S., and Dasgupta, P. K., 1992. 'Flow injection analysis in the capillary format using electroosmotic pumping.' *Analytica Chimica Acta*, **268**, pp. 1–6.

Lyklema, J., 1995. *Fundamentals of Interface and Colloid Science, Vol II.* Academic Press, London.

Markx, G. H., Dyda, P. A., and Pethig, R., 1996. 'Dielectrophoretic separation of bacteria using a conductivity gradient.' *Journal of Biotechnology*, **51**, pp. 175–180.

Markx, G. H., Pethig, R., and Rousselet, J., 1997a. 'The dielectrophoretic levitation of latex beads, with reference to field-flow fractionation.' *Journal of Physics D: Applied Physics*, **30**, pp. 2470–2477.

Markx, G. H., Rousselet, J., and Pethig, R., 1997b. 'DEP-FFF: Field-flow fractionation using non-uniform electric fields.' *Journal of Liquid Chromatography Related Technology*, **20**, pp. 2857–2872.

Masuda, S., Washizu, M., and Iwadare, 1987. 'Separation of small particles suspended in liquid by non-uniform travelling field.' *IEEE Transactions on Industry Applications*, **23**, pp. 474–480.

Masuda, S., Washizu, M., and Kawabata, I., 1988. 'Movement of blood cells in liquid by nonuniform travelling field.' *IEEE Transactions on Industry Applications*, **24**, pp. 214–222.

Melcher, J. R., 1966. 'Traveling wave induced electroconvection.' *Physics of Fluids*, **9**, pp. 1548–1555.

Melcher, J. R., and Firebaugh, M. S., 1967. 'Traveling-wave bulk electro-convection induced across a temperature gradient.' *Physics of Fluids*, **10**, pp. 1178–1185.

Michels, D. A., Hu, S., Schoenherr, R. M., Eggertson, M. J., and Dovivhi, N. J., 2002. 'Fully automated two-dimensional capillary electrophoresis for high sensitivity protein analysis.' *Molecular & Cellular Proteomics*, **1**, pp. 69–74.

Morgan, H., and Green, N. G., 2003. *AC Electrokinetics: Colloids and Nanoparticles.* Research Studies Press, Ltd. Baldock, Hertfordshire, England.

Morgan, H., Green, N. G., Ramos, A., and Sanchez, P. G., 2007. 'Control of two-phase flow in a microfluidic system using AC electric fields.' *Applied Physics Letters* 91, 254107.

Morgan, H., Hughes, M. P., and Green, N. G., 1999. 'Separation of submicron bioparticles by dielectrophoresis.' *Journal of Biophysics*, **77**, pp. 516–525.

Morgan, H., Izquierdo, A. G., Bakewell, D., Green, N. G., and Ramos, A., 2001. 'The dielectrophoretic and travelling wave forces generated by interdigitated electrode arrays: Analytical solution using Fourier series.' *Journal of Physics D: Applied Physics*, **34**, pp. 1553–1561.

Mpholo, M., Smith, C. G., and Brown, A B. D., 2003. 'Low voltage plug flow pumping using anisotropic electrode arrays.' *Sensors and Actuators B*, **92**, pp. 262–268.

Müller, T., Arnold, W. M., Schnelle, T., Hagedorn, R., Fuhr, G., and Zimmermann, U., 1993. 'A travelling-wave micropump for aqueous solutions: Comparisons of 1g and μg results.' *Electrophoresis*, **14**, pp. 764–772.

Nguyen, N. T., and Wu, Z., 2005. 'Micromixers – a review.' *Journal of Micromechanics and Microengineering*, **15**, pp. R1–R16.

Oda, R. P., Clark, R., Katzmann, J. A., and Landers, J. P., 1997. 'Capillary electrophoresis is as a clinical tool for the analysis of protein in serum and other body fluids.' *Electrophoresis*, **18**, pp. 1715–1723.

Okhonin, V., Berezovski, M., and Krylov, S. N., 2004. 'Sweeping capillary electrophoresis: A non-stopped-flow method for measuring biomolecular rate constant of complex formation between protein and DNA.' *Journal of the American Chemical Society*, **126**, pp. 7166–7167.

Pethig, R., 1979. *Dielectric and Electronic Properties of Biological Materials.* John Wiley & Sons Ltd., Chichester, UK.

Pohl, H. A., 1978. *Dielectrophoresis.* Cambridge University Press, Cambridge, UK.

Price, J. A. R., Burt, J. P. H., and Pethig, R., 1988. 'Applications of a new optical technique for measuring the dielectrophoretic behaviour of micro-organisms.' *Biochimica et Biophysica Acta*, **964**, pp. 221–230.

Quirino, J. P., Kim, J. B., and Terabe, S., 2002. 'Sweeping: Concentration mechanism and applications to high-sensitivity analysis in capillary electrophoresis.' *Journal of Chromatography A*, **965**, pp. 357–373.

Ramos, A., Morgan, H., and Green, N. G., 1998. 'Ac electrokinetics: A review of forces in microelectrode structures.' *Journal of Physics D: Applied Physics*, **31**, pp. 2338–2353.

Ramos, A., Morgan, H., Green, N. G., and Castellanos, A., 1999. 'Ac electric-field-induced fluid flow in microelectrodes.' *Journal of Colloid and Interface Science*, **217**, pp. 420–422.

Ramos, A., González, A., Castellano, A., Green, N. G., and Morgan, H., 2003. 'Pumping of liquids with ac voltages applied to asymmetric pairs of microelectrodes.' *Physical Review E*, **67**, pp. 056302.

Ramos, A., Morgan, H., Green, N. G., González, A., and Castellanos, A., 2005. 'Pumping of liquids with traveling-wave electroosmosis.' *Journal of Applied Physics*, **97**, pp. 084906.

Ramos, A., González, A., G-Sánchez, P., and Castellanos, A., 2007. 'A linear analysis of the effect of Faradaic currents on traveling-wave electroosmosis.' *Journal of Colloid and Interface Science*, **309**, pp. 323–331.

Reichmuth, D. S., Chirica, G. S., and Kirby, B. J., 2003. 'Increasing the performance of high-pressure, high-efficiency electrokinetic micropumps using zwitterionic solute additives.' *Sensors and Actuators B*, **92**, pp. 37–43.

Rousselet, J., Markx, G. H., and Pethig, R., 1998. 'Separation of erythrocytes and latex beads by dielectrophoretic levitation and hyperlayer field-flow-fraction.' *Colloids and Surfaces A: Physicochemical and Engineering Aspects*, **140**, pp. 209–216.

Sasaki, N., Kitamori, T., and Kim, H. B., 2006. 'AC electroosmotic micromixer for chemical processing in a microchannel.' *Lab on a Chip*, **6**, pp. 550–554.

Saville, D. A., 1997. 'Electrohydrodynamics: The Taylor-Mechler leaky dielectric model.' *Annual Review of Fluid Mechanics*, **29**, pp. 27–64.

Schmitt-Kopplin, P., and Frommberger, M., 2003. 'Capillary electrophoresis – mass spectrometry: 15 years of developments and applications.' *Electrophoresis*, **24**, pp. 3837–3867.

Schwan, H. P., 1968. 'Electrode polarization impedance and measurements in biological materials.' *Annals of the New York Academy of Sciences*, **148**, pp. 191–209.

Sigurdson, M., Wang, D., and Meinhart, C. D., 2005. 'Electrothermal stirring for heterogeneous immunoassays.' *Lab on a Chip*, **5**, pp. 1366–1373.

Stern, O., 1924. 'The theory of the electrical double-layer.' *Zeitschrift für Elektrochemie*, **30**, pp. 508–516.

Stone, H. A., Strook, A. D., and Ajdari, A., 2004. 'Engineering flows in small devices: Microfluidics toward a lab-on-a-chip.' *Annual Review of Fluid Mechanics*, **36**, pp. 381–411.

Stratton, J. A., 1941. *Electromagnetic Theory.* McGraw Hill, New York.

Studer, V., Pépin, A., Chen, Y., and Ajdari, A., 2004. 'An integrated AC electrokinetic pump in a microfluidic loop for fast and tunable flow control.' *Analyst*, **129**, pp. 944–949.

Sun, T., Green, N. G., and Morgan, H., 2007. 'Analytical solutions of AC electrokinetics in interdigitated electrode arrays: Electric field, dielectrophoretic and travelling-wave dielectrophoretic forces.' *Physical Review E*, **76**, pp. 046610.

Tai, C. H., R.-J., Yang, M. Z., Huang, C. W., Liu, C. H., Tsai, and Fu, L. M., 2006. 'Micromixer utilizing electrokinetic instability-induced shedding effect.' *Electrophoresis*, **27**, pp. 4982–4990.

Takamura, Y., Onoda, H., Inokuchi, H., Asachi, S., Oki, A., and Horiike, Y., 2003. 'Low-voltage electroosmosis pump for stand-alone microfluidics devices.' *Electrophoresis*, **24**, pp. 185–192.

Thormann, W., Caslavska, J., and Mosher, R. A., 2007. 'Modelling of electroosmotic and electrophoretic mobilization in capillary and microchip isoelectric focusing.' *Journal of Chromatography A*, **1155**, pp. 154–163.

Tripp, J. A., Svec, F., Fréchet, J. M. J., Zeng, S., Mikkelsen, J. C., and Santiago, J. G., 2004. 'High-pressure electroosmotic pumps based on porous polymer monoliths.' *Sensors and Actuators B*, **99**, pp. 66–73.

Tritton, D. J., 1988. *Physical Fluid Dynamics.* Oxford University Press, New York.

Urbanski, J. P., Thorsen, T., Levitan, J. A., and Bazant, M. Z., 2006. 'Fast AC electro-osmotic pumps with nonplanar electrodes.' *Applied Physics Letters*, **89**, pp. 143508.

Voldman, J., 2006. 'Electrical forces for microscale cell manipulation.' *Annual Review of Biomedical Engineering*, **8**, pp. 425–454.

Wang, C., Nguyen, N. T., Wong, T. N., Wu, Z., Yang, C., and Ooi, K. T., 2007. 'Investigation of active interface control of pressure driven two-fluid flow in microchannels.' *Sensors and Actuators A*, **133**, pp. 323–328.

Wang, P., Chen, Z., and Chang, H. C., 2006. 'A new electro-osmotic pump based on silica monoliths.' *Sensors and Actuators B*, **113**, pp. 500–509.

Wang, X., Wang, X. B., Becker, F. F., and Gascoyne, P. R. C., 1996. 'A theoretical method of electric field analysis for dielectrophoretic electrode arrays using Green's theorem.' *Journal of Physics D: Applied Physics*, **29**, pp. 1649–1660.

Weissinger, E. M. *et al.*, 2004. 'Proteomic patterns established with capillary electrophoresis and mass spectrometry for diagnostic purposes.' *Kidney International*, **65**, pp. 2426–2434.

Wong, P. K., Chen, C. Y., Wang, T. H., and Ho, C. M., 2004a. 'Electrokinetic bioprocessor for concentrating cells and molecules.' *Analytical Chemistry*, **76**, pp. 6908–6914.

Wong, P. K., Wang, T. H., Deval, J. H., and Ho, C. M., 2004b. 'Electrokinetic in micro devices for biotechnology applications.' *IEEE/ASME Transactions on Mechatronics*, **9**, pp. 366–376.

Wu, C. H., and Yang, R. J., 2006. 'Improving the mixing performance of side channel type micromixers using an optimal voltage control model.' *Biomedical Microdevices*, **8**, pp. 119–131.

Wu, H. Y., and Liu, C. H., 2005. 'A novel electrokinetic micromixer.' *Sensors and Actuators A*, **118**, pp. 107–115.

Wu, J., Ben, Y., and Chang, H. C., 2005a. 'Particle detection by electrical impedance spectroscopy with asymmetric-polarization AC electroosmotic trapping.' *Microfluid Nanofluid*, **1**, pp. 161–167.

Wu, J., Ben, Y., Battigelli, D., and Chang, H. C., 2005b. 'Long-range AC electro-osmotic trapping and detection of bioparticles.' *Industrial Engineering Chemistry Research*, **44**, pp. 2815–2822.

Wu, J., Lian, M., and Yang, K., 2007. 'Micropumping of bio-fluids by alternating current electrothermal effects.' *Applied Physics Letters*, **90**, pp. 234103.

Yang, J., Huang, Y., Wang, X. J., Wang, X. B., Becker, F. F., and Gascoyne, P. R. C., 1999. 'Dielectric properties of human leukocytes subpopulations determined by electrorotation as a cell separation criterion.' *Biophysical Journal*, **76**, 3307–3314.

Yao, S. H., Hertzog, D. E., Zeng, S. L., Mikkelsen, J. C., and Santiago, J. G., 2003. 'Porous glass electroosmotic pumps: Design and experiments.' *Journal of Colloid and Interface Science*, **268**, pp. 143–153.

Ye, M., Hu, S., Schoenherr, R. M., and Dovichi, N. J., 2004. 'On-line protein digestion and peptide mapping by capillary electrophoresis with post-column labelling for laser-induced fluorescence detection.' *Electrophoresis*, **25**, pp. 1319–1326.

Zeng, S. L., Chen, C. H., Mikkelsen, J. C., and Santiago, J. G., 2001. 'Fabrication and characterization of electroosmotic micropumps.' *Sensors and Actuators B*, **79**, pp. 107–114.

Zeng, S. L., Chen, C. H., Santiago, J. G., Chen, J. R., Zare, R. N., Tripp, J. A., Svec, F., and Fréchet, J. M., 2002. 'Electroosmotic flow pumps with polymer frits.' *Sensors and Actuators B*, **82**, pp. 209–212.

Zhao, H., White, L. R., and Tilton, R. D., 2005. 'Lateral separation of colloids or cells by dielectrophoresis augmented by AC electroosmosis.' *Journal of Colloid and Interface Science*, **285**, pp. 179–191.

Zhou, X. F., Markx, G. H., Pethig, R., and Eastwood, I. M., 1995. 'Differentiation of viable and non-viable bacteria biofilms using electrorotation.' *Biochimica et Biophysica Acta*, **1245**, pp. 85–93.

Zhou, X. F., Markx, G. H., and Pethig, R., 1996. 'Effect of biocide concentration on electrorotation spectra of yeast cells.' *Biochimica et Biophysica Acta*, **1281**, pp. 60–64.

Zhou, X. F., Burt, J. P. H., and Pethig, R., 1998. 'Automatic cell electrorotation measurements: Studies of the biological effects of low-frequency magnetic fields and of heat shock.' *Physics in Medicine and Biology*, **43**, pp. 1075–1090.

6

EWOD DROPLET MICROFLUIDIC DEVICES USING PRINTED CIRCUIT BOARD FABRICATION

Chang-Jin 'CJ' Kim and Jian Gong

6.1 Droplet (digital) microfluidics

6.1.1 *Handling of liquids at the sub-millimeter scale*

In microfluidics, a fluid is typically handled by flowing it continuously through channels of inner diameters near or below a millimeter. The flow is usually maintained by a pressure and can be regulated through pumps and valves placed mostly outside the microdevices or by an electrokinetic means regulated through an electric potential. However, recent advances have shown a fluid can also be handled as discrete packets (e.g. droplets) either in a confined geometry or on an open surface. The droplets are usually moved around by actuating them locally and individually or by suspending them in an immiscible carrier fluid that flows continuously. Our interest in microfluidics is the handling of liquids as sub-millimeter droplets by actuating the individual droplets using the mechanism of electrowetting-on-dielectric (EWOD). In particular, we describe a type of EWOD device that allows a full and independent electric access to all the positions on chip with a reasonable cost. While the device is fabricated mostly using the commercial printed circuit board (PCB) manufacturing, a series of post-PCB fabrication processes are described to provide a range of EWOD performances along with the accompanying fabrication complexities (i.e. costs).

6.1.2 *Continuous microfluidics*

In most microfluidic systems, fluids are manipulated inside channels with the feature size (e.g. inner diameter) in micrometers (typically from ~10 μm to ~1 mm), flowing continuously (Gravesen *et al.* 1993; Whitesides and Stroock 2001). Two essential functions are required to manipulate fluids: methods to drive fluids (e.g. pump) and ways to regulate their movement (e.g. valves). The common practice today is to complete the system by connecting the microfluidic device to external parts and equipment (e.g. pumps, valves, tubes, and electronic instrument). Although the microfluidic chip is miniaturized down to millimeters, the system is not. Although this hybrid approach provides an easy solution and is adequate for many applications, efficient and sophisticated fluid manipulations are often challenged because of, for example, the large dead volume and power consumption. To miniaturize the entire system, many investigators have been developing micropumps and microvalves to go along with the microchannels, most ideally on a chip.

Many mechanical and non-mechanical microdevices have been developed over the last two decades to serve as micropumps and microvalves. The earliest microfluidics devices were essentially micron size versions of pumps, valves, channels, and other familiar components used in macro world (Shoji and Esashi 1994). The fluid was pumped in a microchannel connected to reciprocating micromembranes. By controlling the operation of micromembranes, the device could propel the fluid into one direction of microchannels. The microvalve was also built in the same fashion to totally close or open the microchannels and regulate the fluid flows. A variety of driving mechanisms have been explored to operate micromembranes, including pneumatic (Thorsen *et al.* 2002), piezoelectric (Linnemann *et al.* 1998), electrostatic (Zengerle *et al.* 1995), thermopneumatic (van de Pol *et al.* 1990), electromagnetic (Zhang and Ahn 1996), bimetallic (Yang *et al.* 1996), and shape memory alloy (Benard *et al.* 1998). Pumping with membranes is popular because the pumped medium is physically isolated from the actuation mechanism so not restricted by the actuation mechanism. However, performance and reliability (e.g. leakage and wear) are difficult challenges in using micromechanical parts to pump and regulate fluids, not to mention the difficulties and costs of fabricating and controlling such complex mechanical systems.

To avoid the need to develop micromechanical devices and often to be completely free of any moving parts, many investigators explored non-mechanical mechanisms to pump fluids, from the advent of MEMS technology. Unlike mechanical micropumping, which physically compresses the fluid to pressurize it at one end, most non-mechanical pumps use different working principles to generate a fluid motion. Several fluidic driving mechanisms have been demonstrated: electrohydrodynamic (EHD) pumps (Bart *et al.* 1990), electroosmotic flow (EOF) (Selvaganapathy *et al.* 2002), magnetohydrodynamic (MHD) pumps (Lemoff and Lee 2000), electrochemical pumps (Bohm *et al.* 1999), acoustic-wave pumps (Yu and Kim 2002), and more. Researchers have also manipulated liquids in a passive way by creating on the solid surface a pattern of alternate wettabilities (Gau *et al.* 1999; Kataoka and Troian 1999).

6.1.3 *Droplet-based microfluidics*

Recently, droplet-based (often called digital) microfluidic systems have attracted increasing attention. In droplet microfluidics, individual droplets typically in a range between picoliters and microliters are manipulated. Droplet microfluidic systems have several advantages over traditional continuous flow counterparts. By manipulating droplets, the common problem of dead volume is eliminated. Moreover, unlike the continuous microfluidics, mixing is fast when small droplets are merged and can be performed in an aliquot fashion by a precisely prescribed combination of discrete droplets. It has also been shown that a single cell or molecule can be encapsulated in an extremely small (e.g. a picoliter) droplet, so that particular biological events can be induced and monitored (He *et al.* 2005). To manipulate droplets not by a pressure in a microchannel but in an open gap or on an open surface, however, the droplets need to be driven individually on active control sites. Note that there are always interfaces between the droplet and the surrounding environment, whether as liquid–liquid or liquid–gas. Considering the high population of surfaces per given fluid volume to move in micro scale, one can see that using surface tension as the source of force to drive fluidic objects in micro scale would be an attractive proposition (Kim 2000).

6.1.4 *Surface tension at the micro scale*

Surface tension is an inherently dominant force in micro scale. Scaling analysis reveals that surface tension is a unique type of force, which scales directly to the length of the subject. It becomes dominant over pressure (surface force) or mass (body force) when the dimension of interest shrinks down to sub-millimeter range. In the development of microdevices, the disproportionately large surface force has been a serious hindrance against the fabrication of freed elements (stiction problem) and also against the movement of the freed element (adhesion friction).

On the other hand, researchers have been aspiring to take advantage of the dominant surface force in designing devices inherently effective in micro scale (Jun and Kim 1998; Kim 2000; Lee and Kim 2000; Tseng *et al.* 2002). For instance, surface tension could be used to keep small droplets in position against vibration for micromechanical switches (Latorre *et al.* 2002), perform fluidic self-assemblies (Srinivasan *et al.* 2001), regulate liquid flows as a virtual check valve for an inkjet head (Tseng *et al.* 2002), or simply to confine fluids (Lam *et al.* 2002). The recent advancement in superhydrophobic surfaces is another example of surface tension being used productively in micro scale (Choi and Kim 2006).

Although surface tension has been passively used to confine or stabilize fluids effectively in micro scale as summarized above, it has also been used to actuate and create motions despite the increased challenges of implementation. To pump liquid by surface tension for a general engineering purpose, large and reversible changes in the surface tensions on a time scale of seconds or less are necessary (Rosslee and Abbott 2000). An obvious way to change liquid surface tension would be to use surfactants. Researchers have demonstrated that the properties of the surfactants can be controlled by an electrochemical (i.e. oxidation of the surfactants) method (Gallardo *et al.* 1999) or by using UV light (i.e. change of aggregation state of the surfactant) (Shih and Abbott 1999). As a result, the liquid surface tension was reversibly changed. The main disadvantage of chemical and topographical patterns, however, is their static nature, which prevents an active and dynamic control. Considerable work has been devoted to the development of surfaces with controllable wettability—typically coated with self-assembled monolayers. However, the degree of switchability, the speed of switching, the long-term reliability, and the compatibility with a variety of environments that have been achieved so far are not suitable for most practical applications.

The surface tension of a solid surface coated with a proper self-assembled monolayer (SAM) can be controlled by electric potential (Lahann *et al.* 2003) or light (Ichimura *et al.* 2000). Using the surface tension under a temperature gradient, one can also pump a liquid by traveling bubbles or as liquid droplets in a microchannel (Jun and Kim 1998; Sammarco and Burns 1999). The control of surface tension by electric potential, i.e. electrocapillary or electrowetting, has been proposed as an actuation means to pump liquids (Matsumoto and Colgate 1990a) and implemented as MEMS devices (Lee and Kim 2000; Lee *et al.* 2002). Electrowetting has continued to take a significant stride in micropumping and lab-on-a-chip applications because of their ease of implementation in microfluidic devices. Compared to thermocapillary as a method, electrowetting is much more energy efficient and compatible with most biofluids in terms of temperature. It also responds faster to control signals compared with other methods of modulating the surface tension.

6.2 Electrowetting-on-dielectric (EWOD)

6.2.1 *EWOD phenomenon*

Electrocapillarity, or electrowetting, was first described in detail by Gabriel Lippmann (Lippmann 1875). He showed that the capillary depression of mercury, when in contact with electrolyte solutions, could be varied by applying a voltage between the mercury and electrolyte. In the following hundred years, researchers mainly performed their studies with aqueous electrolytes in direct contact with mercury surfaces or mercury droplets in contact with insulators. A major obstacle to broader applications was electrolysis of electrolyte when applying voltages beyond only a few hundred mV between the mercury and electrolyte. It was not until the 1990s that the electrowetting phenomenon was understood to occur even on an insulating material if a high voltage is applied across the dielectric. Since an insulating layer separates the conductive liquid from the metallic electrode, the electrolysis is prevented, increasing the range of wetting variation and thus widening the window of operations. To distinguish it from the original electrowetting with a long history that was on a metal surface, electrowetting on this relatively new material condition has become known as electrowetting-on-dielectric (EWOD) (Lee *et al.* 2001, 2002).

A simple demonstration of EWOD is shown in Fig. 6.1, where the contact angle can be measured as a function of the applied voltage. A sessile water droplet is placed on an electrode that is covered first by a dielectric layer (e.g. silicon dioxide) and then a hydrophobic layer (e.g. Teflon AF1600). The droplet is contacted with a probe needle from the top. The contact angle θ_0 of water on the hydrophobic Teflon is typically around 120°; the droplet is shown to bead on the surface. When an electric potential is applied between the probe and the electrode below the dielectric, the contact angle decreases depending on the voltage (the minimum angle being dependent on the dielectric material, surface quality, and the liquid); the droplet is shown to wet the surface. The conventional electrowetting can also

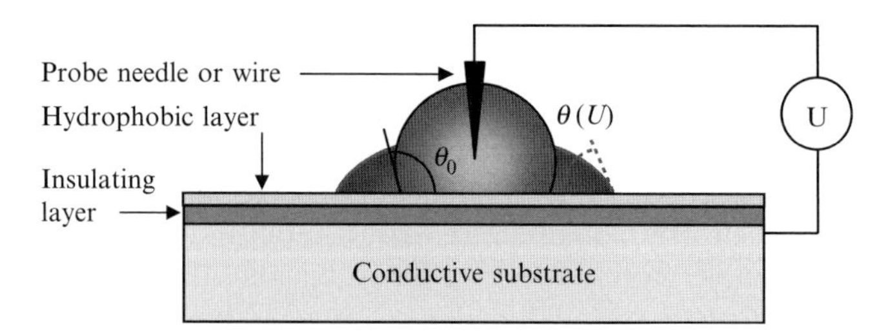

Fig. 6.1: A simple demonstration of EWOD principle with a sessile droplet.

be demonstrated by removing the dielectric layer in the given setup, although the sample needs to be prepared with care to minimize the surface defects. Also, the degree of angle change is much smaller and not as reversible, compared with EWOD.

6.2.2 *Theoretical background of EWOD*

6.2.2.1 *Thermodynamic analysis* To discuss the underlying principle of EWOD, we first define the contact angle and relate the contact angle to the surface property represented by the surface tension on the triple-phase line. When the liquid is not totally wetting the solid surface, it will sit on the surface with a finite angle, known as the contact angle. In the triple-phase region, the gas, liquid, and solid three phases meet together, and three different molecular interactions exist at this point: liquid–gas, solid–liquid, and solid–gas. As shown in Fig. 6.2, Thomas Young (Adamson 1990) proposed to treat the contact angle of a liquid as the result of the mechanical equilibrium of a drop resting on a plane solid surface under the action of three surface tensions: γ_{LG} at the interface of the liquid and gas, γ_{SL} at the interface of the solid and liquid, and γ_{SG} at the interface of the solid and gas.

From a force balance in the horizontal direction, we get:

$$\gamma_{SG} - \gamma_{SL} = \gamma_{LG} \cos\theta. \tag{6.1}$$

Note that Young's equation is an approximation intended for continuum scales. On the molecular scale, equilibrium surface profiles deviate from the

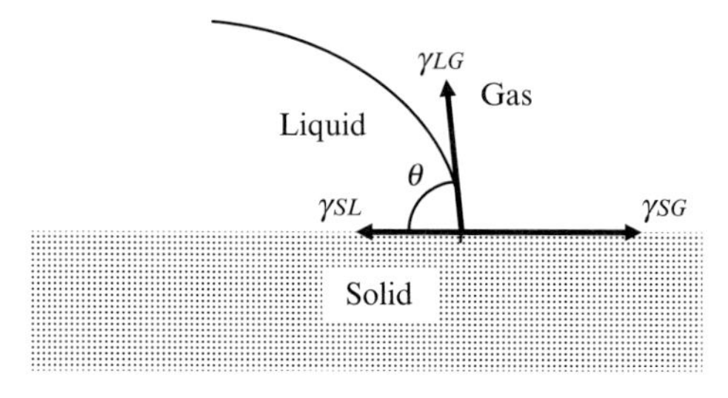

Fig. 6.2: Contact angle of a liquid on a solid surface.

wedge shape in the vicinity of the contact line. Although the profiles that arise are complex, the details are not relevant if one is only interested in the apparent contact angle at the scale much bigger than molecules. On this continuum scale, the contact line can be considered as a one-dimensional object on which the interfacial tensions are pulling. As we will see below, a comparable situation arises in electrowetting.

When an electric potential is applied across a liquid–electrode interface, electrical charges may be accumulated at the interface. Under a constant pressure P and temperature T condition, thermodynamic analysis shows that the electric potential across interface $V_{interface}$ will make the surface tension γ of the interface decrease with a rate of the charge Q per unit area A, or charge density q, as follows. The equation also shows the charge density q in terms of the specific capacitance formed at the interface c and the electric potential $\mathrm{V}_{\mathrm{interface}}$.

$$\left(\frac{\partial\gamma}{\partial V_{\mathrm{interrface}}}\right)_{T,P} = -\frac{Q}{A} = -q = -cV_{\mathrm{interface}} \tag{6.2}$$

6.2.2.2 *Electrowetting on electrode* Although the original electrocapillary experiment of Lippmann considered the surface tension between a *liquid* electrode (mercury) and a liquid electrolyte (Lippmann 1875), the same thermodynamic analysis can be used for a *solid* electrode and a liquid electrolyte (Beni and Hackwood 1981). By applying an electric potential externally, one may vary the charge density of the electric double layer (EDL) at the solid–liquid interface, which in turn changes the surface tension γ_{SL} of the interface. If the external voltage is applied to increase the charge density and thus lower γ_{SL}, the liquid spreads on the solid electrode surface, as illustrated in Fig. 6.3. Since the electric field induces the spreading of liquid (i.e. the liquid wets the surface), this phenomenon is generally

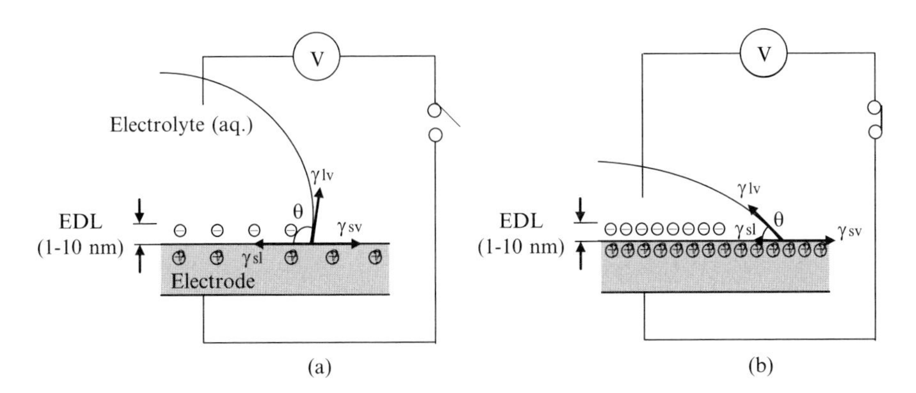

Fig. 6.3: A droplet on a solid electrode spreads when the double layer charge increases by an externally applied electric field (Moon *et al.* 2002). Electrowetting was considered only on metal surfaces until the 1990s.

called 'electrowetting'. Electrowetting has been observed only in this metal-electrolyte configuration until the 1990s.

For the liquid–metal configuration, $V_{\text{interface}}$ is the electric potential across the EDL, c is the specific capacitance of EDL, and q is the specific charge density at EDL. By integrating eqn. (6.2), the following Lippmann equation can be obtained.

$$\gamma - \gamma_0 = -\frac{1}{2}cV_{\text{interface}}^2 = -\frac{1}{2}c(V - V_{PZC})^2 \tag{6.3}$$

where V is the electric potential externally applied to the interface, and γ_O is the surface tension when $V_{interface}$ is made zero by externally applying the potential for zero charge V_{PZC}. By incorporating Young's equation, the above Lippmann's equation can relate to the contact angle θ, resulting in the Lippmann–Young equation, as follows.

$$\cos\,\theta = \cos\,\theta_o + \frac{1}{\gamma_{LG}}\frac{1}{2}cV_{\text{interface}}^2, \tag{6.4}$$

which relates the electric potential at the interface to the contact angle θ at the given voltage and the contact angle θ_o at PZC.

Because only a small voltage drop can be sustained across the EDL, the contact angle change $\Delta\theta$ induced by the conventional electrowetting between liquid–metal is relatively small. Nevertheless, application of this electrowetting mechanism for micropumping (as well as the original liquid–liquid electrocapillary mechanism) has been proposed to the MEMS community in 1990 (Matsumoto and Colgate 1990b) and later successfully reduced to practice as microactuators fabricated by MEMS technologies (Lee and Kim 2000; Lee *et al.* 2001, 2002).

6.2.2.3 *Electrowetting on dielectric (EWOD)* The spread of water on a non-electrode material under a high voltage (over 1000 V) was observed as early as 1980 (Minnema *et al.* 1980), but described as electrowetting in the 1990s (Berge 1993). Electrowetting may occur on an insulating material covering an electrode, as the dielectric layer between the liquid and the electrode emulates the EDL in the conventional electrowetting of liquid–metal. Although a significantly higher voltage is needed to induce the wetting, a thin-film dielectric can sustain much higher voltages than EDL. Overall, more contact-angle changes can be generated reversibly than was possible with the conventional electrowetting on metal. The term electrowetting-on-dielectric (EWOD) is used to clarify that the electrowetting is induced on a dielectric material rather than metal.

Illustrated in Fig. 6.4, EWOD may be considered an electrowetting phenomenon under certain material configurations. The ideal dielectric blocks electron transfer, while sustaining the high electric field at the interface that results in charge redistribution when a potential is applied. By using a hydrophobic dielectric, the large initial contact angle provides room for a large contact angle change $\Delta\theta$ upon electrowetting. Furthermore, by employing a dielectric layer between the

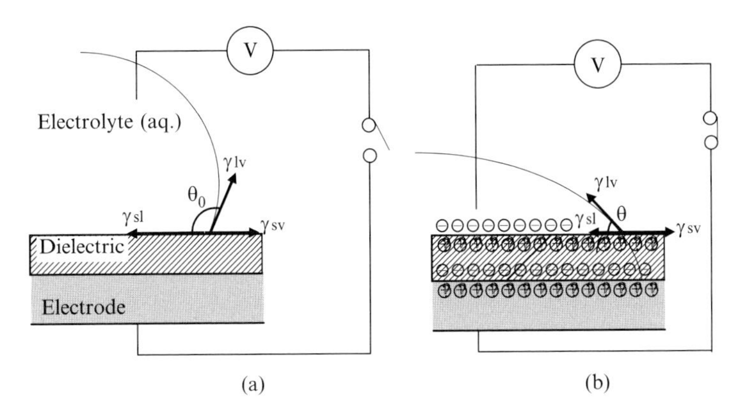

Fig. 6.4: A droplet on solid dielectric spreads when the charge is accumulated on the surface by an externally applied electric field between the liquid and the imbedded electrode. The electrowetting on this materials configuration is called EWOD to emphasize the surface material is a dielectric rather than metal. (Moon *et al.* 2002.)

liquid and electrode, virtually any kind of liquid can be used (barring the change of solid surface by adsorption from the liquid), regardless of the polarization of the interface.

Because the EDL effect is negligible for the EWOD case, the Lippmann–Young equation is now simplified as follows, so that the external voltage can be used directly in the equation as

$$\cos\,\theta = \cos\,\theta_o + \frac{1}{\gamma_{LG}}\frac{1}{2}cV^2, \qquad (6.5)$$

which relates the applied electric potential V to the contact angle θ at the given voltage and the contact angle θ_o at no external potential.

6.2.3 *Moving (pumping) a droplet by EWOD actuation*

To achieve a continuous liquid movement, e.g. liquid pumping and manipulation, an array of driving electrodes is necessary on the surface of an EWOD device to specifically control the surface wettability electrode by electrode. A common configuration of a typical EWOD device contains a liquid droplet placed in the gap between two parallel glass plates for microfluidic operations, as shown in Fig. 6.5. The distance between the plates, or channel height, is determined by the thickness of the spacer placed in between them. The electrodes, made of a conductive material, are placed on both the top and bottom glass plates. The electrode on one plate (top) is an unpatterned reference electrode, covering the whole plate. The reference electric potential (ground) is given to the squeezed droplet by the reference electrode through a thin hydrophobic coating. Driving electrodes are patterned on the other plate (bottom) subsequently coated with a dielectric layer (silicon dioxide, silicon nitride, or polymers) and a thin hydrophobic coating. The reference potential can also be provided from the bottom surface as well by applying both the ground and the actuating voltage on one surface using a coplanar electrode, albeit with a lower actuation force.

When the reference electrode covers the entire area of the top plate, a transparent conductive material, indium-tin-oxide (ITO), is used for the observation of the fluid in the channel. Although the driving electrodes with interdigitated fingers were often used in early days (Pollack *et al.* 2002) for facilitating the droplet movement, the driving electrodes can also be in a simple square shape without losing the droplet pumping performance (Fan *et al.* 2002).

6.2.4 *Basic microfluidic functions*

Using the EWOD device exemplarily shown in Fig. 6.5, we can perform basic microfluidic functions (Cho *et al.* 2003), such as moving merging, cutting, and creating as shown in Fig. 6.6, which are essential for reconfigurable microfluidic circuits. Merging two droplets can be achieved by simply moving two droplets toward each other. Cutting is a more demanding process and requires certain criteria. In brief, with a certain driving voltage and electrode (droplet) size, the channel height should be below a certain value for the cutting to occur. In other words, decreasing the channel height facilitates the cutting process. Creating a droplet from a reservoir is the most challenging among the four and requires various techniques to be effective.

Since electric control of the surface tension is the driving mechanism, discrete pumps, valves, or moving micromechanical parts are not needed, making the system dramatically simple, versatile, and energy efficient (no continuous electric current). Moreover, in this droplet-based microfluidic device, liquid is manipulated between two parallel plates or one single open plate. Microchannels, which usually call for attention to fabricating, bonding and sealing issues, are not required.

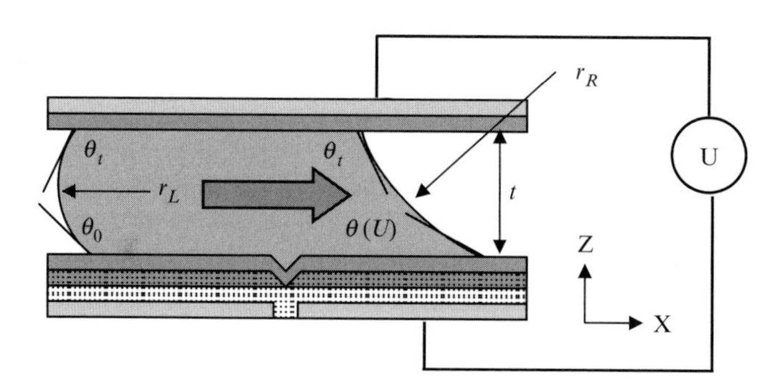

Fig. 6.5: droplet driving mechanism on EWOD device with cross-section view of typical EWOD configuration.

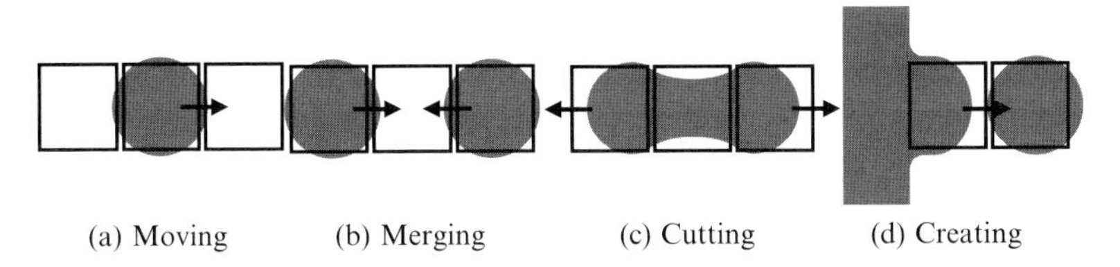

Fig. 6.6: Basic physical microfluidic functions can be performed by EWOD (Cho *et al.* 2003).

In fact, typical fluidic operations such as pumping and mixing can be performed on the same chip by programming electric signals rather than adding more physical structures for each fluidic function.

6.2.5 *Issues with the EWOD phenomenon*

Several important issues affect the performance of EWOD, e.g. how well the droplet moves under a given voltage, how much voltage can be applied before a short-term or long-term failure, etc.

6.2.5.1 *Contact angle saturation* Although the Lippmann–Young equation may suggest that the contact angle can theoretically change to 0° by applying a high enough voltage, the contact angle in reality deviates from the theoretical curve and eventually saturates, as shown in Fig. 6.7. Although the origin of the contact angle saturation is still under investigation by several groups, one should be well aware of

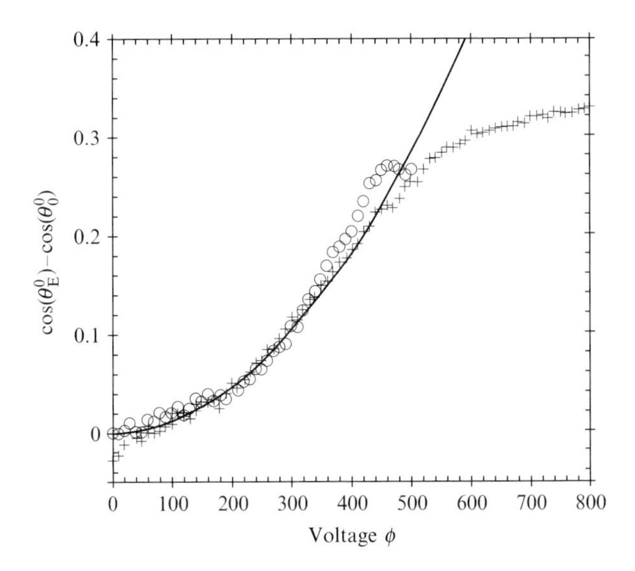

Fig. 6.7: Overlaid plots of contact angle change versus voltage for both AC and DC static contact experiments: (+) AC data; (O) DC data (Blake *et al.* 2000).

the limitation by contact-angle saturation when designing a device actuated by EWOD. The saturation poses a fundamental limitation to the maximum driving force the EWOD actuation can generate.

6.2.5.2 *Materials properties* The dielectric layer is one of the most important material elements in fabricating EWOD surfaces. Recall the Lippmann–Young equation, where the specific capacity is determined by the dielectric constant and thickness of the dielectric. To achieve a large EWOD effect for a given applied voltage, significant efforts have been made for the dielectric materials.

However, the thickness of the dielectric cannot be infinitely small, or else the dielectric strength is not sufficient to prevent an electric breakdown. Eventually a very thin dielectric would approach the case of liquid–metal electrowetting. Another way to increase c is to increase the dielectric constant ε. Moon *et al.* (Moon *et al.* 2002) presented the effect of using different dielectrics, such as SiO_2 ($\varepsilon = 3.8$), Si_3N_4 ($\varepsilon = 7.8$), and BST ($\varepsilon = 180$). By using 700 Å BST as a dielectric, a significant contact angle change (from 120° to 80°) was achieved at a mere 15 V. However, poor deposition quality of these high ε materials made them susceptible to electrolysis (due to many pin holes in the dielectric thin film).

Since there is no perfect dielectric, usually a trade-off needs to be made when designing an EWOD device. When achieving a large contact angle change or when the EWOD driving force is the major concern, a thicker dielectric layer is preferred so that a large voltage can be applied. If a low driving voltage is the main goal, a thinner dielectric should be used with care.

For EWOD performance, the dielectric charging in the dielectric layer and hydrophobic layer is also another important issue. After repeated cycles of high voltage, charges may be accumulated in the dielectric material. These trapped charges induce the electrical field to interact with the droplet and reduce the hydrophobicity of the surface. The EWOD performance would degrade on such a charged dielectric surface. The hydrophobic material such as Teflon®, is a well-known electrets material, which can store a charge for a long time. Although the voltage drop in the Teflon® layer is small, operation over a long time will accumulate a large amount of charge. Driving with AC voltages significantly reduces the charge accumulation and improves the device lifetime. Unfortunately, there is no ultimate solution to resolve the dielectric-charging problem. Systematic research on the charging mechanism on different dielectric materials, hydrophobic coating, and driving voltages should be done in the future.

6.2.5.3 *Surface roughness* The contact angle of the interface where the liquid spreads out, i.e. the advancing contact angle θ_{adv}, is larger than the static contact angle θ, whereas the contact angle where the liquid recedes back, i.e. the receding angle θ_{rec}, is smaller than the static contact angle θ. The difference between the advancing and receding contact angles is called contact angle hysteresis. The inhomogeneous surface conditions are considered the reasons for the hysteresis, surface roughness being an important form of inhomogeneity. As shown in

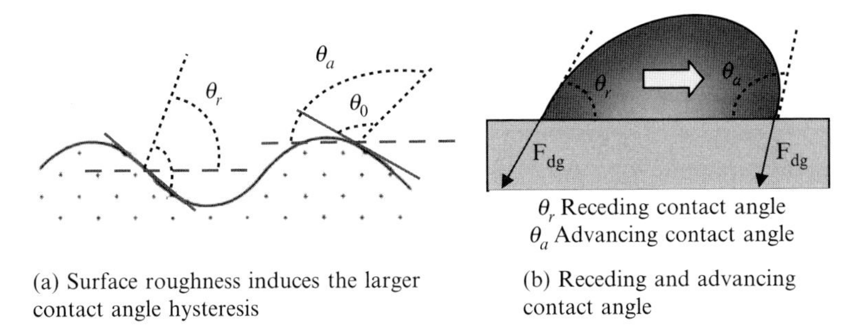

(a) Surface roughness induces the larger contact angle hysteresis

(b) Receding and advancing contact angle

Fig. 6.8: The relationship between contact angle and surface roughness.

Fig. 6.8(a), the rough surface sets the energy barrier for contact line movement. The receding side will be stuck in the valley, and the advancing side will remain in the peak. The surface geometry or the roughness will determine the hysteresis. With increasing surface roughness, the advancing angle increases and the receding angle decreases (Gennes 1985), making the contact angle hysteresis greater—consistent with the general experimental results. Since the static resistance against droplet sliding increases when the difference between the advancing and receding angle increases (essentially proportional to $\gamma(\cos\theta_{adv} + \cos\theta_{rec})$), as illustrated in Fig. 6.8(b), surface roughness is an important factor for droplet actuation. It is important to fabricate an EWOD device with a smooth surface, although the surface defects can be masked by oil (treated or immersed) for a greatly improved result.

6.2.5.4 *Window of driving voltages without dielectric failure* A droplet will move if the driving force is larger than the static resistance. If electrolysis occurs before a droplet starts to move, the droplet cannot be moved by EWOD actuation in the given device. Figure 6.9 shows that below a certain thickness t, for a specific required contact angle change, current leakage (i.e. electrolysis) will occur before the droplet can be moved by EWOD. If the surface has a higher resistance against the droplet movement (e.g. droplets on dry surfaces instead of surfaces covered with oil), larger contact-angle changes are required, and the minimum dielectric thickness should be larger. ('Critical point 2' has a higher voltage than 'Critical point 1', as shown in the figure). For example, PCB substrates have more topography and rougher surfaces than glass or Si substrates, and thus would pose more resistance against droplet movement. As a result, higher operation voltages and thicker dielectric layers are expected for EWOD actuation. In Fig. 6.9, one can consider the Operation Voltage 1 curve is for a glass or Si surface and the Operation Voltage 2 for a PCB surface. For a given surface, however, one can consider Operation Voltage 1 for a dry surface and Operation Voltage 2 for a wet surface (i.e. coated with or immersed in oil). The same fluidic tasks face generally more resistance on a dry surface and require devices with a higher performance. Note that a general goal is to develop an EWOD device

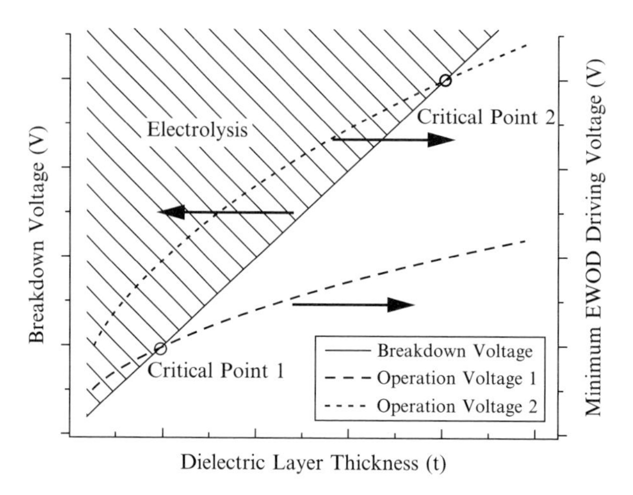

Fig. 6.9: EWOD operation voltage and electrolysis voltage vs. dielectric layer thickness, adapted from Moon *et al.* (Moon *et al.* 2002). Dielectric layer thickness must be larger than critical point to ensure proper EWOD operation. With higher contact angle change needed, the critical dielectric layer thickness increases (such as Critical point 2 > Critical point 1).

capable of operating droplets in air, i.e. of high performance. EWOD chips designed for a droplet-in-air operation also work for a droplet-in-oil operation, but the opposite is not necessarily true.

6.2.5.5 *Driving signals* For EWOD driving signals at a low frequency, the contact angle follows the voltage, while at a high frequency the contact angle follows the RMS voltage. In a theoretical derivation of the EWOD equation, we assumed that the liquid is a perfect conductor. For the DC or low frequency signal, this assumption is true for most aqueous liquids. For a high frequency, however, the droplet behaves more like a capacitor, and the voltage applied on the dielectric layer decrease. Subsequently, the EWOD force or contact angle change will decrease. At a high frequency, however, dielectrophoresis (DEP) plays a role to move the droplet. Jones *et al.* calculated the critical frequency of when the dominant force changes from EWOD to DEP (Jones *et al.* 2004). They also calculated the dependence of the overall force coefficient on frequency. From their calculations, the overall driving force decreases with higher frequency because the EWOD force (line force) is more significant than DEP force (body force). Therefore, DC may be viewed more efficient than AC to drive a droplet. However, from a material point of view, the AC voltage is more robust against a dielectric charging. As a result, overall, a low-frequency AC (around 1 kHz for DI water) is popularly used for EWOD actuation.

In the same surface configuration, one can use both the EWOD and DEP forces by adjusting the driving frequency. Chatterjee *et al.* have explored a droplet

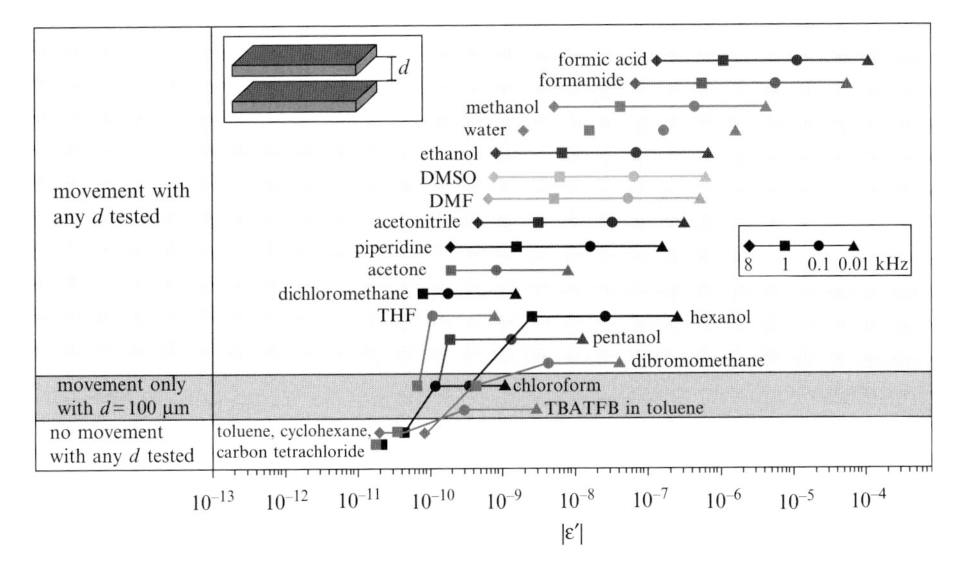

Fig. 6.10: Plot of droplet movement feasibility as a function of the modulus of the complex permittivity ε^* (Chatterjee *et al.* 2006). Liquids in the top group could be moved at the indicated frequencies (symbols) in devices with all spacing tested (d up to 300 μm). Liquid/frequency combinations in the middle group could be moved only in devices with reduced spacing (50 μm $<$ d $<$ 100 μm). Liquid/frequency combinations in the bottom group could not be moved with any spacing tested (50 μm $<$ d $<$ 300 μm). Within each group, the liquids are arranged arbitrarily.

movement on the EWOD configuration with different organic or inorganic chemicals under different parameters such as gap size and frequency (Chatterjee *et al.* 2006), as summarized in Fig. 6.10. They related the feasibility of movement of a certain chemical solution with the function of the modulus of the complex permittivity ε^*:

$$\varepsilon^* = \varepsilon^0 \left(\varepsilon_d - j\frac{\sigma}{\omega\varepsilon_0} \right) \tag{6.6}$$

where ε_0 is permittivity of the vacuum, ε_d is the liquid's dielectric constant, σ is the conductivity, and ω is the driving voltage frequency. They found the solution must have certain complex permittivity to be able to move, but it is not necessary to have contact angle change with voltage, such as m-dichlorobenzene, clearly indicating a DEP response. If the chemical has a certain polarity, it may be actuated by EWOD or DEP, indicating the EWOD device can be used over a wide range of solutions.

6.3 Two-dimensional (2-D) EWOD devices using PCB fabrication

6.3.1 *Droplet microfluidics on a two-dimensional (2-D) playing field*

The main advantages of EWOD-driven droplet microfluidics are in the reconfigurability of the fluidic operation on a given device and the simplicity in the device design and fabrication, compared with the usual pressure-driven continuous microfluidics. Most EWOD devices of today, however, do not exercise their full capability, as they drive droplets along strings of driving electrodes. Although such 1-D configurations fabricated with a single conductive layer for both driving electrodes and connective lines can produce a variety of electrode patterns, each pattern is limited to specific microfluidics protocols, not allowing for one universal device with full reconfigurability. The much-anticipated capability of parallel and multiplexed liquid operations would require driving electrodes laid out in a 2-D configuration, e.g. an M × N array of electrodes. Furthermore, a full operational control of the device requires the ability to electrically access (i.e., reference) each of the electrodes.

For a device fabricated with a single electrode layer, as the number of electrodes in a 2-D pattern increases, the number of conduction lines from the inner electrodes to the external control circuit increases likewise (Fan *et al.* 2002). As shown in Fig. 6.11, these access lines must run in the electrode gaps, making the gap wider. This consequence is not favorable, however, as the driving efficiency of EWOD diminishes when the gap between the neighboring electrodes increases. As a result, in practice, only a small number of electrodes can be designed in a grid array, unless additional layers of electrical conducting lines are introduced. To utilize the full capabilities of the EWOD-driven droplet microfluidics, an innovative chip design and device fabrication are desired.

The most general design for a 2-D electrode array would require a multilayer arrangement of electrical connections, where each of its M × N driving electrodes

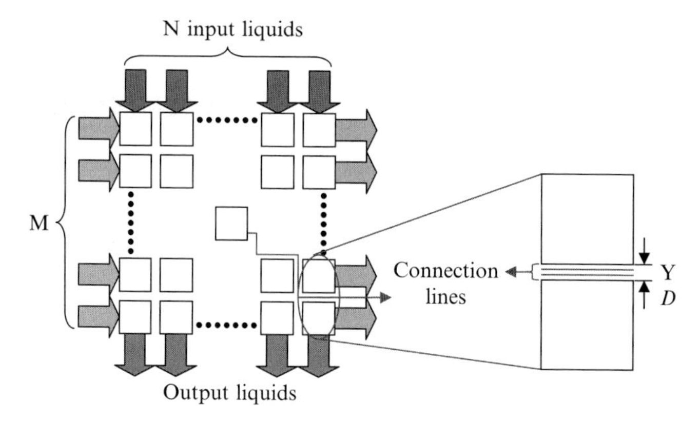

Fig. 6.11: Array size of 2-D driving electrodes is limited for EWOD devices using a single conductive layer (Fan *et al.* 2002).

can be accessed directly and independently through underlying layers of conductive lines. Multiple conducting layer structures can be made using typical integrated circuit (IC) fabrication methods (with special care if high voltage is required) on glass or Si substrates, as demonstrated by Gascoyne *et al.* (2004) with a 32×32 DEP programmable fluidic process chip on a silicon-on-insulator (SOI) IC chip. However, cost is an issue for such microfluidic devices, which require multiple steps of thin-film deposition, lithography, and patterning. Note microfluidic applications require chips with a much larger area and much smaller production volume than the typical IC chips, not being able to enjoy the economy of IC fabrication. Furthermore, since many biomedical applications prefer disposability to avoid cross-contamination, multi-layered chips produced via IC fabrication methods are expected prohibitively expensive except for a few rare applications. In addition, IC-like high-density chips would demand extra costs for fluid or electrical interconnections to complete eventual systems, as no standards exist for packaging. The approach for disposable microfluidic chips, therefore, calls for low-cost chip fabrication methods as well as a system using a convenient and reusable packaging scheme.

To allow for 2-D EWOD operations without fabricating multiple metal layers, Fan *et al.* (2002) have reported a cross-referencing design by orthogonally arranging two parallel plates, each having a single electrode layer, shown in Fig. 6.12(a). By energizing the row and column electrodes with opposite signals, the electrode spots at their intersections become most hydrophilic and thus the droplet moves toward them. However, the simultaneous driving of multiple droplets in this cross-referencing device is limited due to its need for a time-multiplexed driving scheme (Fan *et al.* 2002). The problem of simultaneous driving of droplets on a cross-referencing device is shown in Fig. 6.12(b). Furthermore, since the electrodes on both the top and bottom plates need to be connected to the control circuit, the electrical connections and device packaging are more complex.

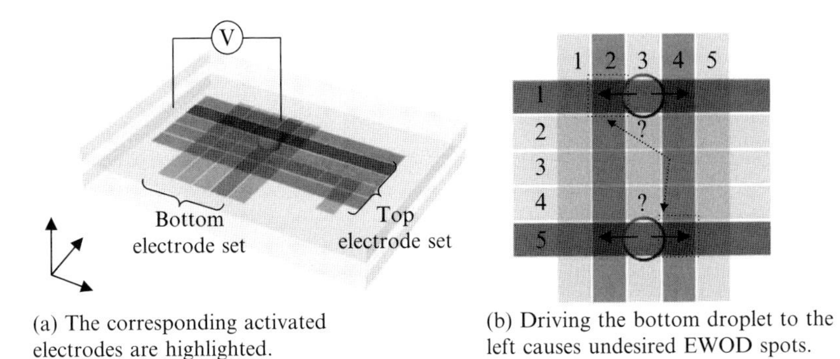

(a) The corresponding activated electrodes are highlighted.

(b) Driving the bottom droplet to the left causes undesired EWOD spots.

Fig. 6.12: A cross-referencing 2-D EWOD device schematically illustrates the driving mechanism and a limitation in driving multiple droplets (Fan *et al.* 2002).

6.3.2 *Multi-layer printed circuit board (PCB)*

The problems of electric addressing as stated above may be solved, for many applications, by using a multi-layer printed circuit board (PCB). It has recently been demonstrated that EWOD droplet microfluidic devices can be fabricated using multi-layer PCB as the substrate. This readily provides the capability of direct referencing as the substrate for 2-D electrode arrays, as first demonstrated by Gong and Kim (2005). PCB, originally built to assemble small electronic components, has made significant advances in recent years to mount IC chips with high density and reliable connections. As shown in Fig. 6.13(a), the copper pads can electrically connect with pins of IC chips by flowing a hot solder on top of them. The fine copper lines interconnect the pads between the IC chips and are prevented to contact with solder by a green solder mask, which is a hard polymer. The typical structure of multi-layer PCB, shown in Fig. 6.13(b), schematically illustrates the multilayer interconnection. Several layers of thin Cu foils are patterned to make fine connection lines in plate and separated by thermal plastic rigid insulating layers. A vertical hole is drilled through the multi-layers, and its inner wall is electroplated with thick Cu to make a vertical connection. Modern commercial PCBs can accommodate up to 30 separate wiring layers and are developing fast to accommodate more layers and smaller feature sizes.

Due to its maturity and low cost, PCB has already been utilized for microfluidic systems. For example, through post-processing steps such as etching, the Cu layer could be machined into diffusers/nozzles, pump chambers (Nguyen and Huang 2001), and thermal flow sensing electrodes (Nguyen *et al.* 2001). Also, hot embossing and multilayer laminations have been used to build entire microfluidic channels, pumps and valves inside the PCB substrate (Fig. 6.14) (Wego *et al.* 2001).

To compare the technologies used for microfluidic device fabrication, Table 6.1 lists typical electrical capabilities (i.e. number of electrode layers), feature sizes, device sizes, and fabrication costs for IC processes, PCB production, a widely used PDMS molding, and in-house (i.e. UCLA Nanolab) cleanroom processes. The in-house cleanroom process is listed here to show the additional cost needed for

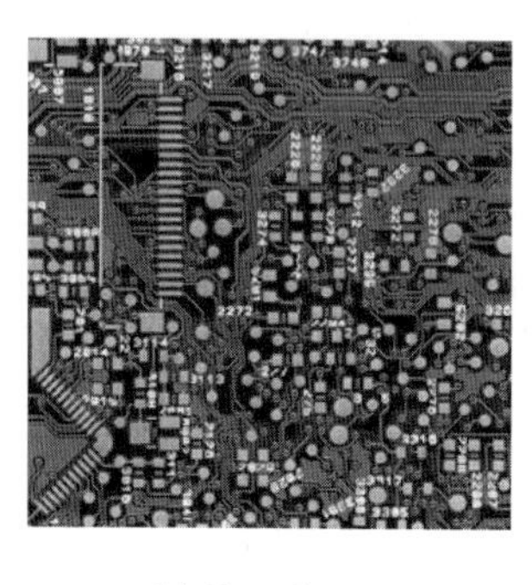

(a) Top view

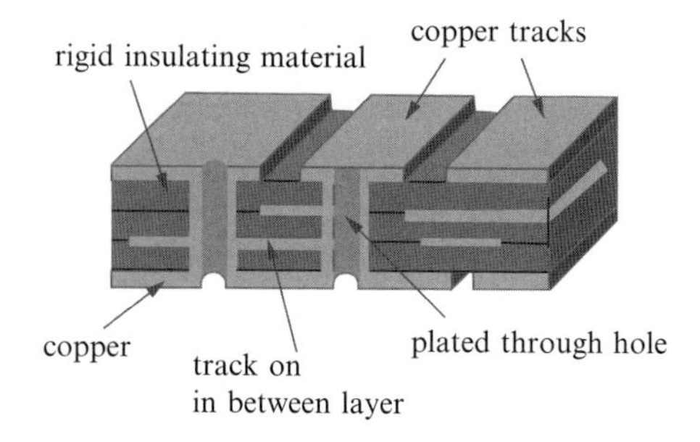

(b) Typical structure of multi-layer PCB (www.ami.ac.uk/courses/ami4809_pcd/unit_01/index.asp)

Fig. 6.13: Multi-layer printed circuit board (PCB).

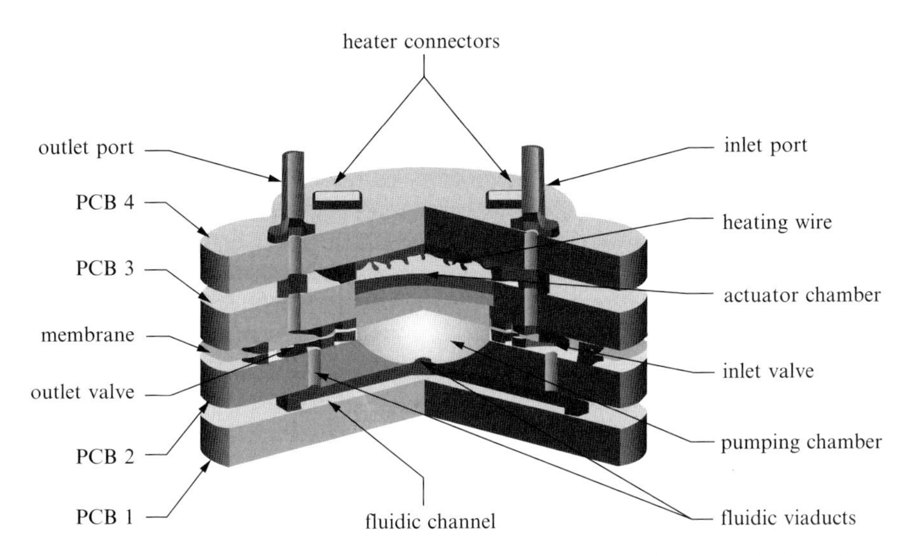

FIG. 6.14: Cross-sectional scheme of the micropumps built with four PCBs (Wego *et al.* 2001).

TABLE 6.1: Comparison of IC, PCB, plastic molding and in-house cleanroom.

Technology	Fabrication method	Typical electrode layers	Typical feature size (μm)	Typical device size (cm)	Typical fabrication cost
IC	Thin-film planarization & lithography	2-10	0.13-1	0.1-1	\$31/cm^2 [a]
PCB	Electroplating & multi-layer lamination	2-30	75-250	1-100	\$21/cm^2 [b]
PDMS molding	Molding & soft lithography	1	10-10000	1-10	Not commercial
In-house clean room	Thin-film planarization & lithography	1-3	2-100	1-10	\$2/cm^2 [c]

[a] Listed for MOSIS AMIS ABN (1.50 μm) two-metal-layer process from http://www.mosis.org/orders/prices/amis/price˙domestic˙amis˙abn.html. For multilayer stacking, the price will increase with the number of layers. Cost for mass production should be lower.
[b] Data for 60000 in^2 production fabrication for four-layer PCB, obtained from http://www.ultimatepcb.com/price˙sample.php.
[c] The price considered the following: 30 min. chemical mechanical polishing (CMP) for thin-film planarization, deposition and patterning of a thin-film metal, and final coating of dielectric layers in the UCLA Nanoelectronics Research Facility (Nanolab) including labor. For multilayer stacking, the price will increase with the number of layers. Cost for mass production should be lower.

post-processing of the PCB substrate to complete EWOD devices. The comparison shows that PCB is a good candidate for microfluidics applications, providing respectable feature sizes at very low fabrication costs. It is worth noting that with the advancement of the PCB technology for IC industry, the existing and fast advancing high-density electrical packaging methods for PCBs will be readily applicable for droplet microfluidics as well.

6.3.3 *Multi-layer PCB as direct-referencing EWOD substrate plate*

Although PCB fabrication cannot provide the feature size and surface quality that the usual IC fabrication does, it instead provides multi-layer conductive patterns as well as via connections between them at a very low cost. While the performance of the PCB-EWOD device strongly depends on the degree of additional cleanroom processes that make the surface smoother and improve the feature size, such post-PCB processes drive the cost of the device higher. The optimum solution is a simplest post-PCB process that provides the EWOD performance needed for a given application. However, the needed performance varies widely, depending on many conditions such as the filler medium (e.g. air, oil), droplet material (e.g. water, blood), and driving signals (e.g. DC, AC). It will be useful if several post-PCB fabrication methods are documented along with the EWOD performance each type provides. This allows a designer to choose the least expensive PCB-EWOD substrate type that satisfies the given application.

6.3.4 *PCB-EWOD substrate: Type 1*

6.3.4.1 *Fabrication* A multilayer PCB can be used directly (i.e. as is) as an EWOD substrate by adapting the surface pads to be the electrodes for EWOD actuation. We have developed and evaluated a four-layer PCB substrate (from a commercial prototype PCB manufacturer) for EWOD application. Fig. 6.15 shows one such device with an 8×8 electrode array composed of 1.5 mm square electrode pads. The gap between electrodes is 75 μm, the connecting via at each electrode is 200 μm in diameter, and the top Cu layer is 25 μm thick (Fig. 6.15 (a,b)). The pattern and arrangement of the EWOD electrodes are shown in Fig. 6.15(d). For electrical connection from the external control circuit, the driving signals are routed from the contact pads surrounding the PCB plate, then through the underlying three metal layers, and then to the driving electrodes on the surface constituting the active EWOD area.

Since the glass transition temperature of FR4 (the polymer used as PCB substrate material) is 185°C, PCBs cannot use high-temperature processes such as plasma-enhanced chemical vapor deposition (PECVD) for silicon dioxide (~300°C), which is often used as the EWOD dielectric layer on glass or Si substrates. Unfortunately, thin films deposited by PECVD processes using lower temperatures tend to exhibit poor dielectric properties, hampering the EWOD performance. Therefore, parylene C, deposited at room temperature, has been chosen as the dielectric material. Parylene has a dielectric constant of 3.2, which is lower than that of silicon dioxide (4.5), but still sufficiently effective for typical EWOD operations. Although parylene itself is hydrophobic (contact angle

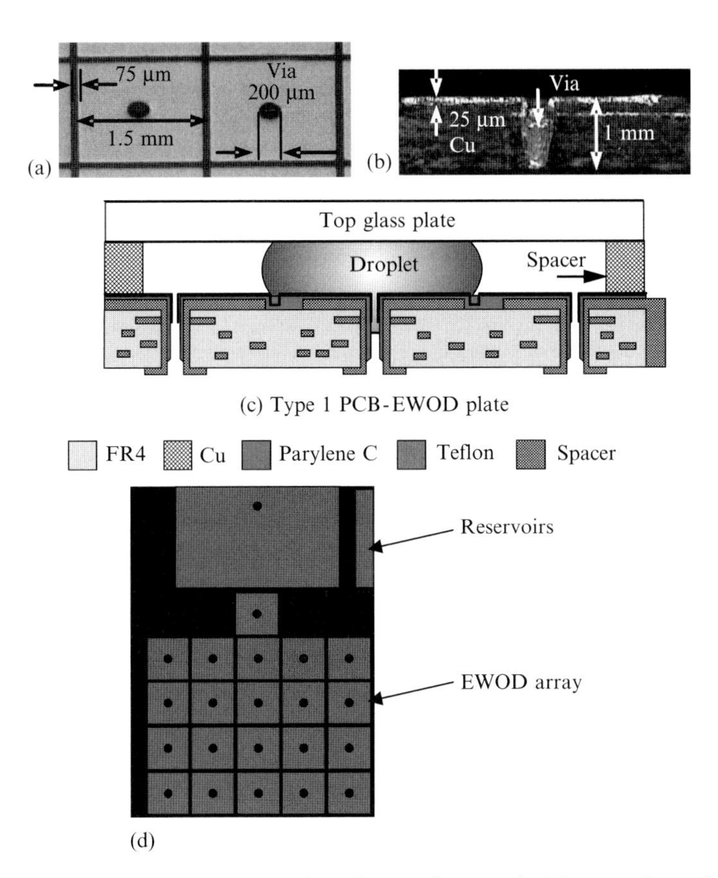

Fig. 6.15: Direct-referencing EWOD device (Type 1) fabricated on four-layer PCB. (a) Optical picture (top view) of the PCB-EWOD substrate, showing two driving electrodes each with a connection via at the center, (b) optical cross-sectional picture of the PCB substrate, showing the top Cu layer and the via, (c) cross-sectional schematic of PCB-EWOD device, showing a Type-1 substrate, a top cover plate in place, and a liquid droplet, (d) optical picture showing a completed PCB-EWOD device, showing a large electrode as liquid reservoir, 5×4 driving electrodes, and a droplet-creation electrode in between.

around 107°), 2000 Å Teflon AF1600 is spin-coated on top of parylene to make the surface more hydrophobic (contact angle around 120°). As shown in Fig. 6.15(c), a glass plate coated with transparent 2000 Å indium tin oxide (ITO) and 2000 Å Teflon is then placed on top of this Type-1 PCB-EWOD substrate to ground the droplets. Appropriate spacers (75–1000 μm) are inserted between the two plates to define the gap height.

Table 6.2: EWOD operation voltage *vs* parylene thickness on Type-1 PCB-EWOD plate. Teflon (2000 Å) is coated on the parylene for all the cases.

Parylene-C thickness	Operation voltage	Transport of droplet along driving electrodes	Electrolysis	Environ-ment
1 μm	70–80 V	Moves between two electrodes but not beyond	Yes	Air
2 μm	~200 V	Moves between two electrodes but not beyond	Yes	Air
4 μm	~400 V	Moves continuously	Some around vias	Air
7 μm	~500 V	Moves continuously	No	Air
7 μm	~200 V	Moves continuously	No	Oil

6.3.4.2 *Experimental results and performance analysis* A series of tests have been conducted for devices made with PCB-EWOD substrate Type 1 to obtain the droplet driving performance in air and oil as shown in Table 6.2. At least 7 μm-thick parylene and 500 V of driving voltage were needed for successful droplet actuation in air. This operation voltage is 10 times larger than that on a glass or Si substrate, on which 2000 Å PECVD oxide is sufficient. Such a high operation voltage may cause electrical shorts on the PCB. Also, it requires a high-voltage source, a special control circuit, and extra safety protection for microfluidic systems. Decreasing the operation voltage is an important challenge for the PCB-EWOD plate. However, the driving voltage can be reduced to around 200 V if the PCB-EWOD plate is immersed in oil. No electrolysis is observed because of the easy movement in oil. This result is reasonably consistent with the oil-filled device that used of 17 μm-thick solder mask as the dielectric (Paik *et al.* 2005).

6.3.4.3 *Limitations and the need for oil* To determine the cause of a very high moving resistance on the PCB substrate, we analyzed the following experimental observations. When we initially placed a droplet between two adjacent electrodes on a PCB-EWOD plate with 1 μm-thick parylene, the droplet could move easily back and forth with 70–80 V of driving voltage, but failed to move further onto the next electrode pad. After careful examination, we learned that the trench between the electrodes prevented the droplet from crossing over. In comparison, electrode pads on a glass or Si substrate, having electrode thicknesses of 1000–2000 Å and gaps of 4–10 μm, allowed a droplet to spread to an adjacent electrode, even without EWOD voltage. This spreading actually enabled EWOD actuation from one electrode to the next. Now, consider the trenches formed in the path of the droplet movement by features on the PCB substrate (i.e. 25 μm thick Cu electrode pads and 75 μm wide gaps between them, as shown in Fig. 6.16). To move between two electrodes, a droplet must first fill in or jump over the trench in order to contact the adjacent electrode, which must then pull the droplet by EWOD. At the same time, the droplet also needs to be pulled off

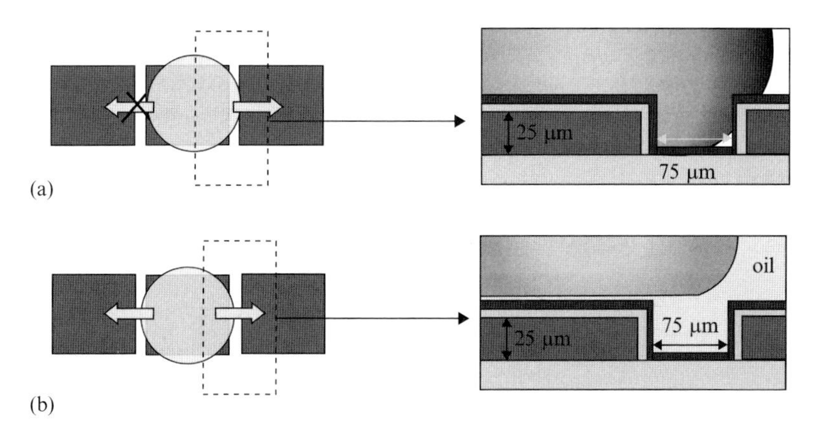

Fig. 6.16: Droplet movement on the PCB-EWOD device (Type 1), illustrating how a large gap and thick electrode impede the droplet moving to the next electrode and how oil masks the problem. (a) Continuous droplet movement was not possible in air with driving voltages as high as 200 V. (b) Continuous droplet movement was possible in oil with driving voltages as low as 60 V.

the previous trench, which poses a significant pullback resistance. As a result, a much larger operation voltage was required to achieve a continuous movement.

Immersing or smearing the device surface with silicone oil can fill trenches and improve EWOD operation performance as shown in Table 6.2 and by Paik *et al.* (2005), but it is not a general solution, as some applications do not allow the use of oil. Interdigitated electrode patterns (Fan *et al.* 2002) would help the droplet transfer between adjacent electrode pads. Unfortunately, any resultant improvement would be quite limited by the large feature size of the PCB fabrication.

6.3.5 *PCB-EWOD substrate: Type 2*

6.3.5.1 *Fabrication* Lower operation voltage being an indication of performance, our approach to improve the PCB-EWOD device is to make the surface flatter and smoother and reduce the gap between the driving electrodes by adding post-PCB microfabrication processes in-house. Since each added step of the microfabrication process significantly increases the cost, we develop three more types of post-PCB processes with increasingly complex fabrication steps, so that users can choose the simplest (i.e. cheapest) type their particular application allows.

The process flow is shown in Fig. 6.17, starting with the PCB substrate as received from the manufacturer. By wet etching (RadioShack® PCB Cu ether: $FeCl_3$), we completely removed the Cu electrodes on top of the PCB, while protecting center holes by patterning PR (AZ 5214) around the center holes and controlling the wet etching time, in order to achieve a flat surface. Then, by

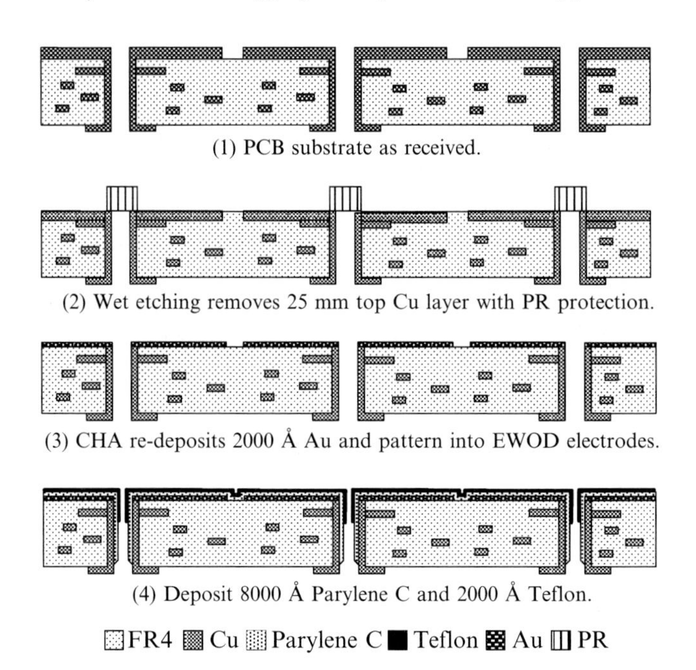

FIG. 6.17: Process flow of post-processing for Type 2 PCB EWOD device.

depositing and patterning a thin metal layer (2000 Å) a thin-film 2-D electrode array was built on top of PCB substrate. Interdigitated electrode patterns were chosen to assist droplet movement. Finally 1 μm parylene and 2000 Å Teflon were deposited to complete the fabrication of PCB-EWOD substrate of Type 2.

6.3.5.2 *Experimental results* The process created a smoother surface with a 0.2 μm step height of the driving electrode (vs. 25 μm for PCB) and smaller electrode gaps of 4 μm by photolithographic patterning (vs. 75 μm for PCB), as shown in Fig. 6.18.

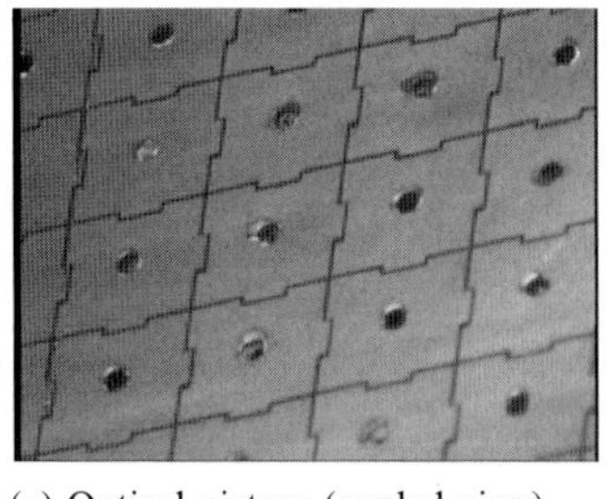

(a) Optical picture (angled view) of Type-2 PCB-EWOD surface after metal (electrode) deposition.

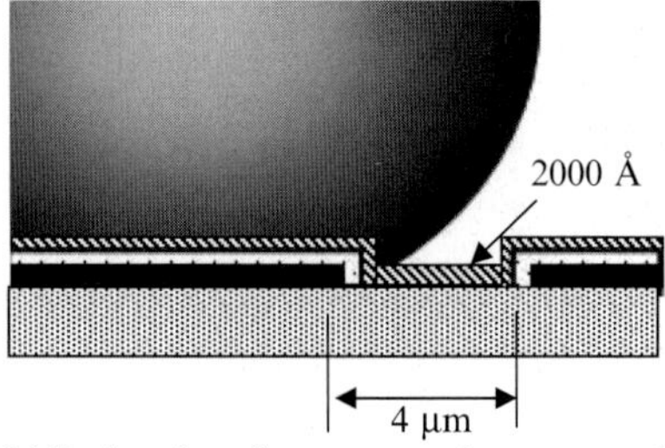

(b) Reduced surface topography compared with Type 1 shown in Fig. 3.7, makes droplet movement easier.

FIG. 6.18: Type-2 PCB-EWOD device surface topography.

A small amount of Cu was still left around the edge of the holes due to the incomplete wet etching, which helped make the connection between the deposited metal layer and Cu inside central holes.

We have tested all of the droplet functions (i.e. creating, moving, cutting, merging droplets) on a Type-2 PCB-EWOD device and found that the driving voltage can be as small as 70 V_{AC} at 1 kHz to move droplets in air. Since the EWOD operation was successful in air, there was no need to test it in oil. However, this improved EWOD performance was still not as good as that on a glass or Si substrate, and cutting and creation of droplets required higher driving voltages and were vulnerable to electrolysis. To improve the performance further, we addressed two more challenges associated with EWOD on PCB substrates, as follows.

6.3.5.3 *Further improvement* First, the connection vias apparently increased the drag against the droplet movement on the surface. During movement, we observed that droplets tend to pin around the vias, substantially increasing the resistance against droplet movement, as shown in Fig. 6.19(a). Furthermore, the via holes, limited by PCB technology to be 200 μm or larger in diameter, are a non-negligible path for liquid to flow into when the distance (i.e. spacer height) between the top and bottom plates is small. The gap height needs to be smaller than a critical value to allow a droplet creation and cutting (Cho *et al.* 2003). The internal pressure of droplets squeezed between the two plates increases as the gap height decreases, and if the gap is too small (i.e. becomes comparable to the via diameter), it may drive the liquid into the via hole.

As shown in Fig. 6.19(b), P_l is the pressure produced by squeezing droplets between two plates to a gap height t, and P_{max} is the maximum (i.e. threshold) pressure above which the droplet fills into the via hole of diameter d. Assuming the advancing contact angles $\cos\theta_a$ are the same at both surfaces and the diameter of the squeezed droplet is much larger than the gap height, the pressure of the droplet is safely below the maximum allowed by:

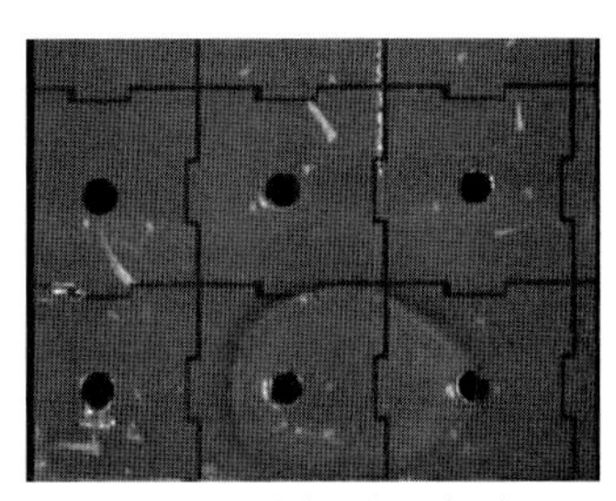

(a) Snapshot of droplet pinning at a via hole at the center of the driving electrode.

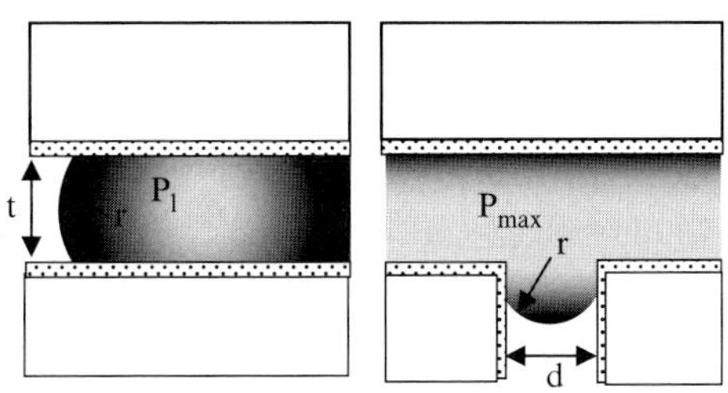

(b) Pressure in the droplet squeezed in the channel (by the gap height) and the maximum pressure before liquid fills into the via hole.

Fig. 6.19: Issues of Type-2 PCB-EWOD device due to the via holes.

$$P_{\max} - P_l = 2\gamma_{\rm lg} \cos \theta_a \left(\frac{1}{t} - \frac{2}{d}\right). \tag{6.7}$$

Note that the cosine value is negative for the current hydrophobic case unless EWOD-actuated. If t is smaller than d/2, the pressure inside the droplet exceeds the maximum allowed, and the droplet would fill into the center hole and tend to pin. In addition to impeding movement, the liquid in the vias caused significant electrolysis because the parylene coating on the corners and inner walls of the vias is not uniform. Considering the difficulties the via holes cause, we conclude that the PCB-EWOD substrate can be improved if the vias are plugged.

Second, the surface of the PCB substrate (general FR4) is rough. After a 2000 Å Teflon coating, the measured surface roughness was around 1–2 μm, which is 1000s of times larger than that of polished glass or Si substrates (only ~1 nm). The advancing and receding contact angles on the Teflon-coated PCB surface were 133° and 96°, respectively, compared with 118° and 107° on Teflon-coated polished glass or Si. In other words, the resistance due to contact-angle hysteresis on the PCB surface was found to be 3–4 times larger than that on the polished glass or Si substrates. Therefore, it is important to improve the surface roughness of PCB-EWOD substrate in order to achieve a higher performance of EWOD operations.

6.3.6 *PCB-EWOD substrate: Type 3*

6.3.6.1 *Fabrication* To improve the performance of PCB-EWOD device further by addressing the above two issues, we developed two additional post-PCB processing methods that involve chemical mechanical polishing (CMP), a widely used technique in IC fabrication to planarize heavily processed surfaces. CMP removes material from uneven topography on a wafer surface until a flat (planarized) surface is created. CMP combines the chemical removal effect of an acidic or basic fluid solution with the 'mechanical' effect provided by polishing with an abrasive material. The CMP system usually has a polishing 'head' that presses the rotating wafer against a flexible pad. The wet chemical slurry containing a micro-abrasive is placed between the wafer and pad. CMP can effectively remove the Cu as tested in the semiconductor industry and has also been used to polish organic polymer materials such as polyarylene (Towery and Fury 1998). A soft polishing pad and a particular polishing slurry should be used to polish the PCB surface. We have tested many combinations of parameters to obtain the best polishing result.

As shown in the process flow of Fig. 16.20, the connection vias on the PCB substrate were filled with silver powder conductive epoxy, as received from the prototype PCB manufacturer. For the first step of post-processing, the Cu layer was lapped down by CMP enough to expose the epoxy filling and PCB surface (FR4). Further, CMP served to polish the PCB surface, now consisting of bare FR4 with silver epoxy dots. After the surface was polished, we deposited and patterned a thin metal layer, and coated a dielectric layer as well as a hydrophobic layer to obtain a high performance for the Type-3 PCB-EWOD plate.

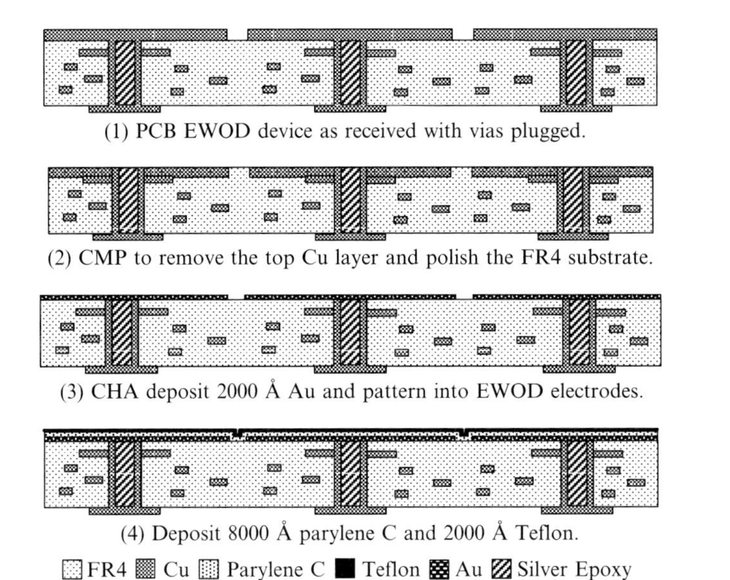

Fig. 6.20: Process flow of post-processing for Type-3 PCB-EWOD device.

6.3.7 *PCB-EWOD substrate: Type 4*

6.3.7.1 *Fabrication* Since FR4 polymer is soft, the polishing process of Type-3 substrate did not work that well. To obtain a highly polished surface, the PCB should be coated with a harder polymer, which then serves as the polished material. Two types of hard polymers were experimented: liquid photo imageable (LPI) solder mask coated by the PCB manufacturer and SU-8 coated in-house. As shown in Fig. 6.21, processing started with CMP lapping of the hard polymer in order to uncover the Cu contact holes, followed by polishing of the resultant Cu and polymer surface. The roughness of a well-polished surface (optimized with different polish slurries and process parameters) was measured as smooth as 100 Åon this Type-4 PCB-EWOD plate. Although this roughness is still ten times larger than that of the polished glass or Si substrates (~10 Å), the contact-angle hysteresis of a water droplet on the Type-4 surface was much less than that of the unpolished devices. Consequently, the driving voltage has been further reduced.

6.3.7.2 *Performance analysis* We made a series of tests on Type-4 PCB-EWOD devices with different dielectric thicknesses and driving voltages to characterize their performances. The droplet movement speeds at different driving voltages were measured with these devices by recording the droplet movement between the EWOD electrodes. As summarized in Fig. 6.22, we have tested five cases: (1) glass substrate with 0.5 μm parylene C in air, (2-4) Type-4 PCB-EWOD plates with 0.5 μm, 0.8 μm and 1.2 μm parylene C in air, and (5) Type-4 PCB-EWOD plate

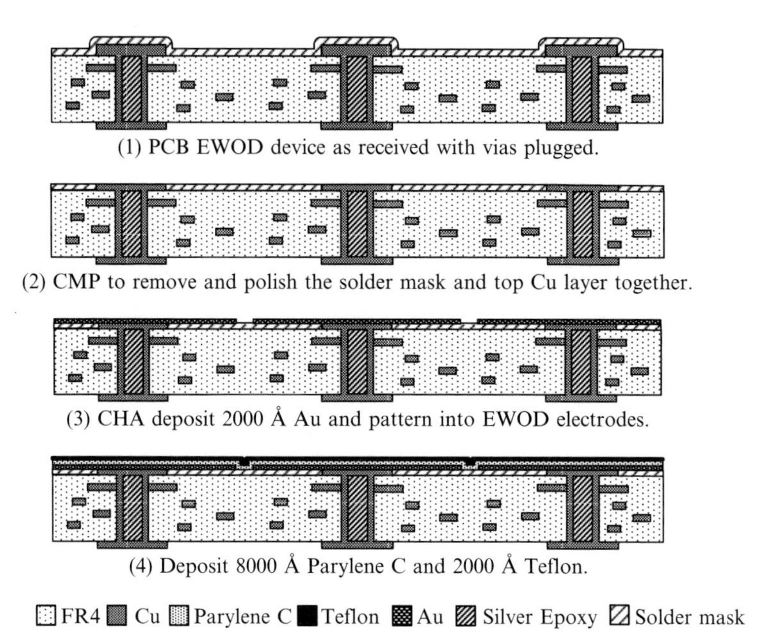

Fig. 6.21: Process flow of post-processing for Type-4 PCB-EWOD device.

with 0.5 μm parylene C immersed in 1 cSt silicone oil (Clearco Products Co.). All the substrates were coated with 2000 ÅTeflon as the top hydrophobic coating. We documented the threshold voltages starting the droplet movement on each substrate. We also confirmed that the speed of the droplet movement increased quickly with voltage, since the EWOD driving force increases parabolically with the voltage. The device had a much smaller driving voltage when immersed in oil, even smaller than glass substrate in air, since silicone oil lubricates the surface and smoothens the rough surface. The results show the Type-4 PCB-EWOD device has a comparable performance with glass substrate with the same dielectric thickness. A device with a thicker dielectric layer would require a higher driving voltage but also reduce the possibility of electrolysis. For the driving voltage range of 50–200 V, the existing high-voltage control circuits for a portable microfluidic system can be used without modification, giving us the freedom to optimize the PCB-EWOD device with the best performance and minimal electrolysis.

6.3.8 *Comparison of PCB-EWOD substrate Types 1–4*

Types 1 to 4 described PCB-EWOD substrates with increasing complications in fabrication processes in return for the increasingly improved EWOD performances. (1) Type 1 is a substrate with a dielectric and a hydrophobic layer deposited directly onto an as-received regular PCB substrate (Fig. 6.15). (2) For Type 2, the thick Cu layer on PCB was removed by wet etching, and then thin electrodes were deposited and patterned, followed by dielectric and hydrophobic layer coatings

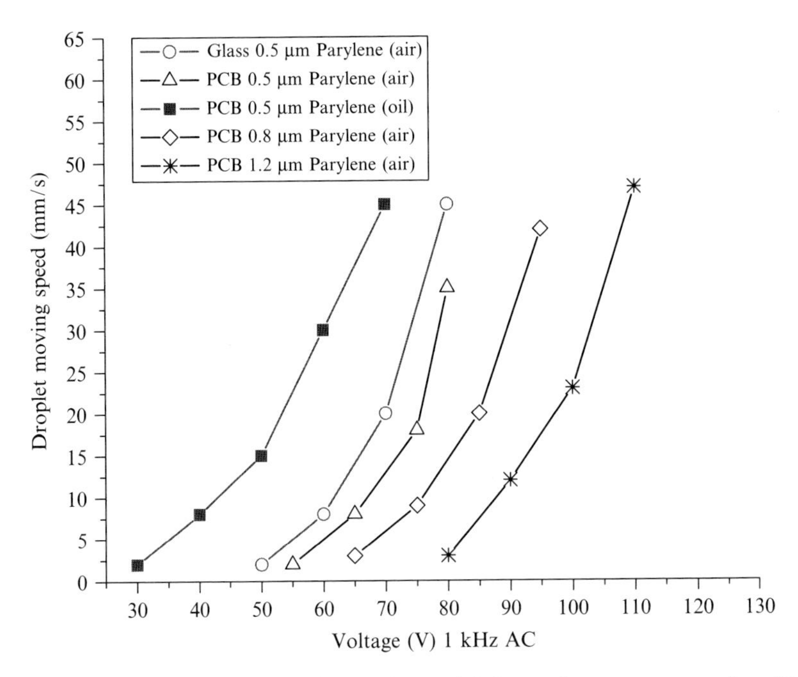

Fig. 6.22: Moving speed of droplet in air vs. driving voltages measured on Type-4 PCB-EWOD devices with different dielectric thicknesses. Tests of Type-4 PCB-EWOD device in oil environment and a glass-EWOD device in air are added for comparison.

(Fig. 6.17). (3) For Type 3, the top Cu layer was removed from a PCB that had its vias filled as received. After polishing the surface by CMP, thin electrodes were deposited and patterned, followed by dielectric and hydrophobic layer coatings (Fig. 6.20). (4) For Type 4, the top Cu layer was removed from a PCB that had filled-vias and a hard polymer coated as received. After polishing the surface by CMP, thin electrodes were deposited and patterned, followed by dielectric and hydrophobic layer coatings (Fig. 6.21). These four methods produced increasingly superior PCB-EWOD substrates (i.e. through flatter surface, reduced roughness, and lower contact-angle hysteresis, all leading to higher EWOD performance), but with increasing fabrication costs.

Table 6.3 lists and compares the surface conditions that the four different methods produced, as well as those of polished glass or Si substrates. While fabrication of the best-performing PCB-EWOD substrate costs the most, we estimate that, if mass-produced, it is still economical even for some disposable applications. Different methods are reported so that one may choose a cheaper method for less demanding microfluidic operations (such as droplet translations only) or droplets in an oil or oil-lubricated environment.

TABLE 6.3: Different direct-referencing EWOD substrates *vs* surface property, EWOD performance and fabrication cost.

Fabrication method	Surface topography	Surface roughness	Contact-angle hysteresis	Driving voltage[a]	Moving speed[b]	Major fabrication cost
As-fabricated PCB (Type 1)	75 μm trench; 200 μm via	15000 Å	34°	500 V[c]	N/A	PCB substrate
Wet etching (Type 2)	Flat surface; 200 μm via	15000 Å	34°	70 V	6 mm/s	PCB + one-layer electrode
CMP + filled vias (Type 3)	Flat surface	5000 Å	28°	60 V	7.8 mm/s	PCB + CMP + one-layer electrode
CMP + filled vias + polymer (Type 4)	Flat surface	100 Å	20°	55 V	13 mm/s	PCB + CMP + one-layer electrode
Glass or Si substrate	Flat surface	10 Å	13°	50 V	20 mm/s	Glass/Si + multilayer electrodes + CMP

[a] Driving voltage is the minimum voltage needed to move droplets on 0.5 μm thick parylene device.
[b] Moving speed is measured on 0.5 μm thick parylene devices while driving with 70 V_{ac} at 1 kHz.
[c] 500 V is the minimum voltage for working devices with 7 μm thick parylene.

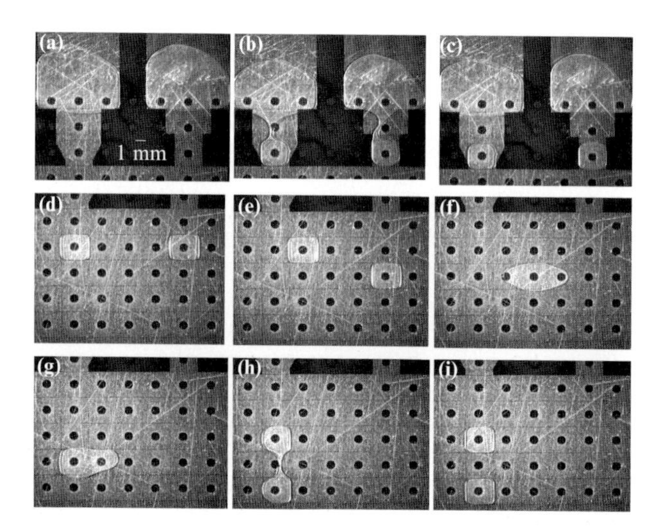

FIG. 6.23: Demonstration of sequential droplet microfluidic operations on 8 × 8 electrode grid of a PCB-EWOD device of Type 4, using 80 V_{AC} at 1 kHz. (a–d) show that two droplets are created from two reservoirs and then transported on the 2-D surface. (e–g) Subsequently they are merged, mixed, and moved as one large droplet together. (h, i) Finally they are cut into two small droplets.

A PCB-EWOD device with an 8 × 8 electrode arrays and on-chip reservoirs was packaged by an LGA socket and evaluated by testing the four essential microfluidic operations of creating, moving, cutting, and merging droplets. Type-4 PCB-EWOD plate as shown in Fig. 6.21 has the smoothest surface and the best EWOD performance. The spacers between the transparent top plate and PCB-EWOD substrate at the bottom defined the gap height of 100 μm to allow droplet creation and cutting. As shown in Fig. 6.23(a–i), with 80 V_{AC} at 1 kHz, comparable to that on glass or Si substrates, multiple droplets were simultaneously created from the on-chip reservoirs, moved on an arbitrary route defined by the user, mixed together and cut into smaller droplets again, confirming droplet microfluidic operations comparable to the regular glass- or Si-based EWOD plates. Type-3 and Type-2 devices were operational albeit with inferior performance. Type-1 device, on the other hand, needed to be immersed in oil for successful EWOD operations.

6.3.9 *Summaries of PCB-EWOD device*

A direct-referencing EWOD plate for digital microfluidics has been developed using a multi-layer PCB substrate. Four different post-PCB processing methods have been developed, resulting in PCB-EWOD plates with varying performance and fabrication cost. On the high-performance PCB-EWOD plates, the microfluidic operations of droplets were comparable to those on polished glass or Si substrates. The low cost of commercial PCB substrate fabrication, and comparable costs for the post processing, make the PCB-EWOD plate viable even for disposable applications, while being two-dimensional (i.e. reconfigurable).

6.4 Reconfigurable droplet (digital) microfluidics

EWOD is an excellent droplet driving mechanism, which can be applied to a wide range of liquids, especially in sub-millimeter scale. Its electrical origin provides an easily accessible control method to design the eventual microfluidic system. The electrolysis and dielectric charging are two main failure mechanisms for EWOD operations. Choosing high quality and high dielectric constant dielectric materials and the use of AC voltages are the keys to achieve long-term EWOD operations. When allowed, filling the device with oil makes the droplet movements much easier. Combining droplet functions such as generation, moving, merging and cutting, into a programmable, versatile microfluidics platform with multiple input and output can be developed based on the EWOD mechanisms.

The state-of-the-art is still not mature enough to legitimately call the devices reported in the field more than *droplet* microfluidics. When fully developed, however, the reconfigurable *digital* microfluidic circuits are expected to present a micro liquid handling technique first envisioned by Lee *et al.* (Lee *et al.* 2001) on one universal chip. To develop an EWOD droplet microfluidic system, one may divide the whole system into several levels of subsystems with the physical and control interfaces between them. As shown in Fig. 6.24, the core of the system is the EWOD chip. The device package around the EWOD chip helps connect it with

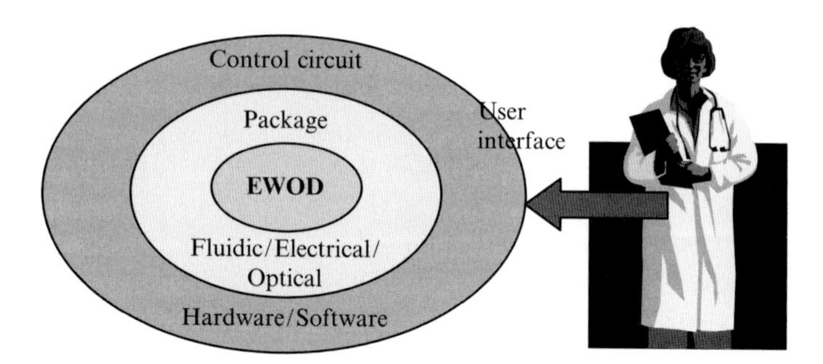

Fig. 6.24: Conceptual description of a reconfigurable droplet (digital) microfluidic system.

the outside world through optical, electrical, and mechanical means. The outer part of the system is the control circuit board with the necessary hardware and software. Last, the user interface of the control board bridges the EWOD microfluidics with the end user.

References

Adamson, A. W., 1990. *Physical Chemistry of Surfaces*, 5th edn., Wiley, New York.

Bart, S. F. *et al.*, 1990. 'Microfabricated electrohydrodynamic pumps.' *Sensors & Actuators A – Physical*, **A21** (1–3), pp. 193–207.

Benard, W. L. *et al.*, 1998. 'Thin-film shape-memory alloy actuated micropumps.' *Journal of Microelectromechanical Systems*, **7**, pp. 245–251.

Beni, G., and Hackwood, S, 1981. 'Electro-wetting displays.' *Applied Physics Letters*, **38** (4), p. 3.

Berge, B, 1993. 'Electrocapillarity and wetting of insulator films by water.' *Comptes Rendus de l'Academie des Sciences Series II*, **317**, pp. 157–163.

Blake, T. D., Clarke, A, and Stattersfield, E. H., 2000. 'An investigation of electrostatic assist in dynamic wetting.' *Langmuir*, **16**, pp. 2928–2935.

Bohm, S, Olthuis, W, and Bergveld, P, 1999. 'An integrated micromachined electrochemical pump and dosing system.' *Biomedical Microdevices*, **1** (2), pp. 121–130.

Chatterjee, D. *et al.*, 2006. 'Droplet-based microfluidics with nonaqueous solvents and solutions.' *Lab on a Chip*, **6**, pp. 199–206.

Cho, S. K., Moon, H, and Kim, C. J., 2003. 'Creating, transporting, cutting, and merging liquid droplets by electrowetting-based actuation for digital microfluidic circuits.' *Journal of Microelectromechanical Systems*, **12** (1), pp. 70–80.

Choi, C. H., and Kim, C. J., 2006. 'Large slip of aqueous liquid flow over a nanoengineered superhydrophobic surface.' *Physical Review Letters*, **96** (6), p. 066001.

Fan, S. K., de Guzman, P. P., and Kim, C. J., 2002. 'EWOD driving of droplet on NxM grid using single-layer electrode patterns.' *Solid-State Sensor, Actuator, and Microsystems Workshop*. Hilton Head Island, SC, pp. 134–137.

Gallardo, B. S. *et al.*, 1999. 'Electrochemical principles for active control of liquids on submillimeter scales.' *Science*, **283**, pp. 57–60.

Gascoyne, P. R. C. *et al.*, 2004. 'Dielectrophoresis-based programmable fluidic processors.' *Lab on a Chip*, **4** (4), pp. 299–309.

Gau, H, Herminghaus, S, and Lenz, P, Lipowsky, R, 1999. 'Liquid morphologies on structured surfaces: From microchannels to microchips.' *Science*, **283**, pp. 46–39.

Gennes, P. G. de, 1985. 'Wetting: Statistics and dynamics.' *Reviews of Modern Physics*, **57**, p. 36.

Gong, J., and Kim, C. J., 2005. 'Two-dimensional digital microfludic system by multi-layer printed circuit board.' *IEEE International Conference on Micro Electro Mechanical Systems*, Miami, FL, pp. 726–729.

Gravesen, P, Branebjerg, J, and Jensen, O. S., 1993. 'Microfluidics – a review.' *Journal of Micromechanics & Microengineering*, **3** (4), pp. 168–182.

He, M. *et al.*, 2005. 'Selective encapsulation of single cells and subcellular organelles into picoliter- and femtoliter-volume droplets.' *Analytical Chemistry*, **77** (6), pp. 1539–1544.

Ichimura, K, Oh, S. K., and Nakagawa, M, 2000. 'Light-driven motion of liquids on a photoresponsive surface.' *Science*, **288**, pp. 1624–1626.

Jones, T. B., Wang, K. L., and Yao, D. J., 2004. 'Frequency-dependent electromechanics of aqueous liquids: Electrowetting and dielectrophoresis.' *Langmuir*, **20**, pp. 2813–2818.

Jun, T. K., and Kim, C. J., 1998. 'Valveless pumping using traversing vapor bubbles in microchannels.' *Journal of Applied Physics*, **83** (11), pp. 5658–5664.

Kataoka, D. E., and Troian, S. M., 1999. 'Patterning liquid flow on the microscopic scale.' *Nature*, **402** (6763), pp. 794–797.

Kim, C. J., 2000. 'Microfluidics using the surface tension force in microscale.' *SPIE Symposium on Micromachining and Microfabrication*. 4177 Santa Clara, CA, pp. 49–55.

Lahann, J. *et al.*, 2003. 'A reversibly switching surface.' *Science*, **299**, pp. 371–374.

Lam, P, Wynne, K. J., and Wnek, G. E., 2002. 'Surface-tension-confined microfluidics.' *Langmuir*, **18** (3), pp. 948–951.

Latorre, L. *et al.*, 2002. 'Electrostatic actuation of microscale liquid-metal droplets.' *Journal of Microelectromechanical Systems*, **11** (4), pp. 302–308.

Lee, J., and Kim, C. J., 2000. 'Surface tension driven microactuation based on continuous electrowetting (CEW).' *Journal of Microelectromechanical Systems*, **9** (2), pp. 171–180.

Lee, J. *et al.*, 2001. 'Addressable micro liquid handling by electric control of surface tension.' *IEEE International Conference on Micro Electro Mechanical Systems*. Interlaken, Switzerland, pp. 499–502.

Lee, J. *et al.*, 2002. 'Electrowetting and electrowetting-on-dielectric for microscale liquid handling.' *Sensors and Actuators A*, **95**, pp. 259–268.

Lemoff, A. V., and Lee, A. P., 2000. 'An AC magnetohydrodynamic micropump.' *Sensors & Actuators B – Chemical*, **63** (3), pp. 178–185.

Linnemann, R. *et al.*, 1998. 'A self-priming and bubble-tolerant piezoelectric silicon micropump for liquids and gases.' *MEMS 98. IEEE*, Heidelberg, Germany, pp. 532–537.

Lippmann, G, 1875. 'Relations entre les phenomenes electriques et capillaires.' *Annales de Chimie et des Physique*, **5**, p. 494.

Matsumoto, H., and Colgate, J. E., 1990a. 'Preliminary investigation of micropumping based on electrical control of interfacial tension.' *IEEE MEMS Workshop 1990.* Napa Valley, CA, pp. 105–110.

Matsumoto, H., and Colgate, J. E., 1990b. 'Preliminary investigation of micropumping based on electrical control of interfacial tension.' *IEEE Micro Electro Mechanical Systems Workshop 1990.* Napa Valley, CA, pp. 105–110.

Minnema, L, Barneveld, H. A., and Rinkel, P. D., 1980. 'An investigation into the mechanism of water treeing in polyethylene high-voltage cables.' *IEEE Transactions on Electrical Insulation*, **EI-15** (6), pp. 461–472.

Moon, H. *et al.*, 2002. 'Low voltage electrowetting-on-dielectric.' *Journal of Applied Physics*, **92** (7), pp. 4080–4087.

Nguyen, N. T., and Huang, X. Y., 2001. 'Miniature valveless pumps based on printed circuit board technique.' *Sensors and Actuators, A*, **88** (2), pp. 104–111.

Nguyen, N. T., Huang, X. Y., and Toh, K. C., 2001. 'Thermal flow sensor for ultra-low velocities based on printed circuit board technology.' *Measurement Science and Technology*, **12** (12), pp. 2131–2136.

Paik, P. Y. *et al.*, 2005. 'Coplanar digital microfluidics using standard printed circuit board processes.' *International Conference on Miniaturized Systems for Chemistry and Life Sciences (microTAS).* Boston, MA, pp. 566–568.

Pollack, M. G., Shenderov, A. D., and Fair, R. B., 2002. 'Electrowetting-based actuation of droplets for integrated microfluidics.' *Lab on a Chip*, **2** (2), pp. 96–101.

Rosslee, C., and Abbott, N. L., 2000. 'Active control of interfacial properties.' *Current Opinion in Colloid and Interface Science*, **5**, pp. 81–87.

Sammarco, T. A., and Burns, M. A., 1999. 'Thermocapillary pumping of discrete drops in microfabricated analysis devices.' *AIChE Journal*, **45** (2), pp. 350–366.

Selvaganapathy, P. *et al.*, 2002. 'Bubble-free electrokinetic pumping.' *Journal of Microelectromechanical Systems*, **11**, pp. 448–453.

Shih, J. Y., and Abbott, N. L., 1999. 'Using light to control dynamic surface tensions of aqueous solutions of water soluble surfactants.' *Langmuir*, **15** (13), pp. 4404–4410.

Shoji, S., and Esashi, M, 1994. 'Microflow devices and systems.' *Journal of Micromechanics & Microengineering*, **4** (4), pp. 157–171.

Srinivasan, U, Liepmann, D, and Howe, R. T., 2001. 'Microstructure to substrate self-assembly using capillary forces.' *Journal of Microelectromechanical Systems*, **10** (1), pp. 17–24.

Thorsen, T, Maerkl, S. J., and Quake, S. R., 2002. 'Microfluidic large-scale integration.' *Science*, **298** (5593), pp. 580–584.

Towery, D., and Fury, M. A., 1998. 'Chemical mechanical polishing of polymer films.' *Journal of Electronic Materials*, **27** (10), p. 7.

Tseng, F. G., Kim, C. J., and Ho, C. J., 2002. 'A high-resolution high-frequency monolithic top-shooting microinjector free of satellite drops – part I: Concept, design, and model.' *Journal of Microelectromechanical Systems*, **11** (5), p. 10.

van de Pol, F. C. M. *et al.*, 1990. 'A thermopneumatic micropump based on micro-engineering techniques.' *Sensors and Actuators A, Physics*, **21**–**23**, pp. 198–202.

Wego, A, Richter, S, and Pagel, L, 2001. 'Fluidic microsystems based on printed circuit board technology.' *Journal of Micromechanics Microengineering*, **11** (5), pp. 528–531.

Whitesides, G. M., and Stroock, A. D., 2001. 'Flexible methods for microfluidics.' *Physics Today*, **54** (6), pp. 42–48.

Yang, Y. *et al.*, 1996. 'A bimetallic thermally actuated micropump.' *Journal of Microelectromechanical Systems*, **DSC-59**, pp. 351–354.

Yu, H., and Kim, E. S., 2002. 'Noninvasive acoustic-wave microfluidic driver.' *IEEE International Conference on Micro Electro Mechanical Systems*, Las Vegas, NV, pp. 125–128.

Zengerle, R. *et al.*, 1995. 'A bi-directional silicon micropump.' *IEEE International Conference on Micro Electro Mechanical Systems*, Amsterdam, Netherlands, pp. 19–24.

Zhang, W., and Ahn, C. H., 1996. 'A bi-directional magnetic micropump on a silicon wafer.' *IEEE Solid-State Sensor and Actuator Workshop*, Hilton Head, SC, pp. 94–97.

7

MICRO PARTICLE VELOCIMETRY

Carl D. Meinhart and Steven T. Wereley

7.1 Introduction

Micron-resolution Particle Image Velocimetry (micro-PIV) refers to the measurement of fluid motion with resolved length scales ranging from 10^{-4}–10^{-7} m using Particle Image Velocimetry (PIV). The first reported effort on the subject was a particle-streak velocimetry approach by Brody (1996). He used epi-fluorescent microscopy to record the streaks of 0.9 um dia. fluorescent particles. Before the development of pointwise measurement techniques, such as micron-resolution PIV, most of the experiments conducted in micro scale geometries consisted of bulk flow measurements, such as flow rate, pressure drop, thrust, etc. (for a review see Gad-el-Hak 1999 and Ho & Tai 1998). Particle image velocimetry has been applied at the macro scale for the past 25 years, and it has been adopted as the gold standard velocimetry technique in fluid mechanics. Extensive reviews of the PIV technique are given by Adrian (1991, 2005), and Raffel *et al.* (2007).

The micro-PIV technique that is commonly used in microfluidics was introduced by Santiago *et al.* (1998) and Meinhart *et al.* (1999), where they used epi-fluorescent illumination to record discrete particle images. Since this early work, there have many variations and extensions of the technique. As a result, the original micro-PIV technique is now commonly referred to as 'classical micro-PIV'. There are currently ~100 publications on the subject per year and as of June 2006 nearly 400 cumulative publications (see Fig. 7.1). Reviews on the subject have been given by Sinton (2004), and Wereley and Meinhart (2005, 2007). The purpose of this review article is to extend the previous reviews on micro-PIV, and to update the reader of recent developments that have occurred over the past several years in micro-PIV.

The most common recording mode of PIV is to record two successive images of flow-tracing particles. The two particle images are separated by a known time delay, Δt. Typically, the two particle image fields are divided into uniformly spaced interrogation regions, which are cross-correlated to determine the most probable local displacement. A first-order estimate of the local velocity of the fluid, $\boldsymbol{u}$, is then obtained by divided the measured displacement, $\boldsymbol{\Delta x}$, by the time delay, such that

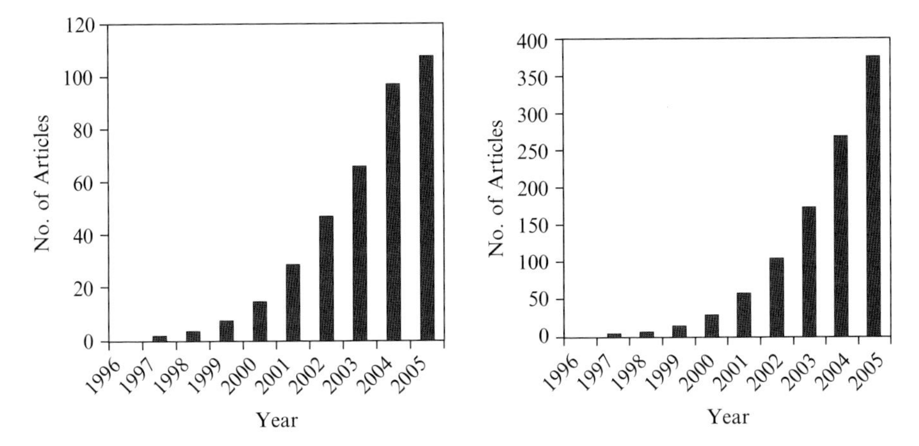

Fig. 7.1: Number of articles published on micro-PIV: (a) number of articles on a per year basis, (b) cumulative number of articles (data as of June 2006). The data was collected based upon journal papers published with title or keywords of micro and PIV.

$$\boldsymbol{u} = \frac{\Delta \boldsymbol{x}}{\Delta t}. \tag{7.1}$$

High spatial resolution is achieved by recording the images of flow-tracing particles with sufficiently small diameters, d_{p}, so that they faithfully follow the flow in microfluidic devices, which often exhibit high velocity gradients near flow boundaries. The particle should be imaged with high diffraction-resolution optics and with sufficiently high magnification so that the particles are resolved with at least 3-4 pixels per particle diameter. Typically, microscope objective lenses range from oil-immersion lenses, $M = 60$, $NA = 1.4$ to air-immersion $M = 10$, $NA = 0.1$.

7.2 Volume illumination theory

The small length scales associated with measuring fluid motion at the micro scale create several challenges that must be addressed by the micro-PIV technique. These challenges are met for judicious choice of seed particles, particle illumination and recording techniques, and particle-image interrogation/post processing techniques. In macro-scale PIV, flow-tracing particles are commonly illuminated by a light sheet, which defines the measurement plane (Adrian 1991). At the micro scale, it is difficult, if not impossible, to form and focus a light sheet of just a few microns. As a result, alternative approaches are incorporated, such as volume illumination or evanescent illumination (Zettner and Yoda 2003).

In volume illumination, the entire test section is illuminated by a volumetric cone of light, and the measurement plane is defined by the depth of field of the recording lens (see Fig. 7.2). Fortunately, microscope objective lenses have large

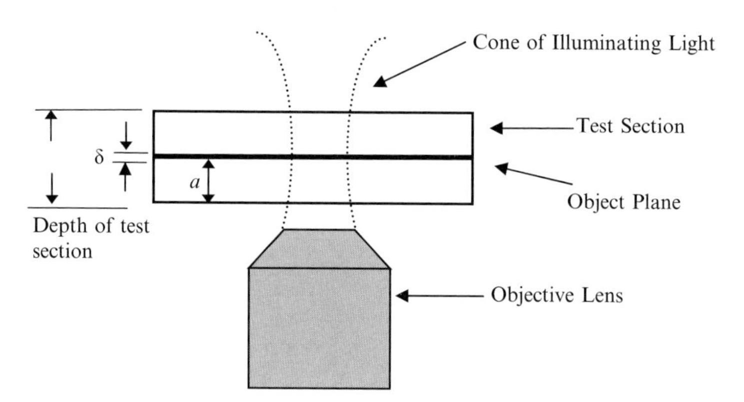

Fig. 7.2: Volume illumination particle image velocimetry geometry. In the epi-fluorescent mode, the objective lens is used to both deliver the illumination light and record the particle images (after Wereley and Meinhart 2005).

apertures with sharply defined objective planes. This allows for particles to quickly transition from being in focus to being out of focus.

Theories describing the measurement depth for volume illumination have been presented by Meinhart *et al.* (2000) and Olsen and Adrian (2000). The theory for correlation depth was developed by Olsen and Adrian (2000). The theory considers particle imaging through air with relatively low numerical apertures. This theory is experimentally verified Bourdon *et al.* (2004).

In order to extend this theory to high numerical aperture imaging systems, a relationship between *f-number* and numerical aperture must be determined. Following Meinhart and Wereley (2003) the *f-number* can be related to NA for an infinity-corrected lens as

$$f_{\infty}^{\#} = \frac{1}{2}\left[\left(\frac{n_0}{\mathrm{NA}}\right)^2 - 1\right]^{1/2}, \tag{7.2}$$

where n_0 is the index of refraction of the immersion medium, NA is the numerical aperture. Assuming the paraxial approximation, equ. (7.2) can be reduced to

$$f_p^{\#} \approx \frac{n_0}{2\mathrm{NA}}. \tag{7.3}$$

Using the paraxial approximation for *f-number*, the depth of correlation, z_{corr}, is given by Bourdon *et al.* (2006)

$$z_{corr} = 2\left\{\frac{(1-\sqrt{\epsilon})}{\epsilon}\left[\frac{n_0^2 d_p^2}{4\mathrm{NA}^2} + \frac{5.95(M+1)^2\lambda^2 n_0^4}{16M^2\mathrm{NA}^4}\right]\right\}^{1/2}. \tag{7.4}$$

Table 7.1: Thickness of the measurement plane for typical experimental parameters, z_{corr} (μm).

Particle Size d_p	Microscope Objective Lens Characteristics $M = 60$ $NA = 1.4$	$M = 40$ $NA = 0.75$	$M = 40$ $NA = 0.6$	$M = 20$ $NA = 0.5$	$M = 10$ $NA = 0.25$
0.01 μm	2.1	2.1	2.1	2.2	2.3
0.10 μm	2.1	2.2	2.2	2.3	2.9
0.20 μm	2.2	2.4	2.6	2.8	4.3
0.30 μm	2.3	2.8	3.1	3.5	5.9
0.50 μm	2.6	3.7	4.3	5.0	9.4
0.70 μm	3.1	4.7	5.7	6.7	13
1.00 μm	3.9	6.4	7.9	9.3	18
3.00 μm	10	18	23	27	55

Here, $\epsilon \approx 0.01$ is a customary value for the threshold weighting function, and M is magnification. According to Bourdon *et al.* (2006), eqn. (7.4) gives good agreement with experimental results for $M=20$, NA $=0.7$; $M=40$, NA $=1.25$; $M=63$, NA $=1.4$ imaging systems (using 1.0 μm dia. fluorescent particles). It is interesting to note that the depth of correlation agrees more closely to experiments when using the paraxial approximation relation between *f-number* and NA (eqn. 7.3), than using the exact relationship given by eqn. (7.2).

Using $\lambda = 560\,\text{nm}$, eqn. 7.4 is used to estimate the depth of correlation for a range of imaging systems and particle sizes are given in Table 7.1.

7.3 Evanescent imaging

Zettner and Yoda (2003) were the first to use an evanescent field to illuminate flow-tracing particles in a microchannel. The evanescent field illuminated an order 200 nm region near a microchannel wall. This increased the maximum spatial resolution in the out-of-plane direction by nearly an order of magnitude to ~100-200 nm, while decreasing the in-plane spatial resolution by an order of magnitude (Zettner and Yoda 2003; Jin *et al.* 2004; Sadr *et al.* 2004, 2005). Typically, the averaging volume required to obtain a valid velocity measurement using evanescent illumination is similar to the averaging volume reported in much of the micro-PIV literature.

Figure 7.3 shows an evanescent wave at the boundary between a waveguide with refractive index n_1 and a fluid with refractive index n_2. The evanescent field intensity decays exponentially with increasing distance from the interface, such that

$$I_z = I_0 e^{-z/d}, \tag{7.5}$$

where I_Z represents the intensity at a perpendicular distance z from the interface and I_O is the intensity at the interface. The characteristic penetration depth d at is given by

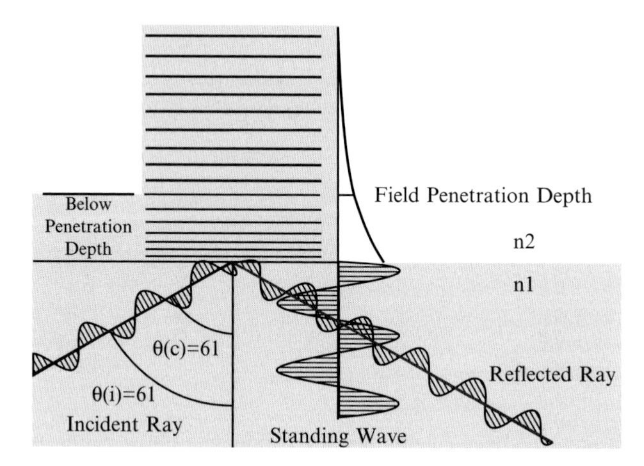

Fig. 7.3: Evanescent field intensity decays exponentially with distance. Altering the beam incident angle or wavelength will change the field penetration depth. Shown is a beam with $\lambda = 403$ *nm* incident on the waveguide at an angle of 61 degrees. [Taken from Axelrod *et al.* 1987].

$$d = \frac{\lambda_0}{2\pi\sqrt{n_1^2 \sin^2 \Theta - n_2^2}}, \tag{7.6}$$

where λ_0 is the vacuum wavelength and Θ is the beam's angle of incidence (Zettner and Yoda 2003).

The field penetration depth decreases as the incident angle increases. This depth is also dependent upon the refractive indices of the media present at the interface and the illumination wavelength. In general, the value of d is on the order of the incident wavelength.

The numerical aperture *NA* of the two refractive indices is found by taking the sin of the largest angle that allows TIR.

$$NA = \sin \Theta_c \leq \sqrt{n_1^2 - n_2^2}, \tag{7.7}$$

when n_1 is greater than n_2, and when Θ exceeds the critical angle Θ_C, TIR occurs within the waveguide. The critical angle is defined by the equation:

$$\Theta_c = \sin^{-1} \frac{n_2}{n_1}. \tag{7.8}$$

For angles less than Θ_C, a majority of the incident light propagates directly through the interface with a refraction angle measured from the normal as defined by Snell's Law (see Fig. 7.4). Even in this situation, some of the incident light is reflected back into the glass. However, for all angles greater than the critical angle, TIR is achieved.

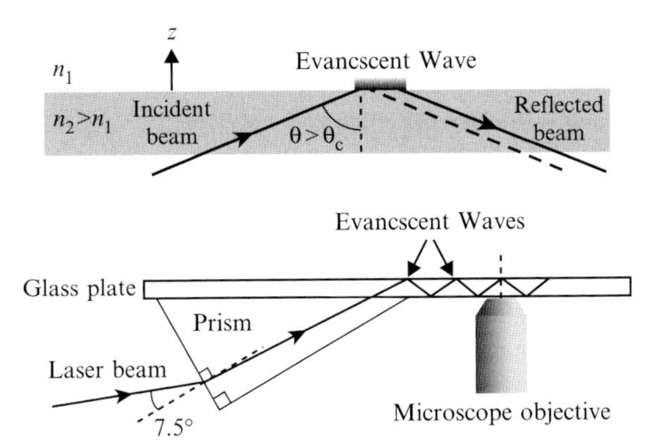

Fig. 7.4: Schematic illustrating evanescent illumination. The incident beam is beyond the critical limit and is subject to geometric total internal reflection. The result are evanescent waves that penetrate ~100 nm into the microchannel. [Image taken from Zettner and Yoda 2003].

7.4 Flow-tracing particles

The most common flow-tracing particles for micro-PIV are 200 nm-2 μm dia. fluorescently-labeled polystyrene particles. Fluorescent particles are readily available from manufactures like Duke Scientific (http://www.dukescientific.com), Bangs Laboratories (http://www.bangslabs.com) and Invitrogen (http://probes.invitrogen.com). These particles allow for good visibility, even down to ~200 nm diameters with high numerical aperture (NA = 1.4) oil-immersion objective lenses.

The size of the flow-tracing particle must be selected such that it is sufficiently small compared to the length scale of the flow, so that it follows the flow faithfully, and will not clog the device (Sharp and Adrian 2005). The particle must also be sufficiently large so that it can readily be imaged, and sufficiently dampen Brownian noise.

A general-purpose equation for the force balance on a dilute suspension of rigid spherical particles in a fluid is given by Maxey and Riley (1983)

$$\begin{aligned} m_\rho \frac{dV_i}{dt} &= m_F \frac{Du_i}{Dt}\bigg|_{\mathbf{Y}(t)} - \frac{1}{2} m_F \frac{d}{dt}\left\{ V_i(t) - u_i[\mathbf{Y}(t),t] - \frac{1}{10} a^2 \nabla^2 u_i\big|_{\mathbf{Y}(t)} \right\} \\ &+ (m_\rho - m_F) g_i + f_i - 6\pi a \mu \left\{ V_i(t) - u_i[\mathbf{Y}(t),t] - \frac{1}{6} a^2 \nabla^2 u_i\big|_{\mathbf{Y}(t)} \right\}, \\ &- 6\pi a^2 \mu \int_0^t d\tau \left(\frac{d/d\tau \left\{ V_i(\tau) - u_i[\mathbf{Y}(\tau),\tau] - (1/6) a^2 \nabla^2 u_i\big|_{\mathbf{Y}(\tau)} \right\}}{\left[\pi\nu(t-\tau)\right]^{1/2}} \right) \end{aligned} \tag{7.9}$$

where m_ρ is the mass of a particle, m_F is the mass of fluid displaced by a particle, $\mathbf{Y}(t)$ is the location of a particle, V_i(t) is the velocity of a particle, u_i(x,t) is the velocity field of fluid, a is the radius of a particle, μ is the dynamic viscosity of the fluid, ν is the kinematic viscosity of the fluid. The terms in eqn. (7.9) represent, from left to right, the particle inertia, the pressure-viscous force, the added mass force, the buoyancy force, an arbitrary external force applied to the particle, the Stokes drag force, and the Basset history force.

Equation (7.9) was derived by making assumptions about the particles and the surrounding flow. Each particle is assumed to be far from any neighboring particle, and also far from any boundaries of the flow. In practice, this restriction translates to the distance from the nearest boundary or neighboring particle being very much larger than the particle's diameter. Effects due to non-zero relative Reynolds number are also ignored. These effects, for the case of steady relative motion, are specifically the Oseen correction to Stokes drag, the modified drag due to particle rotation, and Saffman effect—a lateral force due to shear in the undisturbed flow. All of these effects are small compared to the remaining Stokes drag term with the assumption of low Reynolds number. The overriding criteria that must be satisfied for eqn. (7.9) to accurately model the dynamics of a particle suspended in a flow are that the relative particle Reynolds number, aW_0/ν, be much less than unity and that the quantity $(a^2/\nu)(U_0/L)$ also be very much less than unity. In these parameters W_0 is a measure of the particle velocity relative to the flow and U_0/L is a scale for the velocity gradients in the problem.

For the small-sized particles, eqn. (7.9) can be simplified to the Stokes drag law plus the arbitrary force f_i, which can assume any form or value, depending upon the particular experiment being conducted. For example, f_i could represent an electrostatic force acting on the particles, as in electrophoresis, or it could represent a magnetic body force. The arbitrary force f_i could be a quantity being measured, or it could be a quantity degrading the accuracy of the fluid velocity measurements.

For μPIV, three phenomena primarily act to prevent the seed particles from following the flow. These are:

- Brownian motion—the random thermal vibrations of the seed particles.
- Saffman effect—which can become large in the case of particles very near a boundary.
- Arbitrary external forces.

7.4.1 *Brownian motion*

The error due to Brownian motion relative to the displacement in the x-direction, ϵ_B, is given by Santiago *et al.* (1998)

$$\epsilon_B = \frac{\langle s^2 \rangle^{1/2}}{\Delta x} = \frac{1}{u}\sqrt{\frac{2D}{\Delta t}}, \tag{7.10}$$

where s^2 is the random mean square particle displacement associated with Brownian motion, Δx is the particle displacement, u is the local fluid velocity, and Δt is the time interval between images. The diffusivity of spherical particles, D, is given by the Stokes–Einstein equation

$$D = \frac{k_B T}{3\pi \; \mu \; d_p}, \tag{7.11}$$

where k_B is Boltzmann's constant, and T is the absolute fluid temperature, μ is the dynamic viscosity of the working fluid.

The Brownian error (eqn. 7.10) establishes a lower limit on the measurement time interval Δt since, for shorter times, the measurements are dominated by uncorrelated Brownian motion. These quantities (ratios of the root mean square fluctuation-to-average velocity) describe the relative magnitudes of the Brownian motion and will be referred to here as Brownian intensities. The relative Brownian intensity error decreases as the time of measurement increases. Larger time intervals produce flow displacements proportional to Δt while the root mean square of the Brownian particle displacements grow as $\Delta t^{1/2}$. In practice, Brownian motion is an important consideration when tracing 50 to 500 nm particles in flow field experiments with flow velocities of less than about 1 mm/s. For a velocities on the order of 0.5 mm/s and a 500 nm seed particle, the lower limit for the time spacing is approximately 100 μs for a 20 per cent error due to Brownian motion. This error can be reduced by both averaging over several particles in a single interrogation spot and by ensemble averaging over several realizations. The diffusive uncertainty decreases as $1/\sqrt{N}$, where N is the total number of particles in the average (Bendat and Piersol 1986).

Equation (7.10) demonstrates that the effect of Brownian motion is relatively less important for faster flows. However, for a given measurement, when u increases, Δt will generally be decreased. Equation (7.10) also demonstrates that when all conditions but the Δt are fixed, increasing Δt will decrease the relative error introduced by Brownian motion. Unfortunately, longer Δt will decrease the accuracy of the results because the PIV measurements are based on a first order accurate approximation to the velocity. Using a second order accurate technique called Central Difference Interrogation (CDI) allows for longer Δt to be used without increasing this error (Wereley and Meinhart 2001).

7.4.2 *Saffman effect*

The Saffman effect will cause particles to migrate across streamlines in a deterministic manner. This deterministic effect competes with the random effects resulting from Brownian motion. Saffman (1965) analytically considered the case of a rigid sphere translating in a linear unbounded shear field. He used matched asymptotic expansions to determine an equation for the particle migration velocity that is correct to leading order. Consider a rigid sphere of radius a in an unbounded shear flow of fluid density ρ and kinematic viscosity ν. Assuming that the velocity field depends linearly on one spatial coordinate such that the velocity gradient is G. The particle is initially sitting at the origin (at which point the fluid velocity is identically zero). The particle is assumed to move relative to the fluid at a speed $-V_s$ and rotate with an angular velocity Ω. The pertinent Reynold's numbers in this situation are

$$\mathrm{Re}_G = \frac{4a^2 G}{\nu}, \ \mathrm{Re}_s = \frac{2aV_s}{\nu}, \ \mathrm{Re}_\Omega = \frac{4a^2 \Omega}{\nu}, \tag{7.12}$$

where the quantity Re_s is the slip Reynold's number. If these Reynold's numbers are small compared to unity, Saffman's analysis can be used to estimate a migration velocity V_m, such that

$$\frac{V_m}{V_s} = 0.343a\sqrt{\frac{G}{\nu}}, \tag{7.13}$$

As long as the quotient V_m/V_s is small, the particle migration across streamlines caused by the velocity gradient will have a negligible effect on the velocity measurements.

7.4.3 *Quantum dots*

In recent years, quantum dots (QDs) have been introduced as potential flow-tracing particles (Pouya *et al.* 2005). QDs are nanocrystalline materials that fluoresce at a particular wavelength, depending upon their physical size. The emitted fluorescence results from the quantum confinement effect, where the physical particle size is smaller than the Bohr radius of the electron-hole pair it harbors. (Quantum Dot Corp. 2005). QDs are commonly composed of a cadmium selenide (CdSe) core, and an inorganic coating of zinc sulfide (ZnS) that passivates the QD and increases water solubility (Wu 2003; Murray *et al.* 1993; Dabboursi *et al.* 1997). A polymer shell is used to conjugate functionalized surfaces, so that the QDs can be miscible in polar or non-polar liquids, and conjugated to specific molecules.

Quantum dots have several advantages over their fluorescent polystyrene counterparts. Firstly, they are sufficiently smaller, with hydrodynamic diameters of ~18 nm, do not suffer from photobleaching, and exhibit a relatively large Stokes shift. The small size of QDs causes them to have a high diffusivity and exhibit large Brownian motion.

In their original experiment, Pouya *et al.* (2005) used an argon-ion laser to illuminate an ~100 nm thick evanescent field to excite 10.7 nm diameter QDs. The QDs emitted at a wavelength of ~615 nm. The ~10 nm QDs, exhibited a significant amount of Brownian motion, which contributed to particle dropout between exposures, and noise in the velocity signal. Their flow velocity was estimated by tracking individual QDs. They estimated the mean streamwise velocity to ~18.4 μm/s, with an rms fluctuation of ~10.5 μm/s. The rms velocity was approximately 57 per cent of the mean velocity, indicating the noise level associated with Brownian motion.

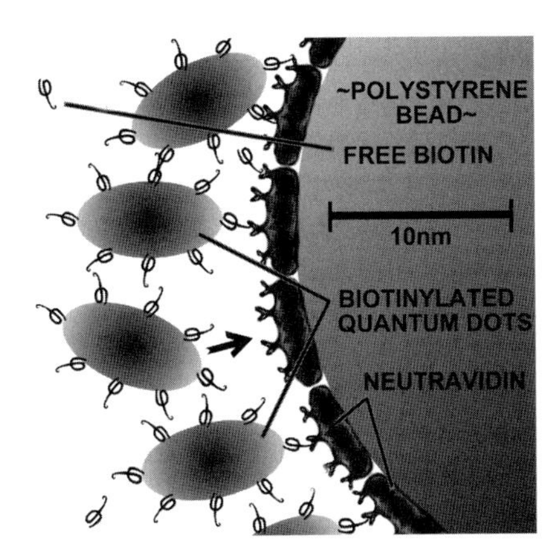

Fig. 7.5: Schematic showing the structure of a Quantum Nanosphere (QN). Polystyrene beads are coated with NeutrAvidin and titrated into a solution containing biotinylated quantum dots (QDs). Approximately 60 QDs are conjugated to a bead, thereby forming a single QN (after Freudenthal *et al.* 2007).

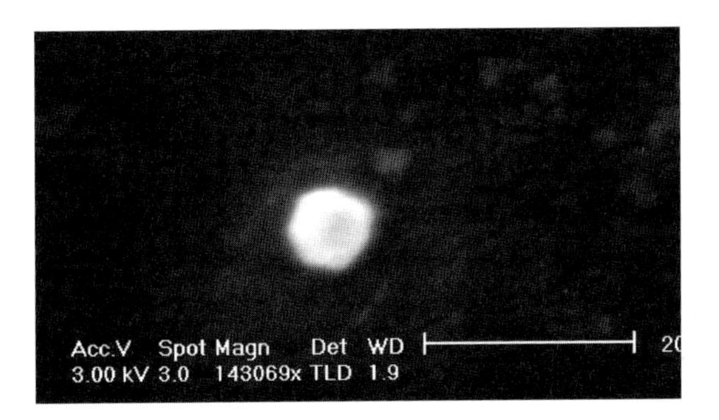

Fig. 7.6: Scanning electron microscope (SEM) image of a single ~75 nm dia. Quantum Nanosphere on a dry slide (after Freudenthal *et al.* 2007).

One approach to mitigate high levels of Brownian motion is to incorporate quantum nanospheres (QNs) developed by Freudenthal *et al.* (2007). A QN is formed by conjugating ~60 QDs to a single polystyrene nanoparticle (see Fig. 7.5). This allows for independent control of the particle diffusivity and emitted fluorescent color. An SEM image of a single QN is shown in Fig. 7.6.

7.5 Special micro-PIV processing methods

Working in the microscopic world is generally more difficult than the macroscopic. That is certainly true when comparing conventional PIV and micro-PIV. However, there is at least one beneficial consequence of the small size of microscopic systems. Generally microscopic systems are low Reynolds number systems (Re<1). This simplifies the fluid mechanics analysis and renders many problems of biomedical interest steady, quasi-steady or periodic. This simplification allows novel types of processing algorithms to be used, such as *ensemble correlation*—also commonly referred to as *correlation averaging.* Correlation averaging was developed in the framework of micro-PIV applications in an effort to reduce the influence of Brownian motion, low seed particle concentration, and low quality images. Rather than obtaining displacement data for each individual image pair, the technique relies on averaging coincident correlation planes from a sequence of images. With increasing frame counts a single correlation peak will accumulate for each correlation plane, reflecting the mean displacement of the flow (Wereley and Meinhart 2005; Devasenathipathy *et al.* 2003; Meinhart *et al.* 2000). Although computationally very efficient, the main drawback of this approach is that all information pertaining to the unsteadiness of the flow is lost (e.g. turbulence intensities and Reynolds stresses, among others). Its use with conventional macroscopic PIV recording has been verified by the authors and seems appropriate for fast calculation of the mean flow. Since this processing is very fast, it has potential as a quasi-online diagnostic tool.

To demonstrate the effectiveness of the ensemble correlation technique, Meinhart *et al.* (2000) compared the three different averaging algorithms applied to a series of images acquired from a steady Stokes flow of water through a 30 μm by 300 μm glass microchannel. The signal-to-noise ratio for measurements generated from a single pair of images was relatively low, because there was an average of only 2.5 particle images located in each 16×64 pixel interrogation window. The velocity measurements are noisy and approximately 20 per cent appear to be erroneous. The three different types of averaging schemes compared are:

- image averaging in which the images themselves averaged to produce an average first image and an average second image that are then correlated;
- correlation field averaging in which the correlation function at each measurement point is averaged across all image pairs;
- velocity field averaging in which a velocity measurement is calculated at each measurement position in each image pair and then averaged across all image pairs.

The relative performance of the three averaging algorithms was quantitatively compared by varying the number of image pairs used in each averaging technique from 1 to 20. The fraction of valid measurements for each averaging technique was determined by identifying the number of velocity measurements in which the streamwise velocity component deviated by more than 10 per cent from the known solution at each point. For this comparison, the known solution was the velocity vector field estimated by applying the average correlation technique to 20 realizations, and then smoothing the flow field. Figure 7.7 shows the fraction of valid measurements for each of the three averaging algorithms as a function of the number of realizations used in the average. The average correlation method

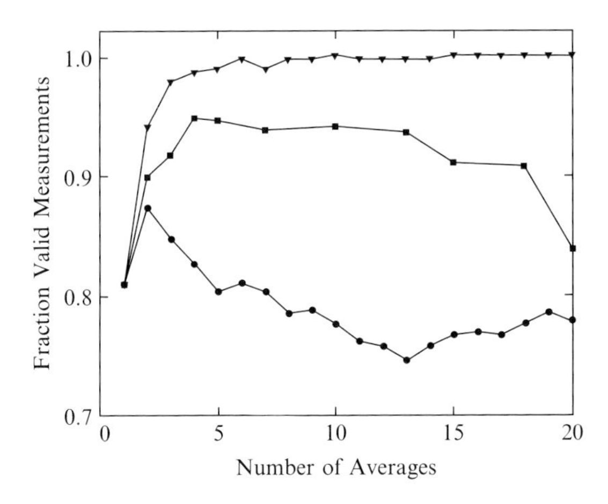

FIG. 7.7: Comparison of three averaging schemes: correlation averaging (triangles), image averaging (squares) and velocity averaging (circles).

clearly performs better than the other two methods, and produces less than 0.5 to 1 per cent erroneous measurements after averaging eight realizations. The average image method produces about 95 per cent reliable velocity measurements, and reaches a maximum at four averages. Further increases in the number of realizations used to average the images decreases the signal to noise ratio of the average particle-image field, and produces noise in the correlation plane due to random correlation between non-paired particle images. The average velocity method reaches a maximum of 88 per cent reliable measurements using two velocity averages. Further increases in the number of averages decreases the fraction of reliable measurements, due to an increase in the probability of an encountering an erroneous measurement.

7.5.1 *Single pixel evaluation*

Once the ensemble correlation algorithm is being used, there is practically no limit to how small the interrogation region can be made. In principle, it can be reduced to a single pixel by replacing the typical spatially-averaged correlation with one averaged in time. Conventional spatially-averaged cross-correlation typically needs two multi-pixel interrogation windows to calculate one valid displacement vector. Ideally, the second window contains all of the information from the first one and no velocity gradient occurs inside the region. Under these conditions a correlation peak indicating the true in-plane displacement will be obtained. The mathematical expression for such a spatially-averaged correlation is written as

$$\Phi_{cc}(m,n) = \sum_{x=1}^{P} \sum_{y=1}^{Q} f(x,y) \cdot g(x+m, y+n), \tag{7.14}$$

where P and Q denote the side lengths in pixels of the interrogation window in x and y direction, respectively. f and g are the intensity functions of the first and the second windows. m and n are the displacement in the correlation domain. The parameters that may influence the accuracy perhaps come from the particle density, image quality, and interrogation window size. Obviously, enlarging the window size is a way to ensure that sufficient spatial information is included. However, large interrogation regions may result in serious bias errors when a strong velocity gradient is present in the flow as well as a reduction in spatial resolution. To solve this problem, a novel algorithm named single pixel evaluation (SPE) was proposed (Westerweel *et al.* 2004) to drastically increase the spatial resolution without decreasing SNR. The cross-correlation function from eqn. (7.14) can be rewritten for an interrogation region consisting of a single pixel averaged over k different time steps can be expressed as follows

$$\Phi_{SPE,\,FDI}(m,n) = \sum_{k=1}^{N} f_k(x,y) \cdot g_k(x+m, y+n), \tag{7.15}$$

where N is the number of image pairs (Gui *et al.* 2002). The main concept is shrinking the interrogation window to its physical limitation given the pixelated format in which the images are recorded: one pixel and collecting the spatial information by searching the second pixel within some specified radius. Based on the ensemble correlation average (Meinhart *et al.* 2000), the SNR can be increased by increasing the number of image pairs in a given sample. For instance, if a particle density with 10 particles over a 32×32 pixel window (9.77×10^{-3} particle/pixel) is used in conventional cross-correlation, at least 1024 image pairs will be needed in SPE in order to reach the same SNR, i.e. particles per pixel.

The accuracy of the SPE technique can be improved by using a more accurate finite difference approximation. The SPE formula given in eqn. (7.15) is a first-order accurate forward difference approximation (hence the FDI subscript on $\Phi_{SPE,FDI}$ above). This can be improved by using a second-order accurate central difference approximation (which will be denoted $\Phi_{SPE,CDI}$). The mathematical expression for the CDI SPE algorithm is

$$\Phi_{SPE,\,CDI}(m,n) = \sum_{k=1}^{N} [f_k(x,y) \cdot g_k(x+m, y+n) + g_k(x,y) \cdot f_k(x-m, y-n)], \tag{7.16}$$

which contains an extra term to account for the backward-time correlation. Besides the accuracy improvement, an additional advantage of this approach is the rapid elimination of the background noise, resulting from the double number of the image pairs. Theoretically, in the case of micro-PIV, an ultimate in-plane resolution of 60 nm could be attainable when a 60 nm particle is imaged with an M=100, NA=1.4 objective lens and a CCD camera with a 6 μm pixel size.

7.6 Two-dimensional and two-component velocimetry

Figure 7.8 is a schematic of a typical μPIV system, originally described by Meinhart *et al.* (1999). It is common to use a pulsed monochromatic light source, such as a pulsed Nd:YAG laser system, which are available from *NewWave Research, Inc.* (47613 Warm Springs Blvd., Fremont, CA). The laser system is specifically designed for PIV applications, and consists of two Nd:YAG laser cavities, beam combining optics, and a frequency doubling crystal. The laser emits two pulses of light at $\lambda = 532$ nm. The duration of each pulse is on the order of 5 ns and the time delay between light pulses can vary from order hundreds of nanoseconds to a few seconds. The illumination light is delivered to the microscope through beam-forming optics, which can consist of a variety of optical elements that will sufficiently modify the light so that it will fill the back of the objective lens, and thereby broadly illuminate the microfluidic device. The beam-forming optics can consist of a liquid-filled optical fiber to direct the light into the microscope. The illumination light is reflected upward towards the objective lens by an antireflective coated mirror (designed to reflect wavelength 532 nm and transmit 560 nm).

A number of investigators have used continuous chromatic light sources, such as a mercury arc lamp or halogen lamp, to provide illumination light (see Santiago *et al.* 1998). In this situation, an excitation filter must be used to allow only a narrow wavelength band of light to illuminate the test section. If a continuous light source is used, then a mechanical or electro-optical shutter must be used to gate the light before reaching the image recording device, so that discrete particle images are formed.

For high resolution measurements, the microscope lens can be an oil immersion, high numerical aperture ($NA = 1.4$), high magnification ($M = 60$, or $M = 100$), low distortion, *CFI Plan Apochromat* lens. Lower resolution and lower magnification microscope lenses, such as an air immersion lens with a numerical aperture $NA = 0.6$ and magnification $M = 40$, can be used, but with decreased measurement resolution.

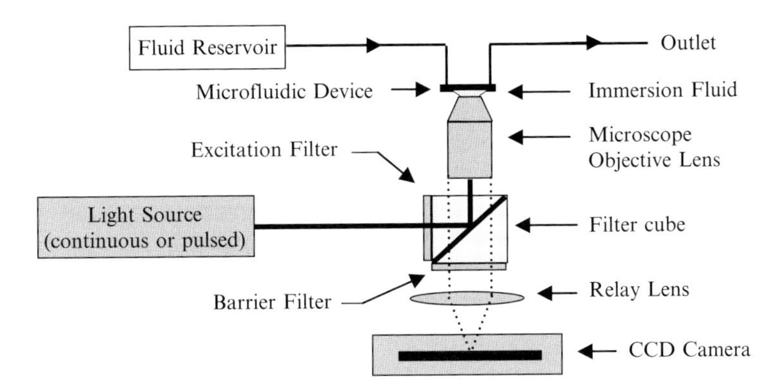

Fig. 7.8: Schematic of an epi-illumination μPIV system using either a continuous or pulsed illumination light source.

The microfluidic test section must have at least one optically transparent wall, so that it can be viewed through the microscope lens. A barrier filter is positioned between the mirror and the relay lens (see Fig. 7.8). The barrier filter is usually a long pass filter that filters out the illumination light that is reflected by the surface of the test section or scattered by the particles.

A sensitive large-format interline-transfer CCD camera is commonly used to record the particle-image fields. Suitable cameras are available from most PIV system manufacturers. A large-format CCD array is desirable because it allows for more particle images to be recorded, and increases the spatial dynamic range of the measurements. The interline transfer feature of the CCD camera allows for two particle-image frames to be recorded back-to-back to within a 200 ns time delay.

7.6.1 *Flow through a rectangular capillary*

The accuracy of μPIV can be determined by comparing measured velocity fields to a known analytical solution. The velocity field of flow through a rectangular duct can be calculated by solving the Stokes equation (the low Reynolds number version of the Navier–Stokes equation), with no-slip velocity boundary conditions at the wall (Meinhart, *et al.* 1999). Given dimensions of a rectangular duct of $2H$ for height and $2W$ for width, assuming that $W \gg H$, the coordinates in the vertical and the wall-normal directions can be made nondimensional by the channel half height H (see Fig. 7.9), such that

$$Y = \frac{y}{H} \text{and} Z = \frac{z}{H}, \tag{7.17}$$

The velocity in the streamwise x-direction, scaled by the bulk velocity in the channel is given as

$$\frac{u}{U} = \frac{3\mu v_x}{H^2(-dP/dx)}, \tag{7.18}$$

The analytical solution to the flow field near the wall for high aspect channels (i.e. $W \gg H$) can be expressed as (Deen 1998)

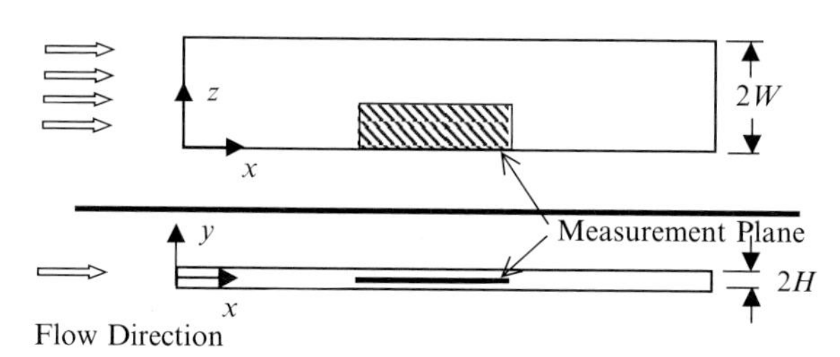

FIG. 7.9: Geometry of the microchannel. The microchannel is 2H high and 2W wide, and is assumed infinitely long in the axial direction. The centerline of the channel is at y = 0. The microscope objective images the test section from below, in the lower figure (Stone *et al.* 2002).

$$\frac{u}{U} = \frac{3}{2}(1 - Y^2) - 6\sum_{n=0}^{\infty}\frac{(-1)^n}{\lambda_n^3}e^{-\lambda_n Z}\cos\lambda_n Y, \tag{7.19}$$

where Eigenvalues are given as $\lambda_n = (n + 1/2)\pi h$. Sufficiently far from the wall (i.e. Z $\gg$ 1) the analytical solution in the Y direction (for constant Z), converges to the well known parabolic profile for flow between infinite parallel plates. In the Z direction (for constant Y) however, the flow profile is unusual in that it has a very steep velocity gradient near the wall ($Z < 1$) which reaches a constant value away from the wall ($Z \gg 1$).

A 30 μm × 300 μm × 25 mm glass rectangular microchannel, fabricated by *Wilmad Industries,* was mounted flush to a 170 μm thick glass coverslip and a microscope slide. The microchannel was horizontally positioned using a high-precision x-y stage, and verified optically to within ~400 nm using epi-fluorescent imaging and image enhancement. The flow in the glass microchannel was imaged through an inverted epi-fluorescent microscope and a *Nikon Plan Apochromat* oil-immersion objective lens with a magnification $M = 60$ and a numerical aperture $NA = 1.4$. The object plane was placed at approximately 7.5 ± 1 μm from the bottom of the 30 μm thick microchannel.

The Plan Apochromat lens was chosen for the experiment, because it is a high quality microscope objective designed with low curvature of field, low distortion, and corrected for spherical and chromatic aberrations.

Since deionized water (refractive index n_w=1.33) was used as the working fluid but the lens immersion fluid was oil (refractive index n_i=1.515), the effective numerical aperture of the objective lens was limited to $NA \approx n_w/n_i = 1.23$. A filtered continuous white light source was used to align the test section with the CCD camera and to test for proper particle concentration. During the experiment, the continuous light source was replaced by the pulsed Nd:YAG laser. A *Harvard Apparatus* syringe pump was used to produce a 200 μl hr^{-1} flow through the microchannel.

The particle images were analyzed using a custom-written μPIV interrogation package described in Wereley and Meinhart (1998). Specifically, the images were analyzed using

- correlation averaging,
- background removal based on the minimum function, and
- central difference interrogation (CDI).

For this experiment, 20 realizations were used to obtain a high quality signal-to-noise ratio, considering a first interrogation window of only 120 × 8 pixels. The high aspect ratio interrogation region was chosen to provide maximal spatial resolution in the wall normal or spanwise direction while sacrificing spatial resolution in the streamwise direction where there are no velocity gradients at all. The signal-to-noise ratio resulting from these interrogation techniques was high enough that there were no erroneous velocity measurements.

Figure 7.10.a shows an ensemble-averaged velocity-vector field of the microchannel. The images were analyzed using a low spatial resolution away from the

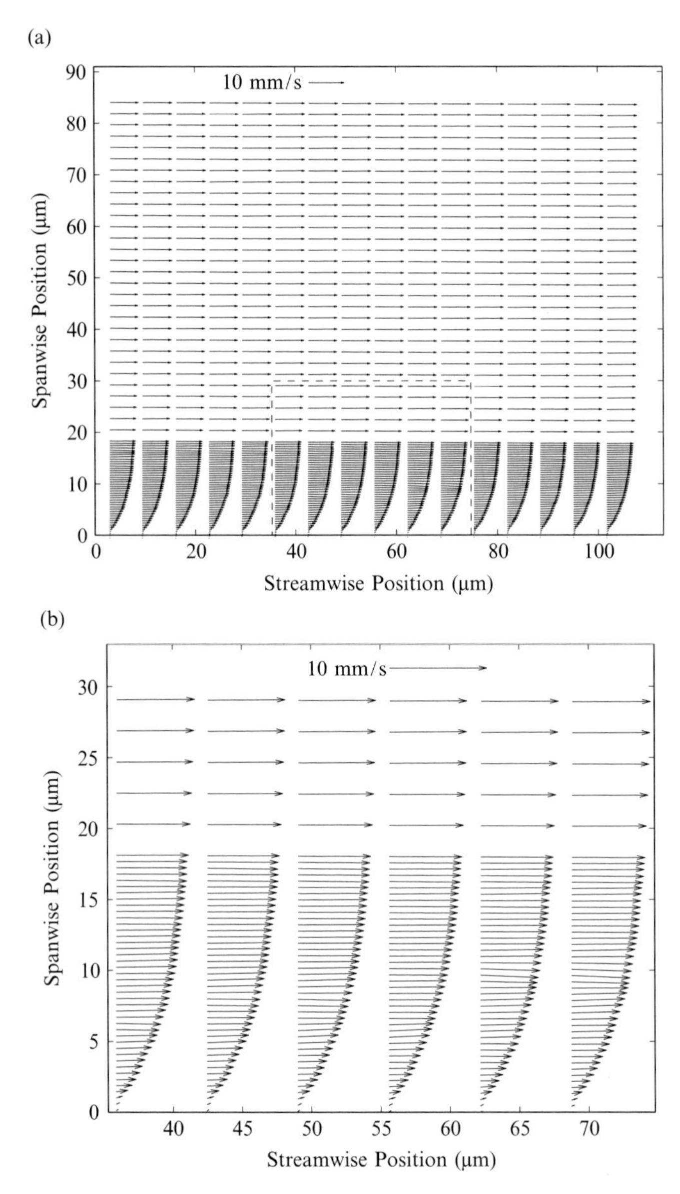

FIG. 7.10.a: Large area view of ensemble-averaged velocity-vector field measured in a 30 μm deep × 300 μm wide × 25 mm channel. The spatial resolution, defined by the interrogation spot size of the first interrogation window, is 13.6 μm × 4.4 μm away from the wall, and 13.6 μm × 0.9 μm near the wall (Meinhart *et al.* 1999). (b) Near wall view of boxed region from Fig. 7.10.a (Meinhart *et al.* 1999).

wall, where the velocity gradient is low, and using a high spatial resolution near the wall, where the wall-normal velocity gradient is largest. The interrogation spots were chosen to be longer in the streamwise direction than in the wall-normal direction. This allowed for a sufficient number of particle images to be captured in an interrogation spot, while providing the maximum possible spatial resolution in the wall-normal direction. The spatial resolution, defined by the size of the first interrogation window, was 120×40 pixels in the region far from the wall and 120×8 pixels near the wall. This corresponds to a spatial resolution of $13.6\ \mu m \times 4.4\ \mu m$ and $13.6\ \mu m \times 0.9\ \mu m$, respectively. The interrogation spots were overlapped by 50 per cent to extract the maximum possible amount of information for the chosen interrogation region size according to the Nyquist sampling criterion. Consequently, the velocity vector spacing in the wall-normal direction was 450 nm near the wall. The streamwise velocity profile was estimated by line-averaging the measured velocity data in the streamwise direction. Figure 7.10.b is a close-up of Fig. 7.10.a, detailing the near-wall region.

Figure 7.11 compares the streamwise velocity profile estimated from the PIV measurements (shown as symbols) to the analytical solution (eqn. 7.19) for laminar flow of a Newtonian fluid in a rectangular channel (shown as a solid curve). The agreement is within 2 per cent of full-scale resolution. Hence, the accuracy of μPIV is at worst 2 percent of full-scale for these experimental conditions. This experiment confirmed for the first time that micro-PIV can be used to measure accurately two-dimensional velocity vector fields in microchannels.

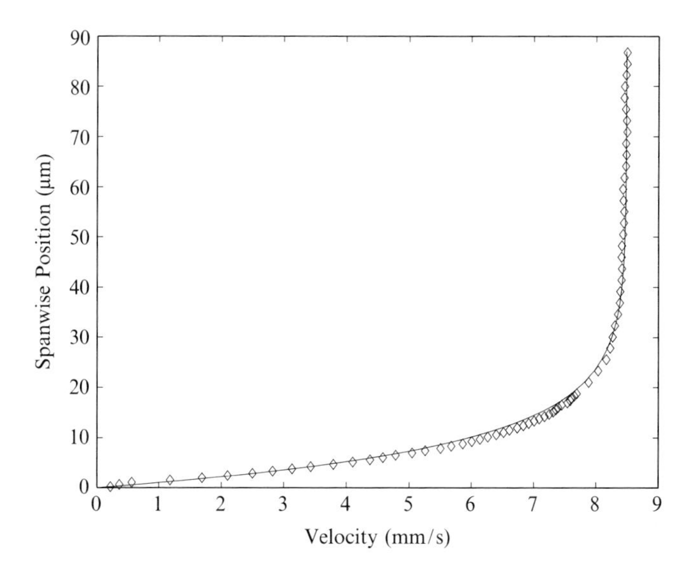

Fig. 7.11: Velocity profile measured in a nominally $30\ \mu m \times 300\ \mu m$ channel. The symbols represent the experimental PIV data while the solid curve represents the analytical solution (Meinhart *et al.* 1999).

7.6.2 *Flow around a blood cell*

A surface tension driven Hele – Shaw flow with a Reynolds number of 3×10^{-4} was developed by placing deionized water seeded with 300 nm diameter polystyrene particles between a 500 μm thick microscope slide and a 170 μm coverslip. Human red blood cells, obtained by auto-phlebotomy, were smeared onto a glass slide. The height of the liquid layer between the microscope slide and the coverslip was measured as approximately 4 μm by translating the microscope objective to focus on the glass surfaces immediately above and below the liquid layer. The translation stage of the microscope was adjusted until a single red blood cell was visible (using white light) in the center of the field of view. This type of flow was chosen because of its excellent optical access, ease of setup, and its 4 μm thickness, which minimized the contribution of out of focus seed particles to the background noise. Also, since red blood cells have a maximal tolerable shear stress, above which hemolysis occurs, this flow is potentially interesting to the biomedical community.

The images were recorded in a serial manner by opening the shutter of the camera for 2 ms to image the flow and then waiting 68.5 ms before acquiring the next image. Twenty-one images were collected in this manner. Since the camera is exposed to the particle reflections at the beginning of every video frame, each image can be correlated with the image following it. Consequently, the 21 images recorded can produce 20 pairs of images, each with the same time between exposures, Δt. Interrogation regions, sized 16×16 pixels, were spaced every 4 pixels in both the horizontal and vertical directions for a 75 per cent overlap. Although technically overlaps greater than 50 per cent over sample the images, they effectively provide more velocity vectors to provide a better understanding of the velocity field. The resulting velocity field is shown in Fig. 7.12.

The images were interrogated using:

- background removal with the minimum function and
- correlation averaging.

The flow exhibits the features that we expect from a Hele–Shaw flow. Because of the disparate length scales in a Hele–Shaw flow, with the thickness much smaller than the characteristic length and width of the flow, an ideal Hele–Shaw flow will closely resemble a 2-D potential flow (Batchelor 1987). However, because a typical red blood cell is about 2 μm while the total height of the liquid layer between the slides is 4 μm, there is a possibility that some of the flow will go over the top of the cell instead of around it in a Hele–Shaw configuration. Since the velocity field in Fig. 7.12 closely resembles that of a potential flow around a right circular cylinder, we can conclude that the flow is primarily a Hele–Shaw flow. Far from the cylinder, the velocity field is uniformly directed upward and to the right at about a 75° angle from horizontal. On either side of the red blood cell, there are stagnation points where the velocity goes to zero. The velocity field is symmetric with respect to reflection in a plane normal to the page and passing through the stagnation points. The velocity field differs from potential flow in that near the red blood cell

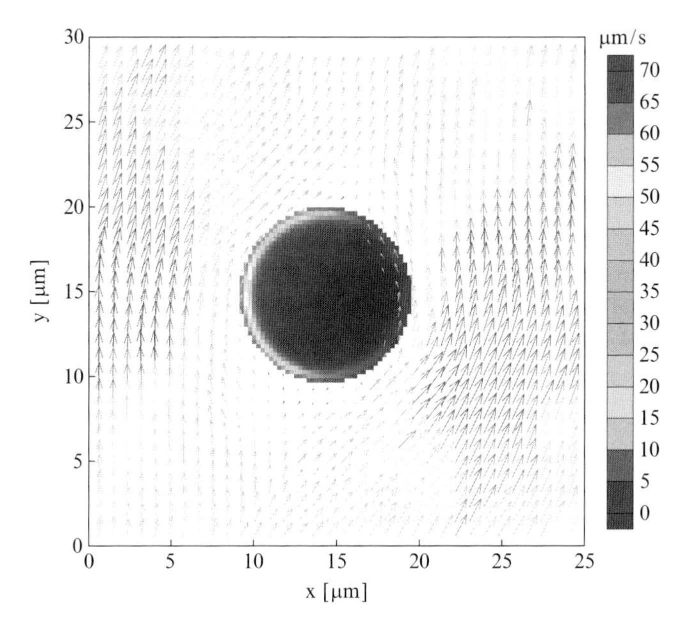

FIG. 7.12: Flow around a single human red blood cell (Wereley and Meinhart 2001).

there is evidence of the no-slip velocity condition. These observations agree well with the theory of Hele–Shaw. Although these measurements are of no particular physiological significance, because the blood cells are affixed to a rigid boundary, measurements such as these could be used to study the reaction of red blood cells or other cells to flow-induced stresses.

7.6.3 *Flow in microfluidic biochip*

Microfluidic biochips are microfabricated devices that are used for delivery, processing, and analysis of biological species (molecules, cells, etc.). Gomez *et al.* (2001) successfully used μPIV to measure the flow in a microfluidic biochip for impedance spectroscopy of biological species. The biochip used for the PIV experiment is fabricated in a silicon wafer with a thickness of 450 μm. It has a series of rectangular cavities connected by channels with a depth of 12 μm. The surface of the chip is covered with a piece of glass of about 0.2 mm thick, so that the images of seeded flows can be taken from the top. During the experiment, water-based suspensions of fluorescein-labeled latex beads with a mean diameter of 1.88 μm were injected into the biochip. The flow is illuminated with a constant intensity mercury lamp. Images are captured with a CCD camera through an epi-fluorescence microscope and recorded at a video rate (30 Hz).

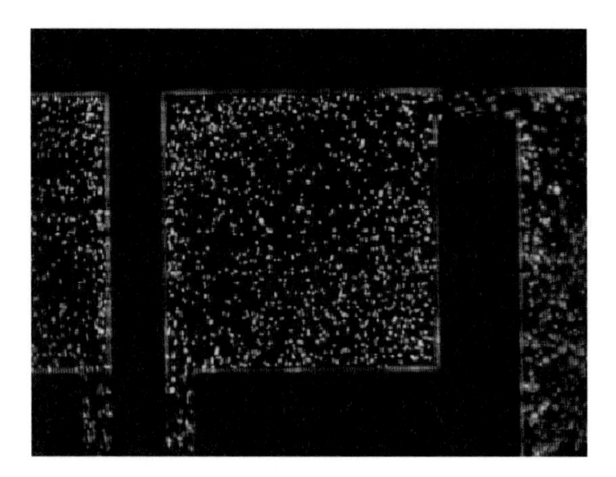

Fig. 7.13: Digital image of the seeded flow in the cavities and channels of the biochip (360×270 pixels, 542×406 μm^2).

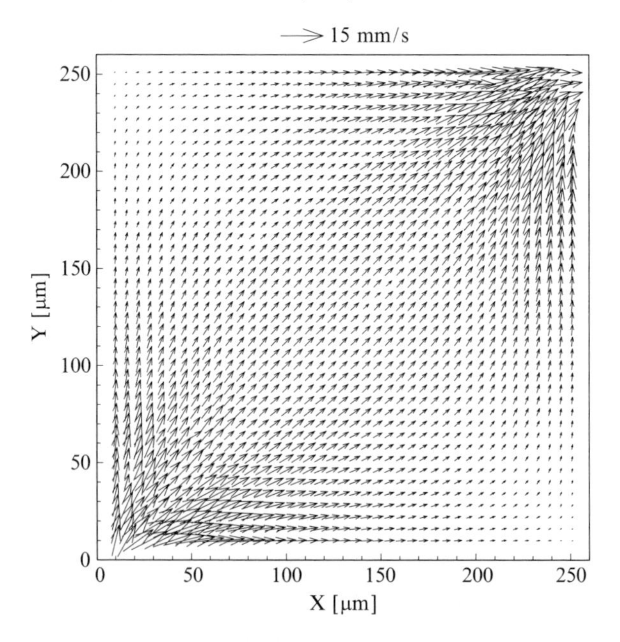

Fig. 7.14: PIV measurement results in a rectangular cavity of the biochip with a spatial resolution of 12×12 μm^2.

One of the PIV images covering an area of 542×406 μm^2 on the chip with a digital resolution of 360×270 pixels is shown in Fig. 7.3. The flow in a rectangular cavity of the biochip is determined by evaluating more than 100 μPIV recording pairs with the ensemble correlation method, CDI and the image correction technique, and the results are given in Fig. 7.14. An interrogation window of 8×8 pixels is chosen for the PIV image evaluation, so that the corresponding spatial resolution is about 12×12 μm^2. The measured velocities in the cavity range from about 100 μm/s to 1600 μm/s.

7.7 Three-dimensional and three-component velocimetry

Over the past five years there has been a significant effort in extending 'classical' micro-PIV beyond the 2-D planar measurements of in-plane two-component (2-C) velocity measurements. The extension from 2-D space into 3-D space is achieved by a number of techniques, digital holography, confocal fluorescent microscopy, and scanning of multiple 2-D planes.

7.7.1 *Digital holography*

Digital holography for microflows was introduced by Yang and Chuang (2005) using a photopolymer plate to record stereoscopic images of flow over a backward facing step. Satake *et al.* (2005, 2006) developed a holographic particle tracking method and applied it to a 92 μm dia. micropipe. Figure 7.15 shows a typical reconstructed holographic image of 1 um dia. particles. From these particle image fields, 104 3-D velocity vectors were recovered and shown in Fig. 7.16.

One of the major drawbacks with holographic imaging is the relatively low particle density that can be recorded. Sheng *et al.* (2006) were able to achieve relatively high image density (of ~1000 particle images) of 0.75 μm and 3.2 μm dia. particles in measurement volumes with depths of 1–10 mm.

An indirect method of using holographic imaging to obtain velocity information was developed by di Leonardo *et al.* (2006). They used holographic techniques to create multiple optical traps in a flowing microchannel. The optical traps were

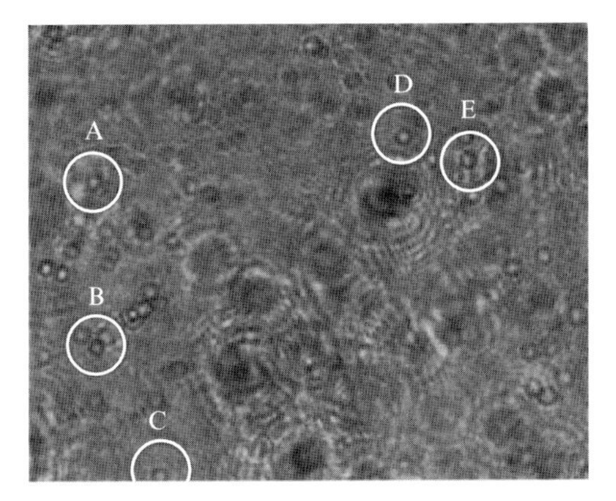

Fig. 7.15: Typical reconstructed holographic images of 1 um dia. particles. Several particle images are highlighted by circles (taken from Satake *et al.* 2006).

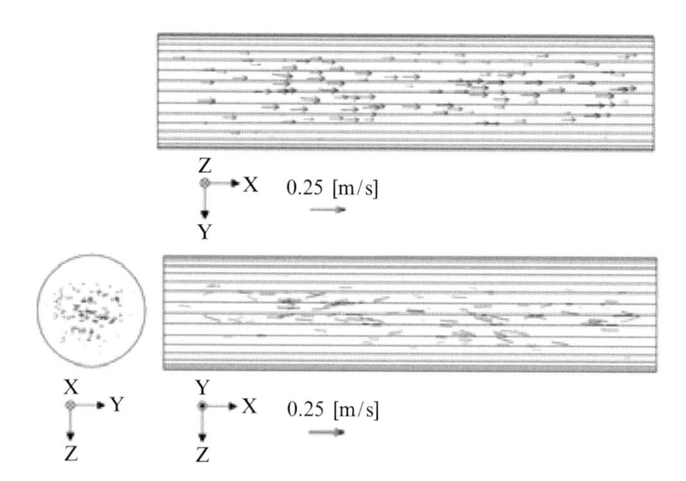

FIG. 7.16: Velocity vectors obtained from reconstructed holographic images; 104 vectors are calculated from 320 particle images (taken from Satake *et al.* 2006).

actuated on and off to trap and release polystyrene beads to probe the velocity at multiple locations in a microchannel flow.

7.7.2 *Stereo imaging*

There have recently been several significant efforts to apply stereo microscope imaging to micro-PIV. This allows one to measure all three components of velocity in a single two-dimensional plane. Stereo imaging has been used successfully and commercialized in macroscale PIV (Arroyo and Greated 1991; Willert 1997; and Prasad 2000).

The primary limitation of stereo micro-PIV, is that there are conflicting requirements between spatial resolution and accuracy of the out-of-plane velocity component. High spatial resolution requires a large imaging aperture. At the same time, a large aperture limits the off axis viewing angle, which is required for accurate stereo imaging.

The first successful attempt of stereo micro-PIV was reported by Lindken *et al.* (2006). They investigated the three-dimensional flow in a 'T-shaped' micromixer with dimensions of $800 \times 200\ \mu m^2$, at a Reynolds number, Re = 120. A typical velocity field is shown in Fig. 7.17. In order to achieve off axis viewing, they used low numerical apertures of NA = 0.14 and 0.28, to achieve a spatial resolution of $44 \times 44 \times 15\ \mu m^3$. The out of plane distance between successive measurement planes was 22 μm.

Bown *et al.* (2006) compared the performance of stereo micro-PIV to computational fluid dynamics simulations of flow over a micro back step. The results indicate that the accuracy of correlation-based measurements is limited by the

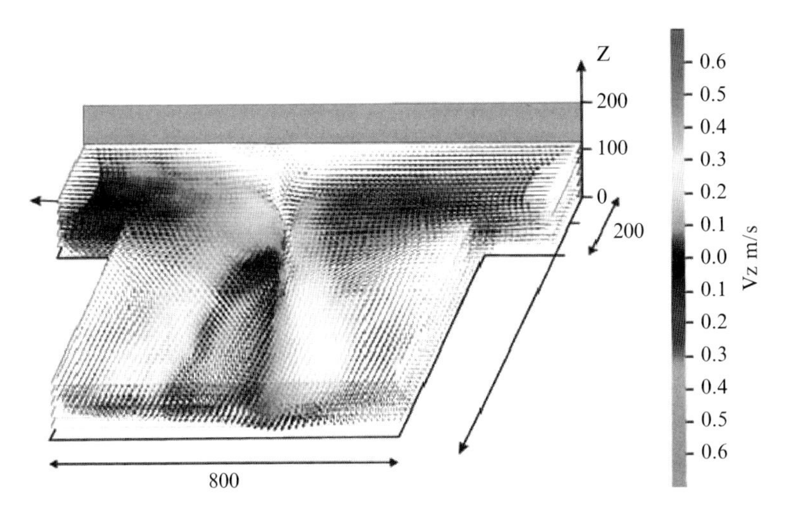

Fig. 7.17: Stereoscopic velocity field of a T-mixer operating a Reynold's number Re = 120 (Image taken from Lindken *et al.* 2006).

degree of overlap between of the two object fields. They suggested that particle tracking is more accurate than correlation-based interrogation and that the accuracy of stereo-PTV is not affected by inexact plane overlap. They present data with a spatial resolution of $10 \times 10 \times 10\ \mu m^3$.

7.7.3 *Defocusing*

The complexity of stereo imaging at the micron scale can be mitigated by employing defocusing digital particle image velocimetry (DDPIV). Here, particle images are recorded through a single collection lens and a three-hole aperture (Willert and Gharib 1992). The DDPIV was applied to a micro backward facing step by Yoon and Kim (2006). Averaging over 2000 vector volumes yielded an average spatial resolution of 5 μm (in plane) and 1 μm (out of plane).

Pereira *et al.* (2007) obtained 3-D velocity measurements of an evaporating water droplet in a $400 \times 300 \times 150$ micron volume with 2 micron diameter particles. Figure 7.18.a shows reconstructed multiple 3-D particle trajectories, depicting the complex flow inside the evaporating droplet. In Fig. 7.18.b, a single particle trajectory is shown is 3-D space, and its respective planar projections.

The concept of defocusing can also be applied to single aperture imaging systems. Building upon the technique of deconvolution microscopy, Park and Kihm (2006) used a single aperture to record images of a 500 nm fluorescent particles flowing over a spherical obstacle. They were able to estimate the out of plane location of individual particles by measuring the particle image location and diameter. They could then estimate the out of plane particle location by comparing the image diameter to the point spread function of the recording optics.

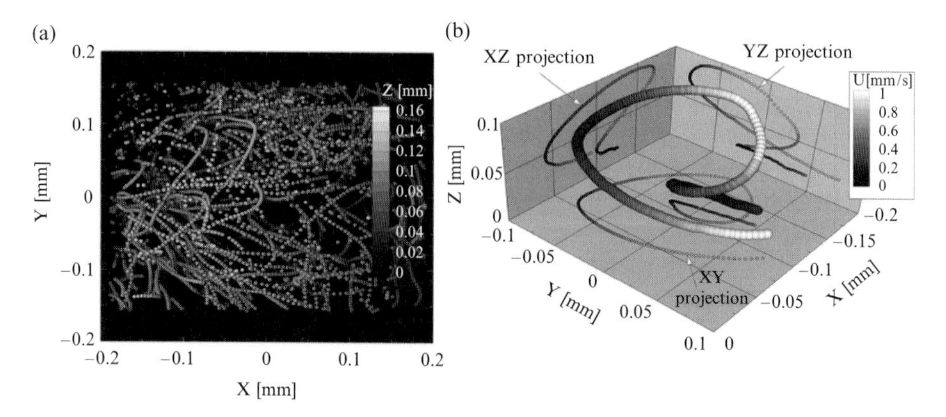

Fig. 7.18: Three dimensional particle trajectories of flow in an evaporating droplet. The measurement volume is $400 \times 300 \times 150$ μm (Images taken from Pereira *et al.* (2007)).

7.7.4 *Confocal imaging*

The application of fluorescent confocal microscopy to micro-PIV was first reported by Park *et al.* (2004). Confocal imaging focuses the illumination and recording optics to a single spot (see Fig. 7.19). This significantly decreases the depth of field and the signal-to-noise of the image. Since confocal can only image a single point simultaneously, it must scan in order to obtain imaging information over a 2-D plane. Figures 7.19 and 7.20 show the optical path through confocal imaging systems. The time duration required to scan a 2-D plane places restrictions on the applicability of confocal to a wide range of microfluidic applications. In practice, high scanning rates are obtained by using a rotating NipKow disk (see Fig. 7.20).

A comparison between standard epi-fluorescence imaging and confocal imaging is presenting by Park *et al.* (2004) and shown in Fig. 7.21. In this case 200 nm dia. fluorescent particles are imaged through a 100 um ID pipe. The top image in Fig. 7.21 is taken using a standard 40x objective lens. It is clear the background light decreases the quality of the in-focus particle image field. In comparison, the bottom image in Fig. 21 is taken using confocal imaging. The in-focus images have higher signal to noise and will provide stronger signal to the correlation, as compared to standard epi-fluorescence imaging.

It should be noted that in this particular situation, improvement in particle-image quality can also be achieved in epi-fluorescent imaging mode by using a higher numerical aperture objective lens, say $M = 60$, $NA = 1.4$. This would provide high-quality images without requiring expensive hardware associated with confocal microscopy, and without the restriction of measuring relatively low-speed velocity fields as a result of the time required for confocal scanning.

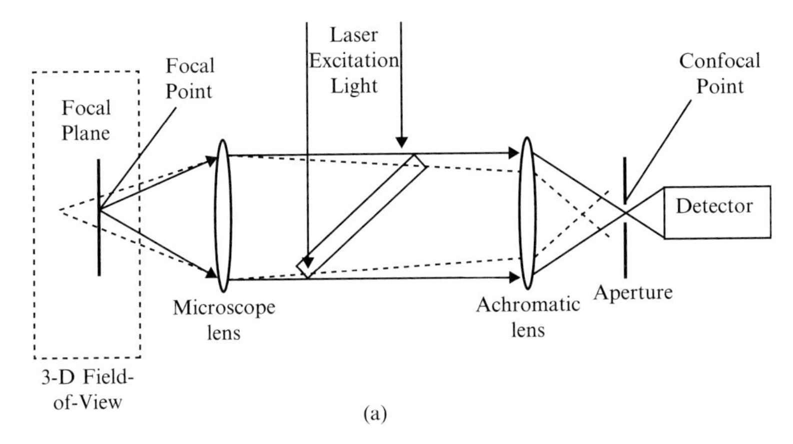

Fig. 7.19: Optical paths through a confocal imaging system (taken from Park *et al.* (2004)).

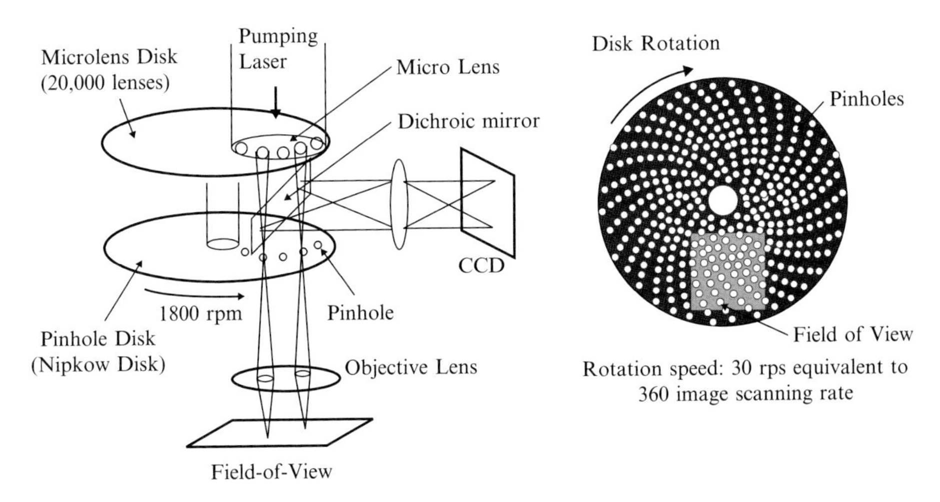

Fig. 7.20: High speed scanning can be achieved using a Nikow disc (taken from Park *et al.* (2004)).

The primary advantages of confocal microscopy are the sharp extinction of background light from out of focus images, and increased resolution in the out of plane direction. This advantage comes at the cost of requiring that the image plane be scanned point-by-point to obtain particle image data. The rate at which a 2-D image plane can be scanned to acquire high-quality images, limits the applicability of the confocal technique to high-speed velocity fields.

The ability of confocal imaging to optically section is a valuable tool for microfluidic diagnostics (Chao *et al.* 2005). It has been exploited for a variety for

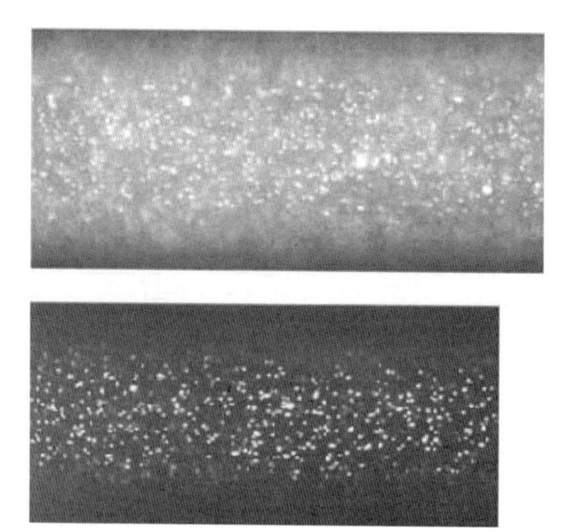

Fig. 7.21: Typical images of 200 nm dia. flow-tracing particles imaged in a 1000 um ID pipe. The top image is taken using a standard epi-fluorescent imaging with a 40x objective lens. The bottom image is obtained using confocal microscopy (Image taken from Park *et al.* 2004).

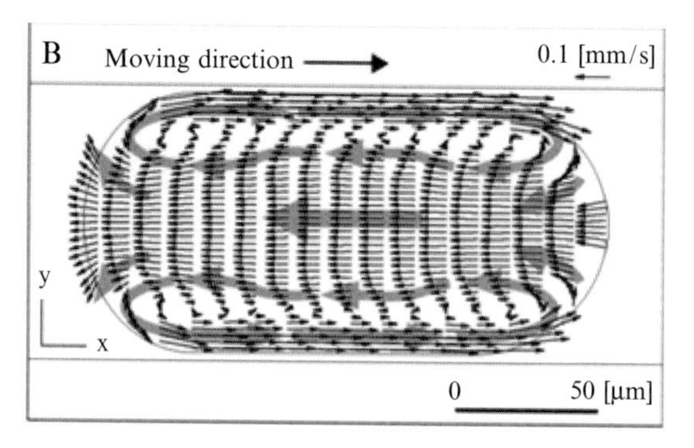

Fig. 7.22: Two-dimensional velocity measurements of an oil droplet surrounded by water flowing through a microchannel. The particle images were imaged using confocal microscopy (Image taken from Kinoshita *et al.* 2007).

imaging flow conditions. For example, it has been applied to study flow of blood cells in a physiological fluid by Lima *et al.* (2006) and used to measure simultaneously velocity fields and pH fields by Ichiyanagi *et al.* (2007).

Digital microfluidics is an important and growing field of microfluidics, where discreet oil/water droplets are formed and manipulated in a microfluidic device. Kinoshita *et al.* (2007) applied confocal micro-PIV to investigate water/glycerol droplets surrounded by silicone oil flowing through a 100×58 μm PDMS microchannel. A typical velocity field of flow inside a water/glycerol droplet at 12 μm from a wall is shown in Fig. 7.22. The flow is in the frame of reference traveling with the droplet. After scanning multiple planes using confocal imaging, the continuity equation was then applied to estimate the out-of-plane velocity component, measuring the full 3-D, 3-C velocity field with a spatial resolution of $9.1 \times 9.1 \times 2$ μm.

7.8 Future directions

As of the writing of this chapter, approximately 10 years after the inception of micro-PIV, many of the variations of microscopic imaging methods have been used successfully to improve μPIV. For example, scanning confocal microscopy, stereo microscopy, fluorescence microscopy, total internal reflection microscopy, and so on, have all been used. Each of these techniques has delivered some improvement to the basic two-dimensional, two-component technique. Future innovations may include other microscopic imaging techniques not yet adapted to micro-PIV. In addition, future improvement may come from using shorter wavelengths, e.g. UV illumination, that could improve the maximum spatial resolution of the μPIV technique from where it stands at approximately 1 μm to approximately 250 nm. Higher frame rate cameras and lasers are allowing the same kind of cinematographic PIV performed at large length scales to be performed at small length scales. One limitation here is the sensitivity and noise levels typically found in CMOS imagers. However, these specifications should be improved in time. Furthermore, different modalities of imaging may allow micro-PIV to be performed in domains previously thought impossible. For example, X-Ray microimaging and magnetic resonance imaging (MRI) are promising techniques that can enable micro-PIV in domains inside living tissues. Micro-PIV should certainly be regarded as a technique that is at the beginning of its useful lifetime and one that will be pervasive in the coming decades.

References

Adrian, R. J., 1991. 'Particle-imaging tecniques for experimental fluid mechanics.' *Annual Review of Fluid Mechanics*, **23**, p. 261.

Adrian, R. J., 2005. '20 years of particle image velocimetry.' *Experiments in Fluids*, **39**, pp. 159–169.

Arroyo, M., and Greated, C, 1991. 'Stereoscopic particle image velocimetry.' *Measurement Science and Technology*, **2**, pp. 1181–1186.

Axelrod, D., Long, J., and Davidson, M., *Olympus Microscopy Resource Center.* http://www.olympusmicro.com/primer/java/tirf/penetration/index.html.

Batchelor, G. K., 1987. *An Introduction to Fluid Dynamics.* Cambridge University Press, Cambridge.

Bendat, J. S., and Piersol, J. G., 1986. *Random Data: Analysis and Measurement Procedures.* Wiley, New York.

Bourdon, C. J., Olsen, M. G., and Gorby, A. D., 2004. 'Validation of an analytical solution for depth of correlation in microscopic particle image velocimetry.' *Measurement Science and Technology*, **15** (2), pp. 318–327.

Bourdon, C. J., Olsen, M. G., and Gorby, A. D., 2006. 'The depth of correlation in micro-PIV for high numerical aperture and immersion objectives.' *Journal of Fluids Engineering*, **128** (4), pp. 883–886.

Bown, M. R., Macinnes, J. M., and Allen, R. W. K., 2006. 'Three-component micro-PIV using the continuity equation and a comparison of the performance with that of stereoscopic measurements.' *Experiments in Fluids*, **42** (2), pp. 197–205.

Brody, J. P., Yager, P., Goldstein, R. E., and Austin, R. H., 1996. 'Biotechnology at low Reynolds numbers.' *Biophysical Journal*, **71**, pp. 3430–3441.

Chao, S. H., Holl, M. R., Koschwanez, J. H. *et al.*, 2005. 'Velocity measurement in microchannels with a laser confocal microscope and particle linear image velocimetry.' *Microfluidics and Nanofluidics*, **1** (2), pp. 155–160.

Dabboursi, B. O., Rodriquez-Viejo, J., Mikulec, F. V., Heine, J. R., Mattoussi, H., Ober, R., Jensen, K. F., and Bawendi, M. G., 1997. '(CdSe)ZnS core-shell quantum dots: synthesis and characterization of a size series of highly luminescent nanocrystallites.' *Journal of Physical Chemistry B*, **101**, p. 9463.

Deen, W. M., 1998. *Analysis of Transport Phenomena*. Oxford University Press, Oxford.

Devasenathipathy, S., Santiago, J. G., Wereley, S. T., and Meinhart, C. D., 2003. 'Particle tracking techniques for microfabricated fluidic systems.' *Experiments in Fluids*, **34**, pp. 504–513.

di Leonardo, R., Leach, J., Mushfique, H., Cooper, J., Ruocoo, G., and Padgett, M., 2006. 'Multipoint holographic optical velocimetry in microfluidic systems.' *Physical Review Letters*, **96**, p. 134502.

Freudenthal, P., Pommer, M., and Meinhart, C. D., 2007. 'Quantum nanosperes for sub-micron velocimetry.' *Experiments in Fluids*, **43**, pp. 525–533.

Gad-el-Hak, M., 1999. 'The fluid mechanics of microdevices – The Freeman scholar lecture.' *Journal of Fluids Engineering*, **121**, pp. 5–33.

Gomez, R., Bashir, R., Sarakaya, A., Ladisch, M. R., Sturgis, J., Robinson, J. P., Geng, T., Bhunia, A. K., Apple, H. L., and Wereley, S. T., 2001. 'Microfluidic biochip for impedance spectroscopy of biological species.' *Biomedical Microdevices*, **3** (3), pp. 201–209.

Gui, L., and Wereley, S. T., 2002. 'A correlation-based continuous window-shift technique to reduce the peak-locking effect in digital PIV image evaluation.' *Experiments in Fluids*, **32**, 506–517.

Ho, C. M., and Tai, Y. C., 1998. 'Micro-electro-mechanical-systems and fluid flows.' *Annual Review of Fluid Mechanics*, **30**, pp. 579–612.

Ichiyanagi, M., Sato, Y., and Hishida, K., AUG 2007. 'Optically sliced measurement of velocity and pH distribution in microchannel.' *Experiments in Fluids*, **43** (2–3), pp. 425–435.

Jin, S., Huang, P., Park, J., Yoo, J. Y., and Breuer, K. S., 2004. 'Near-wall surface velocity using evanescent wave illumination.' *Experiments in Fluids*, **37**, pp. 825–833.

Kinoshita, H., Kaneda, S., Fujii, T. *et al.*, 2007. 'Three-dimensional measurement and visualization of internal flow of a moving droplet using confocal micro-PIV.' *Lab on a Chip*, **7** (3), 338–346.

Lima, R., Wada, S., Tsubota, K. *et al.*, APR 2006. 'Confocal micro-PIV measurements of three-dimensional profiles of cell suspension flow in a square microchannel.' *Measurement Science & Technology*, **17** (4), pp. 797–808.

Lindken, R., Westerweel, J., and Wieneke, B., 2006. 'Stereoscopic micro particle image velocimetry.' *Experiments in Fluids*, **41** (2), pp. 161–171.

Maxey, M. R., and Riley, J. J., 1983. 'Equation of motion for a small rigid sphere in a nonuniform flow.' *Physics of Fluids*, **26**, pp. 883–889.

Meinhart, C. D., and Wereley, S. T., JUL 2003. 'The theory of diffraction-limited resolution in microparticle image velocimetry.' *Measurement Science & Technology*, **14** (7), pp. 1047–1053.

Meinhart, C. D., Wereley, S. T., and Santiago, J. G., 1999. 'PIV measurements of a microchannel flow.' *Experiments in Fluids*, **27**, pp. 414–419.

Meinhart, C. D., Wereley, S. T., and Santiago, J. G., 2000. 'A PIV algorithm for estimating time-averaged velocity fields.' *Journal of Fluids Engineering*, **122**, pp. 285–289.

Murray, C. B., Morris, D. J., and Bawendi, M. G., 1993. 'Synthesis and characterization of nearly monodisperse CdE (E = sulfur, selenium, tellurium) semiconductor nanocrystallites.' *Journal of American Chemistry Society*, **115**, p. 8706.

Olsen, M. G., and Adrian, R. J., 2000. 'Out-of-focus effects on particle image visibility and correlation in microscopic particle image velocimetry.' *Experiments in Fluids*, **29**, S166–S174 (Suppl. S).

Olsen, M. G., and Bourdon, C. J., 2003. 'Out-of-plane motion effects in microscopic particle image velocimetry.' *Journal of Fluids Engineering-Transactions of the ASME*, **125** (5), pp. 895–901.

Park, J. S., Kihm, K. D., 2006. 'Three-dimensional micro-PTV using deconvolution microscopy.' *Experiments in Fluids*, **40** (3), pp. 491–499.

Park, J. S., Choi, C. K., and Kihm, K. D., 2004. 'Optically sliced micro-PIV using confocal laser scanning microscopy (CLSM).' *Experiments in Fluids*, **37** (1), pp. 105–119.

Pereira, F., Lu, J., Castano-Graff, E., Gharib, M., 2007. 'Microscale 3D flow mapping with mDDPIV.' *Experiments in Fluids*, **42** (4), pp. 589–599.

Poiseuille, J. L. M., 1836. *Annales des Sciences Naturelles*, **5**, p. 111.

Pouya, S., Koochesfahani, M., Snee, P., Bawendi, M., and Nocera, D., 2005. 'Single quantum dot (QD) imaging of fluid flow near surfaces'. *Experiments in Fluids*, **39** (4), 784–786.

Prasad, A., 2000. 'Stereoscopic particle image velocimetry.' *Experiments in Fluids*, **29**, pp. 103–116.

Raffel, M., Willert, C., Wereley, S., and Kompenhans, J., 2007. *Particle Image Velocimetry: A Practical Guide*. Springer, New York.

Sadr, R., Yoda, M., Zheng, Z., and Conlisk, A. T., 2004. 'An experimental study of electro-osmotic flow in rectangular microchannels.' *Journal of Fluid Mechanics*, **506**, pp. 357–367.

Sadr, R., Li, H., and Yoda, M., 2005. 'Impact of hindered brownian diffusion on the accuracy of particle-image velocimetry using evanescent-wave illumination.' *Experiments in Fluids*, **38** (1), pp. 90–98.

Saffman, P. G., 1965. 'The lift on a small sphere in a slow shear flow.' *Journal of Fluid Mechanics*, **22**, pp. 385–400.

Santiago, J. G., Wereley, S. T., Meinhart, C. D., Beebe, D. J., and Adrian, R. J., 1998. 'A micro particle image velocimetry system.' *Experiments in Fluids*, **25** (4), pp. 316–319.

Satake, S., Kunugi, T., Sato, K. *et al.*, 2005. 'Three-dimensional flow tracking in a micro channel with high time resolution using micro digital-holographic particle-tracking velocimetry.' *Optical Review*, **12** (6), pp. 442–444.

Satake, S., Kunugi, T., Sato, K. *et al.*, 2006. 'Measurements of 3D flow in a micro-pipe via micro digital holographic particle tracking velocimetry.' *Measurement Science & Technology*, **17** (7), pp. 1647–1651.

Sharp, K. V., and Adrian, R. J., 2005. 'On flow-blocking particle structures in microtubes.' *Microfluidics and Nanofluidics*, **1**, pp. 376–380.

Sheng, J., Malkiel, E., and Katz, J., 2006. 'Digital holographic microscope for measuring... particle distributions and motions.' *Applied Optics*, **45**, pp. 3893–3901.

Sinton, D., 2004. 'Microscale flow visualization.' *Microfluidics and Nanofluidics*, **1**, pp. 2–21.

Stone, S. W., Meinhart, C. D., and Wereley, S. T., 2002. 'A microfluidic-based nanoscope.' *Experiments in Fluids*, **33**, pp. 613–619.

Quantum Dot Corp. 2005. www.qdot.com

Wereley, S. T., and Meinhart, C. D., 2001. 'Adaptive second-order accurate particle image velocimetry.' *Experiments in Fluids*, **31**, pp. 258–268.

Wereley, S. T., and Meinhart, C. D., 2005. 'Micron-resolution particle image velocimetry' in *Micro- and Nano-Scale Diagnostic Techniques*, ed. Kenny Breuer. Springer-Verlag, New York.

Wereley, S. T., and Meinhart, C. D., 2007. Micro-PIV. In *Particle Image Velocimetry: A Practical Guide*, ed. M. Raffel, C. Willert, S. Wereley, and J. Kompenhans. Springer, New York.

Wereley, S. T., Santiago, J. G., Meinhart, C. D., and Adrian, R. J., 1998. 'Velocimetry for MEMS applications.' *Proceedings of ASME/DSC*, **66** (*Microfluidics Symposium*, Nov. 1998, Anaheim, CA).

Westerweel, J., Geelhoed, P. F., and Lindken, R., 2004. 'Single-pixel resolution ensemble correlation for PIV applications.' *Experiments in Fluids*, **37**, pp. 375–384.

Willert, C., 1997. 'Stereoscopic digital particle image velocimetry for application in wind tunnel flows.' *Measurement Science and Technology*, **8**, pp. 1465–1479.
Willert, C. E., and Gharib, M., 1992. 'Three-dimensional particle imaging with a single camera.' *Experiments in Fluids*, **12**, pp. 353–358.
Wu, X., 2003. 'Detecting nuclear antigens using Q-dot streptavidin conjugates.' *Quantum Dot Vis*, **1**, pp. 10–13.
Yang, C. T., and Chuang, H. S., AUG 2005. 'Measurement of a microchamber flow by using a hybrid multiplexing holographic velocimetry.' *Experiments in Fluids*, **39** (2), pp. 385–396.
Yoon, S. Y., and Kim, K. C., 2006. '3D particle position and 3D velocity measurement in a microvolume via the defocusing concept.' *Measurement Science and Technology*, **17**, pp. 2897–2905.
Zettner, C., and Yoda, M., 2003. 'Particle velocity field measurements in a near-wall flow using evanescent wave illumination.' *Experiments in Fluids*, **34** (1), pp. 115–121.

8

NEAR-FIELD OPTICAL AND PLASMONIC IMAGING

Yuan Wang and Xiang Zhang

This chapter reviews the recent development of near-field biomolecule imaging systems. Near-field scanning optical microscopy (NSOM) offers a practical means of optical imaging at a resolution well beyond the diffraction limit of light. Applications are limited, however, due to strong attenuation of transmitted light through the sub-wavelength aperture. A variety of approaches to address this problem, such as apertureless NSOM and plasmonic near-field focusing devices, are covered in this chapter. These approaches provide an enhanced nanoscale imaging tool for cellular visualization, single molecule detection, and many other applications requiring high spatial and temporal resolution.

8.1 Introduction to nanoscale optical imaging for biomedical applications

Biochemical processes of molecular diffusion, association/dissociation, and conformational change are fundamental to all biological phenomena in living cells. Real-time visualization of such processes at the molecular level is increasingly important for current and future research in biology and medicine. There have been extensive efforts to develop a variety of techniques toward the ultimate goal of single molecule detection and imaging, and single molecule detection has been realized within certain constraints (e.g., via highly dilute solutions). To the best of our knowledge, at the time of this writing, there have been no demonstrations of an imaging system capable of single molecule detection in a living cell with natural biomolecule concentrations.

Imaging the position and trajectory of a single molecule, on a cell membrane or within the cytoplasm, is extremely challenging in optical microscopy due to the fundamental half wavelength diffraction limit in spatial resolution. For instance, the resolution limit for green illumination at 532 nm is around 324 nm in air. Any object smaller than this limit appears as an 'airy disc' when using a conventional optical system (Born *et al.* 1993).

Other imaging technologies, such as scanning electron microscopy (SEM) or atomic force microscopy (AFM), allow superior spatial resolution over optical microscopy. SEM uses an electron beam where the associated de Broglie wavelength is much smaller than wavelengths in the visible regime. While the Rayleigh criterion still holds for electron optics, the imaging resolution limit is vastly improved. Unfortunately, SEM is not a feasible imaging technique for living cells; vacuum requirements, SEM sample preparation, and charge accumulation typically damage delicate biological samples before images can be obtained. AFM, on the other hand, scans a small, sharp tip across the surface of the sample. The force between tip and sample is monitored by tracking the deflection of the cantilever. This provides topological information where nanometer and even sub-nanometer resolution is possible. An AFM can also operate in solution, providing a powerful imaging tool for biology and medical research. Nevertheless, important information such as Raman scattering and fluorescence signals cannot be obtained using AFM.

Near-field scanning optical microscopy (NSOM) techniques combine the advantages of scanning probe technology with optical microscopy; thus, both topographical and local optical information are collected. NSOM studies have been intensively applied in the fields of biology, material science, surface chemistry, information storage, and nanofabrication. NSOM is currently one of the most promising tools capable of realizing optical imaging of biological samples with nanometer resolution.

8.2 Aperture NSOM

8.2.1 *History and principle of conventional near-field imaging system*

The fundamental principle underlying a NSOM system is outlined in Fig. 8.1. Visible light illuminates a conducting plate with a sub-wavelength aperture and light passing through the aperture just beneath the plate remains laterally confined to the aperture size (Hsu 2001). Beyond the plate, the radiating field from the aperture diverges in the far-field resulting in an airy diffraction pattern and cannot accurately reproduce the geometrical image of the sub-wavelength aperture.

When the aperture is positioned in close proximity to the sample surface, the light radiating from the aperture can be used to image a specimen before diffrac-

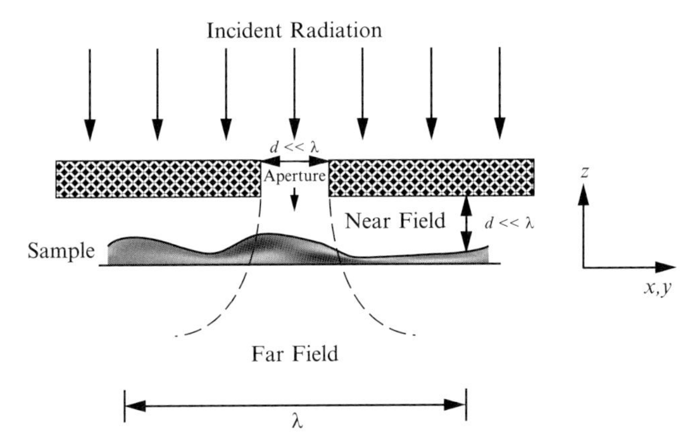

FIG. 8.1: A sketch showing the basic principles of subwavelength resolution in near-field optics (Hsu 2001).

tion degrades the resolution. By scanning this sub-wavelength light source over the object, the scattered light from the object can be detected through the same aperture or from far-field. The detected optical signal can then be used to generate a high-resolution image. Since the resolution is dependent upon the aperture size rather than the wavelength, it should be possible to obtain 50 nm or better resolution if a sufficiently small aperture is used.

Distinguishing NSOM from more conventional diffraction limited techniques is important. An optical microscope can obtain images of isolated nanoscale objects with characteristic length scales in the tens of nanometers via contrast enhancement. The images, however, are at best limited to a size of 100–300 nm by diffraction. Thus determining the real size of nanoscale objects is an impossible task for conventional techniques. Furthermore, when two or more nanoscale objects are separated by less than the diffraction limit (as determined by the Rayleigh criterion), image overlap prevents individual resolution of separate objects. NSOM, in contrast, generates images indicating an object's real size and can still resolve individual objects separated by tens of nanometers, much less than the wavelength.

Initial attempts to implement near-field scanning optical microscopy attest to the difficulty of the technique (Betzig *et al.* 1986). To understand the technical challenges inherent in this form of microscopy, people have performed calculations to obtain an approximation to the pattern of radiation in the near field (Leviatan 1986). These calculations modeled the transmission of light through a small aperture in a perfectly conducting infinite thin screen, and the results showed that the radiation remains collimated up to a distance of at least half the slit width. This suggests that rigid stability requirements are necessary, especially along the z axis.

Experimental evidence of high-resolution imaging using a subwavelength aperture was first demonstrated at microwave frequencies (wavelength $\lambda = 3$ cm) by Ash and Nicholls in 1972 (Ash *et al.* 1972). In their pioneering experiment, periodic features in a metal grating sample were measured with $\lambda/60$ spatial resolution. These exciting results illustrated the feasibility of aperture NSOM first proposed by Synge in 1928 and 1932 and renewed interest for similar investigations using visible radiation (Synge 1928). The results of the microwave experiment could not be immediately extended to the visible regime because the metal screens used in the microwave case are comparatively thicker and nearly completely conducting, whereas the finite conductivity of thin metal films must be considered at visible wavelengths (Dunn 1999). Finally, in the 1980s, initial results of sub-diffraction limit optical measurements were reported by Pohl's laboratory at IBM Zurich (Pohl *et al.* 1984).

8.2.2 *Typical NSOM system configurations*

NSOM designs come in a variety of laboratory setups tailored to the particular needs of the experimentalists. A common NSOM setup, shown in Fig. 8.2, represents an illumination mode NSOM (Dunn 1999). In this design, a NSOM tip on top of an inverted fluorescence microscope illuminates the sample and collects signals

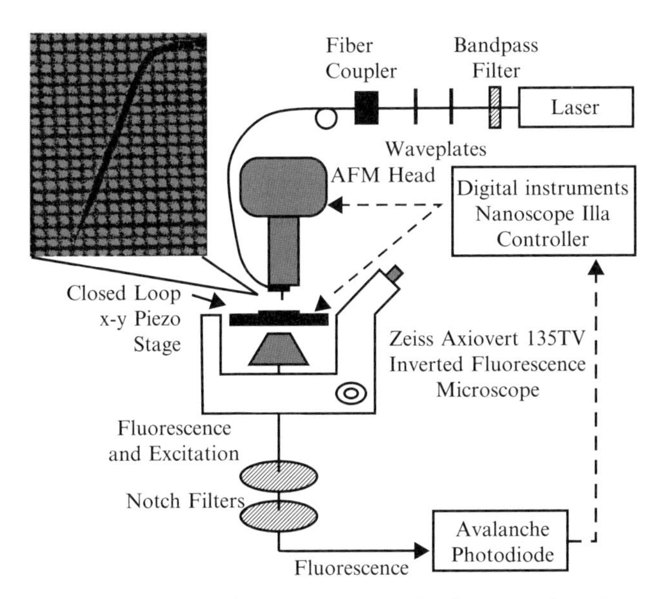

FIG. 8.2: Schematic of a near-field microscope built around an inverted fluorescence microscope (Dunn 1999).

in the far-field—quite appropriate for cell imaging applications. The sample is located and studied through the normal imaging modes of the inverted microscope before performing higher resolution NSOM scans. As shown in Fig. 8.2, the incident light is filtered to a smaller bandwidth before the polarization is adjusted by a variable wave plate. The light is then coupled into an optical fiber and exits at the tapered NSOM tip. The near-field probe tip is mounted on a piezoelectric stage which controls the tip-sample gap distance during a measurement. The transparent sample is mounted on a separate stage that moves, or scans, the region of interest laterally underneath the NSOM probe. The sample fluoresces under light exiting from the NSOM probe and the fluorescence is collected from below using a high numerical aperture microscope objective. Residual excitations are removed by filters and the remaining fluorescence signal is imaged by a high sensitivity detector such as an avalanche photodiode or a photomultiplier tube.

Another NSOM configuration, called collection mode, illuminates the sample from below while the NSOM probe collects the near-field signals. A third mode usually used for opaque samples, reflection mode, measures the reflected signals from probe tip illumination by either fiber-coupling back through the probe or from the side with a long working distance objective.

As with any scanning probe system, the clarity of images obtained depends strongly on the probe. A sharp NSOM tip, capable of exciting samples and/or collecting signals is a necessity for almost all NSOM systems. Tip quality also presents the greatest technical challenge for high performance near-field optical scanning. Thus, tip and aperture quality are perhaps the most important feature for any near-field microscope. Early NSOM tip designs included etched quartz crystals and pulled micropipettes, but these tips generally suffered from low throughput and poor reproducibility.

Tip fabrication typically involves two steps: first producing a sharp tapered probe tip followed by depositing a thin metal film with an aperture at the apex. Two popular tip fabrication processes are laser heating with pulling and chemical etching. Laser heating with pulling utilizes axial tension on the fiber while simultaneously applying heat with a focused laser spot at the desired break point. Accurate control of the heating and pulling parameters allows excellent reproducible tapers and tip diameters, and an example is shown in Fig. 8.3(a) (Lazarev *et al.* 2003b). While laser heating with pulling fabricates tips with a smooth taper surface, chemical etching produces tips with a very short taper. Various etching based methods that produce different configurations of fine optical fiber tips have been proposed in the literature. Several groups have demonstrated the most successful method, HF static etching (Hoffmann *et al.* 1995; Lazarev *et al.* 2003). These tips are at least one order of magnitude more efficient than their pulled counterparts. The tips are etched in a HF solution with an organic solvent on the surface (e.g. silicon oil) and the tip is formed by the meniscus of the etchant along the fiber (Fig. 8.3(b) (Lazarev *et al.* 2003a)).

The design of a tapered fiber optic waveguide coated with a reflective metal film has been used for most successful NSOM tips to date (Pangaribuan *et al.* 1992;

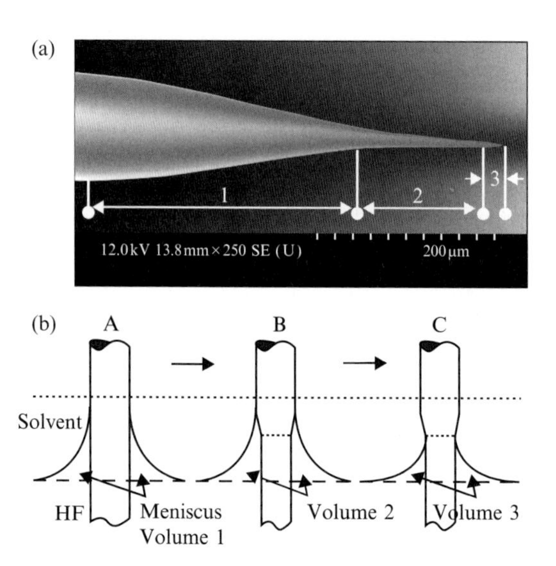

Fig. 8.3: (a) Scanning electron microscope (SEM) image of a typical 'pulled' fiber tip with three distinct taper regions. (b) Meniscus etching method using an organic solvent protection layer over a HF etching solution. (Lazarev *et al.* 2003a; Lazarev *et al.* 2003b).

Shalom *et al.* 1992). However, it involves a complex loss mechanism, and usually suffers from a low transmission coefficient ranging between 10^{-8} and 10^{-4} for apertures with a 30–100 nm diameter. As the diameter of the waveguide decreases in the tapered region (from region 1 to region 2 in Fig. 8.3(a)), the mode confinement also decreases until finally the light intensity dies off exponentially below the cutoff diameter (region 3 in Fig. 8.3(a)). Light leaking from the fiber below the cutoff diameter prevents the formation of a well defined spot, so the sides of the probe must be coated with metal to confine the light. The distance between the cutoff diameter and the apex—characterized by the cone angle—is the most crucial parameter of the tip. Low signal measurements including single molecule spectroscopy, Raman spectroscopy, data storage applications, and high-speed imaging would greatly benefit from even more efficient NSOM probes. As a result, alternative taper formation methods offering more control over the cone angle of the tip are under investigation (Biagioni *et al.* 2005; Mihalcea *et al.* 1996; Suh *et al.* 2000).

Popular near-field scanning microscope modes discussed in literature are constant distance mode (CDM), constant height mode (CHM), and constant optical intensity mode (COM) (Carminati *et al.* 1996; Carminati *et al.* 1997; Hecht *et al.* 1997). CDM uses a feedback loop to maintain a constant tip to sample surface distance. CHM, on the other hand, scans without a feedback loop so the NSOM probe does not follow the sample topography but rather remains at a fixed height relative to the stage, not the sample surface. Alternatively, the feedback loop for COM acts to maintain a constant optical intensity. CHM and COM, believed to be free from topography artifacts, have been demonstrated as equivalent imaging techniques (Carminati and Greffet 1996). These two modes account for purely optical information from the sample including topography. CDM is extensively used in scanning near-field optics and allows the vibration amplitude (and thus the average sample-probe distance) to be held constant during scanning. This notably minimizes the direct influence of the sample topography on the NSOM signal and causes little to no damage to the tip or sample, so CDM is also the most popular mode for bio-sample imaging. In addition, the output signal of the feedback provides a topographic image of the sample. On a final note, NSOM, is a two dimensional technique where molecules under study must lie within a short distance ($<$50 nm) to obtain high spatial resolution. This can be an advantageous depending on the particular application.

8.2.3 *Applications and challenges*

The ability to image samples at high magnification is extremely important in the biological sciences. Nanometer resolution eliminates the averaging effect in conventional diffraction limited images, and this opens new avenues of investigation into fundamental molecular level processes. Single molecule detection, imaging, and improved resolution remain long-time goals of microscopy. In 1993, Betzig and Chichester reported the detection of fluorescence from single molecules with NSOM (Betzig *et al.* 1993). They probed carbocyanine dye (diIC12) molecules spread in a thin coating of poly (methyl methacrylate) (PMMA) on a glass cover slip. These results not only demonstrated single molecule detection with NSOM but also determined the orientation of the transition dipole. This has spurred further efforts to develop NSOM measurements under ambient conditions and even in aqueous surroundings.

Researchers have measured single molecule fluorescence lifetimes and the results revealed complications associated with metal coated NSOM tips (Ambrose *et al.* 1994; Bian *et al.* 1995; Meixner *et al.* 1995). The metal on the tip may reduce the fluorescence signal from a single molecule and several experiments have concluded that the tip-molecule gap indeed modifies the measured fluorescence lifetime of the molecule. A probable explanation includes a radiative and non-radiative decay rate dependence on the metal surface tip distance. Therefore, one can measure variations in the fluorescence lifetime due to the actual distance between a metal-coated NSOM tip and the molecule.

The number and breadth of applications using NSOM to gain insight into sample properties at the nanometer level is expanding at a rapid pace. For example, single molecule diffusion experiments illustrate how high spatial resolution yields new, previously hidden dynamics at interfaces (Ambrose *et al.* 1994; Bian *et al.* 1995; Meixner *et al.* 1995). NSOM measurements on fixed cells under both dry and buffered conditions have become routine, but the extension to free cells remains a major challenge. Overcoming this challenge will present exciting potential for measuring the structure and dynamics of living cells with unprecedented spatial precision.

Near-field microscopy offers a unique, mostly non-invasive technique to determine the position and orientation of a single molecule. Unfortunately, NSOM resolution has been limited to 20 nm at best with little improvement since the first results in 1984 (Pohl *et al.* 1984) because there exists a minimum optical aperture size limit for metallic holes with a diameter less than the skin depth of the surrounding metal. Another near-field imaging configuration to obtain resolution below this limit is presented in the next section.

8.3 Apertureless NSOM

8.3.1 *Field enhancement by metallic tips*

In apertureless NSOM, a metallic probe tip is used to convert evanescent electromagnetic waves into propagating waves. One reason for metallic probe tips instead of a dielectric tip (or sharpened optical fiber) is the higher scattering efficiency of metal compared to dielectrics (Inouye 1994). The technique has been successfully demonstrated for coherent, fluorescence, and non-linear imaging.

Recent studies focus on the key challenge of achieving significant optical near-field enhancement at the end of the NSOM probe. Several theoretical studies have predicted a large field enhancement via plasmon resonance excitations when a component of the electric field is parallel to the tip axis (Martin *et al.* 2001; Novotny *et al.* 1997). These studies also note that NSOM tips are more accurately modeled as semi-infinite rods or semi-infinite cones rather than as finite scatters (e.g. a cylinder or an ellipsoid). New calculations show that resonance conditions are substantially different compared to the finite ellipsoid solid model due to the severe phase retardation (Martin *et al.* 2001).

8.3.2 *Apertureless NSOM systems*

Figure 8.4 shows the configuration of a reflection mode apertureless NSOM system (Inouye 1994). The system is built on a conventional upright optical microscope that simultaneously allows focusing of the laser beam, collection of the reflected light, as well as observation of the sample, tip, and diffraction spot. The sample is placed upon a glass prism with refractive index matching oil. The NSOM probe is an etched metal wire attached to a piezoelectric ceramic which vibrates at a natural resonant frequency in the z-direction. The radius of curvature for the apertureless NSOM tip is between 10 nm and 100 nm. The vibration amplitude of the cantilever is measured by a transverse laser beam to control the tip-sample separation. The stiffness constant and vibration amplitude of the apertureless NSOM probe adapt well to performing an AFM/STM measurement; this ensures that the tip-sample distance is small and allows for high resolution near-field imaging (Bachelot *et al.* 1997; Inouye and Kawata 1994).

An undesirable property of apertureless NSOM with a metallic tip is the noise, or scattered light, from the sample surface. Two methods have been developed to enhance the signal-to-noise ratio and reduce the scattered light. One method uses evanescent illumination with total internal reflection since the intensity of the localized field on the surface decays rapidly along the z-axis. Another method, lock-in detection, uses the z-axis vibration frequency as a reference to demodulate the detected signal. Lock-in amplifiers eliminate undesired signals, such as background light or non-evanescent scattered waves, by selecting only a small bandwidth of frequencies in the detected signal.

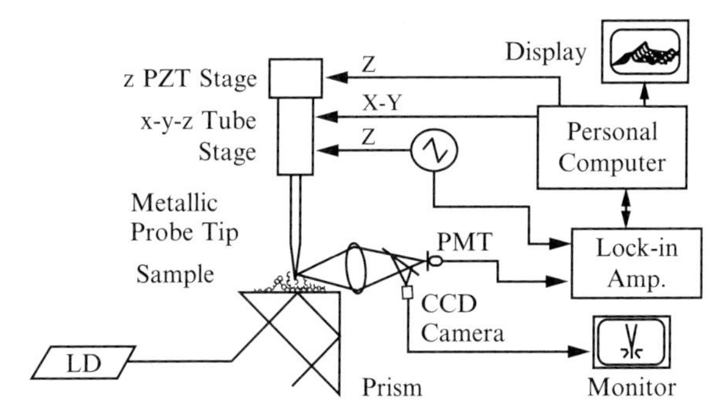

Fig. 8.4: Optical setup of the NSOM with a metallic tip. LD, laser diode; PMT, photomultiplier tube (Inouye *et al.* 1994).

8.3.3 *Applications*

Perhaps the most fascinating application of apertureless NSOM is imaging living biological samples in vitro. Other techniques, such as electron microscopy, image structural information with beautiful detail but are rarely compatible with physiological conditions. Images of a murine myocardial cell using an apertureless NSOM were recently demonstrated (Fig. 8.5) (Inouye 2001). Unlike the topography image, the localized optical contrast in the NSOM image represents the distribution of the effective refractive index of the sample. In particular, narrow channel structures observed in the NSOM image (highlighted by white arrows in Fig. 8.5(a)) are not present in the topography image.

The combination of apertureless NSOM with two photon absorption provides even higher spatial resolution. Photolithography with a spatial resolution as sharp as $\lambda/10$ was recently reported; this represents an improvement of nearly two times the resolution achieved in previous far-field two photon lithography experiments (Yin *et al.* 2002).

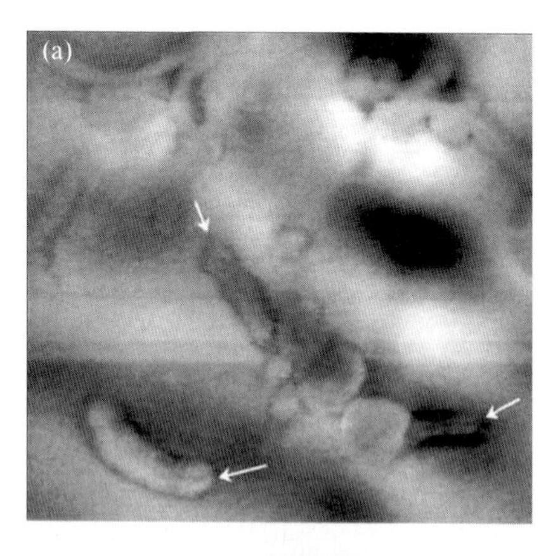

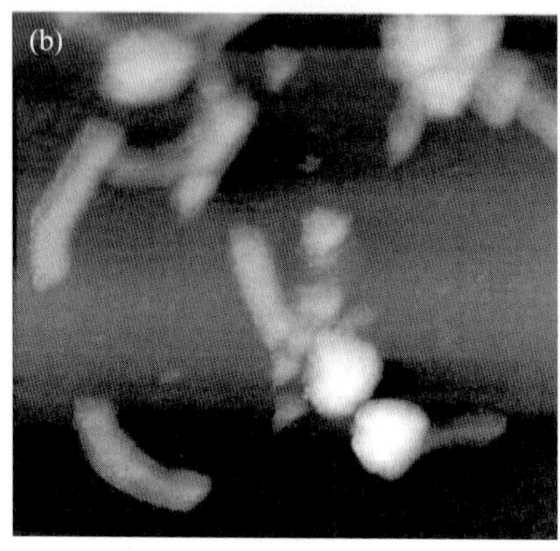

Fig. 8.5: (a) NSOM image and (b) topography of a murine embroynal myocardial cell (Inouye 2001).

8.3.4 *Comparison of aperture and apertureless NSOM*

Apertureless NSOM offers several advantages compared with aperture based NSOM systems. First, higher resolution is obtained by a sharper apertureless NSOM tip. Second, high numerical aperture objectives can efficiently collect the scattered light whereas metal film coatings on aperture NSOM tips strongly inhibit light throughput. Finally, sharp metallic tips provide small contact areas with the sample so that the tip itself coincides with the scattering centers of the evanescent waves. So the near-field optical image using a sharp metal tip corresponds to the topography obtained through the AFM channel. Aperture probes, though, use a thick metal coating which cannot guarantee a single contact point and may lead to scanning artifacts.

8.4 Plasmonic near-field imaging system

8.4.1 *Surface plasmon polaritons*

Surface plasmon polaritons are collective oscillations of electrons coupled to electromagnetic waves and are excited at the interface of a metal and dielectric. Surface plasmons exhibit a maximum intensity at the surface with an exponentially decaying amplitude perpendicular to the interface (Raether 1988). These collectively coupled oscillations (i.e. electrons with electromagnetic waves) may be excited by either electron beams or light incident via attenuated total reflection (ATR). Additionally, corrugated, rough, or periodically textured surfaces such as gratings are typically used to couple electromagnetic waves with surface plasmons and vice versa. Confinement of the electrons near the interface produces a strong enhancement of the electromagnetic near-field in the vicinity of the surface plasmon. It has been well documented that the spectral response of the surface plasmon resonance serves as a sensitive indicator of the optical properties of thin films within a wavelength of the surface. Control and confinement of this strong, localized enhancement prefaced a windfall of collaboration in surface sciences including the fields of surface chemistry, physics, and biology. Accordingly, many surface plasmon based techniques are now applicable and showing great promise: enhanced photoluminescence by surface plasmons, surface enhanced Raman scattering (SERS), non-linear effects such as second harmonic generation (SHG), plasmonic optical devices, etc.

Manipulation and utilization of photons (similar to the degree of control of electrons in electronics), a major objective in photonics related research, has spurred great interests in the potential combination of surface plasmons and optics. For example, extraordinary transmission through sub-wavelength hole arrays in metallic films is orders of magnitude greater than zero-order transmission predicted by standard aperture theory (Thio *et al.* 1999). Theory and experimental evidence show this effect is due to the coupling of incident light with surface plasmons at the interface between the metallic film and the surrounding dielectric media.

8.4.2 *Plasmonic near-field imaging system*

Recent demonstrations of plasmon active substrates include two-dimensional metallic slab waveguides, metallic nanoparticles on substrates, as well as surface plasmon focusing via a plasmonic lens. The nano-scale structured lens guides surface plasmon waves to the focal point of a circular or elliptical shape ending in a slit etched through a metal film (Drezet *et al.* 2005; Liu *et al.* 2005; Yin *et al.* 2005). Light then couples to surface plasmon waves through the sharp edge of the circular slit and these waves focus to a region in the lens smaller than the free space wavelength. This circular plasmonic nanostructure provides a unique method for sub-wavelength manipulation of light that can be adapted to specific applications such as a plasmonic near-field imaging system.

Surface defects and texturing (e.g. periodic gratings) are routinely employed to reflect and scatter waves in surface plasmon focusing studies (Andersen *et al.* 2002; Ditlbacher *et al.* 2002; Liu *et al.* 2005). This method, however, has a low reflection coefficient and the energy transferred is most likely insufficient for many proposed applications. Instead of surface defects or texturing, the sharp edge of a slit milled through a metallic film couples light into surface plasmons and the edge acts as a line of surface plasmon point sources. If the slit width is smaller than half of the incident wavelength of light, a majority of the light is diffracted. This diffracted light gains a wave vector in the direction along the film, allowing a portion of the incident light to excite surface plasmons. The excited surface plasmon wave vector must satisfy the dispersion relation and therefore depends on the frequency of incident light and the dielectric functions of the metal and the surrounding media. The direction of the wave vector, which determines the energy propagation direction, will be normal to the slit.

Excited surface plasmon waves can be directed toward the focal point (Drezet *et al.* 2005; Liu *et al.* 2006; Liu *et al.* 2005; Steele *et al.* 2006; Yin *et al.* 2005). When a circular plasmonic lens is excited by normal incident light, surface plasmon waves concentrate at the center of the lens. For non-normal incident light, the focal point depends on the incident angle. With concentric rings, surface plasmons generated at the outer rings travel inwards to the center where they are transmitted and reflected by the inner ring. Simultaneously, plasmon waves generated by the inner rings interfere with the transmitted surface plasmon waves from the outer rings. By controlling the phase mismatch, or spacing, between neighboring rings, the waves interfere either constructively or destructively at the surface plasmon frequency. Hence, the circular grating effectively becomes an active coupling element.

The electromagnetic near-field of these structures has been studied using NSOM in collection mode with a metal coated tip, as shown in Fig. 8.6 (Steele *et al.* 2006). Concentric rings were cut into thin silver films using a focused ion beam (FIB). The surface plasmons were excited with linearly polarized laser light incident from below the quartz substrate. Figure 8.7 shows the cross-section of the measured NSOM signal through the center of a sample with four rings along with

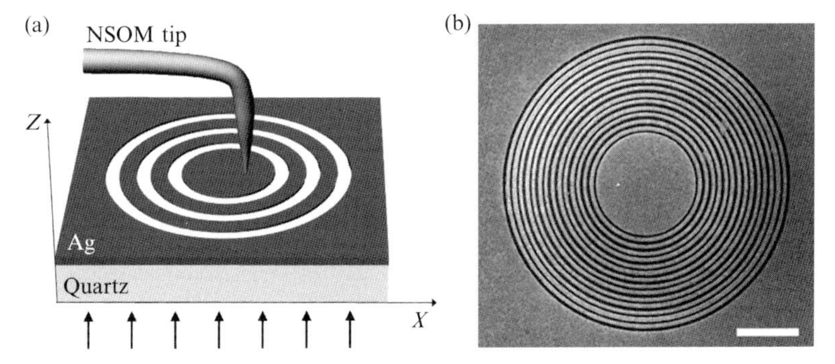

Fig. 8.6: (a) Experimental scheme for near-field measurements. Circular gratings are cut into a silver film deposited on a quartz substrate. Laser light is normally incident from the quartz side, and the electromagnetic nearfield is monitored with a metal coated NSOM tip. (b) SEM image of a sample with 15 rings. The scale bar is 5 microns (Steele *et al.* 2006).

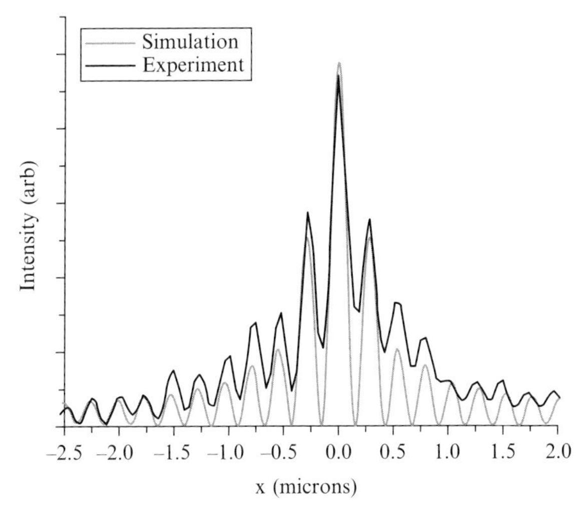

Fig. 8.7: NSOM measurement of the center region of a sample with 4 rings. The rings have a period of 514 nm (Steele *et al.* 2006).

the calculated intensity according to the surface plasmon interference model (Steele *et al.* 2006). Since the NSOM tip preferentially measures the electric field components parallel to the surface, the near-field shows interference fringes that reach a maximum in the center of the circle (Liu *et al.* 2005). By fitting the experimental data with theoretical values, the transmission coefficient is determined to be 0.88 and a phase shift as surface plasmons propagate across a single slit is determined to be 0.07 radians. The high transmission agrees with other experimental studies that launch surface plasmons across sub-wavelength slits. The measured period of plasmonic lens interference pattern (242 nm) is close to the calculated value of 244 nm (or half of the surface plasmon wavelength). This indicates that the pattern is solely due to the interference between surface plasmons excited on the air side of the silver film. The small difference likely arises from the surface roughness of the silver film causing the surface plasmon wavelength to be different from the calculated value for a smooth surface.

8.4.3 *Properties and advantages of plasmonic lens*

One unique advantage of plasmonic lens is the highly enhanced local field. At the focus point, the maximum intensity can be much higher than the incident light. The local enhancement can be raised by increasing the number of rings, so additional surface plasmon waves are generated and converge at the center of the plasmonic lens. Figure 8.8(a) shows the measured intensity at the center of the plasmonic lens as a function of the number of rings for a period corresponding near and far from the plasmon resonance for a silver film with an incident wavelength of 514 nm (Steele *et al.* 2006). The calculated intensity at the center of the circle was obtained using a theoretical model of surface plasmon interference. For a period of 633 nm, additional rings do not increase the intensity at the center of the circle, and in some cases reduce the intensity. This is expected since the surface plasmon waves are destructively interfering to a small degree due to a varying phase mismatch. For a grating period of 514 nm, there is still a small phase mismatch

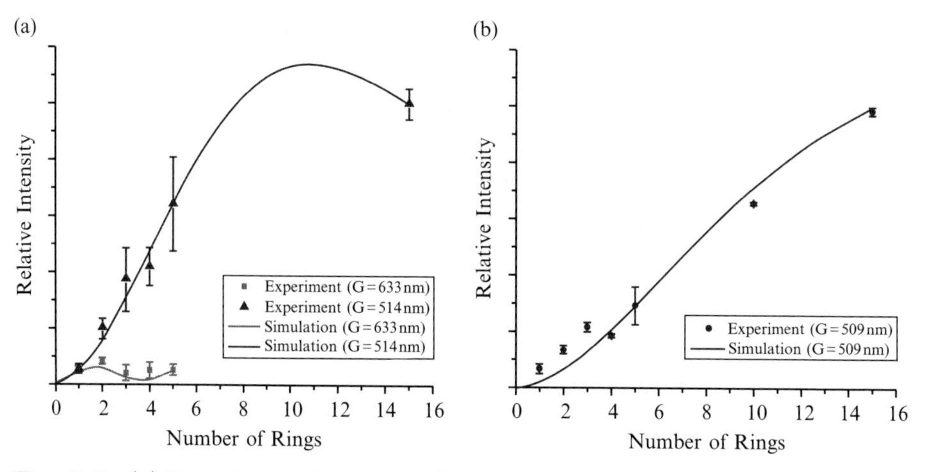

FIG. 8.8: (a) Intensity at the center of the circular grating for increasing number of rings with an incident laser wavelength of 514 nm. Two sets of rings are measured, one with a period G=514 nm and one with G=633 nm. For G=514 nm, the period is close to resonance with the surface plasmon wavelength of 490 nm. (b) Intensity at the center of the circular grating for rings with period G=509 nm exactly resonant with the surface plasmon wavelength for an excitation wavelength of 532 nm (Steele *et al.* 2006).

with a surface plasmon wavelength of 490 nm. This small phase mismatch increases with each ring added to the plasmonic lens. If the surface plasmon waves do not decay along the silver surface, one would expect that the intensity at the center of the plasmonic lens oscillates between a minimum and maximum value as additional rings are added. However, because the surface plasmon waves decay and partially reflect at each slit, the contribution from each additional ring to the intensity at the focal point decreases, damping out the expected oscillations of the focal intensity. In fact, simulations predict that the intensity increases at the center of the inner ring up to approximately 10 rings before leveling off (Steele *et al.* 2006).

In principle, if the period of the rings exactly matches the surface plasmon wavelength and the phase gain at each slit is small, additional rings always increase the intensity of the near-field at the center of the inner ring until the radius of the outermost ring becomes greater than the decay length of the surface plasmon wave. A plasmonic lens with a 509 nm grating period, or the theoretical surface plasmon wavelength at the air/silver interface for an incident wavelength of 532 nm, was fabricated and measured in the same setup as mentioned above. Figure 8.8(b) presents the experimental results (Steele *et al.* 2006). The intensity at the center of the lens increases with the number of rings for more than 15 rings provided a small phase mismatch between the rings. For 15 rings, the intensity at the focus point is an order of magnitude larger than that of a single ring, offering a large near-field intensity desirable in most sensing applications such as enhanced Raman scattering.

The experimental realization of tuning the focal point position of a plasmonic lens by adjusting the angle of the incident light, similar to conventional lenses, was recently demonstrated for fluorescent molecule excitation (Liu *et al.* 2006). For small incident angles, the lens shows well behaved focal point position control depending only on the angle of the incident beam. This tunable plasmonic lens can be used in nanoscale photonics, biological sensing, and biomolecule manipulation. Figure 8.9 shows the simulated total electric field distribution above a silver plasmonic lens (Liu *et al.* 2006). Since coherent surface plasmons inevitably experience interference when they cross each other, the focus profile is modulated by interference fringes. The focus spot size is therefore determined by surface plasmon interference to the order of one-quarter of the surface plasmon wavelength, which is much smaller than the excitation free space wavelength. The focus position shifts about 1 μm from the center when the incident angle is changed from 0° to 20° (Fig. 8.10), which agrees with the estimated calculation from the theoretical model. The surface plasmon focal point is found to be independent of polarization, while overall field distribution depends on the excitation polarization.

8.4.4 *Applications and challenges*

There are already many applications that use the tunable plasmon resonances of nanostructures to operate at specific wavelengths in optical regions (Halas 2005;

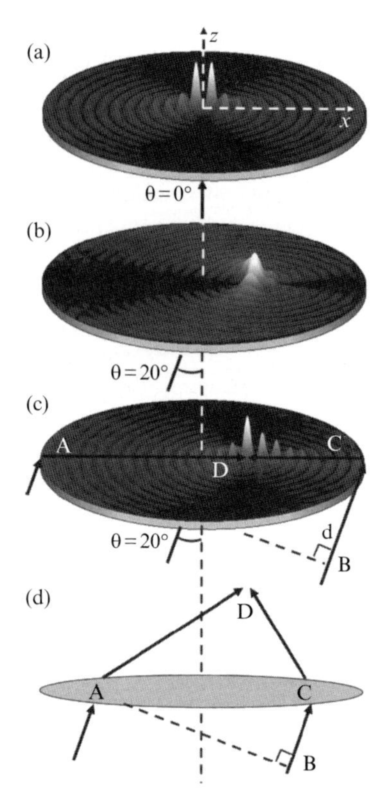

Fig. 8.9: (a)–(c) MWS simulation of the electrical field intensity distributions of a plasmonic lens with different incident angles and incident polarizations. The diameter of the simulated plasmonic lens is 6 μm. The excitation wavelength is 532 nm in a vacuum. (a) Normal incident with x polarization; (b) 20° incident angle with y polarization; (c) 20° incident angle with x polarization; (d) a schematic illustration of light focusing by a conventional optical lens (Liu *et al.* 2006).

Johnson *et al.* 1972). As a new type of nano-scale structure amenable to planar architectures, plasmonic lenses provide enhanced near-field 'hot-spots' upon exposure to a specific wavelength of light allowing for multiple fluorescent monitoring and tracking for separate biological processes. Circular grating plasmon lenses may therefore achieve multicolor sensing on one substrate. In addition, the dependence on grating period allows for the possibility of tuning different areas of a metal film as surface couplers for multicolor localized sensing of different molecular species.

The focusing properties for the plasmonic lens are very similar to those of a conventional optical lens except for two major differences. First, the plasmonic lens is a two-dimensional element where energy transport and the lens itself are in the same plane in contrast to conventional optics. Second, the focused spot size can

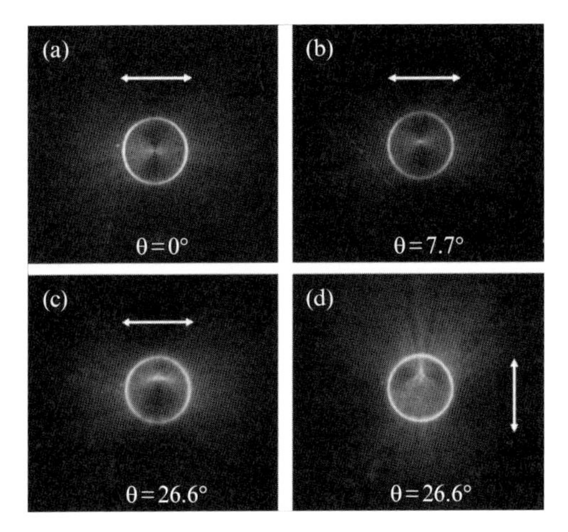

Fig. 8.10: Fluorescence images observed by optical microscope under different incident angles and polarizations. (a) Normal incident with horizontal polarization; (b) 7.7° incident angle with horizontal polarization; (c), (d) 26.6° incident angle with horizontal and vertical polarization, respectively. The polarization directions are indicated by the doubleside white arrows. The incident direction is upwards (Liu *et al.* 2006).

be much smaller than the free space diffraction limit, making it an excellent candidate for ultra small photonics. In fact, the focus size is only limited by the surface plasmon diffraction limit, which is about one-half of the surface plasmon wavelength. In addition, the large field enhancement at the focal point is of special importance for field dependent sensing schemes, such as SERS and SHG.

8.5 Concluding remarks

The goals of near-field based optical microscopy are to develop an imaging tool with nanometer scale spatial resolution that can visualize the biological pathways at the molecular level, to sense the transient responses in real cellular conditions, and to ultimately understand the mechanisms of human physiology in real time. Thus high sensitivity and high resolution observations of sub-cellular activities at the molecular level are critical. The abilities to detect the positions and motions of molecules on the cell membrane and within the cytoplasm are extremely challenging for conventional optical microscopy, not only because of the microscope resolution limit, but also because of the complexity of the molecular components (Hell 2003). Featuring a sub-wavelength optical element, near-field scanning optical microscopes have displayed 1~50 nm resolution along the x–y directions. In particular, apertureless NSOM systems are capable of nearly the same scanning and detection capabilities as other scanning probe systems but also include important information contained in optical signals from biological samples. Another near-field nanoscale optical technique is the plasmonic lens, which presents additional advantages besides high resolution detection/imaging. The large field intensity at the focal point of plasmonic lens has the potential for advanced imaging techniques at the nano scale, such as second harmonic generation, fluorescence bleaching, and optical incisions. This property is essential to explore new avenues in cellular nano scale science with visible light and regular illumination power. Research efforts aimed at imaging the position and motion of single molecules on a cell surface in real time will involve optimizing the plasmonic nanostructure designs with improved understanding of the near-field physics. Near-field optical and plasmonic based imaging may lead to exciting applications in fundamental molecular and cellular biology fields including cancer research, disease pharmacology, stem cell research, and perhaps many other fields yet to be discovered.

References

Ambrose, W. P., Goodwin, P. M., Martin, J. C., and Keller, R. A., 1994. 'Alterations of single-molecule fluorescence lifetimes in near-field optical microscopy.' *Science*, **265** (5170), pp. 364–367.

Andersen, P. C., and Rowlen, K. L., 2002. 'Brilliant optical properties of nanometric noble metal spheres, rods, and aperture arrays.' *Applied Spectroscopy*, **56** (5), pp. 124A–135A.

Ash, E. A., and Nicholls, G., 1972. 'Super-resolution aperture scanning microscope.' *Nature*, **237** (5357), pp. 510–512.

Bachelot, R., Gleyzes, P., and Boccara, A. C., 1997. 'Reflection-mode scanning near-field optical microscopy using an apertureless metallic tip.' *Applied Optics*, **36** (10), pp. 2160–2170.

Betzig, E., and Chichester, R. J., 1993. 'Single molecules observed by near-field scanning optical microscopy.' *Science*, **262** (5138), pp. 1422–1425.

Betzig, E., Lewis, A., Harootunian, A., Isaacson, M., and Kratschmer, E., 1986. 'Near-field scanning optical microscopy (NSOM) – development and biophysical applications.' *Biophysical Journal*, **49** (1), pp. 269–279.

Biagioni, P., Polli, D., Labardi, M., Pucci, A., Ruggeri, G., Cerullo, G., Finazzi, M., and Duo, L., 2005. 'Unexpected polarization behavior at the aperture of hollow-pyramid near-field probes.' *Applied Physics Letters*, **87** (22), pp. 2231121–3.

Bian, R. X., Dunn, R. C., and Xie, X. S., 1995. 'Single molecule emission characteristics in near-field microscopy.' *Physical Review Letters*, **75** (26), pp. 4772–4775.

Born, M., and Wolf, E., 1993. *Principles of Optics.* Pergamon, Oxford.

Carminati, R., and Greffet, J. J., 1996. 'Equivalence of constant-height and constant-intensity images in scanning near-field optical microscopy.' *Optics Letters*, **21** (16), pp. 1208–1210.

Carminati, R., Madrazo, A., NietoVesperinas, M., and Greffet, J. J., 1997. 'Optical content and resolution of near-field optical images: Influence of the operating mode.' *Journal of Applied Physics*, **82** (2), pp. 501–509.

Ditlbacher, H., Krenn, J. R., Schider, G., Leitner, A., and Aussenegg, F. R., 2002. 'Two-dimensional optics with surface plasmon polaritons.' *Applied Physics Letters*, **81** (10), pp. 1762–1764.

Drezet, A., Stepanov, A. L., Ditlbacher, H., Hohenau, A., Steinberger, B., Aussenegg, F. R., Leitner, A., and Krenn, J. R., 2005. 'Surface plasmon propagation in an elliptical corral.' *Applied Physics Letters*, **86** (7), p. 074104.

Dunn, R. C., 1999. 'Near-field scanning optical microscopy.' *Chemical Reviews*, **99** (10), pp. 2891–2927.

Halas, N., 2005. 'Playing with plasmons: Tuning the optical resonant properties of metallic nanoshells.' *MRS Bulletin*, **30** (5), pp. 362–367.

Hecht, B., Bielefeldt, H., Inouye, Y., Pohl, D. W., and Novotny, L., 1997. 'Facts and artifacts in near-field optical microscopy.' *Journal of Applied Physics*, **81** (6), pp. 2492–2498.

Hell, S. W., 2003. 'Toward fluorescence nanoscopy.' *Nature Biotechnology*, **21** (11), pp. 1347–1355.

Hoffmann, P., Dutoit, B., and Salathe, R. P., 1995. 'Comparison of mechanically drawn and protection layer chemically etched optical fiber tips.' *Ultramicroscopy*, **61** (1–4), pp. 165–170.

Hsu, J. W. P., 2001. 'Near-field scanning optical microscopy studies of electronic and photonic materials and devices.' *Materials Science & Engineering R-Reports*, **33** (1), pp. 1–50.

Inouye, Y., 2001. 'Apertureless metallic probes for near-field microscopy.' *Near-Field Optics and Surface Plasmon Polaritons*, **81**, pp. 29–48.

Inouye, Y., and Kawata, S., 1994. 'Near-field scanning optical microscope with a metallic probe tip.' *Optics Letters*, **19** (3), pp. 159–161.

Johnson, P. B., and Christy, R. W., 1972. 'Optical-constants of noble-metals'. *Physical Review B*, **6** (12), pp. 4370–4379.

Lazarev, A., Fang, N., Luo, Q., and Zhang, X., 2003a. 'Formation of fine near-field scanning optical microscopy tips. Part I: By static and dynamic chemical etching.' *Review of Scientific Instruments*, **74** (8), pp. 3679–3683.

Lazarev, A., Fang, N., Luo, Q., and Zhang, X., 2003b. 'Formation of fine near-field scanning optical microscopy tips. Part II: By laser-heated pulling and bending'. *Review of Scientific Instruments*, **74** (8), pp. 3684–3688.

Leviatan, Y., 1986. 'Study of near-zone fields of a small aperture.' *Journal of Applied Physics*, **60** (5), pp. 1577–1583.

Liu, Z. W., Steele, J. M., Lee, H., and Zhang, X., 2006. 'Tuning the focus of a plasmonic lens by the incident angle.' *Applied Physics Letters*, **88** (17).

Liu, Z. W., Steele, J. M., Srituravanich, W., Pikus, Y., Sun, C., and Zhang, X., 2005. 'Focusing surface plasmons with a plasmonic lens.' *Nano Letters*, **5** (9), pp. 1726–1729.

Martin, Y. C., Hamann, H. F., and Wickramasinghe, H. K., 2001. 'Strength of the electric field in apertureless near-field optical microscopy.' *Journal of Applied Physics*, **89** (10), pp. 5774–5778.

Meixner, A. J., Zeisel, D., Bopp, M. A., and Tarrach, G., 1995. 'Superresolution imaging and detection of fluorescence from single molecules by scanning near-field optical microscopy.' *Optical Engineering*, **34** (8), pp. 2324–2332.

Mihalcea, C., Scholz, W., Werner, S., Munster, S., Oesterschulze, E., and Kassing, R., 1996. 'Multipurpose sensor tips for scanning near-field microscopy.' *Applied Physics Letters*, **68** (25), pp. 3531–3533.

Novotny, L., Bian, R. X., and Xie, X. S., 1997. 'Theory of nanometric optical tweezers.' *Physical Review Letters*, **79** (4), pp. 645–648.

Pangaribuan, T., Yamada, K., Jiang, S. D., Ohsawa, H., and Ohtsu, M., 1992. 'Reproducible fabrication technique of nanometric tip diameter fiber probe for photon scanning tunneling microscope.' *Japanese Journal of Applied Physics Part 2-Letters*, **31** (9A), pp. L1302–L1304.

Pohl, D. W., Denk, W., and Lanz, M., 1984. 'Optical stethoscopy: Image recording with resolution lambda/20.' *Applied Physics Letters*, **44** (7), pp. 651–653.

Raether, H., 1988. *Surface Plasmons*. Springer, Berlin.

Shalom, S., Lieberman, K., Lewis, A., and Cohen, S. R., 1992. 'A micropipette force probe suitable for near-field scanning optical microscopy.' *Review of Scientific Instruments*, **63** (9), pp. 4061–4065.

Steele, J. M., Liu, Z. W., Wang, Y., and Zhang, X., 2006. 'Resonant and non-resonant generation and focusing of surface plasmons with circular gratings.' *Optics Express*, **14** (12), pp. 5664–5670.

Suh, Y. D., and Zenobi, R., 2000. 'Improved probes for scanning near-field optical microscopy.' *Advanced Materials*, **12** (15), pp. 1139.

Synge, E. H., 1928. 'A suggested method for extending microscopic resolution into the ultra-microscopic region.' *Philosophical Magazine*, **6** (35), pp. 356–362.

Thio, T., Ghaemi, H. F., Lezec, H. J., Wolff, P. A., and Ebbesen, T. W., 1999. 'Surface-plasmon-enhanced transmission through hole arrays in Cr films.' *Journal of the Optical Society of America B-Optical Physics*, **16** (10), pp. 1743–1748.

Yin, L. L., Vlasko-Vlasov, V. K., Pearson, J., Hiller, J. M., Hua, J., Welp, U., Brown, D E., and Kimball, C. W., 2005. 'Subwavelength focusing and guiding of surface plasmons.' *Nano Letters*, **5** (7), pp. 1399–1402.

Yin, X. B., Fang, N., Zhang, X., Martini, I. B., and Schwartz, B. J., 2002. 'Near-field two-photon nanolithography using an apertureless optical probe.' *Applied Physics Letters*, **81** (19), pp. 3663–3665.

9

OPTOELECTRONIC TWEEZERS

Eric P.Y. Chiou and Ming C. Wu

9.1 Introduction

9.1.1 *Overview of micro-manipulation tools*

Micromanipulation tools that enable trapping, sorting, addressing, and assembly of cells, colloidal particles, nanoparticles, and macromolecules play an important role in the development of bio- and nanotechnologies. Several manipulation tools have been developed and widely used, including optical tweezers (Grier 2003), electrophoresis (Kremser *et al.* 2004), dielectrophoresis (DEP) (Hughes 2002), magnetic tweezers (Lee *et al.* 2004), acoustic traps (Hertz 1995), hydrodynamic manipulators (Sundararajan *et al.* 2004), and optoelectronic tweezers (OET) (Chiou *et al.* 2005c). Each tool has its own unique combinations of advantages and limitations. For example, optical tweezers provide accurate three-dimensional trapping of particles but have a limited throughput; magnetic force offers non-invasive manipulation but requires the targets attached to magnetic beads; electrophoresis and dielectrophoresis offer high throughput but have limited resolution for dynamic traps. Optoelectronic tweezers combine the advantages of optical tweezers and DEP, and allow massively parallel manipulations through optically patterned virtual electrodes. This is the main subject of this chapter, which is organized as follows: first we will briefly review the principles of optical tweezers and DEP. Next we will introduce the concept of OET. The operation principle, device structure, theoretical models, and experimental results will be presented. Recent results of single nanowire manipulation and bio-compatible OET will be described before the concluding remarks.

9.1.2 *Optical tweezers*

Optical tweezers were first proposed by Ashkin in 1986 (Ashkin *et al.* 1986). He showed that by focusing a laser beam with a high numerical aperture (NA) objective lens, a transparent dielectric particle was trapped at the focal point. In 1987, he demonstrated laser trapping of live cells by using near infrared (NIR) lasers to avoid strong laser heating and photodamage of biological samples (Ashkin *et al.* 1987). Since then, optical tweezers have become an important tool for characterizing the forces generated by bio-motors (Wang *et al.* 1998, Bustamante *et al.* 2003), as well as manipulating, sorting, and patterning biological cells. The optical force consists of two components, the radiation force and the gradient force. The radiation force pushes the particle along the light propagation direction and does not trap the particle stably by itself. The optical gradient force attracts particles to regions with stronger light intensity. To form a stable optical trap, a tightly focused laser beam is required so that the gradient force dominates over the radiation force.

The force of the optical trap can be expressed by $F = Q\frac{np}{c}$, where c/n is the speed of light in the medium, P is the optical power, and Q is the trap efficiency (Howard 2001). For a spherical particle with a radius equal to the wavelength of light, the efficiency is ~ 0.1 and the optical force is ~ 0.5 pN/mW. For particles much smaller than the wavelength, the efficiency decreases as the optical force in this size range is proportional to the volume of the particles. Simultaneous trapping of multiple particles is possible by splitting the laser beam with diffractive optical elements. Recently, progresses in holography optical tweezers allow 3-D dynamic optical traps that can be reconfigured in real time (Grier 2003).

Optical tweezers are not generally considered as high throughput tools due to two fundamental limitations. One is the small manipulation area due to the small field of view of high NA objective lenses required for tight focusing. For a $100\times$ oil immersion lens, the effective area is less than $100 \times 100\ \mu m^2$. There have been several attempts to increase the manipulation areas, for example, using near-field optical gradient near a total-internal-reflection surface or using high NA focusing mirrors fabricated on a surface (Garces-Chavez *et al.* 2005, Merenda *et al.* 2007). The other limitation is the high optical power required to stably trap a particle and overcome its Brownian motions. An optical power of ~ 1 mW is needed to stably trap a 1-μm particle (Curtis *et al.* 2002). The optical power required for multiple traps is linearly proportional to the number of traps. The maximum number of traps is therefore limited by the available laser power. The maximum power that a spatial light modulator can tolerate is another limiting factor.

9.1.3 *Dielectrophoresis*

Dielectrophoresis is a phenomenon in which a force is exerted on a dielectric particle when it is subjected to a non-uniform electric field (see Chapter 5 for more detailed discussion). The magnitude of DEP force can be expressed as (Jones 1995)

$$< F_{dep}(t) >= 2\pi a^3 \epsilon_m \mathrm{Re}[K^*(\omega)] \nabla (E_{rms}^2) \tag{9.1}$$

$$K^*(\omega) = \frac{\epsilon_p^* - \epsilon_m^*}{\epsilon_p^* - 2\epsilon_m^*},\ \epsilon_p^* = \epsilon_p - j\frac{\sigma_p}{\omega},\ \epsilon_m^* = \epsilon_m - j\frac{\sigma_m}{\omega} \tag{9.2}$$

where $<F_{dep}(t)>$ represents the time-average of the function $F_{dep}(t)$, E_{rms} is the root-mean-square magnitude of the imposed AC electric field, a is the particle radius, ϵ_m and ϵ_p are the permittivities of the surrounding medium and the particle, respectively, σ_m and σ_p are the conductivities of the medium and the particle,

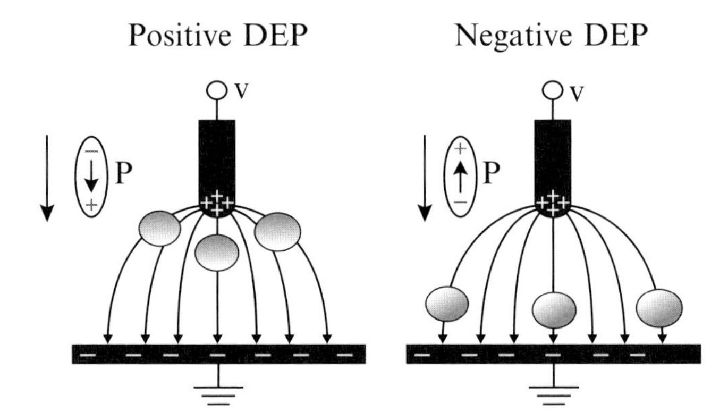

Fig. 9.1: Illustration of positive and negative DEP forces. (a) For positive DEP, the field induced electric dipole on the particle is collinear with the direction of the applied electric field. The particles move towards the strongest electric field region. (b) For negative DEP, the dipole direction is anti-parallel to the applied electric field and the particles are pushed towards the weakest electric field region.

respectively, ω is the angular frequency of the applied electric field, and $K^*(\omega)$ is the Clausius–Mossotti (CM) factor.

The CM factor of a particle represents its frequency response to an external electric field. It is the particle's dielectric signature, which is determined by several factors including the particle size, material composition, structure, and surface charges. The value of the real part of $K^*(\omega)$, $\mathrm{Re}[\mathrm{K}^*(\omega)]$, is between 1 and $-1/2$, depending on the polarizability of the medium and the particle. If $\mathrm{Re}[\mathrm{K}^*(\omega)] > 0$, the induced electric dipole is collinear with the electric field as shown in Fig. 9.1(a). The particle will move towards the strongest electric field region, a phenomenon known as positive DEP. On the other hand, if $\mathrm{Re}[\mathrm{K}^*(\omega)] < 0$, the induced electric dipole is anti-parallel to the electric field as shown in Fig. 9.1(b). The particle moves towards the weakest electric field region. This is known as negative DEP.

Dielectrophoresis has been widely applied to manipulate micro- and nano-scale particles, including trapping, sorting, and patterning a variety of objects such as biological cells (mammalian cells and bacteria), viruses, macromolecules (DNAs and proteins), nanowires and carbon nanotubes, and liquid droplets (Becker *et al.* 1995, Yang *et al.* 2000, Jones 2002, Ermolina *et al.* 2003, Krupke *et al.* 2003, Velev *et al.* 2003, Kadaksham *et al.* 2004, Tang *et al.* 2004, Ying *et al.* 2004). DEP is generally produced by microelectrodes, which can be easily fabricated over a large area for high throughput. A general-purpose DEP manipulator with an $\mathrm{N} \times \mathrm{N}$ array of individually addressable electrodes has been constructed using complementary metal-oxide-semiconductor (CMOS) technology (Manaresi *et al.* 2003). The resolution of the DEP array is limited by the electrode size. With the rapid advances of integrated circuit (IC) technology, CMOS-based DEP chips could potentially be a powerful platform for DEP manipulation, provided the cost is reduced by mass production.

9.2 Optoelectronic tweezers

OET use optically patterned virtual electrodes on a photoconductor surface to sculpt the electric field landscape for DEP manipulation (Chiou *et al.* 2005c). Optical addressing has several advantages over electrical addressing. It is easier to address a large array of electrodes using optical patterns. Furthermore, optical addressing is flexible and easily reconfigurable. Figure 9.2 illustrates the relationship between optical tweezers, dielectrophoresis, and optoelectronic tweezers. Optical tweezers use direct optical forces for particle manipulation. The optical energy is directly converted to mechanical forces, and the trapping force is proportional to the optical power. On the other hand, the optical energy is used to turn on the virtual electrodes in OET. The optically induced DEP force depends on the electrical bias. It is independent of the optical power once the virtual electrode is fully turned on. Using a photoconductor with high optoelectronic gain, the optical power requirement of OET is several orders of magnitude lower than that of optical tweezers. The low optical power requirement is the key to enable massively parallel optical manipulation over a large area using direct optical images.

9.2.1 *Device structure and operation principle*

Figure 9.3 illustrates the device structure of OET. The aqueous solution containing the particles is sandwiched between two electrodes: a transparent conductive electrode made of heavily doped indium tin oxide (ITO) on a glass substrate, and a photoconductive electrode made of multiple featureless layers including an ITO layer, a 10-nm-thick aluminum, a 1.5-μm-thick intrinsic hydrogenated amorphous silicon (a-Si:H), and a 20-nm-thick silicon nitride. The aluminum improves the contact resistance between ITO and the a-Si:H layer. Its thickness is kept below 10 nm to reduce blocking of the incident light. The silicon nitride prevents electrolysis at low frequency. It behaves like a high-pass filter for AC signals. An AC voltage bias is applied between the top and the bottom electrodes. In the absence of light, the majority of the applied voltage drops across the a-Si:H layer since its

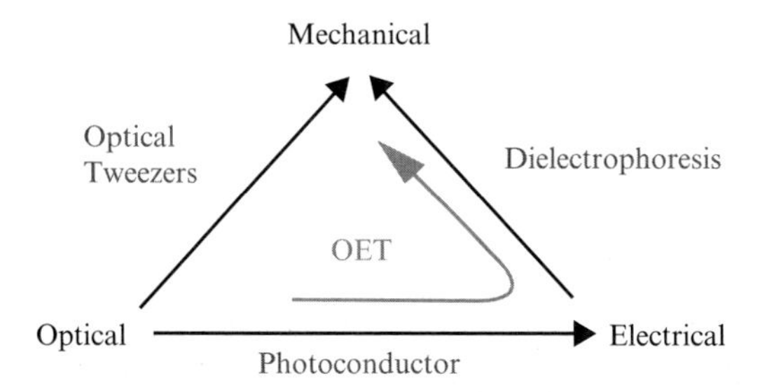

Fig. 9.2: Schematic illustrating the energy conversion process of various manipulation mechanisms. Optical tweezers convert optical energy directly to mechanical force, while DEP converts electrical energy to mechanical force. OET combines the advantages of both optical tweezers and DEP. The optical energy is used to induce DEP on a photoconductive surface.

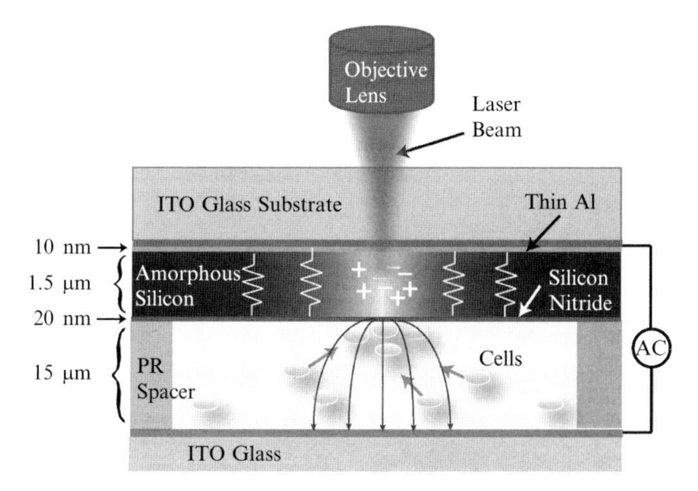

FIG. 9.3: The schematic structure of the OET device.

dark electrical impedance is much higher than that of the aqueous medium. Under optical illumination, the conductivity of the a-Si:H increases due to the photo-generated carriers. With a light intensity of 100 W/cm^2, the photoconductivity increases by about five orders of magnitude (Wei *et al.* 1993). The electrical impedance becomes smaller than that of the aqueous medium. This creates a virtual electrode and shifts the voltage drop to the liquid layer. The electric field near the virtual electrode is highly non-uniform. This results in a DEP force near the a-Si:H surface. For particles experiencing positive DEP forces, they are attracted to the illuminated spot; while for particles experiencing negative DEP forces, they are repelled away. The trapped particles can be continuously moved across the surface by scanning the light beam.

The resolution of the virtual electrode is determined by two factors: the ambipolar diffusion length of the carriers in a-Si:H and the diffraction limit of the optical system. While the ambipolar diffusion length in single crystalline silicon

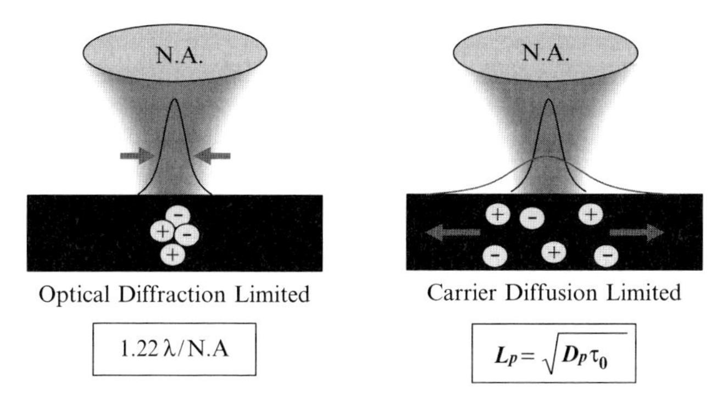

FIG. 9.4: The resolution of light-patterned virtual electrodes in OET devices.

can be as long as hundreds of microns, it is only ~ 100 nm in a-Si:H, much smaller than the diffraction limit of visible light. In this case, the resolution of the virtual electrodes is determined by optical diffraction. According to the Fraunhofer diffraction theory, the diffraction-limited spot size is $1.22\ \lambda/\mathrm{NA}$, where λ is the optical wavelength and NA is the numeric aperture of the objective lens. For example, for $\lambda = 633$ nm and $\mathrm{NA} = 1$, the minimum spot size is $0.77\ \mu$m.

The OET device can be modeled by the equivalent circuit shown in Fig. 9.5(a). The photoconductive layer is essentially a light-controlled resistor. It has a high dark resistance so that most applied voltage drops across the photoconductive layer. The voltage drop across the liquid layer, ΔV, is very small. With illumination, the resistance of the photoconductor reduces significantly, shifting the voltage drop to the liquid layer. This creates a non-uniform electric field that interacts with particle through DEP. The resulting force is proportional to the square of ΔV. The magnitude of ΔV can be solved graphically using the I-V curves of the photoconductor and the load line of the liquid medium, as shown in Fig. 9.5(b). For a-Si:H photoconductor, the maximum ΔV occurs when the liquid conductivity is between 0.001 and 0.1 S/m. For liquid conductivity smaller than 1 mS/m, ΔV is small because the load line has a very small slope (R_{large} in Fig. 9.5(c)). Physically, this means the virtual electrode is already on even without light illumination. On the other hand, if the liquid is too conductive, ΔV is also small because the load line is very steep (R_{small} in Fig. 9.5(c)). The electrode cannot be fully turned on in this case. Thus the standard a-Si:H OET cannot operate in highly conductive

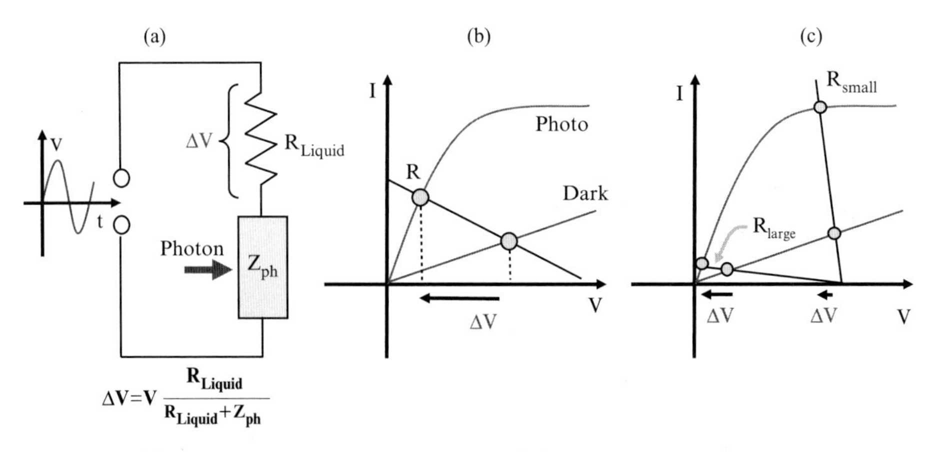

Fig. 9.5: (a) The equivalent circuit of OET. (b) I-V curves of the photoconductor in dark and illuminated states and the load line of the liquid layer with optimum conductivity. Large shift of voltage drop in the liquid layer (ΔV) can be obtained by light illumination. (c) Similar to (b) but with load lines corresponding to high (R_{small}) and low (R_{large}) liquid conductivities. In either case, ΔV is too small to generate appreciable DEP force.

(~1.5 S/m) physiological buffer solutions, which are desirable for most biological applications. Photoconductors with much higher gain are needed for biocompatible OET. In Section IX-5-b, we will describe a phototransistor-based OET that can operate in high salt liquid media (Hsu *et al.* 2007).

9.2.2 *Model and simulation*

Figure 9.6(a) shows the simulated electric field distribution in the liquid for a 17-μm-diameter light spot generated by a focused laser beam under 10 V bias. The conductivity of the liquid is 1 mS/m and the thickness is 15 μm. The photoconductivity of the amorphous silicon layer is assumed to have a Gaussian distribution, following the profile of the incident light with a peak conductivity of 10 mS/m at the center. The three-dimensional electrical field distribution is calculated using the FEMLAB Electrostatic Module.

The electric field strengths at three different heights above the photoconductor are plotted in Fig. 9.6(b). Since the DEP force is proportional to the gradient of E^2, the electric field distribution shows that the OET can generate strong DEP forces within a radius of $\sim$ 20 μm in the lateral direction. There is also a vertical gradient that will attract microparticles towards the photoconductor surface (or repel the particle for negative DEP). Both the lateral and the vertical gradients are strongest near the edge of the laser spot, similar to those generated by physical electrodes.

In OET, the driving force is provided by the applied voltage. Once the virtual electrode is fully turned on, the light-induced DEP force is no longer dependent on optical intensity. Figure 9.7(a) shows the calculated electric field distribution under the illumination of a Gaussian beam for three different light intensities.

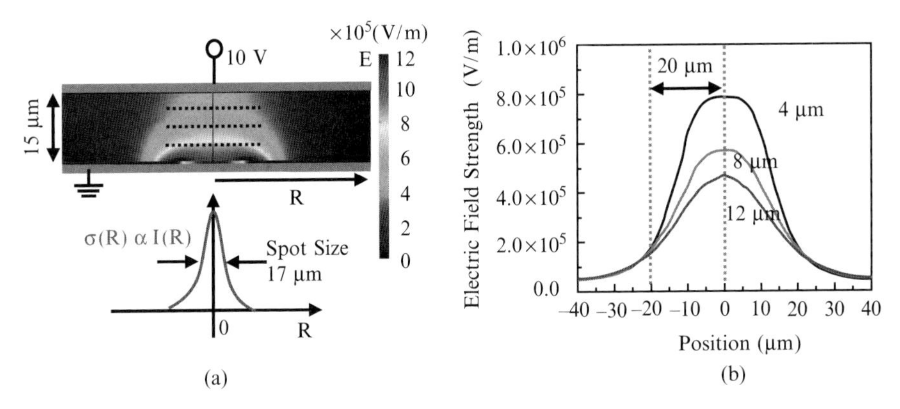

Fig. 9.6: (a) Electric field distribution in the OET device with a liquid layer thickness of 15 μm. (b) The electric field strength at three different heights above the photoconductive surface. This electric field distribution shows that OET generates DEP forces in both the lateral and vertical direction. The lateral range of this OET device is about 20 μm.

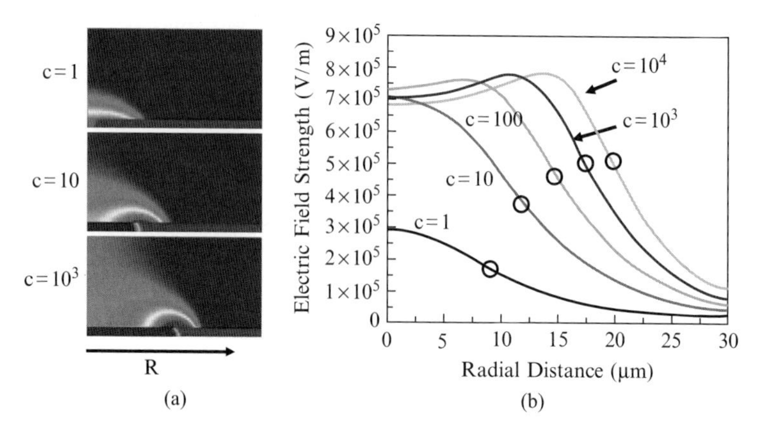

Fig. 9.7. (a) Close-up view of the electric field distribution near the virtual electrodes for various light intensities. The plot is radially symmetric, and only one side of the virtual electrode is shown. (b) The electric field strength distribution for various photo-to-liquid conductivity ratios (c parameter). The field strength is calculated at 5-μm above the photoconductive surface. The virtual electrode is fully turned on when c $>$ 10.

The parameter c is defined as the ratio of the peak photoconductivity to the liquid conductivity. While the effective size of virtual electrode increases with light intensities, the maximum gradients of the electric field (marked by circles in Fig. 9.7(b)) remain almost constant for c $>$ 10.

9.2.3 *Experimental setup*

Figure 9.8 shows the experimental setup of OET on an inverted microscope (Nikon TE 2000) for trapping *E. coli* bacteria. The condenser lens in the top port of the microscope is removed to construct the illumination optics for OET actuation. An 800-μW, 632-nm HeNe laser with a beam width of 0.24 mm is used as the actuation light source. The laser beam is scanned by a 2-D galvanometer mirror (Cambridge Technology, Inc., Model 6210), reflected by a dichroic mirror (Edmund Industrial Optics Inc.), and then focused by a 40$\times$ objective lens to generate a 17-μm light spot on amorphous silicon. A neutral density filter controls the incident optical power. Since the amorphous silicon is not transparent to visible light, the OET device is mounted 'upside down' (i.e. amorphous Si on top and ITO at the bottom) to enable simultaneous fluorescent observation and OET manipulation. The actuation and the observation optics are completely separated in this setup, yielding maximum flexibility for OET operation. The effective manipulation area is not limited to the field of view of the high NA objective lens needed for high magnification observation and detection.

The source for exciting fluorescent signals could also activate OET. This introduces leakage currents and replaces the electric field gradient and DEP force.

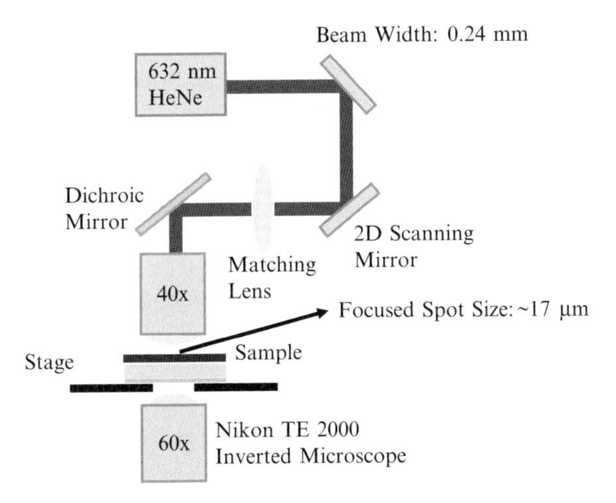

FIG. 9.8: Experimental setup for OET.

Fortunately, the optimal wavelength for OET activation is between 600–680 nm. For wavelengths shorter than 600 nm, they are absorbed within the 0.1 μm in the amorphous silicon (Sze 1981). Their contribution to the overall photoconductivity of the 1$\sim$1.5 μm thick amorphous silicon layer is very small. In most biological fluorescence imaging systems, the excitation wavelengths are shorter than 600 nm, which is compatible with OET operation.

Figure 9.9(a) shows the images of live *E. coli* bacteria under the microscope. The *E. coli* cells have been modified to express green fluorescent protein (GFP) so they can be observed by fluorescent microscope. The conductivity of the aqueous medium is adjusted to 1 mS/m by adding potassium chloride. A 100 kHz, 10 V_{pp} AC bias is applied to OET device. As the electric field is turned on, the rod-shaped *E. coli* bacteria align vertically to the OET surface so the induced dipoles is parallel to the vertical electric fields. The fluorescent image of a group of randomly distributed green spots is shown in Fig. 9.9(b). The *E. coli* bacteria experience positive DEP forces at 100 kHz in this medium. When the laser is turned on, the bacteria within a 20-μm radius are attracted to the illuminated spot and trapped on the OET surface, as shown in Fig. 9.9(c)(d). Scanning the light beam can transport these trapped *E. coli* bacteria to any location on the surface (Fig. 9.9(e)(f)). When the laser and the electric field are turned off, these trapped bacteria swim away from the locations without showing any observable damage.

Light beams with four different laser powers, 8, 120, 400, and 800 μW, were employed to trap *E. coli* cells in this experiment. The corresponding light intensities are 2.4, 36, 120, and 240 W/cm^2, respectively. Figure 9.10(a) shows the measured velocities of the *E. Coli* cells moving towards the OET trap versus the radial distance from the center of the laser beams. At 800 μW, *E. Coli* bacteria as far as 30 μm away are trapped by OET. The moving speed increases sharply as the

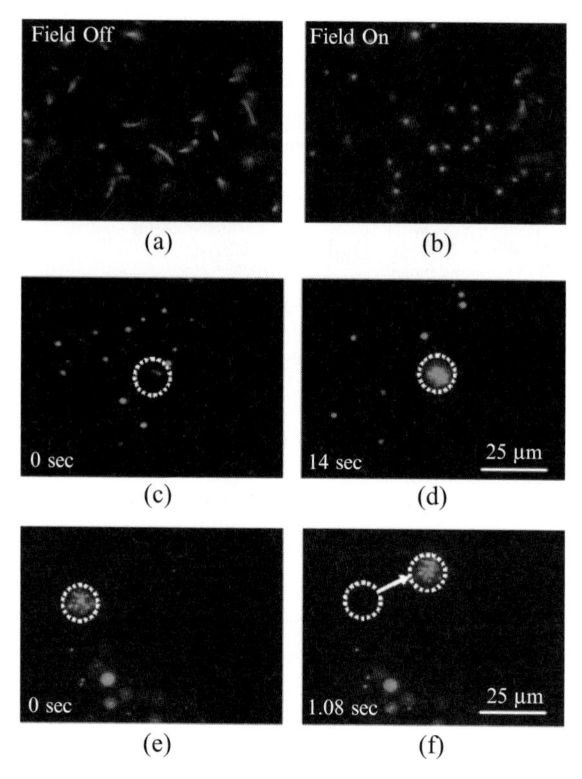

Fig. 9.9: The rod-shaped *E. coli* cells (a) before and (b) after the AC bias voltage is turned on. The *E. coli* cells are aligned with the electrical field and appear as green dots. (c) and (d) the *E. coli* cells are concentrated at the illuminated spot when the light is turned on. (e) and (f) a group of trapped *E. coli* cells are transported to various positions on the OET surface.

E. Coli cells are within 20 μm to the trap center. It reaches a peak velocity of 120 μm/s at 15 μm radial distance. These experimental results match well with the simulated DEP force distribution shown in Fig. 9.6(b). The maximum moving velocity, or the maximum DEP force, is a function of laser optical power. The peak velocity increases from 26 μm/s at 8 μW to 90 μm/s at 120 μW. Above 120 μW, the peak velocity increases slowly, and eventually saturates (Fig. 9.10(a), (b)). This shows that fully turning on a virtual electrode in this OET device requires a light intensity of 36 W/cm^2. Beyond this intensity, the effective size of virtual electrodes still increases with light intensity but the maximum DEP force saturates.

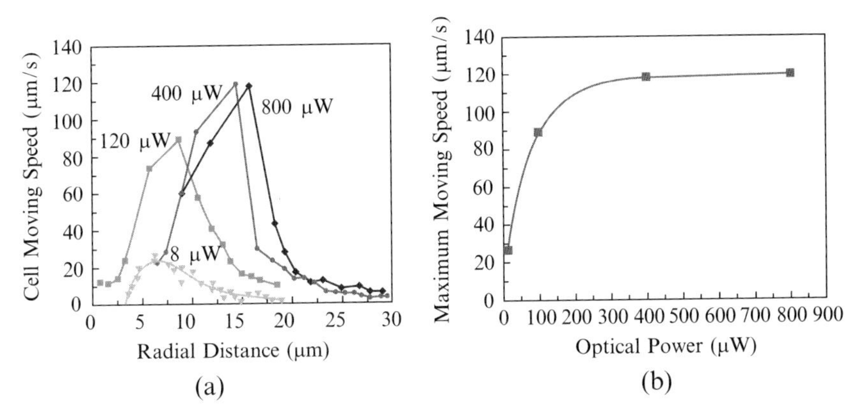

Fig. 9.10: (a) Velocities of *E. Coli* bacteria as they move towards the OET trap. (b) The maximum speed of *E. Coli* bacteria versus the optical power in OET trapping. The maximum speed saturate when the optical power is higher than 400 μW.

9.3 Characterization

9.3.1 *Force*

The OET force is characterized by measuring the maximum traveling speed of trapped particles. When the trap is moved with respect to the media (by either scanning the light pattern or by moving the sample stage), the particle experiences a drag force in addition to the OET force. The particle position will reach equilibrium, at which the OET force is equal to the drag force. The drag force follows the Stoke's Law, $F = 6\pi\eta r v$, where η is the viscosity of the media, r and v are the radius and the velocity of the particle, respectively. As the scan speed increases, the particle eventually cannot follow the trap as the viscous force exceeds the maximum OET force. Thus the maximum speed of a trapped particle is a good measure of the OET force.

Figure 9.11(a) shows the maximum speed of a 25-μm polystyrene bead versus optical power for two bias voltages. The particle experiences a negative DEP force at 100 kHz. At 5 V bias (peak-to-zero), the bead can be actuated with optical power as low as 1 μW (the intensity is 3 nW/μm^2). The maximum speed of the bead is 4.5 μm/s. As the optical power increases, the virtual electrode becomes less resistive. More voltage is shifted to the liquid medium and the particle experience a larger OET force. At 100 μW optical power and 10 V bias, the maximum particle speed is 397 μm/sec, which corresponds to an OET force of 93.4 pN. The OET force at 10 V bias is approximately four times of the force at 5 V bias, which agrees well with the DEP force expression in eqn. (9.1). The maximum voltage that can be applied to the OET is limited by the dark current in amorphous Si, which tends to reduce ΔV and therefore the DEP force, as illustrated in Fig. 9.5. Using a thicker

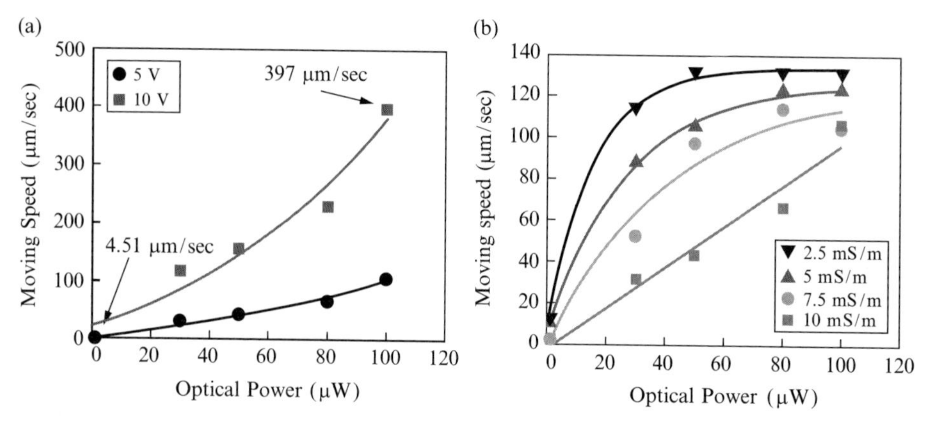

FIG. 9.11: (a) Maximum particle speed under various optical powers and applied AC voltages. The square (■) symbol represents the data for an applied voltage of 10 V and the circular (●) symbol represents the data for 5 V. The FWHM spot size of the laser beam is 17 μm. (b) Liquid conductivity dependent OET actuations.

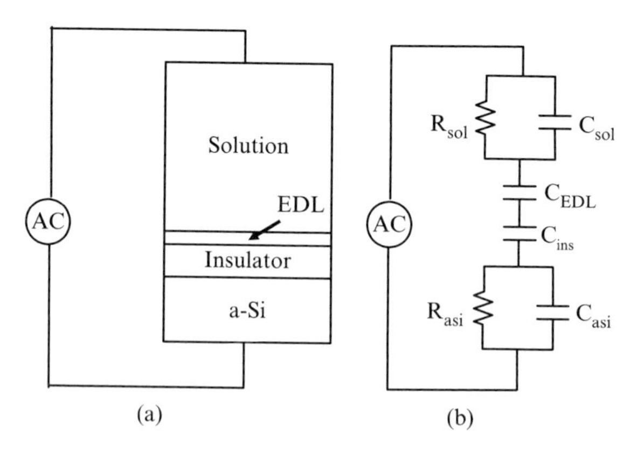

FIG. 9.12: (a) The schematic and (b) the equivalent circuit model of an OET device. The multilayer structure is modeled as serially-connected lumped circuit elements.

amorphous silicon layer can increase the maximum voltage and the induced DEP forces in an OET device.

9.3.2 *Liquid conductivity dependence of OET actuation*

The optical power requirement is a function of the liquid conductivity. Figure 9.11 (b) shows the maximum speed of trapped particles versus optical power for various liquid conductivities. In liquid with a conductivity of 2.5 mS/m, an optical power of 50 μW is sufficient to fully turn on the virtual electrode. The maximum speed saturates after the virtual electrode is fully turned on. As the liquid becomes more conductive, the amount of optical power required to turn on the virtual electrode increases. In the medium with a conductivity of 0.01 S/m, the virtual electrode is not fully turned on at 100 μW, the maximum power applied in this experiment.

9.3.3 *Frequency response of OET devices*

The frequency response of the OET device can be modeled by an AC equivalent circuit. Figure 9.12(a) shows the generic layer structure of an OET device. The corresponding equivalent circuit is shown in Fig. 9.12(b). The amorphous Si is modeled as a variable resistor with a shunting capacitor. Similarly the liquid layer is modeled as a resistor with a shunting capacitor. The dielectric layer (silicon nitride) is repeated by a capacitor. The solid – liquid interface is modeled as an electric double layer (EDL), which behaves as a capacitor. The frequency response of the equivalent circuit can be easily simulated.

Figure 9.13 shows the electric impedances for various elements of the equivalent circuit in OET devices. They are calculated using the following parameters: the relative permittivitives, ϵ_r, of water, silicon nitride, silicon dioxide, and amorphous silicon are 78, 7, 4, and 11, respectively, and the dark conductivity of the amorphous silicon is 10^{-8} S/m. The conductivity of the insulator and the electric double layers is assumed to be negligible. The thickness of the liquid layer is 100 μm and the thickness of the double layer is 4 nm, which corresponds to an electrolyte concentration of 10 mM. The area of the device is 250 μm^2 in this calculation. The liquid conductivity varies from 1 mS/m to 1 S/m.

Due to the low dark conductivity of amorphous silicon, its impedance is dominated by the shunting capacitor at frequencies greater than 10 kHz. The

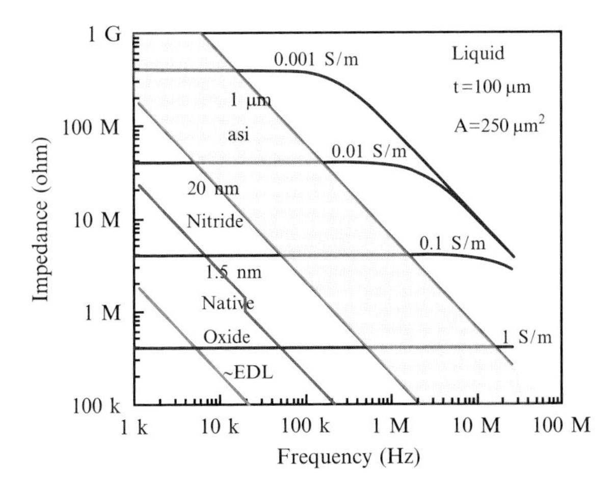

Fig. 9.13: The impedances of various equivalent circuit elements in an OET device. The thickness of the liquid layer used in this calculation is 100 μm and the device area is 250 μm^2. The thickness of the EDL is assumed to be 4 nm.

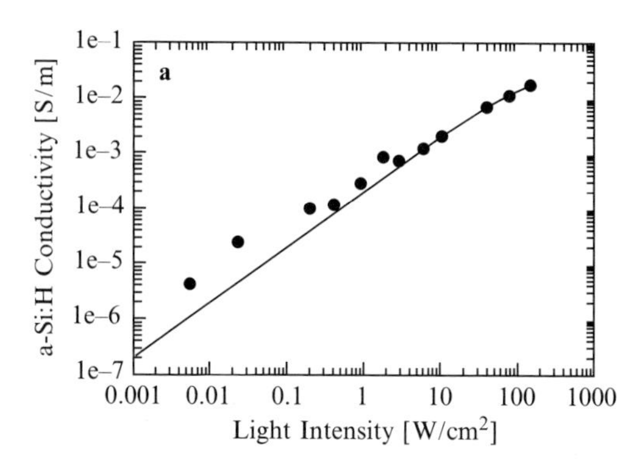

Fig. 9.14: Measured photoconductivity of undoped a-Si:H as a function light intensity at 5V bias.

impedance decreases with frequency, and eventually becomes smaller than the liquid impedance. When this happens, the OET device is no longer operable since the virtual electrode is always on. This implies the OET cannot operate in insulating liquid such as oil. As shown in Fig. 9.13, the high cut-off frequencies for the OET device are 20 kHz, 200 kHz, 2 MHz, and 20 MHz for liquid conductivities of 0.001, 0.01, 0.1, and 1 S/m, respectively, assuming a 1-μm-thick amorphous Si. The low cut-off frequency is determined by the capacitance of the insulating layer.

The OET surface is often coated with a dielectric layer to prevent electrolysis or other self-assembled monolayer for surface functionalization. It is important to keep the dielectric layer as thin as possible so its impedance is smaller than both the liquid and the amorphous Si. For 20-nm-thick silicon nitride, the low cut-off frequencies of the OET device are 500 Hz, 5 kHz, 50 kHz, and 500 kHz for liquid conductivities of 0.001, 0.01, 0.1, and 1 S/m, respectively. The operating frequency of the OET is therefore bound by these two cut-off frequencies, as shown by the gray area in Fig. 9.13. The frequency window can be expanded if the insulation layer is thinned or removed. The OET layer structure can be optimized for different applications.

9.2.4 *Photoconductivity and light intensity*

Figure 9.14 shows the measured photoconductivity of the undoped a-Si:H used in our OET devices. At 5V bias, the a-Si:H film has a conductivity of 0.9×10^{-6} S/m under room light. The conductivity increases steadily with light illumination. At 145 W/cm^2, the photoconductivity increases by three orders of magnitude to 1.7×10^{-2} S/m (Ohta *et al.* 2007).

9.3 Trapping of colloidal particles

The low optical power requirement of OET opens up many opportunities. First, since optical coherence is not required, incoherent light sources such as light-emitting

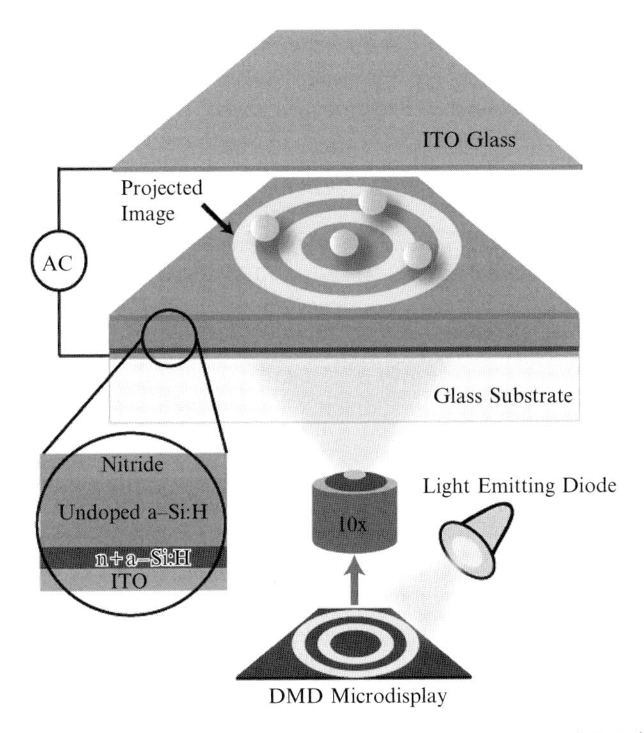

FIG. 9.15: Schematic illustrating direct optical image-driven OET (Chiou *et al.* 2005c). A spatial light modulator projects programmed optical images to create dynamic electric field patterns on OET, which generates reconfigurable DEP forces for particle manipulation.

diodes (LEDs) or even halogen lamps can be used to actuate OET. Second, OET can be actuated using direct optical images from a spatial light modulator (SLM). Parallel dynamic optical traps can be realized over a large manipulation area.

Figure 9.15 shows the schematic setup of direct optical image-driven OET using digital micromirror device (DMD) spatial light modulator from Texas Instruments. The DMD chip (Discovery kit) used has 1024×768 pixels. Each pixel is $13.68 \times 13.68\ \mu\text{m}^2$. After projection onto the OET surface by a $10\times$ objective lens, the optical pixel size is 1.5 μm. A single red LED with a center wavelength of 625 nm is used as the illumination source. After projection, the average optical power per pixel is 25 nW (~ 1.7 W/cm^2), which is strong enough to partially turn on virtual electrodes. The total effective manipulation area is 1.3 mm $\times$ 1.0 mm on the OET plane, limited by the projected area of the entire DMD chip. Instead of the Discovery DMD chip and custom light source, one can also use commercial digital projectors. They are low cost SLMs providing a reasonably high optical intensity of ~ 10 W/cm^2 on OET surface.

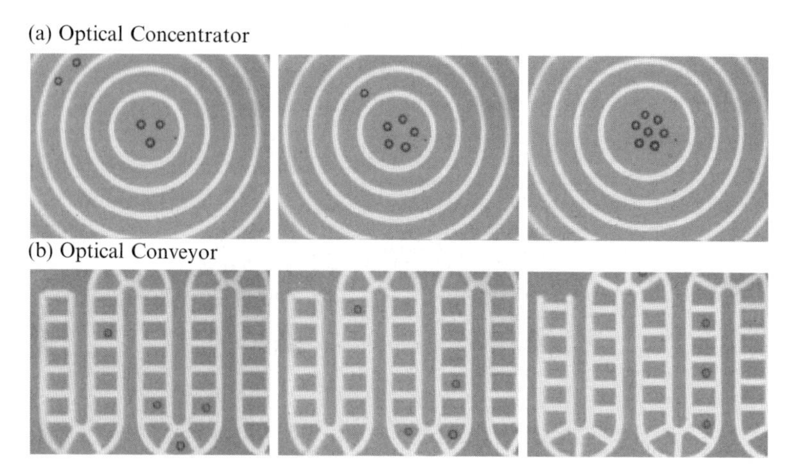

Fig. 9.16: Dynamic optical patterns projected on an OET device to (a) concentrate and (b) transport a group of 20-μm polystyrene beads.

Though DMD is used for OET actuation here, other spatial light modulators such as liquid crystal SLM can also be used. As an alternative to optical projection, the OET device can be actuated by directly placing it on a flat panel liquid crystal display (Choi *et al.* 2007) or a light-emitting diode array. This results in a more compact system, though the pixel size on OET is usually larger without demagnification.

9.3.1 *Particle concentrator and conveyor*

With a spatial light modulator, arbitrary patterns can be easily generated and reconfigured to perform desired functions. For example, a dynamic optical pattern consisting of multiple concentric optical rings concentrates cells or particles to the center of these concentric rings, as shown in Fig. 9.16(a). By shrinking these rings synchronously, a group of 20-μm polystyrene particles experiencing negative DEP forces are concentrated in the center. Since these particles are all polarized by the electric field perpendicular to the OET surface, they are pushed away from each other by the dipole-dipole repulsive forces. The particles form symmetrical colloidal structure in the center of these concentric rings. Other functional dynamic optical patterns, such as an optical conveyor belt that transports multiple particles from the left to the right, have also been demonstrated (Fig. 9.16(b)).

9.3.2 *High density traps*

To demonstrate the high-resolution capabilities of OET, we have created 15,000 OET traps over an area of 1.3×1.0 mm^2, as shown in Fig. 9.17. The particles are trapped in the dark area by the induced negative DEP forces, which push the

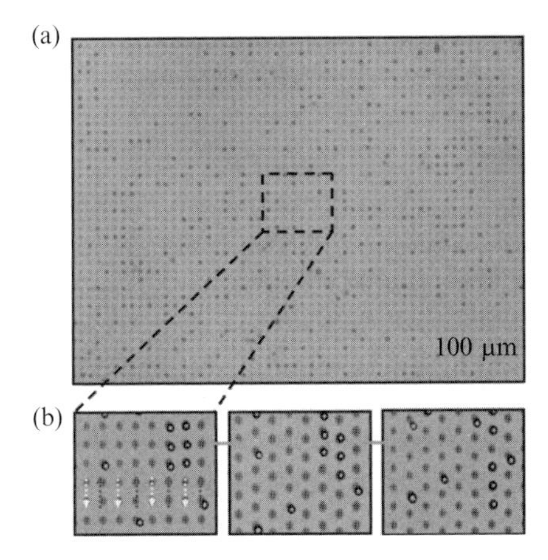

FIG. 9.17: (a) 15,000 OET traps are created over a 1.3 mm × 1.0 mm area. The 4.5 μm-diameter polystyrene beads experiencing negative DEP forces are trapped in the dark area. Each trap has a diameter of 4.5 μm, which is adjusted to fit a single particle. (b) Parallel transportation of single particles. Three snapshots from the captured video show the particle motion in a subset of the manipulation area. The trapped particles in two adjacent columns move in opposite directions, as indicated by the blue and yellow arrows.

beads into the non-illuminated regions where the electric field is weakest. The size of each trap is optimized to capture a single 4.5-μm diameter polystyrene bead. The pitch between particles is 12 μm, corresponding to 8 optical pixels. By programming the projected images, these trapped particles can be individually moved in parallel. Snapshots of the captured video images are presented in Fig. 9.17 to show the particles in adjacent columns moving in opposite directions.

9.3.3 *Real-time interactive microparticle manipulation*

Optical manipulation by direct optical images is a salient feature that enables real-time, interactive control of individual particles without the extensive mathematical calculation needed for holographic optical tweezers. Interactive manipulation can be easily accomplished using a DMD-based digital projector (Ohta *et al.* 2007). The optical traps are drawn using Microsoft's PowerPoint. The pattern is projected onto the OET device through a 10× objective lens. The power at the projector output was measured to be approximately 600 mW. Approximately 7 per cent of this power is collected by the objective lens and focused onto the OET device. Therefore, the optical power incident on the OET surface (not including losses through the bulk optics) is 42 mW, corresponding to an optical intensity of 12 W/cm^2.

Figure 9.18 shows an individually addressable OET trap array for 20- and 45-μm particles. Since the particle experience negative DEP force, each particle is trapped by a square image with size matched to the particle. The traps can be translated in

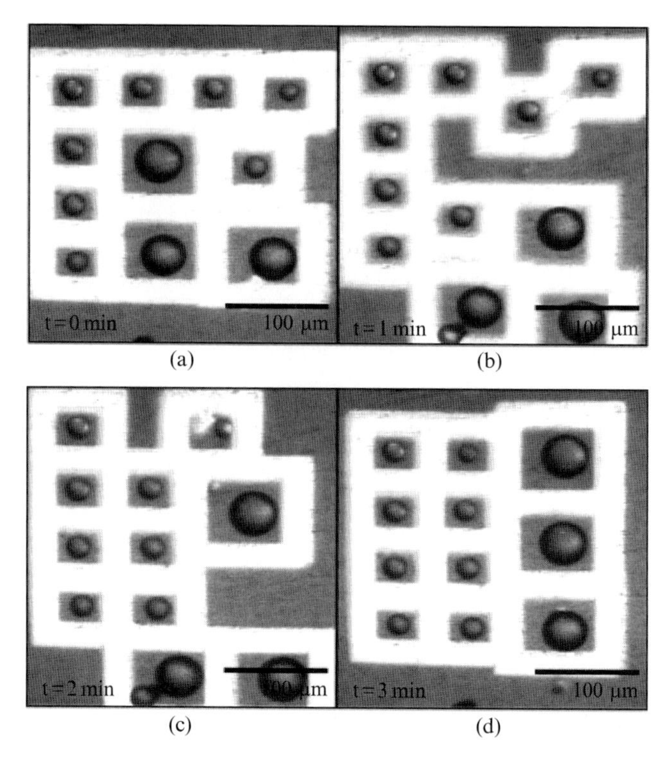

Fig. 9.18: Capture video images showing interactive manipulation of an individually addressable OET trap array for 20- and 45-μm particles. The 45-μm particles are moved to the right while the 20- μm particles are moved to the left.

real-time under user control at velocities up to 25 μm/s. Since each cell of the array is an independent single-particle trap, the array can be dynamically rearranged. As a demonstration, the 20- and 45-μm polystyrene beads are reorganized under operator control. Although the formation of the array was performed manually here, it can be automated by combining OET with an image-analysis feedback-control system (Chiou *et al.* 2005a, Chiou *et al.* 2005b).

9.4 Recent advances in optoelectronic tweezers technique

The OET has become a vibrant research area. New applications are being explored, including cell manipulation in highly conductive media and parallel assembly of nanowires. In this section, we will review the recent developments in OET.

9.4.1 *Characterization of OET operational regimes*

Even though the main operation principle of OET has been the light-induced DEP, there are other operation regimes in which other physical effects are also present, and in some cases, might be dominant over the DEP (Valley *et al.* 2008). A comprehensive understanding of these operational regimes is essential in optimizing the OET performance as well as positively utilizing other regimes. The three main operational regimes of OET include: DEP, electrothermal flow, and light-actuated AC electroosmosis (Chiou *et al.* 2006) (LACE).

The DEP operational regime has been discussed in previous sections. The electro-thermal flow is an effect due to a temperature gradient present in the liquid layer created by optical absorption in the photoconductor. This effect is only observed at very high optical power intensities of more than 100 W/cm^2, which is much larger than typical optical power intensities used for trapping. The electrothermal flow is mostly a parasitic effect and is not capable of trapping any objects; however, due to the very high optical powers necessary to achieve this operational regime, it does not interfere with typical OET operation.

The other effect that has been observed in OET is light-actuated AC electroosmosis or LACE. At low frequencies, the lateral component of the electric field created in the liquid layer interacts with the double-layer charges on the surface of the OET device and can accelerate them laterally, creating a flow vortex around the illuminated area. This effect is typically observed at frequencies lower than

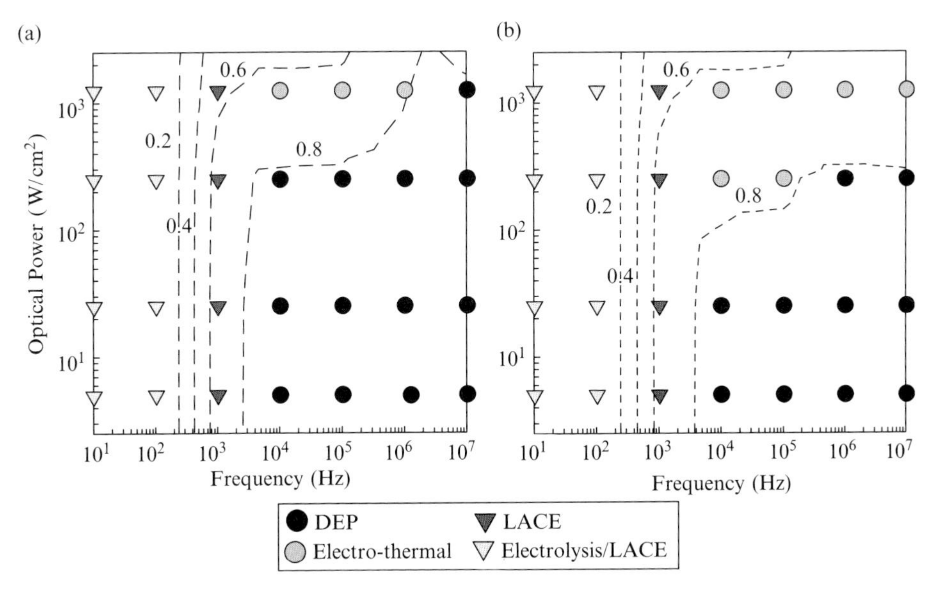

Fig. 9.19: Experimental and theoretical operation regimes of the OET device for a liquid conductivity of 1 mS/m at (a) 20 Vpp and (b) 10 Vpp. The light-induced DEP is dominant at the lower right corner (i.e. high bias frequency, low optical intensity). For bias frequencies below 1 kHz, LACE is dominant. Electro-thermal flow dominates at high optical intensity (> 100 W/cm^2) (Valley *et al.* 2008).

10 kHz and can be used for trapping nano-scale objects such as polystyrene beads as small as 200 nm. Using this method, it has been demonstrated that 31,000 individually addressable traps can be created for particles larger than 1 μm in diameter (Chiou *et al.* 2006).

The different operational regimes can be achieved in OET by tuning the parameters such as optical power and AC bias frequency. Figure 9.19 shows the dominant effects in OET as a function of optical intensity and bias frequency. In general, light-induced DEP is dominant for frequency > 10 kHz and optical intensity < 100 W/cm^2. At low frequencies (< 1 kHz), LACE play a key role, while at high optical intensities (> 100 W/cm^2), electro-thermal flow is the main effect.

9.4.2 *Optical manipulation of biological objects*

One of the most important applications of OET is its ability to manipulate cells and other biological samples in the liquid media with significantly lower optical intensities than optical tweezers, minimizing harmful effects on the cells. Previously, OET had been used for manipulation of red and white blood cells (Chiou *et al.* 2005d, Ohta *et al.* 2006) and separation of live and dead human B cells (Chiou *et al.* 2005). However, conventional OET devices with hydrogenated amorphous silicon as the photoconductive layer have been limited to manipulation of particles in liquids whose conductivities are smaller than 100 mS/m. This limitation arises from the fact that conventional OET is incapable of effectively switching the AC voltage from the photoconductive layer to the liquid layer due to relatively small photoconductivity of the amorphous silicon layer.

The ability to manipulate cells in high-conductivity physiological media is essential for maintaining cell viability (Voldman 2006) and performing other applications such as cell electroporation. To overcome this limitation of the conventional OET device, it is possible to replace amorphous silicon as the photoconductive layer with an N^+PN phototransistor structure which has more than two orders of magnitude larger photoconductivity due to the higher carrier mobility in single crystalline silicon and the current gain in the phototransistor structure. This novel OET device is called phototransistor OET (phOET). Figure 9.20(a) shows a schematic of the phOET structure. It consists of a top transparent ITO electrode and a bottom N^+PN phototransistor structure with an AC bias applied between them. The liquid layer containing the particles of interest is sandwiched between these two surfaces.

The photoconductivity of the phototransistor is 100 times higher than that of a-Si:H, as shown in Fig. 9.20(b). This enables phOET to turn on the virtual electrodes in liquid with a high salt content at an optical intensity as low as 1 W/cm^2. It has been demonstrated that phOET is capable of transporting Hela and Jurkat cells with speeds higher than 30 μm/s in phosphate-buffered saline (PBS) and Dulbecco's Modified Eagle Medium (DMEM) solutions, with

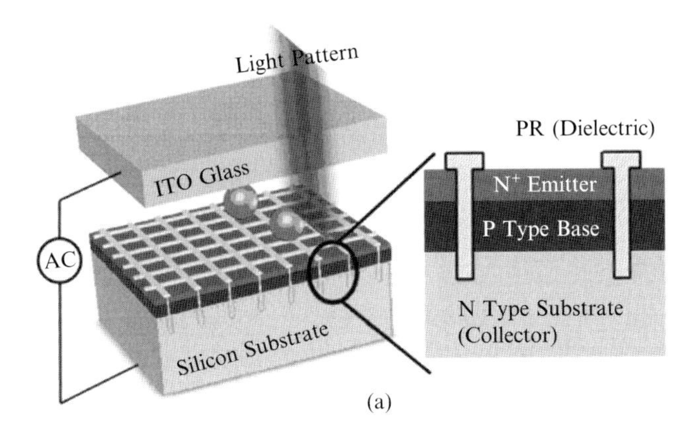

(a)

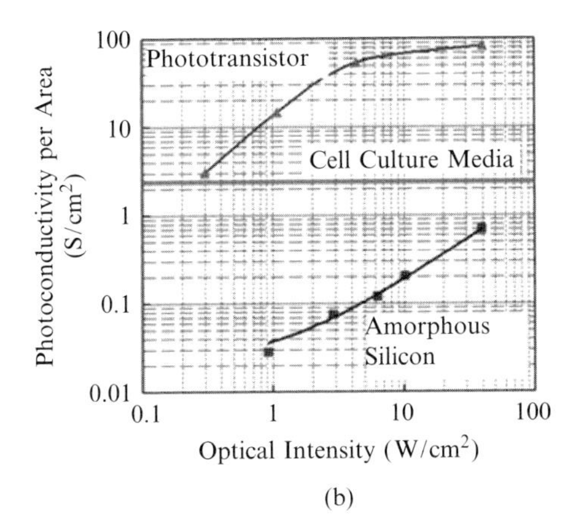

(b)

Fig. 9.20: (a) Schematics of phototransistor-based OET (phOET). The amorphous silicon photoconductor layer is replaced by a bipolar junction phototransistor with N^+PN doping in single crystalline silicon. (b) The measured photoconductivity of phototransistor and amorphous Si photoconductor. The photoconductivity in phototransistor is more than 100 times larger than in amorphous silicon. It requires less than 1 W/cm^2 to turn on the virtual electrodes in the cell culture media (Hsu *et al.* 2007).

conductivities of 1.5 S/m (Hsu *et al.* 2007). PhOET typically operates at frequencies in the mHz range. The trapped cells experience negative DEP force.

9.4.3 *Thermocapillary movement of air bubbles*

The ability to move air bubbles is a useful function for optofluidic systems. Air bubbles have been used for a variety of applications such as mixers (Garstecki *et al.* 2006), valves (Hua *et al.* 2002), pumps (Jun *et al.* 1998), and performing Boolean logic (Prakash *et al.* 2007). The bubbles can be either passively positioned using fluidics or actively positioned using various methods such as DEP (Schwartz *et al.* 2004), electrowetting (Pollack *et al.* 2002), optoelectrowetting (Chiou *et al.* 2003), evaporation (Liu *et al.* 2006), and thermal gradient (Kataoka 1999). Conventionally, thermal gradients have been created using resistive heating elements. However, it has been demonstrated that it is possible to create the necessary temperature gradients for thermocapillary manipulation of a bubble using a light source (Kotz *et al.* 2004).

Optically controlled actuation of the thermocapillary force has several advantages over other conventional methods, including flexible and reconfigurable manipulation capability and the ability to control a large number of bubbles simultaneously. This optically-controlled thermocapillary force can be created in a conventional OET device by using the photoconductive substrate as an absorber of the light energy, which results in the formation of a temperature gradient in the liquid layer. This temperature gradient in turn creates a surface tension gradient in the liquid, leading to a flow pattern from warmer areas towards the colder areas to minimize the energy. Therefore, the bubbles move to high temperature gradient areas and are stably trapped in the illuminated area. Figure 9.21 shows the optically-induced thermocapillary trapping of a 109-μm-diameter air bubble in a silicone oil media. The air bubble follows the position of the laser trap as it is scanned across the stage over a period of 12 s.

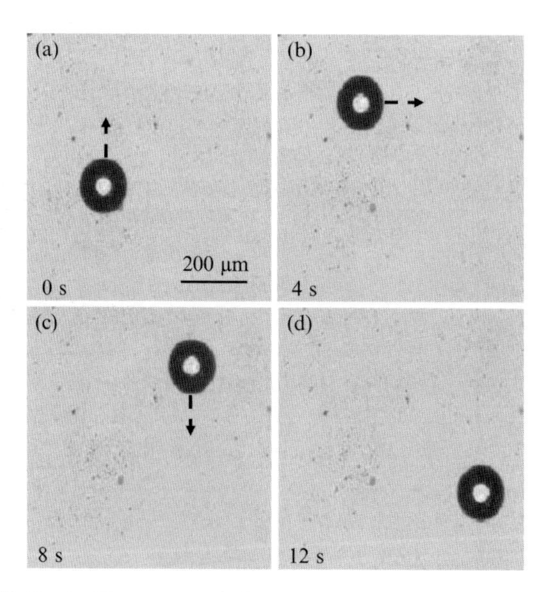

FIG. 9.21: (a–d) Optically actuated thermocapillary transport of an air bubble in silicone oil. A 109 μm-diameter bubble is trapped in the thermal trap created by a laser. The bubble follows the position of the laser spot as it is scanned across the stage (Ohta *et al.* 2007).

This method has been used to transport bubbles with diameters ranging from 33 to 329 μm (corresponding to 19 pl–23 nl). The translation speed of the bubbles is a linear function of optical power intensity used as the laser intensities are directly proportional to temperature gradients. The translation speed for bubbles less than 0.5 nl can reach 1.5 mm/s with 2 kW/cm^2 optical intensity.

9.4.4 *OET for nanoparticle manipulation*

The bottom-up approach is one of the main methods of organizing nano-scale objects. In this method, nano-scale building blocks such as nanowires, gold nanoparticles, and carbon nanotubes are put together to form structures with more complex functionalities. Current advances in the synthesis of nanoscopic materials have created an array of objects with interesting compositions and properties (Yang 2005). Several methods have already been used for manipulation of nanowires and carbon nanotubes, including microelectrode-based DEP (Smith *et al.* 2000, Krupke *et al.* 2003) and optical tweezers (Agarwal *et al.* 2005, Pauzauskie *et al.* 2006). However, in electrode-based DEP, the trapping patterns are fixed, which prevents dynamic and flexible manipulation of trapped nanowires. In the case of optical tweezers, the high optical power intensities required for a stable

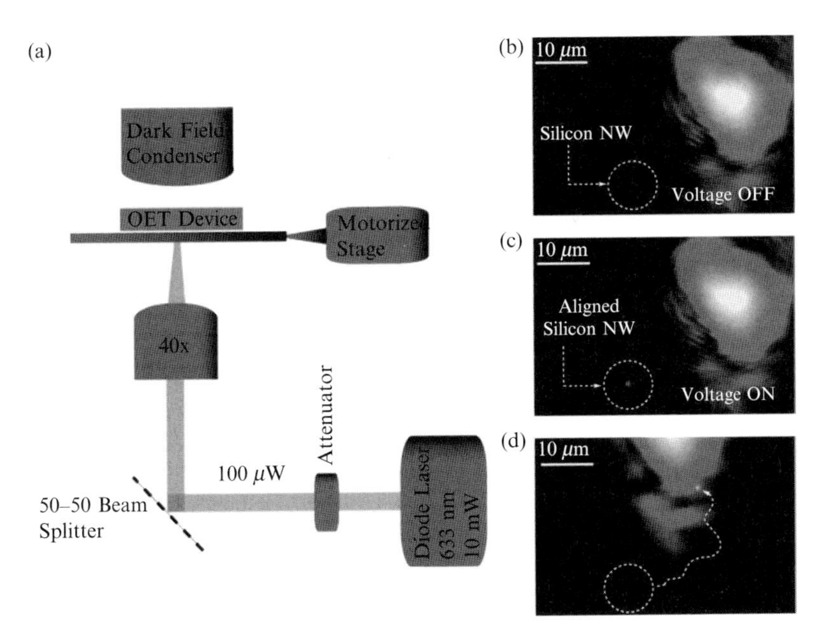

Fig 9.22: (a) Experimental setup for trapping of single nanowires by OET. (b) Snapshot before voltage is applied to the device. (c) After bias voltage is turned on, and the long-axis of the nanowire aligns with the electric field. (d) The nanowire experiences a positive DEP force and follows the movement of the light beam.

trapping limit often damage the nanowires. Optoelectronic tweezers can overcome both limitations by trapping nanostructures using virtual electrode patterns created at low optical power intensities (Jamshidi *et al.* 2008).

As mentioned before, the magnitude of the DEP force scales with the volume of the particle. Therefore, trapping of objects using the dielectrophoresis becomes increasingly challenging as the size of the particle becomes smaller. Previously, the smallest particles that OET was capable of trapping were 1–μm-diameter polystyrene beads. However, in cylindrical nanostructures such as nanowires, the magnitude of the DEP force is enhanced because the length of the wires is in the micrometer scale and the higher polarizability of nanowires along the length direction, which leads to a larger CM factor. As a result, the DEP force experienced by a nanowire is about two orders of magnitude larger than that by a spherical particle with the same diameter. It has been demonstrated that OET is capable of the trapping and transport of nanowires with diameters less than 20 nm and lengths of approximately 5 μm (Jamshidi *et al.* 2008). A wide variety of nanowires have been successfully trapped by OET, including silicon, gallium nitride, and silver.

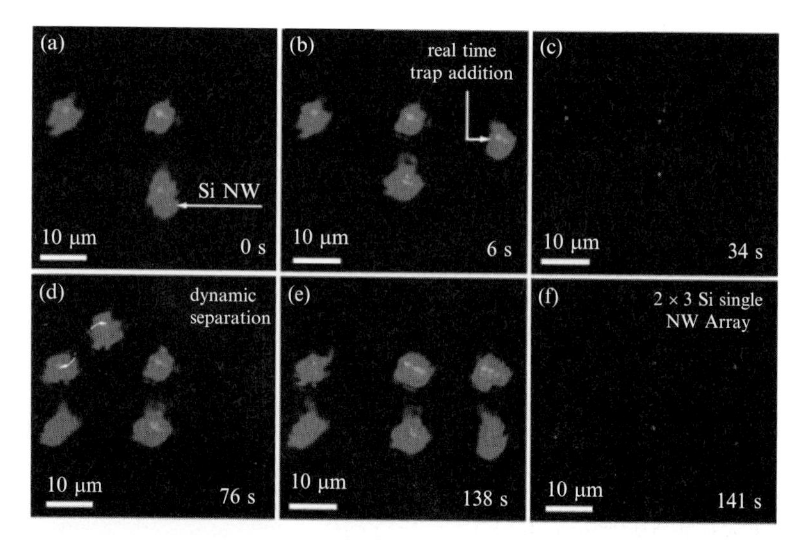

Fig. 9.23: Arrangement of six individual silicon nanowires into a 2 $\times$ 3 array using OET traps created with a DMD. (a) A silicon nanowire moving into the lower laser trap. (b) Addition of a new trap in real time. (c) Nanowires pattern with the laser light filtered. The upper traps contain two silicon nanowires each. (d) To create a single nanowire array, a new trap is added to separate the nanowires in top left corner. (e) The 2 $\times$ 3 single nanowire array. (f) Final pattern of the nanowire array with the laser light filtered. (This figure is a reproduction from Jamshidi *et al.* (2008) (supporting information) with permission © Nature Publishing Group.)

Figure 9.22(a) shows the experimental setup for trapping an individual silicon nanowire. A 10 mW, 633 nm diode laser source is attenuated to 100 μW and focused onto the OET chip using a $40\times$objective lens. A dark-field condenser is used for observation of the scattered light from the nanowires. Figures 9.22(b)–(d) show the process of trapping an individual silicon nanowire using a laser source. In the beginning, there is no voltage applied to the device and the nanowire is undergoing Brownian motion. Once the voltage is applied (Fig. 9.22(c)), the long axis of the nanowire aligns with the electric field, the nanowire experiences a positive DEP force and is attracted to the illuminated area. The trapped nanowire follows the movement of the light beam, as shown in Fig. 9.22(d).

In addition to using a focused laser light, it is possible to use a DMD spatial light modulator to create multiple traps in real time. Figure 9.23 shows the process of creating a 2×3 array of individual silicon nanowires using DMD-projected light patterns. Each trapped nanowire is individually addressable and can be dynamically positioned in real time.

In addition to trapping individual nanowires, the OET is also capable of separating metallic and semiconducting nanowires (Jamshidi *et al.* 2008). Silver nanowires experience a larger OET force because they are more conductive and therefore more polarizable. Using a scanning line pattern, one can find a range of scan speeds such that the silver nanowires stay trapped while silicon nanowires fail to follow the trap as the laser line scans over the sample. This is a simple way to separate metallic and semiconductor nanowires. This is potentially useful for many applications, including the separation of metallic and semiconducting carbon nanotubes, which is one of the main challenges in carbon nanotube nanoelectronics (Krupke *et al.* 2003).

9.5 Conclusion

In this chapter, we reviewed the concept and operation principle of optoelectronic tweezers (OET). Based on light-induced dielectrophoresis (DEP), OET combines the advantages of optical tweezers and DEP. However, unlike optical tweezers, OET can be actuated by low-cost incoherent light sources such as lamps or LEDs. The optical power requirement is also five orders of magnitude smaller than that of optical tweezers. Using a digital-micromirror-device (DMD) spatial light modulator, 15,000 individually addressable particle traps have been created by OET. Actuation using direct optical images offers a great deal of flexibility and versatility that enables OET to perform a wide variety of complex manipulation functions, such as sorting of particles based on size or other visual attributes through pattern recognition and feedback control. We have also discussed the operation regimes of OET and other physical effects present in OET devices, such as light-actuated AC electroosmosis (LACE) and electrothermal flow. By properly selecting the optical power and the bias frequency, OET can operate in predominatly light-induced DEP regime. Phototransistor-based OET (phOET) capable of manipulating and trapping cells in high conductivity physiological solutions has also been described. The trapped cells can be transported at speeds higher than 30 μm/s. Extension of the OET technology to trap nanowires with diameters as small as 20 nm has been achieved, thanks to the enhanced polarizability of nanowires. A scanning light pattern separates metallic and semiconducting nanowires due to the difference of their polarizability. The ability to trap individual nanostructures opens up many new applications in nanoelectronics and nano-optics.

References

Agarwal, R., Ladavac, K., Roichman, Y., Yu, G. H., Lieber, C. M., and Grier, D. G., 2005. 'Manipulation and assembly of nanowires with holographic optical traps.' *Optics Express*, **13**, pp. 8906–8912.

Ashkin, A and Dziedzic, J. M., 1987. 'Optical trapping and manipulation of viruses and bacteria.' *Science*, **235**, pp. 1517–1520.

Ashkin, A., Dziedzic, J. M., Bjorkholm, J. E., and Chu, S., 1986. 'Observation of a single-beam gradient force optical trap for dielectric particles.' *Optics Letters*, **11**, pp. 288–290.

Ashkin, A., Dziedzic, J. M., and Yamane, T., 1987. 'Optical trapping and manipulation of single cells using infrared-laser beams.' *Nature*, **330**, pp. 769–771.

Becker, F. F., Wang, X. B., Huang, Y., Pethig, R., Vykoukal, J., and Gascoyne, P. R. C., 1995. 'Separation of human breast-cancer cells from blood by differential dielectric affinity.' *Proceedings of the National Academy of Sciences of the United States of America*, **92**, pp. 860–864.

Bustamante, C., Bryant, Z., and Smith, S. B., 2003. 'Ten years of tension: Single-molecule DNA mechanics.' *Nature*, **421**, pp. 423–427.

Chiou, P. Y., Moon, H., Toshiyoshi, H., Kim, C. J., and Wu, M. C., 2003. 'Light actuation of liquid by optoelectrowetting.' *Sensors and Actuators*, **104**, p. 222.

Chiou, P. Y., Ohta, A. T., and Wu, M. C., 2005a. 'Microvision activated optical manipulation on microscopic particles.' In *IEEE 18th Annual International Conference on Micro Electro Mechanical Systems (MEMS'05)*, pp. 682–685.

Chiou, P. Y., Ohta, A. T., and Wu, M. C., 2005b. 'Continuous optical sorting of hela cells and microparticles using optoelectronic tweezers.' In *IEEE/LEOS International Conference on Optical MEMS and Their Applications (OMEMS'05)* pp. 83–84.

Chiou, P. Y., Ohta, A. T., and Wu, M. C., 2005c. 'Massively parallel manipulation of single cells and microparticles using optical images.' *Nature*, **436**, pp. 370–372.

Chiou, P. Y., Ohta, A. T., Han, T. H., Liao, J. C., Bhardwaj, U., McCabe, E. R. B., Yu, F., Sun, R., and Wu, M. C., 2005d. 'Manipulation of live red and white blood cells via optoelectronic tweezers.' In *Proceeding of the International Conference on Bio-Nano-Informatics (BNI) Fusion.*

Chiou, P. Y., Ohat, A. T., Jamshidi, A., Hsu, H. Y., and Chou, J. W., M. C., 2006. 'Light-actuated AC electroosmosis for optical manipulation of nanoscale particles.' In *Proceedings of Solid-State Sensor, Actuator, and Microsystems Workshop*, pp. 56–59.

Choi, W., Kim, S. H., Jang, J., and Park, J. K., 2007. 'Lab-on-a-display: A new microparticle manipulation platform using a liquid crystal display (LCD).' *Microfluidics and Nanofluidics*, **3**, pp. 217–225.

Curtis, J. E., Koss, B. A., and Grier, D. G., 2002. 'Dynamic holographic optical tweezers.' *Optics Communications*, **207**, pp. 169–175.

Ermolina, I., Morgan, H., Green, N. G., Milner, J. J., and Feldman, Y., 2003. 'Dielectric spectroscopy of Tobacco Mosaic Virus.' *Biochimica et Biophysica Acta – General Subjects*, **1622**, pp. 57–63.

Garces-Chavez, V., Dholakia, K., and Spalding, G. C., 2005. 'Extended-area optically induced organization of microparticies on a surface.' *Applied Physics Letters*, **86**, p. Art. No. 031106.

Garstecki, P., Fuerstman, M. J., Fischbach, M. A., Sia, S. K., and Whitesides, G. M., 2006. 'Mixing with bubbles: A practical technology for use with portable microfluidic devices.' *Lab on a Chip*, **6**, pp. 207–212.

Grier, D. G., 2003. 'A revolution in optical manipulation.' *Nature*, **424**, pp. 810–816.

Hertz, H. M., 1995. 'Standing-wave acoustic trap for nonintrusive positioning of microparticles.' *Journal Of Applied Physics*, **78**, pp. 4845–4849.

Howard, J., 2001. *Mechanics of Motor Proteins and the Cytoskeleton.* Sinauer Associates, Inc., Sunderland, MA.

Hsu, H. Y., Ohta, A. T., Chiou, P. Y., Jamshidi, A., and Wu, M. C., 2007. 'Phototransistor-based optoelectronic tweezers for cell manipulation in highly conductive solution.' In *IEEE Transducer.* Lyon, France.

Hua, S. Z., Sachs, F., Yang, D. X., and Chopra, H. D., 2002. 'Microfluidic actuation using electrochemically generated bubbles.' *Analytical Chemistry*, **74**, pp. 6392–6396.

Hughes, M. P., 2002. 'Strategies for dielectrophoretic separation in laboratory-on-a-chip systems.' *Electrophoresis*, **23**, pp. 2569–2582.

Jamshidi, A., Pauzauskie, P. J., Schuck, P. J., Ohat, A. T., Chiou, P. Y., Chou, J., Wu, M. C., and Y. P., 2008. 'Dynamic manipulation and separation of individual semiconducting and metallic nanowires.' *Nature Photonics*, **2**, pp. 86–89.

Jones, T. B., 1995. *Electromechanics of Particles.* Cambridge University Press, New York.

Jones, T. B., 2002. 'On the relationship of dielectrophoresis and electrowetting.' *Langmuir*, **18**, pp. 4437–4443.

Jun, T. K., and Kim, C. J., 1998. 'Microscale pumping with traversing bubbles in microchannels'. *Journal of Applied Physics*, **83**, pp. 5658–5684.

Kadaksham, A. T. J., Singh, P., and Aubry, N., 2004. 'Dielectrophoresis of nanoparticles.' *Electrophoresis*, **25**, pp. 3625–3632.

Kataoka, D. E., 1999. 'Patterning liquid flow on the microscopic scale.' *Nature*, **402**, pp. 794–797.

Kotz, K. T., Noble, K. A., and Faris, G. W., 2004. 'Optical microfluidics.' *Applied Physics Letters*, **85**, p. 2658.

Kremser, L., Blaas, D., and Kenndler, E., 2004. 'Capillary electrophoresis of biological particles: Viruses, bacteria, and eukaryotic cells.' *Electrophoresis*, **25**, pp. 2282–2291.

Krupke, R., Hennrich, F., von Lohneysen, H., and Kappes, M. M., 2003. 'Separation of metallic from semiconducting single-walled carbon nanotubes.' *Science*, **301**, pp. 344–347.

Lee, H., Purdon, A. M., and Westervelt, R. M., 2004. 'Manipulation of biological cells using a microelectromagnet matrix.' *Applied Physics Letters*, **85**, pp. 1063–1065.

Liu, G. L., Kim, J., Lu, Y., and Lee, L. P., 2006. 'Optofluidic control using photothermal nanoparticles.' *Nature Materials*, **5**, pp. 27–32.

Manaresi, N., Romani, A., Medoro, G., Altomare, L., Leonardi, A., Tartagni, M., and Guerrieri, R., 2003. 'A CMOS chip for individual cell manipulation and detection.' *IEEE Journal of Solid-State Circuits*, **38**, pp. 2297–2305.

Merenda, F., Rohner, J., Fournier, J. M., and Salathe, R. P., 2007. 'Miniaturized high-NA focusing-mirror multiple optical tweezers.' *Optics Express*, **15**, pp. 6075–6086.

Ohta, A. T., Chious, P. Y., Jamshidi, A., Hsu, H. Y., Wu, M. C., Phan, H. L., Sherwood, S. W., Yang, J. M., and Lau, A. N. K., 2006. 'Spatial cell discrimination using optoelectronic tweezers.' In *Digest of the LEOS Summer Topical Meetings IEEE*. Piscataway, NJ., pp. 23–24.

Ohta, A. T., Chiou, P. Y., Han, T. H., Liao, J. C., Bhardwaj, U., McCabe, E. R. B., Yu, F. Q., Sun, R., and Wu, M. C., 2007a. 'Dynamic cell and microparticle control via optoelectronic tweezers.' *Journal of Microelectromechanical Systems*, **16**, pp. 491–499.

Ohta, A. T., Chiou, P. Y., Phan, H. L., Sherwood, S. W., Yang, J. M., Lau, A. N. K., Hsu, H. Y., Jamshidi, A., and Wu, M. C., 2007b. 'Optically controlled cell discrimination and trapping using optoelectronic tweezers.' *IEEE Journal of Selected Topics in Quantum Electronics*, **13**, pp. 235–243.

Ohta, A. T., Jamshidi, A., Valley, J. K., Hsu, H. Y., and Wu, M. C., 2007c. 'Optically actuated thermocapillary movement of gas bubbles on an absorbing substrate.' *Applied Physics Letters*, **91**, pp. 074103.

Pauzauskie, P. J., Radenovic, A., Trepagnier, E., Shroff, H., Yang, P. D., and Liphardt, J., 2006. 'Optical trapping and integration of semiconductor nanowire assemblies in water.' *Nature Materials*, **5**, pp. 97–101.

Pollack, M. G., Shenderov, A. D., and Fair, R. B., 2002. 'Electrowetting-based actuation of droplets for integrated microfluidics.' *Lab on a Chip*, **2**, pp. 96–101.

Prakash, M., and Gershenfeld, N., 2007. 'Microfluidic bubble logic.' *Science*, **315**, p. 832.

Schwartz, J. A., Vykoukal, J. V., and Gascoyne, P. R. C., 2004. 'Droplet-based chemistry on a programmable micro-chip.' *Lab on a Chip*, **4**, pp. 11–17.

Smith, P. A., Nordquist, C. D., Jackson, T. N., Mayer, T. S., Martin, B. R., Mbindyo, J., and Mallouk, T. E., 2000. 'Electric-field assisted assembly and alignment of metallic nanowires.' *Applied Physics Letters*, **77**, pp. 1399–1401.

Sundararajan, N., Pio, M. S., Lee, L. P., and Berlin, A. A., 2004. 'Three-dimensional hydrodynamic focusing in polydimethylsiloxane (PDMS) microchannels.' *Journal Of Microelectromechanical Systems*, **13**, pp. 559–567.

Sze, S. M., 1981. *Physics of Semiconductor Devices*, 2nd ed. John Wiley & Sons, Inc., New YorK.

Tang, J., Gao, B., Geng, H. Z., Velev, O. D., Din, L. C., and Zhou, O., 2004. 'Manipulation and assembly of SWNTS by dielectrophoresis.' *Abstracts of Papers of the American Chemical Society*, **227**, pp. U1273–U1273,

Valley, J. K., Jamshidi, A., Ohat, A. T., Hsu, H. Y., and Wu, M. C., 2008. 'Operational regimes and physics present in optoelectronic tweezers.' *Journal of Microelectromechanical Systems*, **17**, pp. 342–350.

Velev, O. D., Prevo, B. G., and Bhatt, K. H., 2003. 'On-chip manipulation of free droplets.' *Nature*, **426**, pp. 515–516.

Voldman, J., 2006. 'Electrical forces for microscale cell manipulation.' *Annual Review Of Biomedical Engineering*, **8**, pp. 425–454.

Wang, M. D., Schnitzer, M. J., Yin, H., Landick, R., Gelles, J., and Block, S. M., 1998. 'Force and velocity measured for single molecules of RNA polymerase.' *Science*, **282**, pp. 902–907.

Wei, J. H., and Lee, S. C., 1993. 'Electrical and optical properties of implanted amorphous silicon.' *Journal of Applied Physics*, **76**, pp. 1033–1040.

Yang, J., Huang, Y., Wang, X. B., Becker, F. F., and Gascoyne, P. R. C., 2000. 'Differential analysis of human leukocytes by dielectrophoretic field-flow-fractionation.' *Biophysical Journal*, **78**, pp. 2680–2689.

Yang, P. D., 2005. 'The chemistry and physics of semiconductor nanowires.' *MRS Bulletin*, **30**, pp. 85–91.

Ying, L. M., White, S. S., Bruckbauer, A., Meadows, L., Korchev, Y. E., and Klenerman, D., 2004. 'Frequency and voltage dependence of the dielectrophoretic trapping of short lengths of DNA and dctp in a nanopipette.' *Biophysical Journal*, **86**, pp. 1018–1027.

10

NANOBIOSENSORS

Tza-Huei Wang, Kelvin Liu, Hsin-Chih Yeh, and Christopher M. Puleo

10.1 Introduction

The explosion in nanotechnology has led to an increase of interest in designing sensors engineered at the nano scale from molecular probes and nanomaterials as the basis for next-generation biosensors to achieve even higher sensitivity, more flexible conjugation chemistries, lower cost, and more pervasive application. Although devices such as lab-on-a-chip and DNA microarrays can be engineered to have extremely high sensitivities and throughputs, the disconnect between these surface based detection technologies and solution based biomolecules has created confines that have yet to be fully bridged.

The great breakthrough in nanobiosensors is the ability to tackle the problem from another direction in creating solution based nanoassemblies that are on the same size-scale as the molecules being detected. This has been realized through progress in synthesis chemistries that have allowed scientists to create monodisperse, colloidal dispersions of nanoparticles with well defined shapes, sizes, and compositions, therefore providing a means to accurately control material properties. Further advances in surface functionalization and conjugation chemistry have enabled the application of these unique material properties in novel signal transduction methods to transform each nanoassembly into a fully independent and functional biosensor. Nanomaterial based biosensors have the potential to not only improve upon current techniques but also to operate where current technologies have failed or are lacking.

The fabrication of nanomaterial based sensor assemblies fundamentally differs from that of microscale devices. While typical microscale devices are fabricated from a top-down approach using high-resolution physical processes such as photolithography or e-beam tools to pattern thin film materials, nanoassemblies are created through a bottom-up approach relying on biological and chemical specificity to pre-program the controlled assembly of individual building blocks such as nucleic acids, antibodies, proteins, lipids, polymers, small molecules, and organic linkers into functional sensors.

After a brief introduction of the main concepts in biodetection, this review begins with a discussion of nanobiosensors that incorporate metallic nanoparticles, semiconductor nanocrystals, and nanowire/nanotubes. Finally, single molecule detection through confocal spectroscopy is reviewed as a high sensitivity method for the optical detection of biomolecular sensors. The concluding remarks provide perspective on past developments and future directions.

10.1.1 *Biodetection assay basics*

The general function of an assay is to characterize a specific property of an analyte or an interaction between analytes. Optical, enzymatic, or isotopic labels are typically used in conjunction with molecular probes to quantify measurands. Common biochemical assays include radioimmunoassays (RIA), enzyme-linked immunosorbent assays (ELISA), gel electrophoresis-based assays such as western and northern blots, cellular assays of gene expression, and numerous others. These assays can be used to measure enzymatic activity, characterize protein–protein interactions, sequence DNA, analyze the activity of cellular machinery, deduce protein structure, etc. Analogous assays can be performed in a highly efficient manner using nanobiosensors rather than traditional biochemical techniques. The biodetection assay as it is related to nanobiosensors can be described by the general array of parameters listed in Table 10.1.

The analyte refers to the biomolecule being measured or characterized and is typically the species against which the molecular probe has affinity. Nanomaterials themselves have no biological affinity, preference, or ability to discriminate between biomolecules. Biological affinity to the analyte is therefore generated by probe molecules conjugated to the nanoparticle surface. Bioanalytes that have convenient probes include 1) nucleic acids to which oligonucleotides can be synthesized to hybridize predictably to target DNA/RNA, and 2) proteins, small molecules, and pathogens against which suitable antibodies can be generated. Other methods such as using protein–protein interactions, short peptides generated via phage display libraries, and chemical specificity exist but are far less common.

Once an interaction between the probe and analyte has occurred, the interaction must be transduced into a detectable signal. Nanomaterials provide this signal transduction by acting as optical, electrical, or magnetic labels. Optical detection methods can be based on fluorescence emission, surface plasmon resonance, scattering, colorimetric change, energy transfer, or surface enhanced Raman scattering. Optical detection is by far the most common method of detection due to the high sensitivity of optical systems and due to the legacy from assays developed with molecular fluorophores and dyes. Table 10.2 and Table 10.3 are provided as a concise reference of nanoparticle detection methods.

The assay format describes the general scheme and protocol in which probes and labels are used to detect bioanalytes. Assays using nanosensors are often similar to standard molecular assays but can also differ dramatically. Although the majority of traditional assays are heterogeneous and require a solid support, many nanosensor based assays can be done entirely in the solution phase without the need for a solid surface as the nanoparticle itself functions as the primary support/scaffold. For example, many FRET based nanoparticles assays have no

TABLE 10.1: Key parameters and properties in describing the performance and function of a typical bioassay.

Variable	Bioassay Variables Definition	Examples
Analyte / bioanalyte	Species being analyzed	DNA, proteins, pathogens, small molecules
Affinity generation	Method of providing biological specificity	DNA hybridization, immunochemistry, chemical specificity
Detection mode	Method of transduction into measurable signal	Optical, electrical, magnetic
Heterogeneous / Homogeneous assay	Whether separation of bound / unbound species is required	Separation by stringency wash (thermal, salt, rinsing, denaturing), magnetic separation
Bulk vs. single molecule	Whether measurements are made on ensemble of molecules at once or individually	Bulk spectroscopy—spectroscopy of bulk solutions such as in fluorimeters, plate readers, etc Single molecule spectroscopy—spectroscopy with single molecule sensitivity such as confocal microscopy
Sensitivity	Lowest detectable amount of analyte	Molar concentration (nM), weight concentration (ng/mL), total molar amount (zmol),
Selectivity	Ability to differentiate between analyzed species	Selectivity of perfect match *vs.* mismatch DNA targets, selectivity of bacteria A *vs.* bacteria B
Equilibrium / real-time	Time scale at which an assay performs measurement	Real-time—SPR Equilibrium—Microarray, gel shift assays,
Assay format	Method in which probes and labels are utilized to perform assay	Sandwich—2x probes form sandwich with analyte in middle Competitive—analyte competes with a second substrate for probe binding
Sample volume	Volume of sample required for assay	Given in volume (mL, μl, etc)

comparable molecular counterparts and function freely in solution without the need for a solid surface or anchor.

Depending on the assay format and signal transduction method, separation of bound *vs.* unbound probes and analytes may be required. The majority of traditional assays and microdevices that employ solid supports and surfaces are heterogeneous and require stringency washes to rinse away unbound components in an effort to increase selectivity and reduce background. Among the chief benefits of many nanobiosensors is that they reside in the solution phase, and have signal transduction methods that allow high sensitivity and high selectivity assays to be performed in a homogenous format, greatly simplifying assay protocol. The elimination of stringency washes also widens the window for in-vivo application.

An additional approach that has been enabled by the evolution of colloidal nanoparticles and progress in high sensitivity photonic detectors is the field of single molecule detection. With high sensitivity signal transduction methods, it is now possible to detect even single nanoparticle labels, pushing detection limits lower and enabling study of single molecule dynamics. Still, the large majority of assay techniques rely on bulk measurements in which ensemble averages of solution properties are taken.

Sensitivity, selectivity, and sample volume are perhaps the most widely touted measures of an assay since they are easily quantified. Although the exact definitions can vary depending on the specific assay, the general meanings remain consistent. Sensitivity refers to the lowest concentration or the smallest absolute amount of analyte that is detectable above the baseline. Selectivity quantifies the ability of an assay to distinguish between a specific analyte and the background, and is typically expressed as a ratio. The sample volume defines the minimum volume of sample that a particular assay requires for analysis.

Assays can also be distinguished by the time scales during which measurements can be taken. Heterogeneous assays requiring stringency washes must reach a steady-state equilibrium between probe–analyte binding before measurements are taken to ensure accurate results. In homogeneous assays, it is possible that measurements can be taken real-time as the assay is occurring to obtain kinetics information.

10.1.2 *General applications of nanoscale materials and structures*

Functionalized nano-scale materials and structures such as nanoparticles and nanowires have been applied to the detection of nearly all forms of biological analytes. Early applications of many sensor technologies are often focused on nucleic acid detection because the interactions of DNA strands through hybridization as well as the molecular structure and conjugation of DNA are well characterized and easily predicted. Oligonucleotides comprise the most predictable form of probe for biosensor applications since they are easily synthesized and very stable. Analysis of DNA has been performed in applications such as the detection of sequence specific DNA, mutation status, gene expression, and pathogens. Assays involving DNA analysis have been performed in many formats depending on the specific detection method utilized.

Genomic analysis of gene mutation and gene status has been largely limited to the research laboratory due to the labor-intensive protocols and costly equipment and reagents. Many nanoparticle assays aim to enable low cost detection of mutations such as SNPs and specific alleles to bring genetic analysis from the research lab into the clinical environment for personalized medicine.

While DNA is the blueprint that controls cellular function, proteins are the actual functional molecules synthesized based on DNA sequences. Characterization of the interactions of proteins with other proteins, small molecules, or nucleic acids is crucial in understanding cellular and molecular biology. Although proteins

TABLE 10.2: Summary of recent work in metallic nanoparticle based nanosensors arranged by signal transduction method.

Metallic Nanoparticle Detection Methods		
Detection method	Description	References
Cross-linking induced aggregation	Analytes induce aggregation via cross-linking of nanoparticle probes	(Mirkin, C.A. *et al.* 1996; Elghanian, R. *et al.* 1997; Storhoff, J.J. *et al.* 1998; Cao, Y.W. *et al.* 2001; Chakrabarti, R. and Klibanov, A.M. 2003; Hirsch, L.R. *et al.* 2003; Jin, R. C. *et al.* 2003; Roll, D. *et al.* 2003; Storhoff, J.J. *et al.* 2004; Schofield, C.L. *et al.* 2007)
Non-cross-linking induced aggregation	Analytes induce aggregation via electrostatic or entropic effects without cross-linking of nanoparticle probes	(Sato, K. *et al.* 2003; Li, H.X. and Rothberg, L. 2004; Li, H.X. and Rothberg, L.J. 2004)
SERS—Surface Enhanced Raman Scattering	Raman dyes conjugated to nanoparticle surface as for use labels in assays	(Kneipp, K. *et al.* 1997; Nie, S.M. and Emery, S.R. 1997; Cao, Y.W.C. *et al.* 2002; Cao, Y.C. *et al.* 2003; Grubisha, *et al.* 2003; Faulds, K. *et al.* 2004)
LSPR—Localized Surface Plasmon Resonance	Nanoparticle surface plasmon resonance spectra is sensitive to the local dielectric properties of the immediate medium	(Haes and Van Duyne 2002; McFarland, and Van Duyne 2003)
Electrochemical stripping	Acidic dissolution of nanoparticles for electrical detection via stripping voltammetry	(Authier *et al.* 2001; Wang *et al.* 2001; Wang *et al.* 2001; Wang *et al.* 2002; Ozsoz *et al.* 2003; Wang *et al.* 2003)
RLS—Resonant Light Scattering	Application of nanoparticles as optical tags in assays primarily using their high scattering coefficients	(He *et al.* 2000; Taton *et al.* 2000; Niemeyer and Ceyhan 2001; Bao, P. *et al.* 2002; Storhoff *et al.* 2004)
Electrical impedance	Direct measure of impedance across electrodes to detect nanoparticle labels	(Park *et al.* 2002)
BCA—Bio-bar-Code Amplification	Indirect detection analyte is used to capture nanoparticle probes labeled with bar-code DNA which is subsequently detected via a chip-based nanoparticle sandwich assay	(Nam *et al.* 2003; Nam *et al.* 2004)
Quenching	Use of nanoparticles as quenchers of fluorescence emission	(Dubertret *et al.* 2001; Maxwell, *et al.* 2002; Ray *et al.* 2006; Huang *et al.* 2007)

TABLE 10.3: Summary of recent work in quantum dot based nanosensors arranged by signal transduction method.

	Quantum Dot Detection Methods	
Detection Method	Description	References
Fluorescent label	Use of QD as a label in imaging	(Bruchez *et al.* 1998; Chan and Nie 1998; Akerman *et al.* 2002; Dubertret *et al.* 2002; Parak *et al.* 2002; Dahan *et al.* 2003; Wu *et al.* 2003; Gao *et al.* 2004)
Fluorescent tag	Use of QD as a tag in an optical tag in an assay or array	(Mattoussi *et al.* 2000; Pathak *et al.* 2001; Gerion *et al.* 2002; Goldman *et al.* 2002; Gerion *et al.* 2003; Robelek *et al.* 2004)
Optical coding	Multiplex coding based on intensity and spectral distribution of QDs within a polymer microbead scaffold	(Han *et al.* 2001; Xu *et al.* 2003)
Energy transfer / luminscence modulation	Nonradiative energy transfer from a donor fluorophore to an acceptor fluorophore, e.g. FRET, or quenching of QD photoluminscence	(Willard *et al.* 2001; Medintz *et al.* 2003; Gueroui and Libchaber 2004; Medintz *et al.* 2004; Dyadyusha *et al.* 2005; Hohng and Ha 2005; Levy *et al.* 2005; Zhang *et al.* 2005; Aryal and Benson 2006; Ho *et al.* 2006; So *et al.* 2006; Huang *et al.* 2007; Liu *et al.* 2007; Shi *et al.* 2007; Yao *et al.* 2007)
Colocalization / coincidence	2D spatial or temporal overlap of distinct QD tags indicates local confinement by binding	(Ho *et al.* 2005; Yeh *et al.* 2005; Yeh *et al.* 2006)

themselves are cumbersome to label, protein detection can be accomplished fairly easily due to the widespread availability of antibodies that are highly specific and due to the ease with which antibodies can be conjugated and modified. They can be applied to the detection of disease markers and quantification of specific gene products, among other uses.

Antibodies can be generated for a wide range of antigens not limited to proteins such as virus and bacteria for pathogen detection and small molecules. Pathogen detection has been demonstrated for homeland security applications as well as food-borne illness applications (Zhao *et al.* 2004; Ho *et al.* 2005).

The use of nanomaterial-based sensors is not limited to the detection of analytes but can also be applied to the characterization of biomolecular interactions between two or more species. In addition to sensor technologies, nanomaterials can be incorporated as optical and magnetic labels for imaging and spectroscopy. Both quantum dots and metallic nanoparticles are used as optical labels in microscopy and imaging of cells, tissues, and even whole animals.

A thorough walkthrough of nanobiosensor applications is provided in the following sections.

10.2 Metallic nanoparticles

Metallic nanoparticles were among the first nanomaterials to be used as optical tags and scaffolds in nanosensors. They typically consist of either gold or silver and have diameters between 10–100 nm. The unique properties of noble metal nanoparticles include 1) the incredible efficiency at which they scatter light due to plasmon resonance, 2) the great local enhancement of electric field in the immediate vicinity of the nanoparticle, 3) the well developed conjugation chemistry, 4) size, shape, and composition dependent plasmon resonance, and 5) environment dependent plasmon resonance effects.

These properties are used to varying degrees in five main methods by which detection of nanoparticles can be performed: 1) nanoparticle aggregation, 2) Raman scattering, 3) Rayleigh scattering, 4) electrochemical, and 5) energy transfer.

10.2.1 *Surface coatings and functionalization*

The straightforward chemistry for conjugating Au nanoparticles has promoted their rapid development in nanosensors. A common Au nanoparticle synthesis scheme is the citrate reduction of $HAuCl_4$. This leaves the Au nanoparticles with citrate ions adsorbed to the surface that can easily be functionalized through a thiol-exchange mechanism (Grabar *et al.* 1995). Thus, most nanoparticle functionalization schemes utilize alkanethiols as linker molecules between the Au surface and biomolecule. However, many other linker molecules can be used so long as a thiol-functionalization resides at one end to interface the nanoparticle. Ag nanoparticles can also be linked through thiol chemistry though the bond is less stable (Cao *et al.* 2001). Alternative functionalization chemistries such as electrostatic interactions (Li and Rothberg 2004; Li H.X. and Rothberg, L.J. 2004) and surface adsorption (Maxwell *et al.* 2002) have been used in select applications.

10.2.2 *Applications of metallic nanoparticles*

The earliest applications of metallic nanoparticles were aimed at detecting and analyzing DNA. Once DNA detection is demonstrated, the natural path progresses towards protein and pathogen detection and multiplexing. Although in some applications metallic nanoparticles are simply used as fluorescent analogs, in many other applications the nanoparticles function as scaffolds in nanoassemblies that are quite advanced when compared to those designed in current quantum dot based nanosensors.

10.2.2.1 *Cross-linking induced nanoparticle aggregation* The first nanoparticle sensors used cross-linking induced nanoparticle aggregation as a sensing mechanism. The bulk absorbance and scattering spectra of gold colloids in solution are proportional to particle size, shape, composition, and distribution. As individual nanoparticles are brought into close-proximity to each other, they display a reversible shift in optical properties that can be detected by both absorption and scattering. This red shift in the nanoparticles' surface plasmon resonance band causes a colorimetric change from red to purple which can be exploited as a signal transduction mechanism using cross-linking induced nanoparticle aggregation.

Mirkin first demonstrated that 13 nm Au colloids functionalized with short oligonucleotides could be cross-linked by a DNA duplex containing exposed 'sticky ends' (Mirkin *et al.* 1996). When the nanoparticles are linked to form a macroscopic aggregate, the solution changes from red to pink in color, providing a mechanism for the detection of DNA. Eleghanian extended this work to the detection of perfect match DNA targets with the addition of a thermal stringency wash (Elghanian *et al.* 1997). Nanoparticle induced aggregation created a sharp melting curve that could be exploited for distinguishing perfect complement and mismatch targets. Hybridization could be detected as an easily visualized colorimetric change when the nanoparticle solution was spotted on a thin layer chromatography (TLC) plate.

In order to increase detection sensitivity over standard absorbance based measurements, Storhoff demonstrated scattering-based detection by exciting the nanoparticles through evanescent scatter using a waveguide (Storhoff *et al.* 2004). Detection sensitivity was increased by four orders of magnitude over TLC plate spotting.

Detection of proteins in addition to nucleic acids has also been achieved through aggregation induced colorimetric change. Hirsch used gold nanoshells (silica core surrounded by Au or Ag shell) functionalized with PEG and PEG-antibodies to detect rabbit IgG in saline, serum, and whole blood (Hirsch *et al.* 2003). Roll used nanoparticle aggregation to demonstrate the detection of protein–small molecule interactions (Roll *et al.* 2003). Schofield recently demonstrated the use of lactose conjugated Au nanoparticles, rather than more typical antibody recognition, for the colorimetric detection of cholera toxin (Schofield *et al.* 2007).

10.2.2.2 *Non-cross-linking induced nanoparticle aggregation* While cross-linking induced aggregation relies on biological affinity between the molecular probe and target molecule link nanoparticles into large networks, aggregation can also be induced by alternative mechanisms. Sato used Au nanoparticles functionalized with only a single oligonucleotide probe that would hybridize and aggregate in the presence of complimentary DNA without cross-linking (Sato *et al.* 2003). Rather, hybridization in the presence of target DNA was thought reduce the repulsive forces between nanoparticles thus allowing aggregation to occur at high salt concentrations. Li demonstrated a similar mechanism but in an even simpler assay that relied on differences in the propensity of ssDNA versus dsDNA in adsorbing to Au nanoparticle surfaces (Li and Rothberg 2004). This mechanism was demonstrated in the detection of PCR-amplified genomic DNA and in SNP detection using a simple colorimetric assay (Li and Rothberg 2004).

10.2.2.3 *Surface enhanced Raman scattering (SERS)* Nanoparticle SERS has great potential in creating optical tags for multiplexing. The extremely large Raman cross-section of metallic nanoparticles greatly amplifies the spectroscopic signatures of covalently-linked surface dyes such that dye-functionalized nanoparticles can be easily distinguished due to their unique spectroscopic signatures. When compared to fluorophores, the particles are easily detected due to their high scattering cross-sections, and extremely stable due to the lack of photodecomposition (Kneipp *et al.* 1997; Nie and Emery 1997).

Nanoparticle SERS has been used for the detection of many types of biomolecules as well as multiplex detection. Cao used dye labeled Au nanoparticles combined with silver development in microarray-based multiplex detection of DNA targets and RNA SNP identification (Cao *et al.* 2002). Figure 10.1 shows a schematic illustration of the scheme used. The use of six distinct Raman dyes allowed unambiguous optical identification of the targets due to the unique spectroscopic signatures. Unlike fluorescence, there are no issues of spectral overlap or the need for multiple source excitation. Cao later applied the same technique

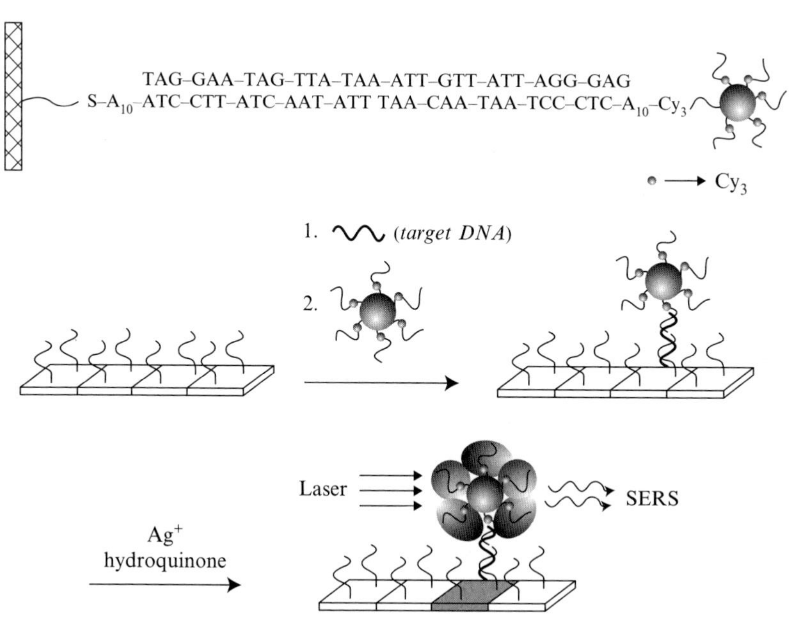

Fig. 10.1: Schematic illustration of nanoparticle SERS for DNA detection. Raman labeled nanoparticles are linked to the surface by target DNA. Subsequent catalytic silver deposition serves to amplify the SERS signal (Cao *et al.* 2002).

to look at protein–protein and protein–small molecule interactions in a microarray format (Cao *et al.* 2003). Quantitative analysis of DNA using Raman-dye labeled Ag nanoparticles was demonstrated by Faulds in comparing the scattering intensity of eight different Raman dyes over concentrations ranging four orders of magnitude (Faulds *et al.* 2004). Grubisha reported Raman reporter labeled Au nanoparticles as a tag in sandwich immunoassay to detect free prostate specific antigen (PSA) in human serum down to fM levels (Grubisha *et al.* 2003).

10.2.2.4 *Local surface plasmon resonance (LSPR)* The localized surface plasmon resonance (LSPR) spectra of a nanoparticle solution are dependent on the dielectric properties of its local environment as well its size, shape, distribution, and composition. By monitoring the wavelength shift in the peak of the Rayleigh scattering spectra, McFarland used this mechanism to directly sense the adsorption of small molecule analytes to the surface of a single Ag nanoparticle (McFarland and Van Duyne 2003). The response was not only proportional to the amount of analyte adsorbed to the surface but also to the size of the analyte, the solvent refractive index, and nanoparticle shape. In addition, the resonance shift could be actively monitored to provide a real-time detection platform for applications. LSPR spectroscopy of Ag nanoparticles derivatized with biotin was used by Haes to study streptavidin–biotin antibody–biotin interactions (Haes and Van Duyne 2002). They showed that the LSPR peak shift was proportional to the amount of bound SA and to the size of the analyte.

10.2.2.5 *Electrical/electrochemical detection* Electrochemical detection of metallic nanoparticles has been demonstrated as an alternative to optical detection to improve speed, cost, and ease of use. Electrochemical stripping detection of captured Au nanoparticles is a common method of electrical detection. Authier immobilized PCR-amplified human cytomegalovirus (HCMV) target DNA to a microwell that served to capture Au nanoparticle labeled oligonucleotide probes (Authier *et al.* 2001). The Au nanoparticles were then subject to acid dissolution, which enabled detection of the released Au^{III} ions via carbon electrodes. Although direct electrochemical detection of DNA hybridization can be performed, stripping detection of Au nanoparticles functions to amplify signal and increase sensitivity.

Wang demonstrated electrochemical stripping for the detection of the breast cancer gene BRCA1, but used magnetic nanoparticles for separation of hybridized DNA rather than a solid phase support (Wang *et al.* 2001). Wang then simplified the assay by replacing acid-dissolution of the nanoparticle tag with solid-state electrochemical stripping via oxidative dissolution (Wang *et al.* 2002). Wang later performed multiplex electrochemical stripping detection of three different breast cancer related DNA targets using three nanoparticle labels, ZnS, CdS, and PbS, conjugated to three separate oligonucleotide probes (Wang *et al.* 2003). Electrochemical stripping detection of the ZnS, CdS, and PbS probes lead to distinct voltammetric signals for each probe. Park used direct impedance measurements rather than electrochemical stripping to detect DNA (Park *et al.* 2002).

Oligonucleotide labeled Au nanoparticles were captured and bound to the surface between two microelectrodes by target DNA. Silver development of captured Au nanoparticle probes enabled detection of hybridization by the measurement of impedance across the electrode gap.

10.2.2.6 *Signal enhancement* Various amplification and alternative detection schemes have been developed in order to increase detection efficiency. While gold is easier to conjugate and more stable, silver has a 10 time larger scattering cross-section and a different plasmon resonance band. Cao took advantage of this combination in using Ag/Au core-shell nanoparticles derivatized with oligonucleotide probes to detect DNA (Cao *et al.* 2001). The core-shell nanoparticles displayed an aggregation induced colorimetric change distinct from that of typical Au nanoparticles and demonstrated the potential of engineering metal nanoparticles with varying compositions to obtain the desired physical, chemical, or optical properties. An alternative approach, referred to as silver development, uses the Au catalyzed reduction of Ag to enhance signal. Niemeyer used gold nanoparticles functionalized with antibody probes via DNA linker molecules in a sandwich immunoassay for the detection of mouse IgG (Niemeyer 2001).

Nam developed a creative approach for high-sensitivity detection of DNA termed Bio-Bar-Code amplification (BCA) combining gold nanoparticles, magnetic microparticles, and silver development to improve the detection limits (Nam *et al.* 2004). The Au nanoparticle probe and magnetic microparticle are functionalized with oligonucleotide probes such that the presence of target DNA hybridizes the two nanoparticles together. The Au nanoparticle is also co-functionalized with bar code DNA in a 100:1 ratio to probe DNA. The hybridized magnetic microparticle-Au nanoparticle assembly is magnetically separated and dehybridized to release the bar code DNA. The bar code DNA is then detected by a chip-based detection system using the sandwich capture of Au nanoparticle probes and silver reduction for further amplification, i.e. scanometric detection. The process is schematically depicted in Fig. 10.2 (Nam *et al.* 2004).

Signal enhancement using Au nanoparticles can be achieved by linking multiple Au labels onto a single DNA target. This can be accomplished by creative design of a self-assembly scheme and of the oligonucleotide probes conjugated to the nanoparticle surface. Mo used nanoamplicons, Au nanoparticles derivatized with random tetramers, as mass labels to greatly increase the sensitivity of a QCM to rival PCR (Mo and Wei 2006). As many as 5000 nanoamplicons could be linked to a single 7.25 kb target. Hazarika used an alternate approach where multiple difunctional DNA-gold nanoparticles were used to form multilayer gold nanoparticle structures for protein detection (Hazarika *et al.* 2005). Difunctional nanoparticles are functionalized with two different oligonucleotide probes and can self assemble in the presence of protein targets to greatly amplify the nanoparticle absorption signal, as shown in Fig. 10.3.

Another approach to increase selectivity and sensitivity involves modifying the functional biological components of the sensor nanoassemblies. Rather than use

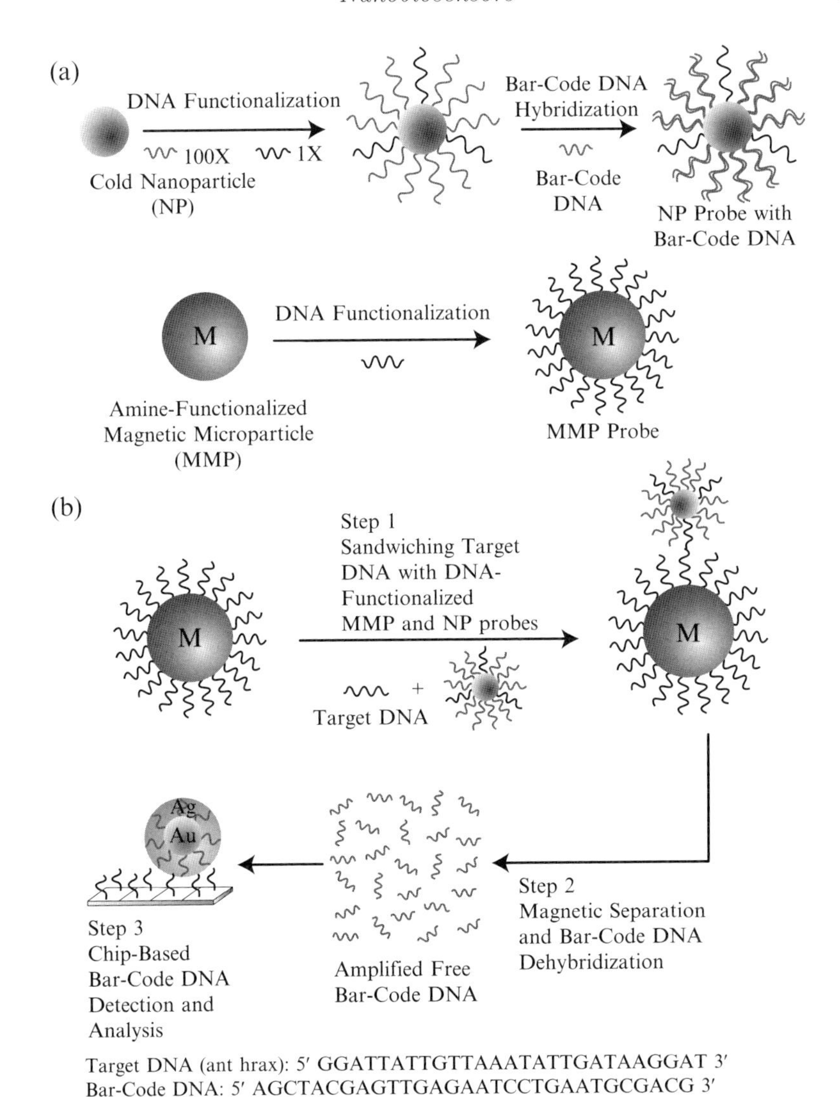

Fig.10.2: Schematic diagram of BCA assay. a) Preparation of Au nanoparticle with bar-code DNA and probe DNA in 100:1 ratio (top) and magnetic microparticle with second DNA probe (bottom). b) Hybridization with target DNA links MMP with NP into a nanoassembly which can be magnetically separated. Bar code DNA is dehybridized and analyzed by scanometric detection. Reprinted with permission from Nam *et al.* (2004). © 2004, American Chemical Society.

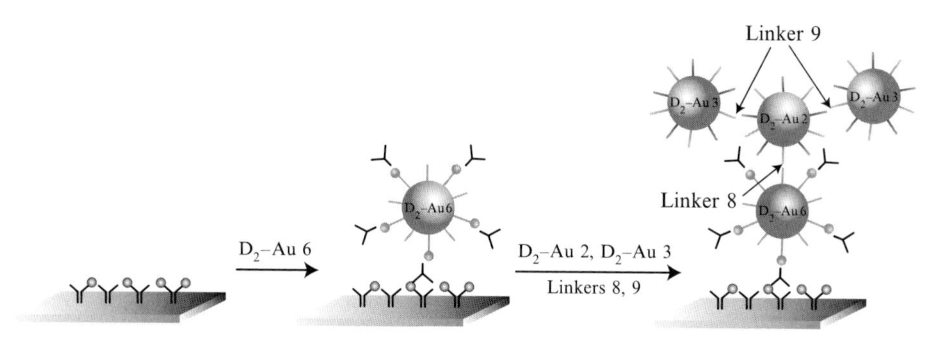

FIG.10.3: Schematic illustration of the assembly of multiple difunctional DNA - Au nanoparticles. The presence of a target ligand binds Au nanoparticle D_2-Au6 to the surface. Difunctional oligonucletodes attached to the surface of D_2-Au6, D_2-Au2, and D_2-Au3 allows a large aggregate to form to amplify signal (Hazarika *et al.* 2005).

traditional synthetic oligonucleotides as the basis for DNA probes, Chakrabarti used peptide nucleic acids (PNAs) to increase stability and base-pair mismatch sensitivity (Chakrabarti and Klibanov 2003). Recently aptamers have become popular as an alternative to antibodies for molecular probes. Aptamers have great potential as probe molecules since they can be randomly generated and selected to have specificity for many types of biomolecules. Huang has combined Au nanoparticles functionalized with aptamers and nanoparticle quenching as a detection mechanism for protein detection (Huang *et al.* 2007). Figure 4 shows a schematic diagram of the aptamer conjugation scheme and signal transduction mechanism.

10.2.2.7 *Other sensor technologies* Gold nanoparticles have also been used to enhance the sensitivity of more established sensor technologies such as molecular beacons, surface plasmon resonance (SPR) devices, and quartz crystal microbalances (QCM).

Gold nanoparticles can be extremely efficient at quenching fluorescence emission and has been used as a quencher in many molecular probes. Dubertret used 1.4 nm Au nanoparticles as quenchers in molecular beacons and found that nanoparticles offered higher quenching efficiencies and greater selectivity in distinguishing between perfect match and mismatch targets than traditional molecular quenchers (Dubertret *et al.* 2001). Maxwell used 2.5 nm Au nanoparticles as a nanoscaffold and fluorescent quencher in a nanobiosensor alternative to molecular beacons (Maxwell *et al.* 2002). As shown in Fig. 10.5 (Maxwell *et al.* 2002), in closed form, the fluorophore is spontaneously adsorbed to the nanoparticle surface and quenched. Upon hybridization with target DNA, the probe becomes stiffer, and the fluorophore is separated from the nanoparticle restoring fluorescence emission. Au nanoparticle quenching was also explored by Ray to monitor the enzymatic activity of DNA nucleases (Ray *et al.* 2006).

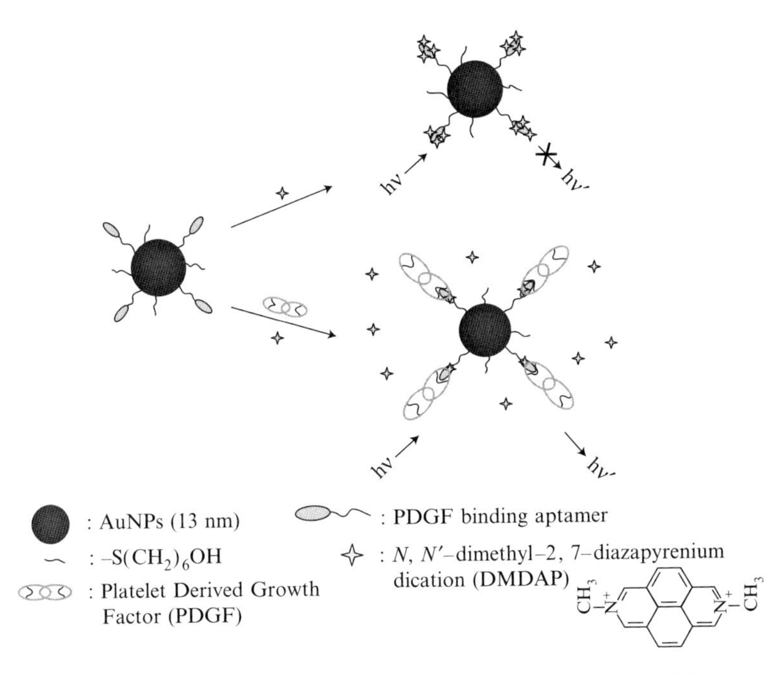

Fig.10.4: Schematic diagram of an aptamer functionalized gold nanoparticle sensor for protein detection. In the absence of PDGF, DMDAP freely intercalates into the PDGF aptamer resulting in quenched fluorescence emission. Binding of PDGF to the aptamer, releases DMDAP and restores fluorescence emission (Huang *et al.* 2007).

Au nanoparticles have also been used to amplify signals in conjunction with microsensor technologies. He used Au nanoparticle tags to increase the sensitivity of SPR-based detection by 1000-fold (He *et al.* 2000). The Au nanoparticle greatly amplifies the SPR angle shift by increasing mass loading, the local index of refraction, and electromagnetic interactions. Taton combined a DNA microarray with Au nanoparticle probes, silver development, and scanometric detection to distinguish single base mismatches in DNA (Taton *et al.* 2000). Catalytic silver deposition of the captured nanoparticles increased signal such that detection of bound nanoparticles could be accomplished with a flatbed camera or CCD and reached an ultimate sensitivity of 50 fM. Weizmann reported the use of Au nanocrystals as mass labels to increase the detection sensitivity of a QCM device in the detection of DNA to 1 fM (Weizmann *et al.* 2001). Catalytic silver deposition on captured Au nanoparticles further served to amplify frequency shift signals. In this application, the optical properties of Au nanoparticles are not used; rather, the nanoparticle is used as a mass label and seed catalyst only.

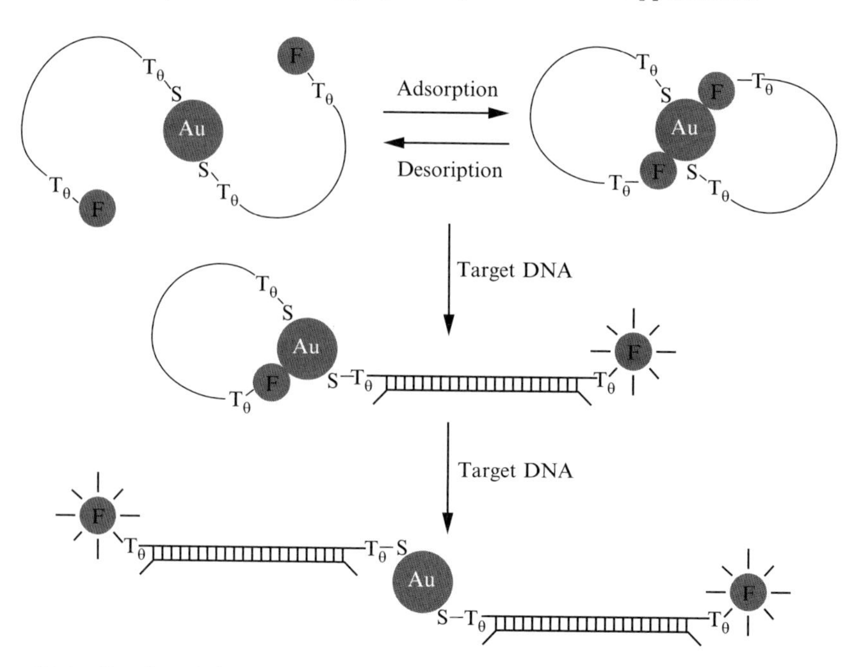

FIG.10.5: In closed form, the fluorophore is adsorbed to the Au nanoparticle surface leading to quenched fluorescence emission. In the presence of target DNA, hybridization occurs and the open form is assumed resulting in restored fluorescence emission. Reprinted with permission from Maxwell *et al.* (2002). © 2002, American Chemical Society.

10.2.3 *Metallic nanoparticle conclusion*

Metallic nanoparticles have been used in increasingly complicated conjugation schemes and assembly methods to demonstrate the potential of these unique nanoparticles. The most advanced nanoassemblies and sensors now use multiple types of nanoparticles and new types of functional probes to achieve increasingly flexible and complex assemblies. Perhaps their true potential can only be reached when they are combined with other types of nanoparticles such as quantum dots and nanowires as reviewed in the following sections.

Table 2 lists the main detection methods used in metallic nanoparticle based nanosensors.

10.3 Semiconductor nanocrystals—quantum dots (QD)

Semiconductor nanocrystals, colloquially known as quantum dots (QD), straddle the line between condensed matter and atomic physics. In a semiconductor nanocrystal, all three dimensions of the crystal are limited to less than the exciton radius of the material, such that discrete energy levels arise due to quantum confinement effects and the spacing of which can be controlled by manipulation of crystal size. This effect leads to the well published properties of quantum dots such as 1) size tailorable emission spectra, 2) broad absorption spectra, and 3) narrow and symmetric emission spectra (Alivisatos 1996). Other commonly exploited benefits include 4) high brightness, 5) highly stable photoluminescence, 6) high surface area to volume ratio, and 7) well documented conjugation chemistries. Generally less desirable features such as fluorescent intermittency, tendency to aggregate, photobrightening, and as synthesized hydrophobic surfaces have led to the widespread use of band gap engineered, core-shell quantum dots and various surface functionalization schemes.

10.3.1 *Surface coatings and functionalization*

The predominant method for synthesizing monodisperse quantum dots is accomplished in inorganic solvents, leaving nanocrystals with a hydrophobic trioctylphosphine/trioctylphosphine oxide TOP/TOPO cap (Murray *et al.* 1993; Peng, X.G. *et al.* 1998). Many techniques have been developed to render the hydrophobic TOP/TOPO cap more hydrophilic through cap exchange, substitution, or addition of components with the choice primarily being dependent upon subsequent bioconjugation requirements.

A straightforward approach involves exchanging the organic TOP/TOPO cap with a mercaptohydrocarbonic acid such as mercaptoacetic acid (Chan and Nie 1998). Dihydrolipoic acid (DHLA) has also been exchanged with TOP/TOPO to provide a negatively charged cap with two anchoring thiol groups for increased cap stability. Alternative methods have used hydrophobic interactions to form amphiphilic copolymer shells (Gao et al. 2004) or phospholipid micelles (Dubertret *et al.* 2002; Gueroui and Libchaber 2004) around the intrinsic TOP/TOPO cap. In these methods the copolymer or phospholipid is engineered with various functional groups to allow subsequent covalent attachment of biomolecules once the shell is formed. Cross-linked silica shells formed through silanization of the nanocrystal surface have been reported to provide stable caps as a result of the increased chemical and physical stability afforded by the cross-linking mechanism (Gerion *et al.* 2001).

A common conjugation strategy takes advantage of streptavidin–biotin interactions to create a universal linkage in which a wide variety of molecules can be mutually linked. Quantum dots can be covalently bound to either streptavidin or biotin while a biomolecule such as an antibody or oligonucleotide is bound to the conjugate enabling a robust linkage to be created and forming a tool kit that allows rapid and nearly universal conjugation.

Although the conjugation of biomolecules is crucial in gaining biological affinity, non-specific adsorption must also be minimized to reduce aggregation and unwanted interactions. In order to reduce non-specific adsorption, many of these schemes have included polyethylene glycol (PEG) passivation among the substitutional groups to good success (Dubertret *et al.* 2002; Gerion *et al.* 2002; Parak *et al.* 2002; Zheng *et al.* 2003). Hydroxyl-terminated surfaces have also been used to reduce non-specific adsorption (Pathak *et al.* 2001).

A mention must be made of the core-shell architecture prevalent among nanocrystal applications. Band gap engineering has led to the development of the core-shell quantum dot in which a core material with a lower band gap (e.g. CdSe: 1.74 eV) is enclosed by a shell of material with a higher band gap (e.g. ZnS: 3.67 eV) (Lide 1995). The shell serves mainly two-fold, in the passivation of the surface against oxidation and degradation and confining charge to within the core resulting in greater photostability and higher quantum yields.

10.3.2 *Applications of quantum dots*

Applications of quantum dots can be divided into two categories in which they function as simple luminescent probes or actively as scaffolds and participants in biosensing. Early applications revolve around using quantum dots to replace molecular fluorophores in the hopes of taking advantage of their unique properties to improve assay performance. Subsequent applications take advantage of biological specificity and self-assembly to form more complex nanosensor assemblies.

10.3.2.1 *Fluorescent labels* In addition to functioning as a fluorescent probe in cellular and molecular imaging, quantum dots have also been explored as fluorescent tags in molecular assays, microarrays, and various other analyte detection technologies. Quantum dots were first used as substitutes for molecular fluorophores in many traditional types of biochemical assays. Among the first demonstrations of quantum dots was Chan in using a CdSe-ZnS quantum dot covalently attached to IgG in a mock immunoassay (Chan and Nie 1998). The quantum dot functioned as a scaffold to induce aggregation in the presence of specific polyclonal antibodies. Medintz then demonstrated a QD-protein conjugate in a competition assay to sense maltose (Mattoussi *et al.* 2000). Maltose was titrated to competitively displace mannose binding protein (MBP)-QD conjugates from an amylose affinity resin resulting in proportionally decreased fluorescence intensity. This sensor concept would later be extended to a more elegant assay involving quantum dot fluorescence resonance energy transfer (Medintz *et al.* 2003).

Goldman also used electrostatic conjugation to link toxin specific antibody-biotin conjugates to DHLA-capped CdSe-ZnS quantum dots. These QD conjugates were used in a sandwich immunoassay for the simultaneous dual-color detection of the protein toxins, cholera toxin and staphylococcal enterotoxin B (SEB) (Goldman *et al.* 2002).

Quantum dots have also been incorporated into more established sensor technologies in order to enhance sensitivity and to multiplex. In these applications, their narrow emission bandwidths, brightness, and resistance to photobleaching make quantum dots particularly suitable replacements for molecular fluorophores. Gerion demonstrated them as tags in a DNA microarray where 4 different oligo-nanocrystal probes bound selectively to complementary ssDNA anchored on a patterned gold surface (Gerion *et al.* 2002). This work was later extended to a DNA microarray where SNP detection of the p53 oncogene and allele specific detection of hepatitis B and hepatitis C virus was accomplished using similar nanocrystal oligo-probes (Gerion *et al.* 2003).

Robelek combined QD-DNA conjugates with surface plasmon resonance (SPR) to perform surface plasmon enhanced spectroscopy (SPFS) and microscopy (SPFM), in which SPR is used to excite chromophores bound near the metal–solution interface (Robelek *et al.* 2004). The two techniques have been used in conjunction with a DNA microarray format to measure hybridization reactions and to probe for the presence of oligonucleotide targets. SPFM enabled identification of target DNA by imaging of the SPR enhanced fluorescence emission of the captured QD probes. SPFS showed that the QD-labels increased SPR reflectivity by ten times and allowed real-time monitoring of DNA hybridization.

Pathak demonstrated the use of QD-oligonucleotide probes in fluorescence *in situ* hybridization (FISH) assays with human sperm cells (Pathak *et al.* 2001).

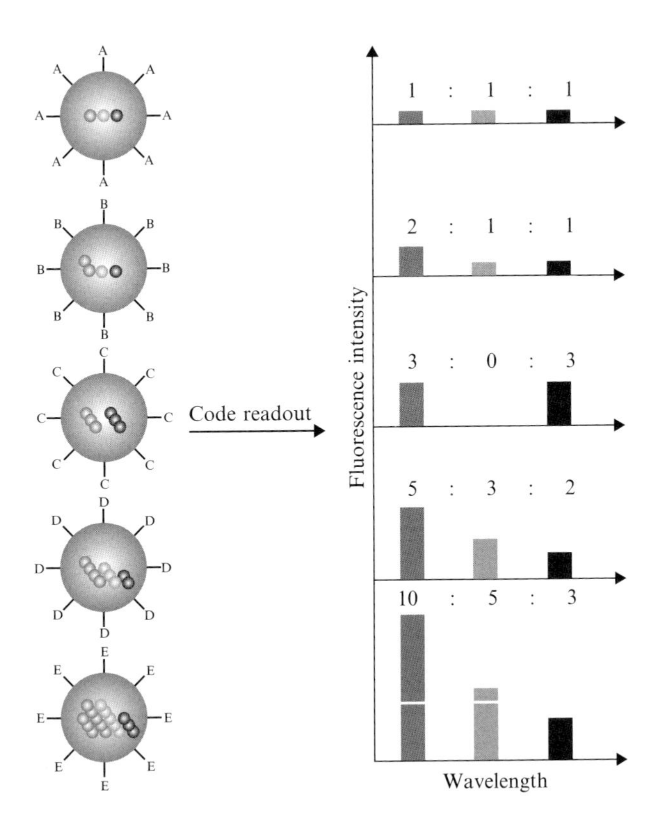

FIG.10.6: Schematic illustration of a multiplex method using quantum dot encoded microspheres. Multicolor quantum dots are embedded into microspheres in different ratios to create unique optical codes of varying color and intensity (Han *et al.* 2001).

This was achieved using hydroxyl-terminated quantum dots conjugated to oligonucleotide probes rather than the more typical carboxyl-terminated quantum dots due to concerns about stability, loading efficiency, and non-specific binding. The QD probe conjugates were specific to the Y-chromosome target while controls with random sequences or hydroxyl-terminations only showed very little non-specific binding.

Quantum dots are well suited to use as optical labels in multiplex detection due to their narrow emission spectra and size tunable emission. Relying solely on spectral separation for multiplexing would limit reliable label identification to a few tens of channels at the most. In order to increase multiplex throughput, Han *et al.* have embedded quantum dots of various colors and ratios in 1.2 μm polystyrene beads to obtain optical coding in both wavelength and intensity as depicted in Fig. 10.6. It is theoretically possible to generate (n^m-1) unique codes where n is the number of intensity levels and m is the number of unique colors. Practical limits will be far less than this as spectral and intensity overlap come into play. Xu applied this multiplex approach in a SNP analysis with two colors and three intensity levels (Xu *et al.* 2003). Data analysis was performed via flow cytometry where the SNP was identified by its encoded microsphere. Multiplex detection of 10 SNPs was performed at nearly 100 percent accuracy.

10.3.2.2 *Quantum dot FRET* In recent advances, quantum dots have taken a more active role not only as luminescent probes but also as scaffolds upon which more complex hybrid inorganic/organic biosensors are built. Signal transduction in these biosensors is most often accomplished through energy transfer, such as fluorescence resonance energy transfer (FRET) and gold nanoparticle quenching, but fluorescence coincidence has also been utilized. Using quantum dots as nanoscaffolds has a unique benefit. Many of the QD-based biosensors can be used in a homogenous, separation-free assay format. Elimination of a stringency wash not only simplifies assay protocol but may also reduce many of the issues faced with surface chemistries such as limitations due to planar diffusion and steric hindrance.

When properly used as a donor in a FRET pair, quantum dots offer benefits of eliminating direct acceptor excitation, reducing spectral overlap between donor and acceptor emission to reduce cross talk, maximizing donor emission and acceptor absorbance overlap, and allowing detection far away from the excitation wavelength to reduce background scattering and autofluorescence.

It has been used in a wide variety of applications from measuring static and dynamic conformational information, to biosensing and detection as well as binding studies. Quantum dot FRET (QD FRET) is a powerful tool for determining structural information since FRET efficiency is such a strong function of intermolecular distance. Hohng studied the dynamics of a DNA Holliday junction by monitoring FRET between one DNA duplex labeled with a QD donor and another DNA duplex labeled with Cy5 acceptor and compared this to results obtained with a Cy3/Cy5 donor-acceptor pair (Hohng, S. and Ha, T. 2005). Although the QD/

Cy5 pair had lower FRET efficiency than the Cy3/Cy5 pair, likely due to the large size of the QD nanoassembly, the conformational dynamics obtained from both pairs of probes were nearly identical. Medintz, on the other hand, has used FRET phenomena to study the orientation of MBP in QD-MBP conjugates (Medintz *et al.* 2004). MBP was electrostatically coupled to DHLA-capped quantum dots through a pentahistidine tail while by analyzing FRET efficiency between the quantum dot and Rhodamine dye in each of the six labeled proteins, the orientational conformation of MBP conjugated quantum dot was modeled.

Other than in determining structural information QD FRET has been used as the basis for probes in genetic analysis. QD FRET-based sandwich hybridization probes were designed as an alternative to molecular beacons and Taqman probes for point mutation detection by Zhang (Zhang *et al.* 2005). A biotinylated capture probe and a Cy5 labeled reporter probe hybridize to target DNA to form a sandwich structure. Hybridized targets can then be captured by streptavidin coated quantum dots. Detection of Cy5 emission indicates that FRET phenomena has occurred and indicates the presence of hybridization products (Zhang *et al.* 2005). Unbound QD FRET nanoprobes were found to produce near-zero background fluorescence but generated a very distinct FRET signal upon binding to even a small amount of target DNA ($\sim$ 50 copies or less).

Medintz has reported a maltose sensor based on QD FRET (Medintz *et al.* 2003). Two sugar sensing nanoassemblies were explored based on either competition with a quencher or competition using a two-step FRET mechanism. In quenching mode, competition between maltose and displaceable fluorescence quencher give rise to concentration dependent photoluminescence. In the two-step FRET nanoassembly a second relay dye, Cy3, is covalently attached to MBP while a displaceable Cy3.5 conjugate occupies the binding site (Fig. 10.7) (Medintz *et al.* 2003). Competition between the Cy3.5 conjugate and maltose give rise to a concentration dependent Cy3.5 fluorescence. The use of a relay dye helps extend

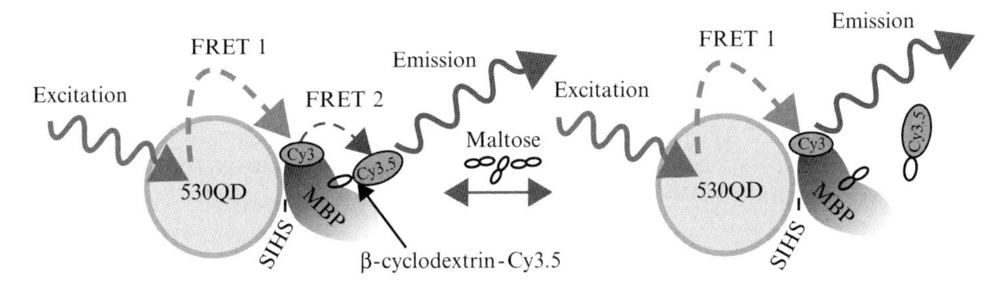

Fig.10.7: In the absence of maltose, excitation of QD530 leads to FRET with the Cy3 relay dye which undergoes a second FRET process to excite the displaceable Cy3.5 conjugate. When maltose is present, β-cyclodextrin-Cy3.5 is competitively displaced, during which Cy3.5 emission is replaced by Cy3 emission. Reprinted with permission from Medintz *et al.* (2003), © 2004, Nature Publishing Group.

the range of FRET phenomena across length of the nanoassembly. One step FRET mechanisms failed in sensing maltose, possibly due to the large size of the nanoassembly.

Studies of protein–ligand interactions have also been performed using QD FRET based nanosensors. Willard used CdSe-ZnS quantum dots covalently linked to biotinylated bovine serum albumin (QD-bBSA) and tetramethylrhodamine-labeled streptavidin (SAv-TMR) to study streptavidin-biotin interactions (Willard *et al.* 2001). QD FRET can also be used to study enzyme activity. Shi linked a quantum dot donor and rhodamine acceptor using a cleavable peptide linkage to study the activity of trypsin by monitoring FRET activity between the pair (Shi *et al.* 2007).

Since QD FRET can be used in a homogenous assay format, it is particularly suitable for in-vivo studies. Ho has taken advantage of this and used QD FRET to monitor intracellular trafficking and unpacking of polymeric gene delivery vehicles as illustrated by Fig. 10.8 (Ho *et al.* 2006). Using Cy5 labeled chitosan and QD labeled plasmid DNA, not only could the particle be tracked as it progressed through the cell, but the unpacking of the nanocomplex could also be monitored through FRET. So presented bioluminescent quantum dot conjugates for in vivo

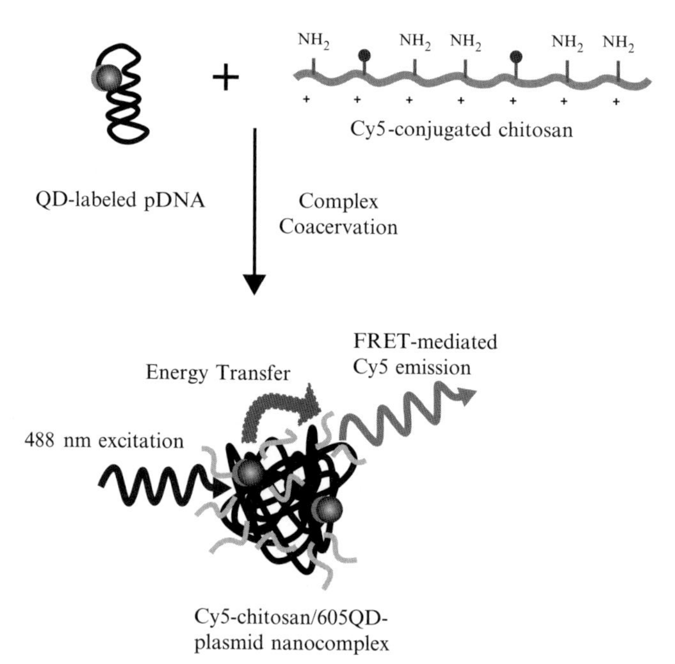

Fig.10.8: Schematic illustration of plasmid polymer nanocomplex using QD FRET to monitor unpacking. FRET between QD labeled plasmid DNA and Cy5-labeled chitosan polymers only occurs when the complex is tightly packed. Cellular trafficking as well as unpacking status can be monitored by tracking the QD and Cy5 signals as the nanocomplex traverses the cell. (Ho *et al.* 2006).

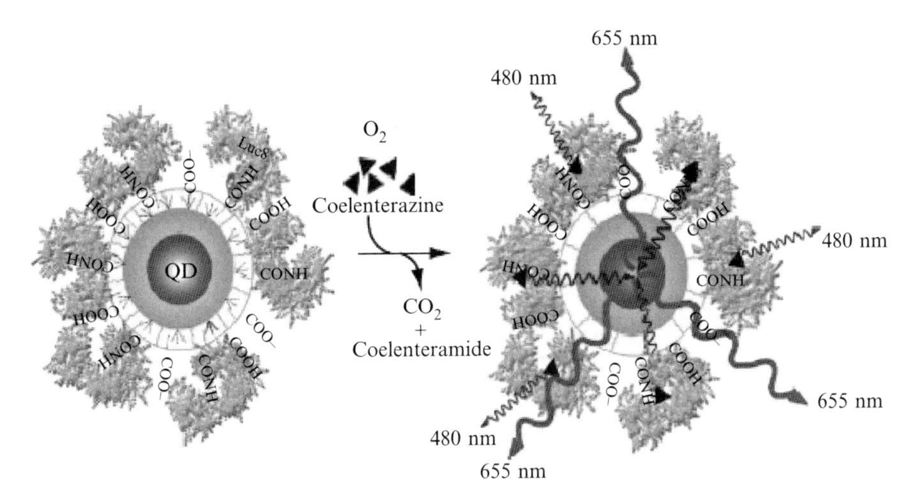

FIG.10.9: Schematic diagram of a BRET nanoassembly. *Renilla reniformis* luciferase (Luc8) is covalently linked to the quantum dot surface and undergoes BRET to the quantum dot when coelenterazine is oxidized (So *et al.* 2006).

imaging that are excited through bioluminescence resonance energy transfer (BRET) rather than an external light source (So *et al.* 2006). This was accomplished by conjugating *R. reniformis* luciferase to a quantum dot such that energy transfer occurred during the luciferase catalyzed oxidation of coelenterazine in the manner shown in Fig. 10.9. Yao applied this technology for the detection to the detection of matrix metalloproteinases (MMPs) in an assay that could potentially be used in vivo where external illumination is inconvenient (Yao *et al.* 2007).

10.3.2.3 *Fluorescence quenching* QD FRET provides an increased photoluminescence with decreased spatial separation. In some applications it may be appropriate for photoluminescence to behave conversely. In such cases, another form of energy transfer, Au nanoparticle quenching, has been used as a signal transduction technique.

Au nanoparticles have been shown to be effective fluorescent quenchers and were used by Gueroui to explore quenching efficiency as a function of distance (Gueroui and Libchaber 2004). Phospholipid-PEG encapsulated QD micelles were conjugated to DNA linkers of various lengths with an Au nanoparticle attached to the opposite end. The Au nanoparticle quenched quantum dot photoluminescence was shown to be a strong function of DNA linker length, demonstrating a potential tool for measuring intermolecular distances.

Dyadyusha used Au nanoparticle quenching of CdSe-ZnS quantum dot emission to demonstrate its potential for use as a probe in biodetection (Dyadyusha *et al.* 2005). Rather than conjugate the Au nanoparticle and quantum dot to opposite ends of an oligonucleotide strand, they were attached to the same end

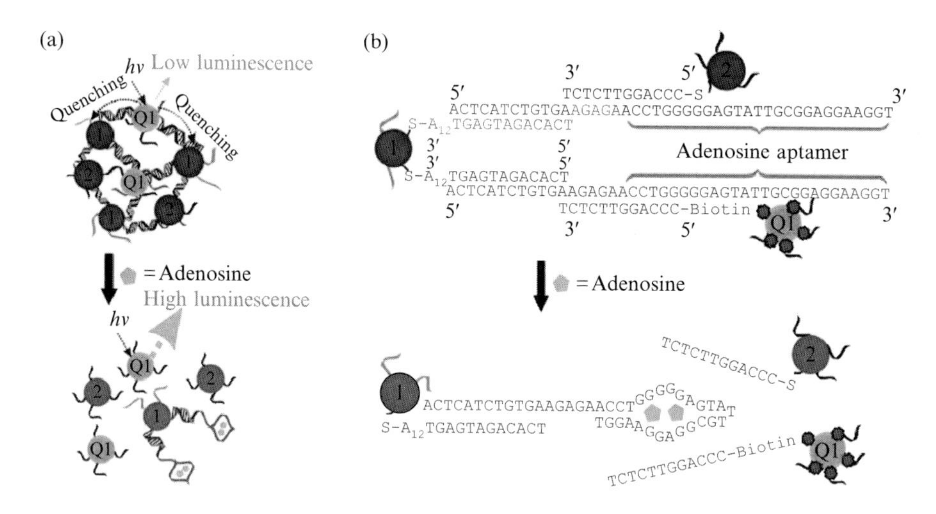

Fig.10.10: a) In the absence of adenosine, the quantum dots (Q1) and Au nanoparticles, 1 and 2, assemble to form large aggregates and quench fluorescence emission. b) Adenosine binds the aptamter region of the oligonucleotide probes, causing disassembly, and restoring fluorescence emission (Liu *et al.* 2007).

of dsDNA strand to improve quenching efficiency. Quenching was shown to be present only when the conjugate oligonucleotide strands were present, thus, providing a mechanism for nucleic acid detection that could be extended to other biomolecules with the correct probes. Liu has developed a more intricate assembly of quantum dots and Au nanoparticles in combination with aptamer probes for the simultaneous detection of cocaine and adenosine (Liu *et al.* 2007). In the absence of ligand, the aptamers help link the two different Au nanoparticles and quantum dots into large aggregates. When adenosine or cocaine are present, aptamer–ligand binding leads to dissociation of the nanoassembly which can be detected through both absorbance and fluorescence. The process is shown in Fig. 10.10.

Levy has designed a quantum dot aptamer beacon for the diction of thrombin. In the absence of target, the quantum dot conjugated aptamer and an oligonucleotide conjugated quencher hybridize to quench quantum dot emission. Binding of thrombin to the aptamer, releases the quencher conjugate and restores fluorescence (Levy *et al.* 2005).

Quantum dots exhibit photoluminescent properties that are dependent on their local environments. Huang used the pH dependence of quantum dot photoluminscent intensity and the enzymatic degradation of urea as an assay for urea concentration (Huang *et al.* 2007). Aryal used solvent effects on quantum dot photolumiscence intensity to probe the salvation status of a fatty acid binding protein (Aryal and Benson 2006).

10.3.2.4 *Coincidence detection* In addition to FRET and quenching based signal transduction, quantum dots have been used as a nanoscaffold to link relevant biomolecules into a nanoscale sensing domain such that detection can be accomplished through fluorescent coincidence detection.

Yeh has demonstrated two quantum dot nanoassemblies based on fluorescent coincidence detection (Yeh *et al.* 2005). In the first, two color CdSe-ZnS QDs were attached to oligonucleotide probes that could hybridize to form sandwich assemblies and cross-link in the presence of target sequences. The quantum dots acted as a scaffold for a larger nanoassembly formation. In the second method, oligonucleotide probes are conjugated to quantum dots and organic fluorophores that hybridize in the presence of target. A single quantum dot functions as a scaffold to capture multiple sandwich assemblies. Detection in both methods is accomplished by dual color coincidence detection. Validation of the techniques was achieved by multiplex detection of three distinct DNA targets.

In order to increase specificity, Yeh has combined coincidence detection and a ligation assay in a unique single molecule, homogenous, separation-free assay format to detect point mutations in the KRAS gene of genomic DNA (Yeh *et al.* 2006). A QD-oligonucleotide discrimination probe and OG488-oligonucleotide reporter probe (Oregon green) are enzymatically ligated together in the presence of a perfect match DNA target. Detection of ligation products was indicated by dual-color fluorescence coincidence of the quantum dot label and the OGG488 molecular fluorophore as seen by a laser induced fluorescence (LIF) spectroscopic system with single fluorophore sensitivity.

Ho has demonstrated an alternative approach to multiplex detection of DNA in a homogenous, separation-free format (Ho *et al.* 2005; Yeh *et al.* 2006). Dual color QD conjugated oligonucleotide probes were hybridized with target DNA to form sandwich nanoassemblies. Detection is accomplished via colorimetric measurements using a CCD where hybridized nanoassemblies appear with mixed colors due to the diffraction-limited resolution of the optical system. Unbound probes retain their original color, obviating the need for separation of bound and unbound molecules. Multiplex analysis of three anthrax related targets for pathogen detection has been demonstrated as a proof-of-concept.

10.3.3 *Quantum dot concluding remarks*

Early ideas have focused on emulating techniques achieved with organic fluorophores. Only papers published within the last few years have begun to explore quantum dots not as fluorescent labels but primarily as nanoscaffolds and the basis for nanosensors. Table 10.3 summarizes the principal detection methods used in various quantum dot nanoassembly applications.

The large preponderance of studies on quantum dot nanosensors have focused on proof-of-principal experiments and have reported unoptimized sensitivities. Currently, quantum dot-based nanosensors operate predominantly in the μM-fM range. It is without doubt that great progress will be made in the application of quantum dots into new assay formats that will push detection limits further into the fM and lower range. Perhaps more importantly though, breakthrough applications of quantum dots will not be realized through their application as analogs to molecular fluorophores, but as nanoscaffolds in new ways in which traditional labels cannot function.

10.4 One-dimensional, nanostructured materials in biosensing platforms

Thus far our discussion has centered on nanoparticle biosensors and their unique transduction mechanisms that yield superior detection limits and multiplexing capabilities, as compared to larger platforms. In this section, we discuss another class of nanosensors, fabricated from so-called one-dimensional (1-D) nanomaterials, including nanowires, nanotubes, and nanorods. Again, the small size of these sensing materials makes them uniquely suited for interfacing the biological world to macroscopic instruments. In addition, distinctive properties of these nano-scale sensing elements make them well-suited for pushing the limits of affinity-based biosensing, therefore providing real, label free alternatives to fluorescent detection. In this section, we will examine several specific examples of the integration and utilization of 1-D nanomaterials within biosensing platforms, where the exceptional properties of the nano-scale sensing element was harnessed to provide improvements in sensor sensitivity, selectivity, addressability, and/or simplicity of signal transduction.

10.4.1 *Silicon nanowire-based FET sensors*

A variety of methods have been devised to prepare metallic, semiconductor, magnetic, inorganic, organic, polymer, dielectric, and magnetic 1-D nanostructures, providing novel optical, electrical, and magnetic properties for biosensor components. Silicon nanowires (NWs) are one of the best-characterized examples of these materials, exhibiting highly reproducible and tunable electrical properties (Patolsky *et al.* 2006). In addition, years of research devoted to chemical modification of planar biological sensors provides a library of methods for linking molecular receptors to silicon NW biosensors (Patolsky *et al.* 2006). Field effect transistor (FET) biosensors are one such set of devices based on standard, planar semiconductor technologies that were first developed decades ago (Kimura, J. and Kuriyama, T. 1990). Standard FET devices rely on electric field control from one 'gate' terminal to control the conductivity and thus electron flow between a semiconductor 'channel', separating a 'source' and 'drain' electrode. This configuration can be made chemically sensitive by modifying the 'gate' with molecular recognition sites. Binding of charged species results in alterations in carrier concentrations within the transistor and allows direct monitoring and label-free molecular detection (Patolsky *et al.* 2006). However, the limited sensitivity of these planar devices restricted widespread use.

The first integration of NWs into field effect devices was demonstrated in 2001, when a p-type, oxidized Si-NW was functionalized with a self-assembled monolayer (SAM) containing two distinct terminal groups, an amine group and a silanol group, respectively (Cui *et al.* 2001). These functional groups responded differently to solvent pH surrounding the devices with protonation of the amine group at low pH and deprotonation of the silanol groups at high pH. Thus, as in traditional planar FET devices, accumulation of excess protons on the surrounding SAM depleted holes in the Si/SiO_2 device, while deprotonation resulted in charge carrier accumulation and an increase in conductance. However, unlike the planar devices, in which depletion and/or accumulation of charge carriers occurred at the surface, the molecular binding induced electric field change affected the bulk of the nanometer-diameter silicon wire. This effectively increases the sensitivity of FET biosensors and has been used in medical diagnostic applications, through functionalization with specific proteins/antibodies or nucleotide target sequences (Wang *et al.* 2005; Zheng *et al.* 2005; Patolsky *et al.* 2006; Patolsky *et al.* 2006). Recently, the benefits of these nanoscale FET devices were demonstrated further by arraying NW-FET components for multiplexed detection of cancer antigens (Fig. 10.11) (Zheng *et al.* 2005). By coupling this arrayed device to a microfluidic device for

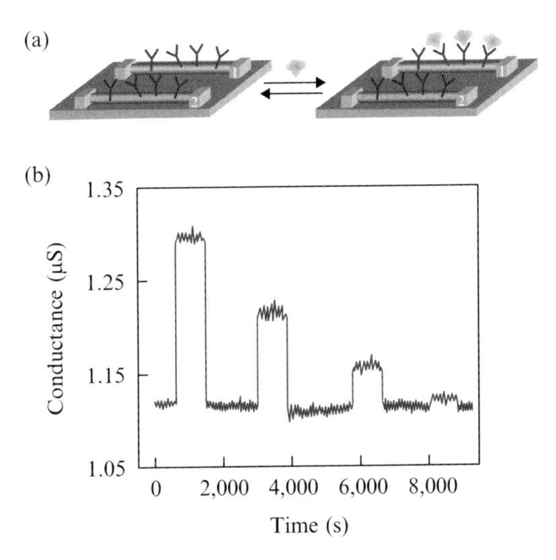

Fig.10.11: Schematic of multiplexed detection of cancer antigens using a NW-FET device. (a) Nanowires within an array of modified with different antibodies against cancer antigens. Cancer marker binding produces a conductance charge on its specific nanowire target only, which is dependent on the surface charge of the protein. (b) Changes in conductance versus time in p-type silicon NWs modified with PSA (Prostate Specific Antigen)—ab1 upon alternating delivery of PSA (0.9 ng/ml, 9 pg/ml, 0.9 pg/ml, 90 fg/ml in chronological order) and buffer solutions using a microfluidic channel. Adapted from Zheng *et al.* (2005). With permission from Nature Publishing Group.

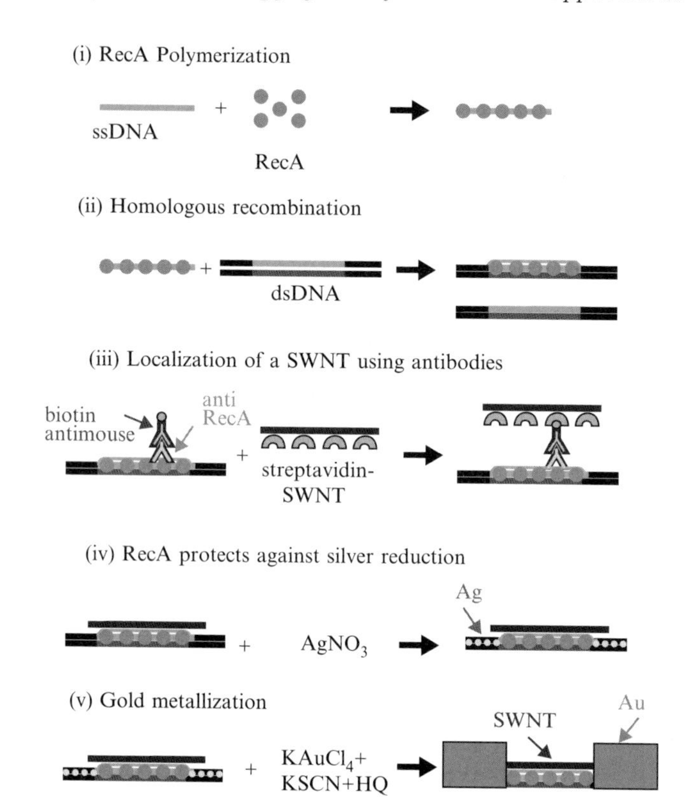

FIG.10.12: DNA-templated FET. (i) RecA monomers polymerize on a ssDNA strand to form a nucleoprotein filament. (ii) Homologous recombination reaction leads to binding of the nucleoprotein filament at specific address on an aldehyde-derivatized ssDNA strand. (iii) Primary antibodies to the protein portion of the filament are used to localize a strepavidin-funcationalized SWNT. (iv) Silver clusters are formed on the unprotected regions that are unprotected by the protein segment by incubation in $AgNO_3$. (v) The silver clusters act as nucleation sites for electrodeless deposition of gold, resulting in DNA-templated gold contacts for the SWNT. Adapted from Keren *et al.* (2003). With permission from AAAS.

sample introduction, the authors were able to both decrease non-specific binding to the NW surface and increase response time from complex serum samples, spiked with low-abundant target molecules.

Several important practical and fabrication considerations arise in the development of the NW-FET technology. First, the diameters of the NWs used in these devices typically range from 20–50 nm and are controlled by the size of the gold nanoparticle catalysts used in the vapor–liquid–solid (VLS) synthesis scheme (Patolsky *et al.* 2006). Further downscaling of the NW diameter does not lead to increases in device sensitivity, due to increased shot noise at lower cross-sectional areas (Reza *et al.* 2006). In addition, detection sensitivities within these field-effect devices are dependent on the ionic strength of the surrounding solution. Thus, testing complex serum samples, as discussed above, requires preconditioning steps, such as desalting (Zheng *et al.* 2005). The common bottom-up VLS growth of NWs also contributes to several disadvantages in NW-FET systems, including the inability to accurately assemble arrayed devices and ohmic contacts. However, improving fabrication technologies in nanoelectronic and nanosensing systems continues to be an exciting area of research, and challenges such as these are being addressed (Patolsky *et al.* 2006).

Figure 10.12 shows the DNA-template guided fabrication of a FET device, in which a DNA scaffold molecule provides localization of a semiconducting nanomaterial and template driven extension of metallic contacts (Keren *et al.* 2003). This method provided the first means for direct, self-assembly of single-wall carbon nanotubes (SWNT) components, which were used as the semiconductor material and will be discussed in the next section.

10.4.2 *Carbon nanotubes in biosensing applications*

While silicon nanowire devices have been buoyed by years of research in silicon-based microelectronics, applications of carbon nanotube (NT) have been limited while researchers investigate methods of selective synthesis and separation of NT with specific structural properties (Patolsky *et al.* 2006). In the past, as-prepared NTs were heterogeneous in structure, varying in length, diameter, and chiral index; in addition, optical and electrical properties of NTs varied with these structural properties (Krupke *et al.* 2003; Peng *et al.* 2007). For instance, altered chirality of SWNTs resulted in electrical properties ranging from a metallic state to semiconducting states (Krupke *et al.* 2003). Still, NT devices have great potential in that they are truly 1-D conductors, with virtually all atoms on their surface, and extensive research into selective synthesis and separation methods have been conducted (Krupke *et al.* 2003; Peng *et al.* 2007).

Application of these methods have led to increased use of NTs in nanoeletronic and sensing applications (Kong *et al.* 2000; Gruner, G. 2006; Star *et al.* 2006). In one such application, a network of NTs was used in a FET sensor to overcome the large shot noise associated with NTs by increasing the number of charge carriers within the device (Star *et al.* 2006). Some such devices have achieved comparable detection limits with metal oxide nanowire systems (Star *et al.* 2006). Still, another challenge in using NTs as biosensing elements remains the lack of flexibility in surface modification techniques compared to more traditional materials. However, several successful applications of NTs in electrochemical and biosensor exist that utilize novel surface functionalization methods (Keren *et al.* 2003; Star *et al.* 2006; Peng *et al.* 2007). For instance, ssDNA has been observed to form stable complexes with individual SWNTs (Zheng *et al.* 2003) and dsDNA has-also been projected to interact with SWNTs as major groove binders (Lu *et al.* 2005).

10.4.3 *Non-electrical sensing platforms*

As discussed above, 1-D nanostructures have also been found to possess unique optical and magnetic properties that can be harnessed in biosensing applications. For instance, SWNTs have been found to fluoresce at near infrared wavelengths and are therefore important in biological applications to minimize absorption by blood and tissue (Barone *et al.* 2005; Jeng *et al.* 2006). Figure 10.13 shows work that combined this optical property of NTs with the aforementioned ability to adsorb molecules on the nanotube surface to detect hybridization of specific DNA sequences (Jeng *et al.* 2006). Hybridization of the complementary strand was detected as an optical modulation in the SWNT fluorescence, resulting in a photostable and label-free detection scheme.

The size of nanostructured materials also allows spatial arraying of multiple functionalities to be incorporated in individual biosensing components. Previously, we discussed nanoparticles that were used as scaffolds for barcoded arrangements

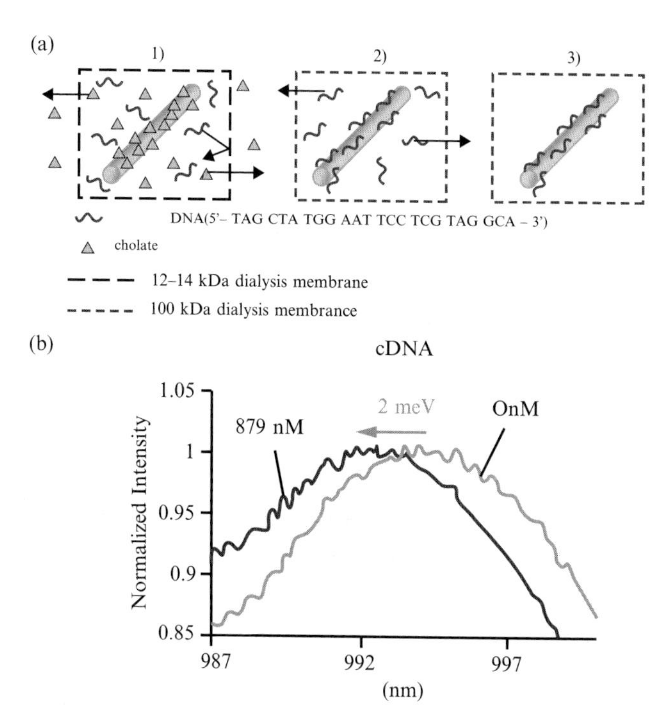

Fig.10.13: Optical detection of DNA hybridization on SWNT. (a) Nanotubes are kept dispersed in 2 wt percent cholate and DNA-SWNT are assembled through dialysis against TRIS buffer in the presence of ssDNA. (b) Addition of complementary DNA strands causes a shift in the fluorescence peak, which is absent in the presence of ssDNA only. Adapted from Jeng *et al.* (2006). With permission from ACS.

of target and probe oligonucleotides. This barcode or multilayered scheme can also be utilized in fabrication techniques for 1-D nanostructures, such as nanowires. Indeed, pulsed-electrodeposition within nanoporous templates has been used to create magnetic nanowires with tunable architecture and magnetic properties (Lee *et al.* 2007). Selective functionalization of these barcoded nanowires may allow direct manipulation, positioning, and selection of target molecules in biosensing applications (Reich *et al.* 2003).

10.4.4 *Multifunctional and composite nanosensing platforms*

For most of the examples above the use of 1-D nanostructures for detection of specific biological molecules was predated by similar detection schemes for simple chemical species (Kong *et al.* 2000; Patolsky *et al.* 2006). Expanding the capabilities of these devices to biomolecular detection usually required methods for functionalization with molecular receptors, which can be made more difficult by the fact that traditional methods, such as covalent functionalization, may disrupt the 1-D electronic structure or optical properties of the nanostructured material (Wang *et al.* 2005; Zheng *et al.* 2005; Jeng *et al.* 2006; Star *et al.* 2006; Peng *et al.* 2007). The use of conducting polymers in biosensing applications has followed a similar path, where materials that exhibited high sensitivity to chemical interactions did not respond well to biosensing applications that required neutral solutions (Huang *et al.* 2003). Doping methods for conductive polymers, such as polyaniline, allowed retention of electrical conductance in neutral solutions; however, these materials did not exhibit the same electrochemical activity or mechanical stability (Wei *et al.* 1996; Bartlett and Astier 2000). Recently, composite self-doped polyaniline/SWNT films have been shown to enhance the biosensing capability of the conducting polymer, by enhancing the stability of the polymer film and acting as conducting doping agents within the polyaniline film (Ma *et al.* 2006). Unlike the examples above, in which the nanostructured material provided the primary sensing surface, this work shows that nanowires can also be used to improve the sensing capabilities of bulk materials and detection platforms.

In addition to augmenting the properties of bulk materials, composite nanoparticle–nanotube structures have recently been investigated. These hybrid nanomaterials, dubbed 'peapod' structures, are created by self-assembly of atoms or molecules within the 'core' of nanotube or nanowire structures (Hornbaker *et al.* 2002; Wu *et al.* 2002; Hu *et al.* 2006). These structures have been synthesized with several materials including all carbon peapod structures within SWNTs (Hornbaker *et al.* 2002), gold peapodded silica (Hu *et al.* 2006), and gold silicide embedded silicon oxide nanowires (Wu *et al.* 2002). These nanohybrids offer the unique prospect of creating nano-scale materials with tunable structures that can be tailored for specific electrical and optical properties (Fig. 10.14) (Wu *et al.* 2002; Hu *et al.* 2006). Synthetic methods with this level of control may be applied to improving the nanostructured building blocks of sensing platforms that we have discussed in this section.

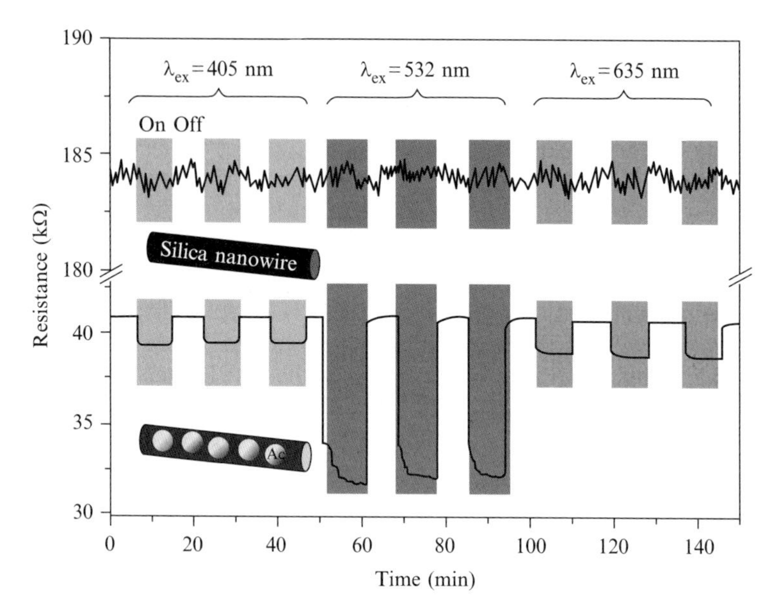

Fig.10.14: Resistance response to silica nanowires in ensemble devices to several wavelengths of light. The traces show that the peapodded structure shows a wavelength dependent photoresponse. The enhanced sensitivity at 532 nm is thought to be due to a pronounced SPR absorption at that wavelength, while the large decrease in background resistance for the peapod structure is thought to be due to enhanced electrical conduction in the carbon-doped silicon oxide. Adapted from Hu *et al.* (2006). With permission from NPG.

10.5 Single biomolecule detection

Recent advances in fluorescence and optics technologies enable measurements of DNA, RNA, and protein at the single molecule level. Such capability has facilitated development of a variety of assay methods capable of detecting low abundant biomolecular targets without using amplification techniques such as PCR. This circumvents technical difficulties associated with PCR such as false-positive and -negative results due to contamination and labor intensive, time consuming processes.

A variety of optical single molecule detection (SMD) techniques (both in solutions and on solid surfaces under ambient conditions) have been developed over the past decade. SMD in solution is mainly based on the use of confocal fluorescence spectroscopy, which detects the emission from a single fluorescent molecule as it travels (by diffusion or active transport within a fluid flow) through a small, confocal detection volume (Fig. 10.15). A laser beam is focused into a diffraction-limited spot in solution through a microscope objective of high-numerical aperture (NA) (typically NA >1.2). Fluorescence is often collected by the same

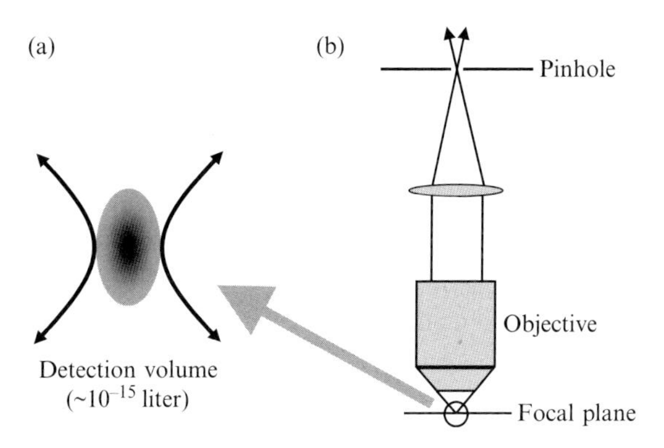

Fig.10.15: A conceptual schematic of confocal optics.

objective and is split from the illumination light with a dichroic mirror. Finally, a pinhole (typically 50–100 μm in diameter) is used to confine the detection volume along the optical axis and an ultra-sensitive photoelectron detector, e.g. photomultiplier tubes and single-photon avalanche photodiodes, is used to detect emitted fluorescence photons. The small detection volume (on the order of 1 fL) of confocal setup significantly reduces the background noise. When measuring fluorescent molecules of <nM, the detected fluorescence signals become digital because the molecular occupancy (the average number of molecules residing in the detection volume at any given time) is smaller than unity. Fluorescence bursts are detected only when single fluorescent molecules pass through the detection volume. Since the advent of SMD, this technology has been largely used for the fundamental study of molecular interactions. Recently, by incorporating new probes or probe strategies, such as fluorescence resonance energy transfer (FRET)-based probes and dual-color fluorescence coincidence analysis, SMD has been applied to genomic analysis including detection of DNA of specific sequences and genetic variations (Eigen and Rigler 1994; Castro and Williams 1997; Knemeyer *et al.* 2000; Wabuyele *et al.* 2003; Goodwin *et al.* 2004; Foldes-Papp *et al.* 2005; Zhang *et al.* 2005; Marme *et al.* 2006; Yeh *et al.* 2006) and gene expression analysis at both the RNA and protein levels.(Korn *et al.* 2003; Nolan *et al.* 2003; Neely *et al.* 2006; Huang *et al.* 2007)

When applying SMD for quantitative analysis, the number of molecules in a sample is correlated to the number of single molecules detected. Since the intensity of the single molecule fluorescence bursts remain unchanged despite decreasing the concentration of the molecules, SMD enables an ideal platform for quantitative analysis of biological samples that contain low concentration biomolecular targets. The small detection volume associated with SMD also confers the unique advantage of requiring small sample volumes for genomic analysis. Indeed, the femtoliter detection element allows the use of one nanoliter sample volumes, while retaining high statistical confidence in measurements (Yeh *et al.* 2006). Still, this small detection volume results in low molecular detection efficiency during measurement and brings about the need for a mechanism to guide molecules within a sample to this volume element. Manipulation of biomolecules using electrokinetic (Haab and Mathies 1999; Anazawa *et al.* 2002; Wang *et al.* 2005) or hydrodynamic forces (Nguyen *et al.* 1987; Castro *et al.* 1993; de Mello and 2007) affords the opportunity to increase mass transport efficiency in micro- and nanofluidic technologies. Therefore, the high detection sensitivity of SMD integrated with the high mass transport efficiency enabled by the use of microfluidic devices makes possible the accurate and quantitative analysis of rare and low abundant target molecules.

10.5.1 *Single molecule detection of nucleic acids targets with molecular FRET probes*

Fluorescence resonance energy transfer (FRET) is a non-radiative energy transfer process from an excited chromophore (donor) to another chromophore (acceptor) in close proximity (usually < 10 nm), through a long-range, dipole-dipole interaction (Lakowicz 1999; Ha 2001). It is sometimes referred to simply as 'resonance energy transfer' due to its radiationless nature. This process is extremely distance dependent and typically occurs at a length scales less than 10 nm. The energy transfer efficiency (E) is inversely proportional to the sixth-power of the separation distance (r) between a donor and an acceptor:

$$E = \frac{R_0^6}{R_0^6 + r^6}$$

where the Forster distance (R_0) is the donor-acceptor distance at which the energy transfer efficiency is 50 percent.

FRET has been employed in a number of molecular probes, such as Taqman (Holland *et al.* 1991) and molecular beacons (Tyagi and Kramer 1996) that rely on changes in intermolecular distances between fluorophores to detect specific nucleic acid sequences (Cardullo *et al.* 1988; Holland *et al.* 1991; Tyagi and Kramer 1996; Knemeyer *et al.* 2000; Dubertret *et al.* 2001; Marme *et al.* 2006). Upon interaction with target sequences these probes undergo conformational changes, thereby allowing detection of the target molecules through changes in the energy transfer efficiency between the excited donor and the acceptor fluorophore. The use of FRET-based probes can eliminate complicated steps in detection protocols, such as the removal or separation of unbound probes, which is usually carried out through immobilization and washing. These steps become unnecessary and DNA detection is conducted in a homogenous format; therefore, one can expect more efficient probe-target binding kinetics and improved detection speed and throughput.

By combining the use of two molecular beacons and confocal spectroscopy, a two-color SMD method has been developed for comparative quantification of nucleic acids targets. In this method, two molecular beacons of different color were used to carry out a comparative hybridization assay for simultaneous quantification of both the target and control strands. Two-color fluorescence techniques are commonly used for comparative gene expression analysis. However, traditional dual-color fluorescence analysis has been plagued by the disparity of the photophysical or photochemical properties between different fluorophores. In addition, when analyzing low-concentration targets, the quantification accuracy is limited by the reduced SNR and the relatively large variations in signal levels. In contrast, the dual-color SMD method obviates such technical complications through quantification schemes that rely on the counting of high SNR single molecule fluorescent bursts in different wavelengths. As this method retains high SNR even with low target concentration, comparative quantification of low-abundant (picomolar or lower) nucleic acid targets is possible, allowing discrimination of as small as two-fold differences in target quantity at these concentration levels (Zhang *et al.* 2005).

As mentioned in previous Section 10.3.2, quantum dots (QD) can be used to make a FRET-based DNA nanosensor (Zhang *et al.* 2005). As shown in Fig. 10.16, each QD-FRET nanosensor is comprised of two target-specific oligonucleotide

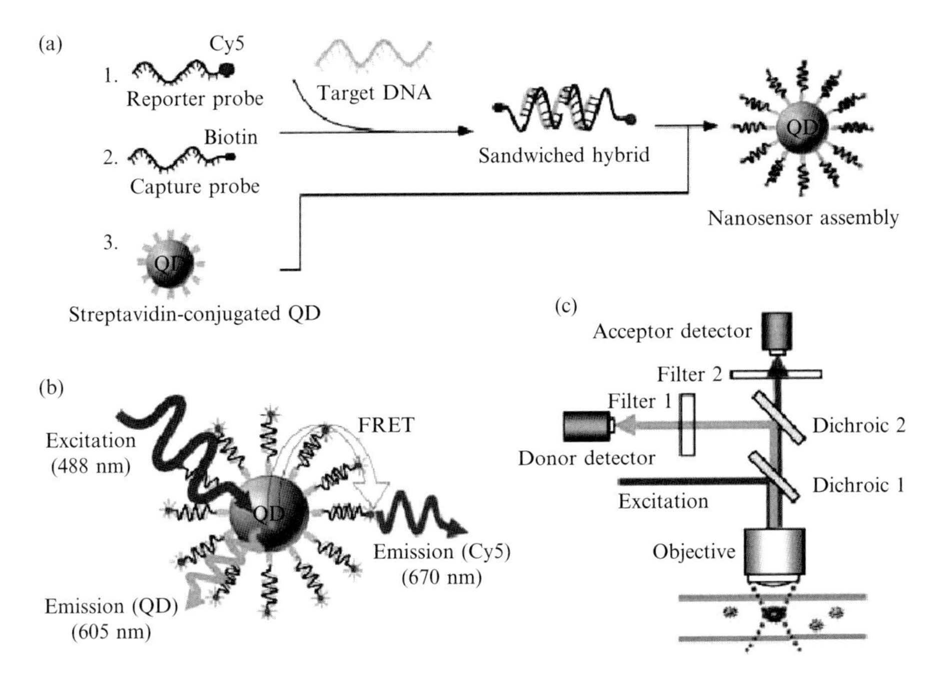

Fig.10.16: Schematic of single QD-FRET DNA nanosensors. a) The conceptual diagram shows the formation of a nanosensor assembly in the presence of targets. b) Fluorescence emission from Cy5 on illumination of QD donor is seen due to FRET between QD (donor) and Cy5 (acceptor) in a nanosensor assembly. (c) A custom-made confocal spectroscopic setup is used for single-QD detection. Zhang *et al.* (2005) Copyright 2005 Nature Publishing Group.

probes. One probe is biotinylated and the other is labeled with an organic fluorophore, Cy5. A streptavidin-coated QD serves as both a FRET donor and a target concentrator. When a target DNA strand is present, it is sandwiched by the two probes and the resulting hybrid is captured by a QD through biotin/streptavidin binding. Multiple hybrids can bind a QD, resulting in a strong emission from the Cy5 *via* resonance energy transfer upon illumination of the QD.

10.5.2 *Analysis of single biomolecules using fluorescence correlation spectroscopy (FCS)*

In the last decade, several assay methods for analysis of biomolecules at the single molecule level based on the fluorescence correlation spectroscopy (FCS) technique have been reported (Eigen and Rigler 1994; Edman *et al.* 1996; Schwille *et al.* 1996; Schwille *et al.* 1997; Wennmalm *et al.* 1997; Kettling *et al.* 1998; Rigler *et al.* 1998; Foldes-Papp *et al.* 2001; Korn *et al.* 2003; Nolan *et al.* 2003; Camacho *et al.* 2004; Winter *et al.* 2004; Foldes-Papp *et al.* 2005). The principle of FCS lies in temporal and spatial fluorescence fluctuations due to molecules diffusing in and out of the small detection volume. The characteristic diffusion time constant (τ_d) is typically determined through autocorrelation analysis of the fluorescence signals, followed by its fitting to the autocorrelation function $G(\tau)$ derived based on a 3D diffusion model:

$$G(\tau) = \frac{1}{N(1 + \frac{\tau}{\tau_d})(1 + (\frac{\omega_0}{z_0})^2 \cdot \frac{\tau}{\tau_d})^{1/2}}$$

where τ is the lag time; ω_0 and z_0 are half axes of the detection volume (Eigen and Rigler 1994). When coupled with the use of a target-specific fluorescent probe, FCS can be used for biomolecule detection by monitoring the change in diffusivity caused by target-probe binding. Rigler and co-workers have combined FCS with a confocal setup for DNA detection and sorting (Eigen and Rigler 1994). This technique of one-color autocorrelation analysis has also been used in the study of DNA conformational fluctuation (Edman *et al.* 1996; Wennmalm *et al.* 1997) and DNA hybridization kinetics (Schwille *et al.* 1996). One complication of one-color FCS analysis is that it is difficult to differentiate binding and non-binding events that involve only small changes in diffusion time constants (Li *et al.* 2003). Therefore, one-color FCS is limited to the analysis of the targets that are substantially larger than the fluorescent probes. Such limitation is overcome by dual-color FCS analysis as it characterizes molecular binding from the change in cross-correlation functions rather than in the diffusion properties. In addition, simpler mathematical evaluation and shorter readout time can also be achieved (Kettling *et al.* 1998). Dual-color FCS (Eigen and Rigler 1994; Schwille *et al.* 1997) has been demonstrated for the detection of amplified DNA target sequences in multiplexed polymerase chain reactions (PCR), (Rigler *et al.* 1998) and mRNA (Korn *et al.* 2003; Nolan *et al.* 2003; Camacho *et al.* 2004; Winter *et al.* 2004). The typical concentration range of FCS measurements is around 1~100 nM and an average of one to hundreds of fluorescent molecules are under interrogation within the detection volume at any time.

Like other fluorescence-based detection methods, FCS is limited by the background fluorescence of the unbound fluorescent probes that are in large excess to the targets. Recently, several schemes for fluorescence background reduction based on FRET have been reported. The use of hairpin molecular beacon probes instead of linear probes in dual-color FCS has been shown to decrease background by a factor of 40 to 100 (Foldes-Papp *et al.* 2005). In addition, background reduction of greater than 100 fold in dual-color FCS has also been reported by the incorporation of two additional quencher-labeled probes that quench the free unbound probes (Nolan *et al.* 2003). The significant reduction in background makes dual-color FCS amenable to the analysis of low-abundance, unamplified targets.

10.5.3 *Single molecule fluorescence burst coincidence detection*

The diffraction-limited minute detection volume of a SMD spectroscope also allows analysis of coincident single molecule fluorescence signals that occur when two or more fluorescent molecules simultaneously pass through the detection volume. This technique is referred to as single molecule fluorescence burst coincidence detection, which has been applied to detect low-abundance biomolecular targets including nucleic acids (Castro and Williams 1997; Castro and Okinaka 1999; Marina and Castro 2004; Neely *et al.* 2006) and proteins (Li *et al.* 2004).

As shown in Fig. 10.17a, the coincidence detection method uses two probes, each labeled with a different fluorescent tag, which bind a specific target. Emission of the two different fluorescent tags is then detected by dual-color confocal spectroscopy. Single molecule fluorescence coincidence detection is typically performed at a concentration level of sub-nanomolar or lower. At this concentration level, the average number of probe molecules residing in the confocal detection volume ($\sim$ fL), i.e. the molecular occupancy, is smaller than unity at any time. The discrete fluorescence bursts are then detected as individual probes or hybrids flow through the detection volume (Fig. 10.17b and Fig. 10.17c). When targets are not present in the solution and thus the two probes move independently, the digital single-probe fluorescence bursts, detected in the two separate emission channels, are not correlated (Fig. 10.17b). On the other hand, when mixed with a sample containing the specific targets, the two probes bind to a target, forming a doubly-bound molecular hybrid. Simultaneous fluorescence bursts can be seen in the two emission channels as the individual doubly-bound hybrids flow through the detection volume (Fig. 10.17c).

The probability of having more than one unbound probe simultaneously present in the detection volume due to stochastic events can be estimated using Poisson statistics:

$$P_x = \frac{\exp(-\lambda) \cdot \lambda^x}{x!}$$

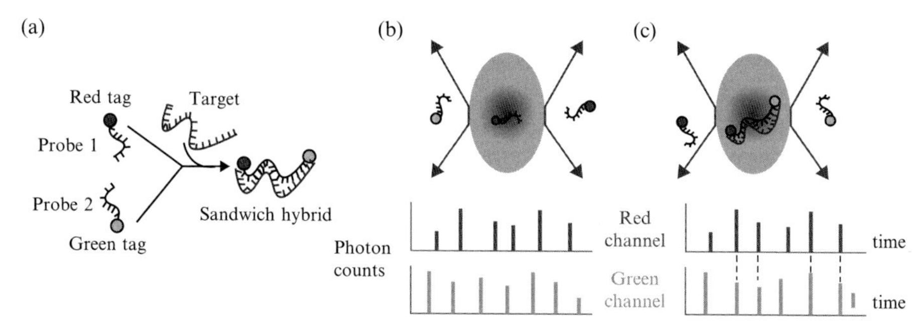

FIG.10.17: Conceptual schematic of two-color coincidence analysis. a) Two different fluorescently labeled probes are designed to bind a target strand, forming a sandwich hybrid. b) In the absence of targets, the fluorescent bursts from the two emission channels do not correlate with each other. c) In the presence of targets, coincident fluorescent bursts from sandwich hybrids are detected.

where P_x stands for the probability of finding exactly x molecules in the focal volume at a given time. λ is the average number of molecules in the detection volume. At a probe concentration level of 1 pM, λ is $\sim 10^{-4}$. A simple calculation shows the probability of finding one (P_1) and two (P_2) molecules in the detection volume is 10^{-4} and 5.0×10^{-9}, respectively. P_2/P_1 is thus 5.0×10^{-5}, implying a low probability of having false coincidences due to the stochastic occurrence of two different color, unbound probes simultaneously residing in the detection volume. Therefore, there is no need to separate the unbound probes from the hybrids due to the low false coincidence probability while the requirement for two probes in coincidence detection increases the detection specificity.

Detection specificity in single molecule coincidence detection can be further improved by using peptide nucleic acid (PNA) probes. PNAs are base sequences attached to an N-(2-aminoethyl)-glycine backbone that is linked by peptide bonds. This DNA analog contains an uncharged backbone, which confers stronger bonding between complementary PNA/DNA sequences, compared to DNA/DNA counterparts. Castro and coworkers demonstrated SMD of specific DNA sequences in unamplified genomic DNA samples, utilizing this stronger binding property of PNA probes (Castro and Williams 1997; Castro and Okinaka 1999). Li and coworkers investigated alternative methods of probing the limits of detection sensitivity in coincidence detection by integrating a dual-laser excitation confocal setup and detecting within samples having a large excess of singly-covalently-labeled DNA molecules (Li *et al.* 2003). Using this dual-laser excitation with its femtoliter-sized detection volume further reduced both background noise and spectral cross-talk.

One challenging aspect of dual-color coincidence detection is to minimize cross-talk induced false coincidences that result from using fluorescent tags with overlapping emission. The small Stokes' shift of organic fluorophores (~20–30 nm)

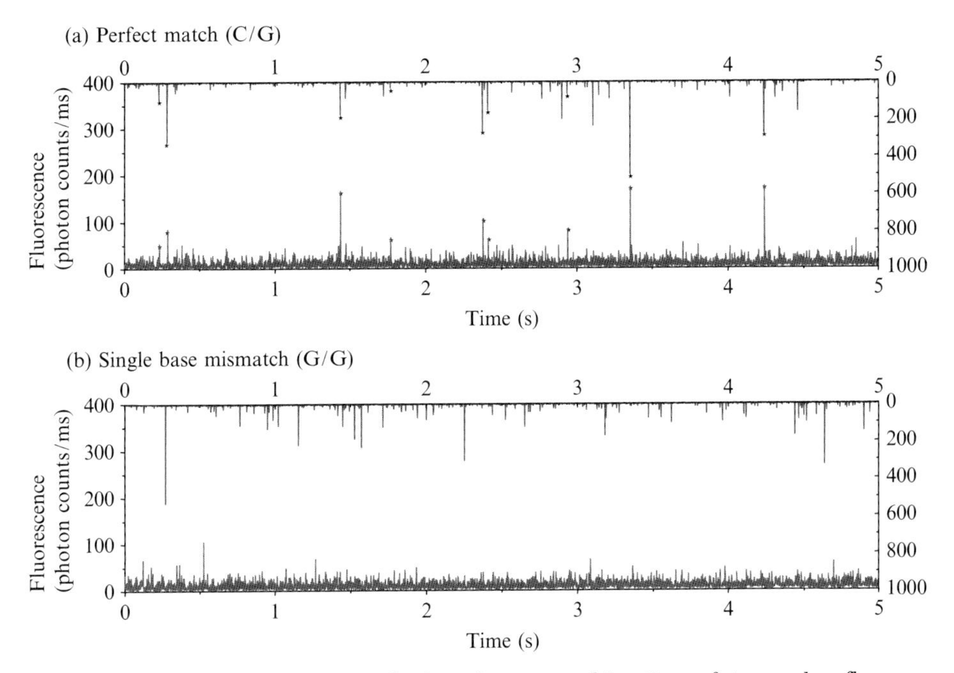

FIG.10.18: Point mutation analysis using a combination of two-color fluorescence coincidence detection and ligation reaction. a) In the presence of perfect-match targets, ligation products are generated that are detectable as they produce coincident fluorescent bursts. b) In the presence of mismatch targets and thus absence of ligation products, non-correlated fluorescence bursts are detected. (Yeh *et al.* 2006).

creates further complications since two fluorophores that can be excited by the same laser will have close peak emission wavelengths. The complication of cross-talk can be overcome by using dual-excitation confocal spectroscopy as it allows use of two fluorophores with distinct excitation and emission spectra. Still, good coincidence detection results from dual-excitation setups necessitates good overlap between the illumination volumes of both lasers, which requires complicated optical alignment and highly corrected, chromatic aberration-free objectives (Schwille *et al.* 1997).

However, dual-color coincidence detection of DNA targets has been demonstrated using single-laser excitation and QDs as fluorescent tags. As aforementioned, QDs have great photophysical properties such as size-tunable spectra, broad absorption and narrow emission spectra, and large Stokes' shifts. As a result, QDs are prime candidates for fluorescent tags that can be efficiently used in multiplexed detection and excited by single laser sources. Figure 10.18 shows an application of coincidence detection in combination with DNA ligation for analysis of tumor-relevant point mutations (Yeh *et al.* 2006). In this method, a pair of allele-specific probes, separately labeled with a QD605 (peak emission 605 nm) and an Oregon Green 488 (peak emission 525 nm), were used in ligation reactions. When perfect match targets were present, the two probes were ligated and coincident emissions of the two fluorescent tags occured upon illumination, thereby revealing the genome type of the target. The combined use of QD and organic fluorophores eased the requirement of instrumentation for coincidence analysis. Thus, the analysis was performed using a single-excitation-wavelength confocal spectroscope. Another advantage of this method is that QDs also served as concentrators that amplified the Oregon Green 488 signal in the presence of ligation products, improving the accuracy in two-color coincidence analysis. This assay method has been demonstrated capable of detecting zeptomoles of targets and achieving an allele discrimination selectivity factor $>10^5$.

10.6 Concluding remarks

Although the development of functional nanomaterials tailored for biosensing applications is still in its infancy, a variety of nanoscale biosensors as reviewed in this chapter have demonstrated superior performance unmatched by conventional biosensors. For example, nanobiosensors derived from semiconductor and metallic nanoparticles have been demonstrated with ultrahigh detection sensitivity at the femtomolar level. In addition, biomolecular detection with a high degree of multiplexing through the implementation of barcoded nanorods and quantum dots in biosensing assays has been reported. We expect that the next ten years will incur a period of rapid maturation and growth in current applications and undoubtedly entail the discovery of new properties and applications that will define the future of nanomaterials in biology and medicine.

Miniaturization in addition to the extraordinary physical properties of materials at the nano scale have enabled new sensors to transduce signals large enough in amplitude and discrete enough in identity to be positively identified at the single molecule level. The combination of nanotechnology and single molecule detection is expected to enable the development of simple, cost-effective, and high-performance biological detection platforms for routine tests in clinically relevant laboratory settings. Indeed, rapid, quantitative, and high-throughput SMD analysis has been achieved by leveraging advances in microfabrication and microfluidic technologies. In the near future, critical challenges in genomic analysis, such as amplification-free detection of rare DNA targets and genetic defects, accurate quantification of low-abundance targets, and differentiation of minute differences in molecular quantity, are poised to be resolved through innovative bioanalytical platforms based on integrated SMD, microfluidics, and nanobiosensor technology.

While the majority of study on the development of nanobiosensors focuses on enhancing the efficiency of signal transduction from a biological event into a detectable physical signal, it has been suggested that the bottleneck in ultrasensitive nanoscale detection resides not in signal transduction but in mass transport (Sheehan and Whitman 2005). Indeed, the multiple highly sensitive nanobiosensors that are reviewed in this chapter commonly presented a concentration detection limit of femtomolar. This may not be just a coincidence; the diffusion-limited interactions of nanosensors and biomolecular targets of extremely low concentration may account for this common detection limit. Directed methods of molecular transport to the sensor surfaces involving micro/nanofluidic, electrostatic, magnetic and/or optical manipulation to increase analyte–sensor interactions will need to be integrated with the high sensitivity nanotransducer/biosensors to surpass this fundamental limit.

References

Akerman, M. E., Chan, W. C. W. *et al.*, 2002. 'Nanocrystal targeting in vivo.' *Proceedings of the National Academy of Sciences of the United States of America*, **99** (20), pp. 12617–12621.

Alivisatos, A. P., 1996. 'Perspectives on the physical chemistry of semiconductor nanocrystals.' *Journal of Physical Chemistry*, **100** (31), pp. 13226–13239.

Anazawa, T., Matsunaga, H. *et al.*, 2002. 'Electrophoretic quantitation of nucleic acids without amplification by single-molecule imaging.' *Analytical Chemistry*, **74** (19), pp. 5033–5038.

Aryal, B. P., and Benson, D. E., 2006. 'Electron donor solvent effects provide biosensing with quantum dots.' *Journal of the American Chemical Society*, **128** (50), pp. 15986–15987.

Authier, L., Grossiord, C. *et al.*, 2001. 'Gold nanoparticle-based quantitative electrochemical detection of amplified human cytomegalovirus DNA using disposable microband electrodes.' *Analytical Chemistry*, **73** (18), pp. 4450–4456.

Bao, P., Frutos, A. G. *et al.*, 2002. 'High-sensitivity detection of DNA hybridization on microarrays using resonance light scattering.' *Analytical Chemistry*, **74** (8), pp. 1792–1797.

Barone, P. W., Baik, S. *et al.*, 2005. 'Near-infrared optical sensors based on single-walled carbon nanotubes.' *Nature Materials*, **4** (1), pp. 86–92.

Bartlett, P. N., and Astier, Y., 2000. 'Microelectrochemical enzyme transistors.' *Chemical Communications*, **2**, pp. 105–112.

Bruchez, M., Moronne, M. *et al.*, 1998. 'Semiconductor nanocrystals as fluorescent biological labels.' *Science*, **281** (5385), pp. 2013–2016.

Camacho, A., Korn, K. *et al.*, 2004. 'Direct quantification of mRNA expression levels using single molecule detection.' *Journal of Biotechnology*, **107** (2), pp. 107–114.

Cao, Y. W. C., Jin, R. *et al.*, 2001. 'DNA-modified core-shell Ag/Au nanoparticles.' *Journal of the American Chemical Society*, **123** (32), pp. 7961–7962.

Cao, Y. W. C., Jin, R. C. *et al.*, 2002. 'Nanoparticles with Raman spectroscopic fingerprints for DNA and RNA detection.' *Science*, **297** (5586), pp. 1536–1540.

Cao, Y. W. C., Jin, R. C. *et al.*, 2003. 'Raman dye-labeled nanoparticle probes for proteins.' *Journal of the American Chemical Society*, **125** (48), pp. 14676–14677.

Cardullo, R. A., Agrawal, S. *et al.*, 1988. 'Detection of nucleic-acid hybridization by nonradiative fluorescence resonance energy-transfer.' *Proceedings of the National Academy of Sciences of the United States of America*, **85** (23), pp. 8790–8794.

Castro, A., and Okinaka, R. T., 1999. 'Ultrasensitive, direct detection of a specific DNA sequence of Bacillus anthracis in solution.' *Analyst*, **125** (1), pp. 9–11.

Castro, A., and Williams, J. G. K., 1997. 'Single-molecule detection of specific nucleic acid sequences in unamplified genomic DNA.' *Analytical Chemistry*, **69** (19), pp. 3915–3920.

Castro, A., Fairfield, F. R. *et al.*, 1993. 'Fluorescence detection and size measurement of single DNA molecules.' *Analytical Chemistry*, **65** (7), pp. 849–852.

Chakrabarti, R., and Klibanov, A. M., 2003. 'Nanocrystals modified with peptide nucleic acids (PNAs) for selective self-assembly and DNA detection.' *Journal of the American Chemical Society*, **125** (41), pp. 12531–12540.

Chan, W. C. W., and Nie, S. M., 1998. 'Quantum dot bioconjugates for ultrasensitive nonisotopic detection.' *Science*, **281** (5385), pp. 2016–2018.

Cui, Y., Wei, Q. *et al.*, 2001. 'Nanowire nanosensors for highly sensitive and selective detection of biological and chemical species.' *Science*, **293** (5533), pp. 1289–1292.

Dahan, M., Levi, S. *et al.*, 2003. 'Diffusion dynamics of glycine receptors revealed by single-quantum dot tracking.' *Science*, **302** (5644), pp. 442–445.

de Mello, A. J., and Edel, J. B., 2007. 'Hydrodynamic focusing in microstructures: Improved detection efficiencies in subfemtoliter probe volumes.' *Journal of Applied Physics*, **101** (8), p. 084903.

Dubertret, B., Calame, M. *et al.*, 2001. 'Single-mismatch detection using gold-quenched fluorescent oligonucleotides.' *Nature Biotechnology*, **19** (4), pp. 365–370.

Dubertret, B., Skourides, P. *et al.*, 2002. 'In vivo imaging of quantum dots encapsulated in phospholipid micelles.' *Science*, **298** (5599), pp. 1759–1762.

Dyadyusha, L., Yin, H. *et al.*, 2005. 'Quenching of CdSe quantum dot emission, a new approach for biosensing.' *Chemical Communications* **25**, pp. 3201–3203.

Edman, L., Mets, U. *et al.*, 1996. 'Conformational transitions monitored for single molecules in solution.' *Proceedings of the National Academy of Sciences of the United States of America*, **93** (13), pp. 6710–6715.

Eigen, M., and Rigler, R., 1994. 'Sorting single molecules: Application to diagnostics and evolutionary biotechnology.' *Proceedings of the National Academy of Sciences of the United States of America*, **91** (13), pp. 5740–5747.

Elghanian, R., Storhoff, J. J. *et al.*, 1997. 'Selective colorimetric detection of polynucleotides based on the distance-dependent optical properties of gold nanoparticles.' *Science*, **277** (5329), pp. 1078–1081.

Faulds, K., Smith, W. E. *et al.*, 2004. 'Evaluation of surface-enhanced resonance Raman scattering for quantitative DNA analysis.' *Analytical Chemistry*, **76** (2), pp. 412–417.

Foldes-Papp, Z., Demel, U. *et al.*, 2001. 'Ultrasensitive detection and identification of fluorescent molecules by FCS: Impact for immunobiology.' *Proceedings of the National Academy of Sciences of the United States of America*, **98** (20), pp. 11509–11514.

Foldes-Papp, Z., Kinjo, M. *et al.*, 2005. 'A new ultrasensitive way to circumvent PCR-based allele distinction: Direct probing of unamplified genomic DNA by solution-phase hybridization using two-color fluorescence cross-correlation spectroscopy.' *Experimental and Molecular Pathology*, **78** (3), pp. 177–189.

Gao, X. H., Cui, Y. Y. *et al.*, 2004. 'In vivo cancer targeting and imaging with semiconductor quantum dots.' *Nature Biotechnology*, **22** (8), pp. 969–976.

Gerion, D., Pinaud, F. *et al.*, 2001. 'Synthesis and properties of biocompatible water-soluble silica-coated CdSe/ZnS semiconductor quantum dots.' *Journal of Physical Chemistry B*, **105** (37), pp. 8861–8871.

Gerion, D., Parak, W. J. *et al.*, 2002. 'Sorting fluorescent nanocrystals with DNA.' *Journal of the American Chemical Society*, **124** (24), pp. 7070–7074.

Gerion, D., Chen, F. Q. *et al.*, 2003. 'Room-temperature single-nucleotide polymorphism and multiallele DNA detection using fluorescent nanocrystals and microarrays.' *Analytical Chemistry*, **75** (18), pp. 4766–4772.

Goldman, E. R., Balighian, E. D. *et al.*, 2002. 'Avidin: A natural bridge for quantum dot-antibody conjugates.' *Journal of the American Chemical Society*, **124** (22), pp. 6378–6382.

Goodwin, P. M., Nolan, R. L. *et al.*, 2004. 'Single-molecule spectroscopy for nucleic acid analysis: A new approach for disease detection and genomic analysis.' *Current Pharmaceutical Biotechnology*, **5** (3), pp. 271–278.

Grabar, K. C., Freeman, R. G. *et al.*, 1995. 'Preparation and characterization of Au colloid monolayers.' *Analytical Chemistry*, **67** (4), pp. 735–743.

Grubisha, D. S., Lipert, R. J. *et al.*, 2003. 'Femtomolar detection of prostate-specific antigen: An immunoassay based on surface-enhanced Raman scattering and immunogold labels.' *Analytical Chemistry*, **75** (21), pp. 5936–5943.

Gruner, G., 2006. 'Carbon nanotube transistors for biosensing applications.' *Analytical and Bioanalytical Chemistry*, **384** (2), pp. 322–335.

Gueroui, Z., and Libchaber, A., 2004. 'Single-molecule measurements of gold-quenched quantum dots.' *Physical Review Letters*, **93** (16), p. 166108.

Ha, T., 2001. 'Single-molecule fluorescence resonance energy transfer.' *Methods*, **25** (1), pp. 78–86.

Haab, B. B., and Mathies, R. A., 1999. 'Single-molecule detection of DNA separations in microfabricated capillary electrophoresis chips employing focused molecular streams.' *Analytical Chemistry*, **71** (22), pp. 5137–5145.

Haes, A. J., and Van Duyne, R. P., 2002. 'A nanoscale optical blosensor: Sensitivity and selectivity of an approach based on the localized surface plasmon resonance spectroscopy of triangular silver nanoparticles.' *Journal of the American Chemical Society*, **124** (35), pp. 10596–10604.

Han, M. Y., Gao, X. H. *et al.*, 2001. 'Quantum-dot-tagged microbeads for multiplexed optical coding of biomolecules.' *Nature Biotechnology*, **19** (7), pp. 631–635.

Hazarika, P., Ceyhan, B. *et al.*, 2005. 'Sensitive detection of proteins using difunctional DNA-gold nanoparticles.' *Small*, **1** (8–9), pp. 844–848.

He, L., Musick, M. D. *et al.*, 2000. 'Colloidal Au-enhanced surface plasmon resonance for ultrasensitive detection of DNA hybridization.' *Journal of the American Chemical Society*, **122** (38), pp. 9071–9077.

Hirsch, L. R., Jackson, J. B. *et al.*, 2003. 'A whole blood immunoassay using gold nanoshells.' *Analytical Chemistry*, **75** (10), pp. 2377–2381.

Ho, Y. P., Kung, M. C. *et al.*, 2005. 'Multiplexed hybridization detection with multicolor colocalization of quantum dot nanoprobes.' *Nano Letters*, **5** (9), pp. 1693–1697.

Ho, Y. P., Chen, H. H. *et al.*, 2006. 'Evaluating the intracellular stability and unpacking of DNA nanocomplexes by quantum dots-FRET.' *Journal of Controlled Release*, **116** (1), pp. 83–89.

Hohng, S., and Ha, T., 2005. 'Single-molecule quantum-dot fluorescence resonance energy transfer.' *Chemphyschem*, **6** (5), pp. 956–960.

Holland, P. M., Abramson, R. D. *et al.*, 1991. 'Detection of specific polymerase chain-reaction product by utilizing the 5′-]3′ exonuclease activity of thermus-aquaticus DNA-polymerase.' *Proceedings of the National Academy of Sciences of the United States of America*, **88** (16), pp. 7276–7280.

Hornbaker, D. J., Kahng, S. J. *et al.*, 2002. 'Mapping the one-dimensional electronic States of nanotube peapod structures.' *Science*, **295** (5556), pp. 828–831.

Hu, M. S., Chen, H. L. *et al.*, 2006. 'Photosensitive gold-nanoparticle-embedded dielectric nanowires.' *Nature Materials*, **5** (2), pp. 102–106.

Huang, B., Wu, H. K. *et al.*, 2007a. 'Counting low-copy number proteins in a single cell.' *Science*, **315** (5808), pp. 81–84.

Huang, C. C., Chiu, S. H. *et al.*, 2007b. 'Aptamer-functionalized gold nanoparticles for turn-on light switch detection of platelet-derived growth factor.' *Analytical Chemistry*, **79** (13), pp. 4798–4804.

Huang, C. P., Li, Y. K. *et al.*, 2007c. 'A highly sensitive system for urea detection by using CdSe/ZnS core-shell quantum dots.' *Biosensors & Bioelectronics*, **22** (8), pp. 1835–1838.

Huang, J., Virji, S. *et al.*, 2003. 'Polyaniline nanofibers: Facile synthesis and chemical sensors.' *Journal of the American Chemical Society*, **125** (2), pp. 314–315.

Jeng, E. S., Moll, A. E. *et al.*, 2006. 'Detection of DNA hybridization using the near-infrared band-gap fluorescence of single-walled carbon nanotubes.' *Nano Letters*, **6** (3), pp. 371–375.

Jin, R. C., Wu, G. S. *et al.*, 2003. 'What controls the melting properties of DNA-linked gold nanoparticle assemblies?' *Journal of the American Chemical Society*, **125** (6), pp. 1643–1654.

Keren, K., Berman, R. S. *et al.*, 2003. 'DNA-templated carbon nanotube field-effect transistor.' *Science*, **302** (5649), pp. 1380–1382.

Kettling, U., Koltermann, A. *et al.*, 1998. 'Real-time enzyme kinetics monitored by dual-color fluorescence cross-correlation spectroscopy.' *Proceedings of the*

National Academy of Sciences of the United States of America, **95** (4), pp. 1416–1420.

Kimura, J., and Kuriyama, T., 1990. 'FET biosensors.' *Journal of Biotechnology*, **15** (3), pp. 239–254.

Kneipp, K., Wang, Y. *et al.*, 1997. 'Single molecule detection using surface-enhanced Raman scattering (SERS).' *Physical Review Letters*, **78** (9), pp. 1667–1670.

Knemeyer, J. P. M. N. *et al.*, 2000. 'Probes for detection of specific DNA sequences at the single-molecule level.' *Analytical Chemistry*, **72** (16), pp. 3717–24.

Kong, J., Franklin, N. R. *et al.*, 2000. 'Nanotube molecular wires as chemical sensors.' *Science*, **287** (5453), pp. 622–625.

Korn, K., Gardellin, P. *et al.*, 2003. 'Gene expression analysis using single molecule detection.' *Nucleic Acids Research*, **31** (16), p. e89.

Krupke, R., Hennrich, F. *et al.*, 2003. 'Separation of metallic from semiconducting single-walled carbon nanotubes.' *Science*, **301** (5631), pp. 344–347.

Lakowicz, J. R., 1999. *Principles of Fluorescence Spectroscopy*. Kluwer Academic/Plenum, New York.

Lee, J. H., Wu, J. H. *et al.*, 2007. 'Iron-gold barcode nanowires'. *Angewandte Chemie* (International edition in English), **46** (20), pp. 3663–3667.

Levy, M., Cater, S. F. *et al.*, 2005. 'Quantum-dot aptamer beacons for the detection of proteins.' *ChemBioChem*, **6** (12), pp. 2163–2166.

Li, H. T., Ying, L. M. *et al.*, 2003. 'Ultrasensitive coincidence fluorescence detection of single DNA molecules.' *Analytical Chemistry*, **75** (7), pp. 1664–1670.

Li, H. T., Zhou, D. J. *et al.*, 2004. 'Molecule by molecule direct and quantitative counting of antibody-protein complexes in solution.' *Analytical Chemistry*, **76** (15), pp. 4446–4451.

Li, H. X., and Rothberg, L., 2004a. 'Colorimetric detection of DNA sequences based on electrostatic interactions with unmodified gold nanoparticles.' *Proceedings of the National Academy of Sciences of the United States of America*, **101** (39), pp. 14036–14039.

Li, H. X., and Rothberg, L. J., 2004b. 'Label-free colorimetric detection of specific sequences in genomic DNA amplified by the polymerase chain reaction.' *Journal of the American Chemical Society*, **126** (35), pp. 10958–10961.

Lide, D. R., 1995. *CRC Handbook of Chemistry and Physics*. CRC Press, Boca Raton, FL.

Liu, J. W., Lee, J. H. *et al.*, 2007. 'Quantum dot encoding of aptamer-linked nanostructures for one-pot simultaneous detection of multiple analytes.' *Analytical Chemistry*, **79** (11), pp. 4120–4125.

Lu, G., Maragakis, P. *et al.*, 2005. 'Carbon nanotube interaction with DNA.' *Nano Letters*, **5** (5), pp. 897–900.

Ma, Y., Ali, S. R. *et al.*, 2006. 'Enhanced sensitivity for biosensors: Multiple functions for DNA-wrapped single-walled carbon nanotubes in self-doped polyaniline nanocomposites.' *Journal of Physical Chemistry B*, **110**, pp. 16359–16365.

Marina, O., and Castro, A., 2004. 'Applications of single-molecule detection to the analysis of pathogenic DNA.' *Current Pharmaceutical Biotechnology*, **5** (3), pp. 279–284.

Marme, N., Friedrich, A. *et al.*, 2006. 'Identification of single-point mutations in mycobacterial 16S rRNA sequences by confocal single-molecule fluorescence spectroscopy.' *Nucleic Acids Research*, **34** (13), e90.

Mattoussi, H., Mauro, J. M. *et al.*, 2000. 'Self-assembly of CdSe-ZnS quantum dot bioconjugates using an engineered recombinant protein.' *Journal of the American Chemical Society*, **122** (49), pp. 12142–12150.

Maxwell, D. J., Taylor, J. R. *et al.*, 2002. 'Self-assembled nanoparticle probes for recognition and detection of biomolecules.' *Journal of the American Chemical Society*, **124** (32), pp. 9606–9612.

McFarland, A. D., and Van Duyne, R. P., 2003. 'Single silver nanoparticles as real-time optical sensors with zeptomole sensitivity.' *Nano Letters*, **3** (8), pp. 1057–1062.

Medintz, I. L., Clapp, A. R. *et al.*, 2003. 'Self-assembled nanoscale biosensors based on quantum dot FRET donors.' *Nature Materials*, **2** (9), pp. 630–638.

Medintz, I. L., Konnert, J. H. *et al.*, 2004. 'A fluorescence resonance energy transfer-derived structure of a quantum dot-protein bioconjugate nanoassembly.' *Proceedings of the National Academy of Sciences of the United States of America*, **101** (26), pp. 9612–9617.

Mirkin, C. A., Letsinger, R. L. *et al.*, 1996. 'A DNA-based method for rationally assembling nanoparticles into macroscopic materials.' *Nature*, **382** (6592), pp. 607–609.

Mo, Z. H., and Wei, X. L., 2006. 'Toward hybridization assays without PCR using universal nanoamplicons.' *Analytical and Bioanalytical Chemistry*, **386** (7–8), pp. 2219–2223.

Murray, C. B., Norris, D. J. *et al.*, 1993. 'Synthesis and characterization of nearly monodisperse Cde (E = S., Se, Te) semiconductor nanocrystallites.' *Journal of the American Chemical Society*, **115** (19), pp. 8706–8715.

Nam, J. M., Thaxton, C. S. *et al.*, 2003. 'Nanoparticle-based bio-bar codes for the ultrasensitive detection of proteins.' *Science*, **301** (5641), pp. 1884–1886.

Nam, J. M., Stoeva, S. I. *et al.*, 2004. 'Bio-bar-code-based DNA detection with PCR-like sensitivity.' *Journal of the American Chemical Society*, **126** (19), pp. 5932–5933.

Neely, L. A., Patel, S. *et al.*, 2006. 'A single-molecule method for the quantitation of microRNA gene expression.' *Nature Methods*, **3** (1), pp. 41–46.

Nguyen, D. C., Keller, R. A. *et al.*, 1987. 'Detection of single molecules of phycoerythrin in hydrodynamically focused flows by laser-induced fluorescence.' *Analytical Chemistry*, **59** (17), pp. 2158–2161.

Nie, S. M., and Emery, S. R., 1997. 'Probing single molecules and single nanoparticles by surface-enhanced Raman scattering.' *Science*, **275** (5303), pp. 1102–1106.

Niemeyer, C. M., and Ceyhan, B., 2001. 'DNA-directed functionalization of colloidal gold with proteins.' *Angewandte Chemie-International Edition*, **40** (19), pp. 3685.

Nolan, R. L., Cai, H. *et al.*, 2003. 'A simple quenching method for fluorescence background reduction and its application to the direct, quantitative detection of specific mRNA.' *Analytical Chemistry*, **75** (22), pp. 6236–6243.

Ozsoz, M., Erdem, A. *et al.*, 2003. 'Electrochemical genosensor based on colloidal gold nanoparticles for the detection of Factor V Leiden mutation using disposable pencil graphite electrodes.' *Analytical Chemistry*, **75** (9), pp. 2181–2187.

Parak, W. J., Boudreau, R. *et al.*, 2002. 'Cell motility and metastatic potential studies based on quantum dot imaging of phagokinetic tracks.' *Advanced Materials*, **14** (12), pp. 882–885.

Parak, W. J., Gerion, D. *et al.*, 2002. 'Conjugation of DNA to silanized colloidal semiconductor nanocrystalline quantum dots.' *Chemistry of Materials*, **14** (5), pp. 2113–2119.

Park, S. J., Taton, T. A. *et al.*, 2002. 'Array-based electrical detection of DNA with nanoparticle probes.' *Science*, **295** (5559), pp. 1503–1506.

Pathak, S., Choi, S. K. *et al.*, 2001. 'Hydroxylated quantum dots as luminescent probes for in situ hybridization.' *Journal of the American Chemical Society*, **123** (17), pp. 4103–4104.

Patolsky, F., Zheng, G. *et al.*, 2006a. 'Fabrication of silicon nanowire devices for ultrasensitive, label-free, real-time detection of biological and chemical species.' *Nature Protocols*, **1** (4), pp. 1711–1724.

Patolsky, F., Zheng, G. *et al.*, 2006b. 'Nanowire-based biosensors.' *Analytical Chemistry*, **78** (13), pp. 4260–4269.

Peng, X., Komatsu, N. *et al.*, 2007. 'Optically active single-walled carbon nanotubes.' *Nature Nanotechnology*, **2**, pp. 361–361–365.

Peng, X. G., Wickham, J. *et al.*, 1998. 'Kinetics of II-VI and III-V colloidal semiconductor nanocrystal growth: Focusing of size distributions.' *Journal of the American Chemical Society*, **120** (21), pp. 5343–5344.

Ray, P. C., Fortner, A. *et al.*, 2006. 'Gold nanoparticle based FRET asssay for the detection of DNA cleavage.' *Journal of Physical Chemistry B*, **110** (42), pp. 20745–20748.

Reich, D. H., Tanase, M. *et al.*, 2003. 'Biological applications of multifunctional magnetic nanowires.' *Journal of Applied Physics*, **93** (10), pp. 7275–7275–7280.

Reza, S., Bosman, G. *et al.*, 2006. 'Noise in silicon nanowires.' *IEEE Transactions on Nanotechnology*, **5** (5), pp. 523–523–529.

Rigler, R., Foldes-Papp, Z. *et al.*, 1998. 'Fluorescence cross-correlation: A new concept for polymerase chain reaction.' *Journal of Biotechnology*, **63** (2), pp. 97–109.

Robelek, R., Niu, L. F. *et al.*, 2004. 'Multiplexed hybridization detection of quantum dot-conjugated DNA sequences using surface plasmon enhanced fluorescence microscopy and spectrometry.' *Analytical Chemistry*, **76** (20), pp. 6160–6165.

Roll, D., Malicka, J. *et al.*, 2003. 'Metallic colloid wavelength-ratiometric scattering sensors.' *Analytical Chemistry*, **75** (14), pp. 3440–3445.

Sato, K., Hosokawa, K. *et al.*, 2003. 'Rapid aggregation of gold nanoparticles induced by non-cross-linking DNA hybridization.' *Journal of the American Chemical Society*, **125** (27), pp. 8102–8103.

Schofield, C. L., Field, R. A. *et al.*, 2007. 'Glyconanoparticles for the colorimetric detection of cholera toxin.' *Analytical Chemistry*, **79** (4), pp. 1356–1361.

Schwille, P., MeyerAlmes, F. J. *et al.*, 1997. 'Dual-color fluorescence cross-correlation spectroscopy for multicomponent diffusional analysis in solution.' *Biophysical Journal*, **72** (4), pp. 1878–1886.

Schwille, P., Oehlenschlager, F. *et al.*, 1996. 'Quantitative hybridization kinetics of DNA probes to RNA in solution followed by diffusional fluorescence correlation analysis.' *Biochemistry*, **35** (31), pp. 10182–10193.

Sheehan, P. E., and Whitman, L. J., 2005. 'Detection limits for nanoscale biosensors.' *Nano Letters*, **5** (4), pp. 803–807.

Shi, L. F., Rosenzweig, N. *et al.*, 2007. 'Luminescent quantum dots fluorescence resonance energy transfer-based probes for enzymatic activity and enzyme inhibitors.' *Analytical Chemistry*, **79** (1), pp. 208–214.

So, M. K., Xu, C. J. *et al.*, 2006. 'Self-illuminating quantum dot conjugates for in vivo imaging.' *Nature Biotechnology*, **24** (3), pp. 339–343.

Star, A., Tu, E. *et al.*, 2006. 'Label-free detection of DNA hybridization using carbon nanotube network field-effect transistors.' *Proceedings of the National Academy of Sciences of the United States of America*, **103** (4), pp. 921–926.

Storhoff, J. J., Elghanian, R. *et al.*, 1998. 'One-pot colorimetric differentiation of polynucleotides with single base imperfections using gold nanoparticle probes.' *Journal of the American Chemical Society*, **120** (9), pp. 1959–1964.

Storhoff, J. J., Lucas, A. D. *et al.*, 2004. 'Homogeneous detection of unamplified genomic DNA sequences based on colorimetric scatter of gold nanoparticle probes.' *Nature Biotechnology*, **22** (7), pp. 883–887.

Storhoff, J. J., Marla, S. S. *et al.*, 2004. 'Gold nanoparticle-based detection of genomic DNA targets on microarrays using a novel optical detection system.' *Biosensors & Bioelectronics*, **19** (8), pp. 875–883.

Taton, T. A., Mirkin, C. A. *et al.*, 2000. 'Scanometric DNA array detection with nanoparticle probes.' *Science*, **289** (5485), pp. 1757–1760.

Tyagi, S., and Kramer, F. R., 1996. 'Molecular beacons: Probes that fluoresce upon hybridization.' *Nature Biotechnology*, **14** (3), pp. 303–308.

Wabuyele, M. B., Farquar, H. *et al.*, 2003. 'Approaching real-time molecular diagnostics: Single-pair fluorescence resonance energy transfer (spFRET) detection for the analysis of low abundant point mutations in K-ras oncogenes.' *Journal of the American Chemical Society*, **125** (23), pp. 6937–6945.

Wang, J., Polsky, R. *et al.*, 2001a. 'Silver-enhanced colloidal gold electrochemical stripping detection of DNA hybridization.' *Langmuir*, **17** (19), pp. 5739–5741.

Wang, J., Xu, D. K. *et al.*, 2001b. 'Metal nanoparticle-based electrochemical stripping potentiometric detection of DNA hybridization.' *Analytical Chemistry*, **73** (22), pp. 5576–5581.

Wang, J., Xu, D. K. *et al.*, 2002. 'Magnetically-induced solid-state electrochemical detection of DNA hybridization.' *Journal of the American Chemical Society*, **124** (16), pp. 4208–4209.

Wang, J., Liu, G. D. *et al.*, 2003. 'Electrochemical coding technology for simultaneous detection of multiple DNA targets.' *Journal of the American Chemical Society*, **125** (11), pp. 3214–3215.

Wang, T. H., Peng, Y. H. *et al.*, 2005. 'Single-molecule tracing on a fluidic microchip for quantitative detection of low-abundance nucleic acids.' *Journal of the American Chemical Society*, **127** (15), pp. 5354–5359.

Wang, W. U., Chen, C. *et al.*, 2005. 'Label-free detection of small-molecule-protein interactions by using nanowire nanosensors.' *Proceedings of the National Academy of Sciences of the United States of America*, **102** (9), pp. 3208–3212.

Wei, X. L., Wang, A. Z. *et al.*, 1996. 'Synthesis and physical properties of highly sulfonated polyaniline.' *Journal of the American Chemical Society*, **118**, pp. 2545–2545–2555.

Weizmann, Y., Patolsky, F. *et al.*, 2001. 'Amplified detection of DNA and analysis of single-base mismatches by the catalyzed deposition of gold on Au-nanoparticles.' *Analyst*, **126** (9), pp. 1502–1504.

Wennmalm, S., Edman, L. *et al.*, 1997. 'Conformational fluctuations in single DNA molecules.' *Proceedings of the National Academy of Sciences of the United States of America*, **94** (20), pp. 10641–10646.

Willard, D. M., Carillo, L. L. *et al.*, 2001. 'CdSe-ZnS quantum dots as resonance energy transfer donors in a model protein-protein binding assay.' *Nano Letters*, **1** (9), pp. 469–474.

Winter, H., Korn, K. *et al.*, 2004. 'Direct gene expression analysis.' *Current Pharmaceutical Biotechnology*, **5** (2), pp. 191–197.

Wu, J. S., Dhara, S. *et al.*, 2002. 'Growth and optical properties of self-organized Au2Si nanospheres pea-podded in a silicon oxide nanowire.' *Advanced Materials*, **14**, pp. 1847–1847–1850.

Wu, X. Y., Liu, H. J. *et al.*, 2003. 'Immunofluorescent labeling of cancer marker Her2 and other cellular targets with semiconductor quantum dots.' *Nature Biotechnology*, **21** (1), pp. 41–46.

Xu, H. X., Sha, M. Y. *et al.*, 2003. 'Multiplexed SNP genotyping using the Qbead (TM) system: A quantum dot-encoded microsphere-based assay.' *Nucleic Acids Research*, **31** (8), p. e43.

Yao, H. Q., Zhang, Y. *et al.*, 2007. 'Quantum dot/bioluminescence resonance energy transfer based highly sensitive detection of proteases.' *Angewandte Chemie International Edition*, **46** (23), pp. 4346–4349.

Yeh, H. C., Ho, Y. P. *et al.*, 2005. 'Quantum dot-mediated biosensing assays for specific nucleic acid detection.' *Nanomedicine: Nanotechnology, Biology and Medicine*, **1** (2), pp. 115–121.

Yeh, H. C., Ho, Y. P. *et al.*, 2006a. 'Homogeneous point mutation detection by quantum dot-mediated two-color fluorescence coincidence analysis.' *Nucleic Acids Research*, **34** (5), pp. e35.

Yeh, H. C., Puleo, C. M. *et al.*, 2006b. 'A microfluidic-FCS platform for investigation on the dissociation of Sp1-DNA complex by doxorubicin.' *Nucleic Acids Research*, **34** (21), pp. e144.

Zhang, C. Y., Chao, S. Y. *et al.*, 2005a. 'Comparative quantification of nucleic acids using single-molecule detection and molecular beacons.' *Analyst*, **130** (4), pp. 483–488.

Zhang, C. Y., Yeh, H. C. *et al.*, 2005b. 'Single-quantum-dot-based DNA nanosensor.' *Nature Materials*, **4** (11), pp. 826–831.

Zhao, X. J., Hilliard, L. R. *et al.*, 2004. 'A rapid bioassay for single bacterial cell quantitation using bioconjugated nanoparticles.' *Proceedings of the National Academy of Sciences of the United States of America*, **101** (42), pp. 15027–15032.

Zheng, G., Patolsky, F. *et al.*, 2005. 'Multiplexed electrical detection of cancer markers with nanowire sensor arrays.' *Nature Biotechnology*, **23** (10), pp. 1294–1301.

Zheng, M., Davidson, F. *et al.*, 2003a. 'Ethylene glycol monolayer protected nanoparticles for eliminating nonspecific binding with biological molecules.' *Journal of the American Chemical Society*, **125** (26), pp. 7790–7791.

Zheng, M., Jagota, A. *et al.*, 2003b. 'DNA-assisted dispersion and separation of carbon nanotubes.' *Nature Materials*, **2** (5), pp. 338–342.

11

SURFACE MOLECULAR PROPERTY CONTROL

Robin L. Garrell and Heather D. Maynard

11.1 General introduction to designed surfaces

Synthetic materials are used in an astonishing range of biomedical applications, from sensors and diagnostics, to inserted and implantable materials and devices. The importance of controling both the chemistry and topology of surfaces that come into contact with biological tissues and fluids has been understood for decades. On the nano scale, molecules can adsorb, adhere, denature, and lose their biological activity. Cells also adsorb and adhere, and can release molecules that initiate a cascade of events that are manifested at the tissue and organism level, including inflammation, immune responses, cellular infiltration, and scar tissue formation. The materials themselves are influenced by these interactions. A surface may become passivated and less reactive, or its texture and lubricity may be altered. It may degrade, or become more prone to colonization by microbes. All of these phenomena result from the complex and time-dependent interactions between the surface and its environment.

It is common for materials or surfaces to be described as 'biocompatible'. The exact meaning of this term is highly context dependent. It generally conveys the idea that the material itself does not cause adverse response(s), such as inflammation, but it tells us nothing about protein and cell interactions with the surface. In some applications, such as contact lenses and microfluidic devices used for separations, the goal is to minimize protein and cellular adsorption. In others, such as many types of biosensors, the goal is to capture and retain specific molecules, cells, or organisms. In other instances, protein adsorption creates a passivating layer that is beneficial because other species (analytes) are then less likely to adsorb. Whether the goal is to minimize or maximize biomolecule–surface interactions, what these diverse applications have in common is the need to design the surface in order to *control* what binds to the surface, how much binds, and for how long.

The wide range of materials and surface modification methods that are now available makes this possible. Nano- and microfluidic devices can be fabricated of materials that intrinsically have the desired surface characteristics, including composition, topology, and micromechanical properties. The surfaces can also be subsequently modified by plasma treatments, adsorption of self-assembled monolayers, application of temporary or permanent coatings, or covalent grafting of synthetic and biological molecules.

Current technologies to modify surfaces and control surface properties are reviewed in this chapter, with particular emphasis on alterations for biomedical and bioanalytical applications. Section 11.2 provides a brief tutorial on mechanisms of biomolecular adsorption. The relationship between surface wettability and biomolecular adsorption, along with methods for modifying wettability, are described in Section 11.3. Materials that minimize or prevent fouling discussed in Section 11.4, and techniques for assessing protein fouling and the effectiveness of surface modifications in Section 11.5. 'Designer surfaces', including both dynamic coatings and covalent modifications, are highlighted in Section 11.6, followed by approaches that lead to the direct and indirect capture of proteins and peptides in Section 11.7. In Section 11.8, some examples of integrating live cells with microfluidic devices are presented. Brief conclusions and future directions are provided in Section 11.9.

11.2. Mechanisms of biomolecular adsorption

Biomolecules adsorb to surfaces through the same types of physicochemical interactions and chemical bonds that form between other types of molecules. When these interactions lead to adsorption of diverse molecules or even organisms, the process is called non-specific adsorption or bio-fouling. Surfaces can also be designed to incorporate molecular recognition, so that only specific proteins, nucleic acid sequences, or organisms can bind. These specific adsorption and capture methods are discussed in Section 11.7.

The mechanisms, thermodynamics and kinetics of protein adsorption have been extensively reviewed (Brash and Horbett 1987; Haynes and Norde 1994; Horbett and Brash 1995; Horbett *et al.* 1996; Norde 1995; Malmsten 1998). The non-specific adsorption problem takes on special significance for sensor and microfluidic devices, because typically the surface-to-volume ratio is high and the samples are small (Locascio *et al.* 2003; Mukhopadhyay 2005). It is generally desirable to minimize or suppress non-specific adsorption, as it leads to loss of analyte and hence poorer detection limits and selectivity, as well as decreased device lifetimes (Schneider *et al.* 2000). It can also result in sample cross-contamination if the same region of the device is re-used. Non-specific protein adsorption is generally irreversible (Horbett *et al.* 1996). The extent of fouling by biomolecules and cells can increase over time, rapidly degrading device performance (Peterson *et al.* 2005; Popat and Desai 2004).

Historically, and for simple systems, adsorption at gas–solid and liquid–solid interfaces has been categorized along an energy continuum. If the adsorbate–surface interactions are relatively weak (less than $\sim$40 kJ/mol) the phenomenon is considered physisorption, while stronger interactions lead to chemisorption (Hiemenz and Rajagopalan 1997; Irene 2008; Masel 1996). This distinction is not particularly useful when discussing biomolecular adsorption, however, because the strength of the interactions typically increases over time as biomolecules change conformation and flatten at the interface (Locasio *et al.* 2002; Yoon and Garrell 2003). It is more useful to focus instead on the physicochemical nature of the interactions, which can then be controlled through choice of solvent, surface materials and surface modifications.

The weakest interactions between biomolecules and surfaces are van der Waals interactions: a term that includes dipole–dipole, dipole–induced dipole, and induced dipole–induced dipole attractive interactions, as well as the Born repulsion, which is effective only at very short distances. The van der Waals attractive forces are electrostatic in origin, but short range, scaling with $1/r^p$ where r is distance and $p \geq 3$ (Dill and Brombert 2003). Because induced dipole moments scale with molecular polarizability (and hence molecular size), these interactions can become significant for biomacromolecules such a proteins.

Hydrogen bonds are particularly important interactions between water and biomolecules and between biomolecules and many types of surfaces. These interactions can be considered a special type of dipole–dipole interaction, although in many cases there is covalent bonding character (sharing of electrons) as well (Schneider 1997). They are on the order of 8–20 kJ/mol in strength. Because there may be many such interactions between a single biomacromolecule and a surface, hydrogen bonds can contribute significantly to adsorption and adhesion.

Electrostatic (Coulombic) interactions scale with 1/r, and hence are much longer range. Proteins and DNA both contain charged groups that play significant roles in their interfacial behavior, affecting not only adsorptivity, but molecular orientation and surface coverage (Norde 1995; Lubarsky *et al.* 2005). Like hydrogen bonds, electrostatic interactions contribute significantly to biomolecular adsorption and adhesion (Lubarsky *et al.* 2005). They can be modulated by varying solution pH (and hence the charge on the biomolecule) and surface composition, as well as by applying an external voltage to control the surface charge and the structure of the electrochemical double layer at the interface. Many of the polymer materials used in microfluidics have a net negative charge; as a result, small anions, negatively charged dyes and DNA exhibit low non-specific adsorption to these surfaces (Locascio *et al.* 2003). Because the charge density is low, however, hydrophobic analytes may still adsorb (Locascio *et al.* 2003).

Solvents play a critical role in biomolecular adsorption. The interaction between water and soluble species is called hydration, a solvation process that is favorable both enthalpically and energetically ($\Delta H < 0$; $\Delta G < 0$). By contrast, many species, such as hydrocarbons and silicon oils, are insoluble in water. The hydrophobic effect is entropic in origin. Water forms hydrogen bonded cage structures around such molecules in order to maximize hydrogen bonding interactions with neighboring water molecules, resulting in a net *decrease* in entropy. (Dill and Brombert 2003). Because transferring a non-polar solute into water results in an enthalpy change $\Delta H \approx 0$, the net free energy change ΔG is > 0, and dissolution does not occur. Water also structures near planar surfaces, driven by the tendency to maximize hydrogen bonds (Dill and Brombert 2003). At a large planar surface, the maximum number of hydrogen bonds that can form is three, compared with four around a small, spherical solute. Thus, inserting a non-hydrogen-bonding planar surface has an enthalpic cost (Southall and Dill 2000).

The hydrophobic effect plays an important role in protein structure and folding, adsorption, and adhesion. The protein surface that is in contact with water tends to be relatively rich in hydrophilic (both neutral and charged) residues that can form hydrogen bonds to water, while the interior is relatively rich in hydrophobic residues. Protein aggregation in solution and protein–surface hydrophobic interactions are maximized at the isoelectric point of the protein (pI), while electrostatic interactions are minimized at that pH (Yoon and Garrell 2008). When a protein molecule diffuses to a hydrophobic surface by Brownian motion, local conformational changes can eventually lead to unfolding (denaturation) (Haynes and Norde 1994). The process is enthalpically favorable, driven by solvent exclusion through which the hydrophobic interior of the protein increases its contact area with the hydrophobic surface and the extruded water molecules become part of the bulk solvent (Dill and Brombert 2003; Yoon and Garrell 2008). A study of the interaction of albumin with polysiloxanes of different surface energies confirmed the importance of the hydrophobic effect in protein adsorption (Janocha *et al.* 2001). Proteins readily adhere to silicon wafers (Sheller *et al.* 1998). and hydrophobic surfaces, such as octadecyltrichorosilane self-assembled monolayers (SAMs) (Ge *et al.* 1998). and many polymers (Locascio *et al.* 2003).

The preceding discussion described the interactions that contribute to protein adsorption. An alternative way to look at the problem is to consider the *processes* that contribute to making protein adsorption exergonic. These include: (1) the redistribution of charged groups at the protein surface and in the electrical double layer at the liquid–solid interface; (2) changes in the hydration of the protein and substrate, and (3) structural rearrangements in the protein molecule (Norde 1995). For proteins with stable secondary and tertiary structures, adsorption is driven by electrostatic interactions and hydrophobic dehydration. As Norde summarizes, (Norde 1995). electrostatic attractions are required for adsorption on hydrophilic surfaces, while hydrophobic dehydration dominates for most hydrophobic surfaces. When proteins are more flexible or structurally unstable, there can be a considerable gain in conformational entropy upon adsorption/denaturation, contributing to a favorable free energy change.

11.3 Surface wettability and biomolecular adsorption

11.3.1. *The role of surface wettability*

Wettability is a qualitative term that describes the propensity of a solvent, usually water, to spread on a horizontal surface. The interactions of solvents, particularly water, with surfaces are key to understanding biomolecular adsorption and conformational stability at liquid–solid interfaces. The hydrophilicity or hydrophobicity of a surface influences not only the behavior of water at the interface (notably, the contact angle and contact angle hysteresis), but also the propensity for biomolecules to adsorb from solution. For most hydrophilic surfaces, biomolecular adsorption is driven by hydrogen bonding interactions and by electrostatic interactions between oppositely charged groups on the surface and adsorbate. In aqueous solutions, hydrogen-bonding interactions are generally dynamic and relatively weak per bond, but there may be quite a few such bonds between an adsorbed biomolecule and the substrate. As noted above, electrostatic interactions can be controlled by choice of buffers and by external forces such as applied electrical potentials (voltages). This is particularly useful in capillary electrophoresis and other electrokinetic separations, as described elsewhere in this volume. For hydrophobic surfaces, biomolecular adsorption is driven by the hydrophobic effect.

The foregoing discussion has focused primarily on the energetics and dynamics of biomolecular adsorption from solution onto pristine surfaces. In reality, the picture is more complicated, and interactions between biomolecules in solution and those already on the surface, intermolecular interactions within the nascent adlayer, and conformational changes in the adsorbates must be taken into account to gain a complete understanding of the structure and dynamics of the interface. Nevertheless, it is clear that prescribing (and assessing) surface wettability represents a simple and general first-line approach for controlling the non-specific adsorption of biomolecules.

Surface wettability is usually assessed by measuring the contact angle θ at the three-phase line of a droplet on a planar surface, and specifically the advancing and receding contact angles, (Chen *et al.* 1999; Gao and McCarthy 2008), as shown in Fig. 11.1. The wettability of a surface depends on the composition and topology of the surface, the composition of the liquid, and the composition of the ambient medium, which might be air or another liquid that is immiscible with the first. Methods for measuring contact angles and interfacial tensions are described in a number of monographs, (Drelich *et al.* 2000; and Hartland 2004), and include contact angle goniometry, and measurements with a de Nouy ring or Wilhelmy plate apparatus. These methods have been systematically compared and evaluated in several short articles that provide excellent entry points to this area of measurement science (Lander *et al.* 1993; and Krishnan *et al.* 2005).

The Young equation shows how the static contact angle depends on the surface tensions at the liquid–vapor (LV), solid–vapor (SV) and solid–liquid (SL) interfaces: $\gamma_{SL} = \gamma_{SV} - \gamma_{LV} \cos\theta$. Contact angle hysteresis is the difference between the advancing and receding contact angles, and is usually measured for a sessile droplet on a tilted substrate or by the Wilhelmy plate method. The advancing

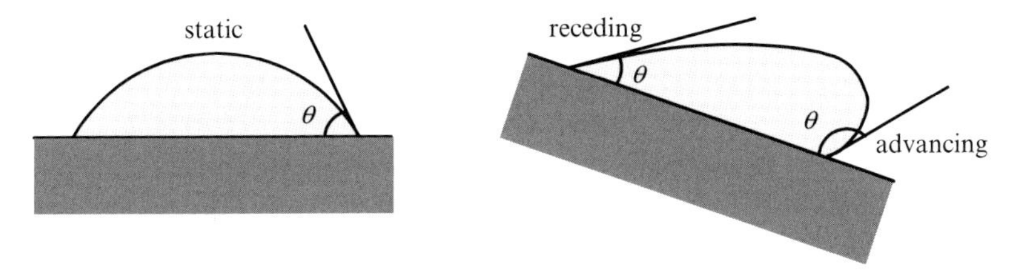

Fig. 11.1: Static, advancing, and receding contact angles. The advancing contact angle may also be measured by pushing liquid into or drawing liquid out of a drop on a horizontal surface.

contact angle $\theta_{advancing}$ is typically larger than the static contact angle measured for an unperturbed droplet on a flat surface, because surface heterogeneity and/or roughness give rise to stiction.

The terms 'hydrophilic' and 'hydrophobic' are qualitative, and Gao and McCarthy have commented on how the definitions of these and related terms such as 'superhydrophobic' have become quite muddled (Gao and McCarthy 2008). In the literature, hydrophilic surfaces are typically considered to be those with water contact angles below 90 ° and hydrophobic surfaces those with contact angles >90 °, but a 65 ° threshold has also been used (Vogler 1999) . It has been suggested that these terms be used only as qualitative or comparative descriptors (Gao and McCarthy 2008). Charged and polar groups contribute to making surfaces hydrophilic, while non–polar groups, particularly perfluorinated ones, make surfaces more hydrophobic.

Wettability is a function not only of the surface composition, but also of the surface topology. In general, roughness, which can be random or periodic, increases the water droplet static contact angle (Cassie and Baxter 1944; Wenzel 1949), with certain surface morphologies giving rise to contact angles exceeding 170° (Chen *et al.* 1999; Bico *et al.* 1999; Ma and Hill; Oner and McCarthy 2000). The reduced liquid–surface contact area can significantly decrease flow friction (Cottin-Bizonne *et al.* 2003; Davies *et al.* 2006). While this may be a desirable feature for nano- and microfluidic applications, the hydrophobicity and high surface area of these materials make them particularly susceptible to biofouling.

11.3.2 *Hydrophobic and hydrophilic surfaces*

Table 11.1 presents a summary of reported static and advancing contact angles for a wide range of materials commonly used in microfluidic devices (Yoon and Garrell 2008). The repeat units of several of these are shown in Fig. 11.2.

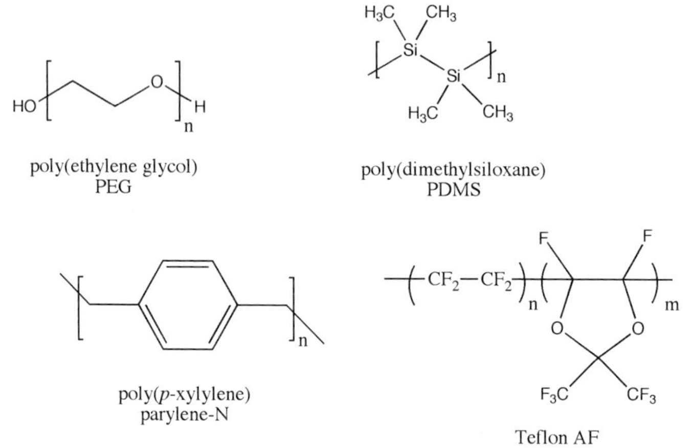

poly(ethylene glycol)
PEG

poly(dimethylsiloxane)
PDMS

poly(*p*-xylylene)
parylene-N

Teflon AF

Fig. 11.2: Repeat units of polymeric materials frequently used in microfluidics and biosensors.

Table 11.1: Advancing and static water contact angles on common materials used for microfluidic device surfaces. Table used from reference with kind permission of Springer Science and Business Media (Yoon and Garrell 2008).*

Materials	θ_{static} (deg)	$\theta_{advancing}$ (deg)
Glass (clean)	0	0
Glass (ordinary)	14, 20, 51	60
Gold (clean)		0
Gold (ordinary)	61, 62	
Si wafer	22	
SiO_2		27 (Chen *et al.* 1999)
Poly(ethylene glycol) [PEG]	18	39–6
Albumin film	64, 70–82	70–100
Lysozyme film	58	
Polycarbonate [PC]	70	86
Poly(methyl methacrylate) [PMMA]	70, 73, 74	
Polystyrene [PS]	87, 90, 91	97
Alkanethiolate or alkylsilane self-assembled monolayers [SAMs]	108, 110, 112, 115	110, 116
Poly(dimethylsiloxane) [PDMS]	108, 113	118
Fluorohydrocarbon	115	
Teflon	112, 113, 115, 118	125

*Refer to the table source for literature citations for each entry.

Glass, common oxides such as silicon dioxide (SiO_2), poly(ethylene glycol), and polymers such as parylene exposed to oxygen plasmas (Nowlin Smith 1980; Pruden *et al.* 2003) or UV light in air (Pruden *et al.* 2003) are hydrophilic. Although freshly prepared, atomically smooth metal surfaces exhibit contact angles approaching zero (i.e. they are completely wettable), metal surfaces prepared by typical vapor deposition conditions and exposed to air and ambient contaminants give contact angles closer to 60°. Hydrocarbon polymers such as polyethylene, polypropylene and polystyrene are hydrophobic, as are surfaces coated with alkanethiolate SAMs. Materials that are intermediate between hydrophobic and hydrophilic include polyesters such as poly(methyl methacrylate) and polycarbonates. Poly(dimethylsiloxane), or PDMS, as well as perfluorocarbons such as Teflon and perfluorocarbon/perfluoroether copolymers such as Teflon AF, are hydrophobic, with static water contact angles exceeding 90°. This makes them susceptible to non-specific adsorption (fouling) (Mukhopadhyay 2005). Interestingly, anions such as hydroxide also physisorb on hydrophobic materials such as Teflon (Aangi and Engberts 2005).

11.3.3 *Modifying surface wettability to control nonspecific adsorption*

The wettability of many device materials can be modified globally (i.e. without spatial control) by a variety of techniques, such as by UV and plasma treatments (Locascio *et al.* 2003). Poly(*p*-xylylene), widely known as parylene or parylene N, is an exemplar. It and its halogenated derivatives, parylene C and D, are dielectric materials commonly used in microfluidic devices and sensor applications. Ultrathin films can be deposited in a way that leaves the chain ends available to react with atmospheric oxygen or other reagents (Senkevich *et al.* 2003). Treatment with oxygen plasma (Nowlin Smith 1980) or UV light in the presence of oxygen (Pruden *et al.* 2003) causes photooxidation. Although these treatments can slightly degrade the thermal stability and dielectric properties of parylene, (Fortin and Lu 2001) they are simple methods for introducing hydroxyl, aldehyde, and carboxylic acid groups that render the surface significantly more hydrophilic.(Pruden *et al.* 2003; Fortin and Lu 2001; Bera *et al.* 2000) Plasma-treated surfaces are more resistant to protein adsorption, and cell adhesion is comparable to that on other tissue culture substrates such as polystyrene (Chang *et al.* 2007). Oxidized and functionalized parylenes provide reactive sites that can improve biocompatibility (Lahann *et al.* 1999), enhance the adhesion of metal or polymer overcoats, (Nowlin and smith 1980), or be used for subsequent grafting of biomolecules (Lahann *et al.* 2001b). Note that these reactive coatings are compatible with soft lithography processes used to pattern DNA, proteins, and cells (Lahann 2006). Electrophilic aromatic substitution reactions can be used to incorporate other functional groups throughout the bulk or confined to the surface, depending on the reaction conditions (Herrera-Alonso and McCarthy 2004). A particularly useful modification utilizes aminomethylation to create a positively charged surface terminated by $-NH_3^+$ groups. Acylation and alkylation reactions have also been demonstrated.

Just as with parylene, the wettabilities of many of the materials in Table 11.1 can be adjusted by incorporating relatively hydrophobic or hydrophilic co-monomers within the polymer chains, or by grafting modifying species onto the fabricated surface. Functional groups or capture agents (e.g. antibodies or aptamers) can also be incorporated that can specifically bind biomolecules of interest. Both strategies can be accomplished with spatial control, and are described in greater detail in later sections of this chapter.

The preceding discussion has focused on the general surface properties of device materials and their relationship to biomolecular adsorption. A phenomenon that is often overlooked in this context is the tendency for solvents to be absorbed into the device or coating material. This can have significant consequences. The solvent can swell the coating, changing its dimensions and wettability, as well as its mechanical properties. For example, hydrocarbons (C_1 to C_{12}) and perfluorocarbons (C_1 to C_7) are absorbed from the gas phase into Teflon AF and permeate the material, acting as plasticizers (Alentiev *et al.* 2002). Teflon AF exhibits selective solvent uptake and permeability. The solvent-saturated film then acts as a supported liquid membrane, into which solutes diffuse and are partitioned (Zhao *et al.* 2004). PDMS also swells in the presence of many solvents, particularly amines, saturated and aromatic hydrocarbons (Lee *et al.* 2003). These solvents extract oligomer contaminants, but also alter the surface properties of PDMS. Polar solvents such as water, DMSO, ethylene glycol, and acetonitrile, as well as perfluorohydrocarbons, are less soluble in PDMS and so do not intercalate as readily (Lee *et al.* 2003).

11.4. Surface materials that minimize or prevent fouling

11.4.1 *Poly (ethylene glycol) and related materials*

Poly(ethylene glycol) (PEG), poly(ethylene oxide) (PEO) and polyoxyethylene are all names for the polymer containing the–(CH_2CH_2O)–repeat unit. Short chain (oligomeric) analogs are called oligo(ethylene glycol) (OEG) or oligo(ethylene oxide) (OEO). The covalent attachment of PEG or OEG to molecules or surfaces is commonly called PEGylation.

The resistance of poly(ethylene glycol) (PEG) and end-tethered PEO to protein adsorption has been known for many years (Harris 1992; Unsworth *et al.* 2008). The adhesion of platelets and bacteria is also inhibited on surfaces treated with PEG, or with the related PEO/PPO/PEO triblock copolymers known as Pluronics (Amiji and Park 1992; Marsh *et al.* 2002). These water-soluble polymers are widely used to prevent protein adsorption *from* solution onto diverse surfaces (Harris 1992), including microchannels (Luk *et al.* 2008). Methods for modifying surfaces with PEG and PEG analogs are described in Section 11.6.

The protein resistance of PEG- and OEG-treated surfaces depends on a set of interrelated factors, including chain length, chain density, hydration, conformation, and the functional group at the distal chain end (Unsworth *et al.* 2008). The mechanisms for protein resistance have been widely discussed and debated. For higher molecular weight (longer chain) PEO, the resistance is commonly attributed to steric barriers, while for shorter chain OEG moieties, 'hydration' barriers are invoked (Unsworth *et al.* 2008; Li *et al.* 2005). There are large differences in the reported minimum EO (or OEG) chain lengths needed to confer protein resistance, ranging from 1–3 to 35–100 (Unsworth *et al.* 2008). Whether the distal group influences protein adsorption in oligomeric systems has also been a matter of debate (Unsworth *et al.* 2008).

Unsworth, Sheardown, and Brasch (2008) have recently shown that there is an optimal surface chain density, and thus an optimal hydration state, associated with maximum protein resistance. The resistance may result from the formation of a weak equilibrium network between PEO and water, and also from water associated beyond the first hydration shell (Heuberger *et al.* 2005). At chain densities up to a critical value of 0.5 chains/nm^2, neither chain length nor distal chemistry were significant factors in protein resistance. Distal chemistry begins to be important at chain densities > 0.5 chains/nm^2, influencing both the rate of adsorption and the ultimate adsorbed quantity (Unsworth *et al.* 2008). The greater adsorption on–OCH_3 than–OH–terminated surfaces may be related to the relative hydrophobicity of the former, or to stronger hydration of the latter. Unsworth *et al.* have concluded that the basic requirement for inhibiting protein adsorption is the formation of a brush-like layer. This is consistent with trends identified for OEO-terminated alkanethiolate SAMs (Harder *et al.* 1998; Herrwerth *et al.* 2003; Vanderah *et al.* 2004).

Other investigators have shown that the amount of protein that adsorbs depends not only on the PEG thickness and grafting density, but also on the charge on the protein and surface (Pasche *et al.* 2005). The ionic strength of the

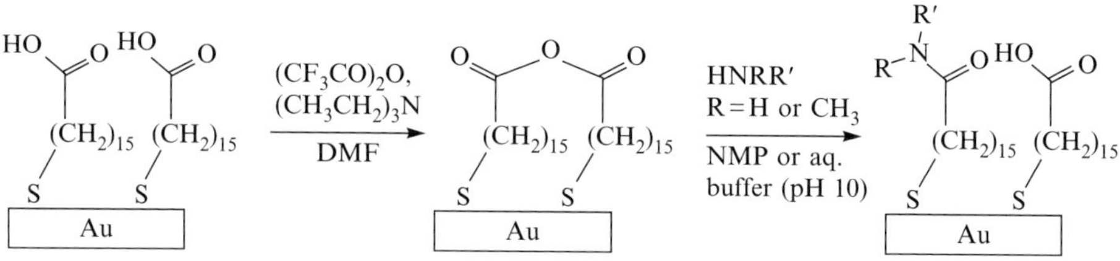

Fig. 11.3: Schematic illustrating the anhydride method for synthesizing mixed SAMs that present a 1:1 mixture of –CONRR' and CO_2H/CO_2^- groups. Reproduced with permission from *J. Am. Chem. Soc.* Copyright 2000 American Chemical Society (Chapman *et al.* 2000).

solution is also important, as it determines the electrical double-layer thickness and hence the extent to which the surface and protein charges are screened from one another (Pasche *et al.* 2005).

11.4.2 *Other protein-resistant materials*

A disadvantage of OEG derivatives is that they tend to autooxidize (Chapman *et al.* 2000). Chapman *et al.* screened SAMs capped with more than 50 different functional groups for resistance to protein adsorption, benchmarking these against OEG-terminated SAMs. They identified several functional groups that conferred protein resistance when incorporated into SAMs at a 1:1 ratio with $-CO_2H/CO_2^-$ groups. The common characteristics of protein-resistant SAMs were that they contained polar functional groups, incorporated hydrogen bond accepting groups, did not contain hydrogen bond donating groups, and had no net charge. Examples of such moieties that are commercially available (or are easy to synthesize) and can be used to prepare SAMs by the anhydride method (Scheme 1) include $HN(CH_3)CH_2CON(CH_3)_2$, $HN(CH_3)CH_2(CH(OCH_3))_4CH_2OCH_3$, $HN(CH_3)CH_2CH_2N(CH_3)PO(N(CH_3)_2)_2$, and $HN(CH_3)CH_2CH_2N(CH_3)COCH_3$ (Chapman *et al.* 2000). Phospholipid and polysaccharide coatings have also attracted attention, but they are less resistant to biomolecular and bacterial adhesion than PEG (Kingshott and Griesser 1999).

11.5 Techniques for assessing protein fouling and the effectiveness of surface modifications

Key to developing surfaces that are resistant to bio-fouling is the availability of a wide range of tools for quantifying protein adsorption. Providing a comprehensive review of these methods is beyond the scope of this chapter, but the sample references included here provide entry points to the literature and introduce the reader to many of the leading investigators in the field.

Some of the earliest experiments to quantifying protein adsorption utilized radiolabeling and related radiotracer techniques (Grant *et al.* 1977). This approach continues to be used (Dyer *et al.* 2007; Benhabbour *et al.* 2008), despite the effort involved in preparing labeled proteins and the costs associated with acquiring isotopes and disposing of radioactive waste. Contact angle measurements (Yoon and Garrell 2003; Chapman *et al.* 2000) and tensiometry represent simple (Miller *et al.* 2004), albeit indirect assays of surface fouling by biomolecules. Widely used optical methods include UV-vis spectroscopy (Prakash *et al.* 2008), solution fluorescence (Holden and Cremer 2005), total internal reflection fluorescence (TIRF) (Golander *et al.* 1990; Sapsford and Ligler 2004), fluorescence recovery after photobleaching (Yang *et al.* 1999), fluorescence microscopy (Holden and Cremer 2005), optical waveguide lightmode spectroscopy (OWLS) (Pasche *et al.* 2005; Tie *et al.* 2003), attenuated total reflection-infrared (ATR-IR) (Mcclellan and Franses 2005), ellipsometry (McClellan and Franses 2005), surface plasmon resonance (SPR) (Chapman *et al.* 2000; Pavey and Olliff 1999; McGurk *et al.* 1999; Green *et al.* 2000; Smith and Corn 2003), and sum-frequency generation (SFG) (Holden and Cremer 2005). Non-optical methods include time-of-flight secondary ion mass spectrometry (TOF-SIMS) and X-ray photoelectron spectroscopy (XPS) (Wagner *et al.* 2002; Michel *et al.* 2005), quartz crystal microbalance or microgravimetry (QCM) (Belengrinou *et al.* 2008; Matsuno *et al.* 2007), and atomic force microscopy (AFM) (Lubarsky *et al.* 2005; Prakash *et al.* 2008). Protein adsorption can also be inferred by assaying enzyme activity in solution (Lenghaus *et al.* 2003).

11.6 Designer surfaces

The choice of surface material for sensor and nano/microfluidic applications should be based on criteria such as ease of fabrication, solvent compatibility, hydrophobicity and resistance to non-specific adsorption, zeta potential and the associated electroosmotic flow mobility, and the need for functional groups that can be used for subsequent chemical modification or analyte capture (Locascio *et al.* 2003). A basic way to establish a specific chemistry is to make the device out of a particular material. Polymers that are commercially available and routinely employed for device fabrication include PDMS, PMMA, polycycloolefin (PCOC), and PC. (Locascio *et al.* 2003; Liu and Lee 2006) (See Table 11.1 for definitions.) Some research groups have fabricated devices out of new materials tailored for a specific application, but, more commonly, traditional materials are modified. The surface chemistry of the materials most frequently used in microfluidics and general strategies for covalent and noncovalent surface modifications have been reviewed by Locascio and co-workers. (2003)

The most straightforward scheme used to alter device surfaces is by simple adsorption of a molecule or polymer. This strategy, called dynamic coating, has the advantage of being easy to carry out. Often a surface-active compound is simply added to the run buffer or provided in a rinsing step directly after device fabrication, prior to use of the device (Belder and Ludwig 2003). The disadvantage is that most coatings are not particularly stable and can deteriorate over time.

Covalent modifications lead to stable constructs, although these involve multiple steps, and thus are not as simple. For example, photolithography can be used to pattern a PEG-like polymer onto substrates such as silicon, oxide, nitride, gold, and platinum (Hanein *et al.* 2001). As described above, PDMS is often oxidized to render the surface hydrophilic and to expose functional groups. Alternatively or in conjunction with oxidation, small molecules or polymers are bound to the surface. Attaching poly(*L*-lysine)-graft-poly(ethylene glycol) in this way gives a surface that is highly resistant to nonspecific protein adsorption (Lee and Voeroes 2005). Polymers can also be grown directly from the surface to achieve high brush densities. Hydrogel materials can be assembled within the device to provide a three-dimensional scaffold for certain applications such as cell seeding. These strategies have been extensively reviewed (Liu and Lee 2006; Belder and Ludwig 2003; Dolnik 2004; Makamba *et al.* 2003). Thus, the particular examples discussed below are meant to be representative and are not comprehensive.

11.6.1 *New materials for device fabrication*

New materials have been devised and employed for fabrication of microfluidic devices. One strategy involves preparing the device from a PEGylated monomer, because PEG resists unwanted protein adsorption. A microchip was fabricated utilizing PEG dimethacrylate (PEGDMA) or PEG diacrylate (PEGDA) (Fig. 11.4) (Kim *et al.* 2006). Materials made from low molecular weight PEG did not swell significantly in aqueous solution and with a supporting layer were suitable for biomedical applications. Devices exhibited significant reductions in cell adhesion and in non-specific adsorption of proteins bovine serum albumin (BSA), fibronectin (FN), and immunoglobulin G (IgG) compared to PDMS/glass and PDMS/PEG-coated glass (Fig. 11.5). Using a similar technique, a copolymer system was prepared from PEGDA, PEG methacrylate, and MMA (Liu *et al.* 2007) and the resultant device applied in peptide and protein separations.

Several groups have fabricated devices from hydrogel materials. Typically the motivation is to prepare microfluidics that support cell adhesion. In some cases,

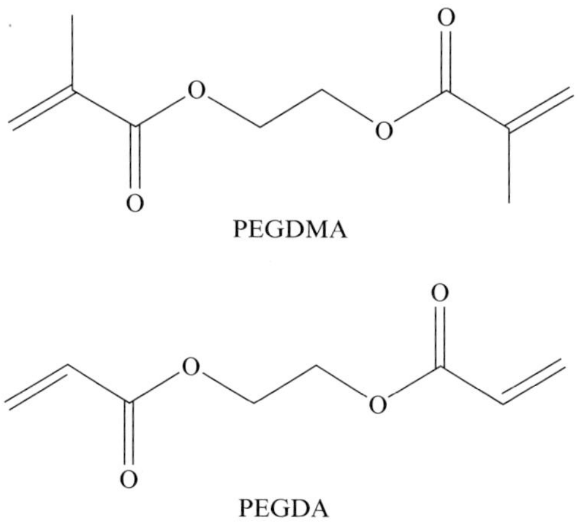

PEGDMA

PEGDA

Fig. 11.4: Monomers utilized to fabricate PEGylated microchips.

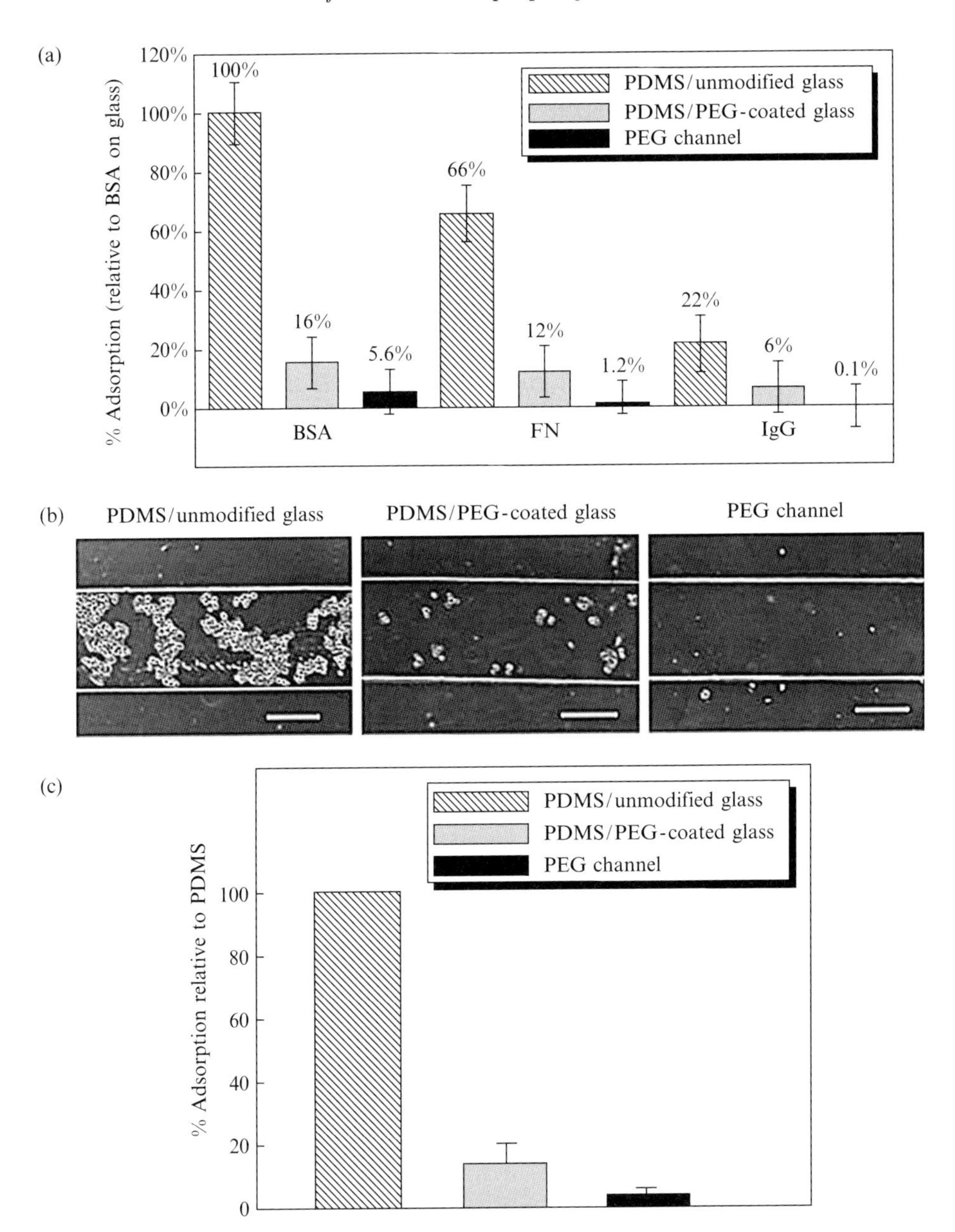

FIG. 11.5: (a) A quantitative analysis of the fluorescent images for protein adsorption where BSA, FN, and IgG were flowed inside three types of 200 μm channels with 80 μm height (PDMS/glass, PDMS/PEG-coated glass, and PEG channels/unmodified glass, respectively). (b) Optical micrographs for the adhesion of NIH-3T3 murine embryonic fibroblasts using the same channels. (c) A quantitative analysis indicates that the PEG channels were the most resistant against cell adhesion ($<\sim2\%$ compared to the PDMS channel).' Caption and figure reproduced with permission of The Royal Society of Chemistry (Kim *et al.* 2006).

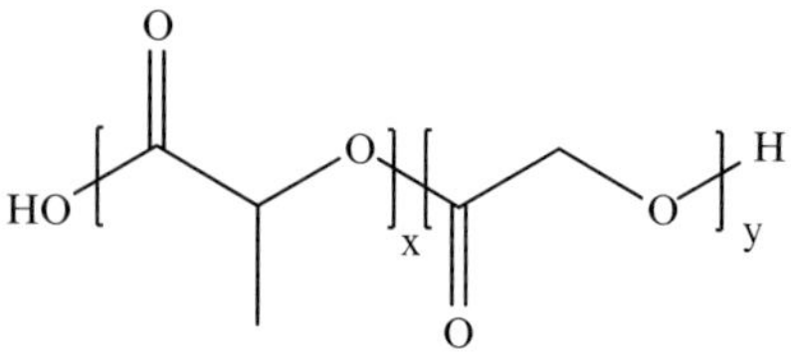

Fig. 11.6: Structure of PLGA used to prepare a degradable device for tissue engineering.

the polymers are also degradable. A device was made from a degradable polymer, poly(DL-lactic-*co*-glycolide) (PLGA) (Fig. 11.6) (King *et al.* 2004), and the material sealed using a thermal fusion bonding process. PLGA is FDA approved for many medical devices and is commonly used in tissue engineering, sutures, and other implantable materials. An important feature for biomaterials utilized as cell scaffolds or drug delivery devices is that nutrients are readily delivered by perfusion. The authors found that fluids passed through the networks without leaks. They suggested that this particular strategy may provide an implantable scaffold with cavities that mimic microvasculature. Another group prepared a device made entirely of calcium alginate (Cabodi *et al.* 2005). The hydrogel was sufficiently stable to form the device, but also permeable to diffusion of molecules. Thus, it would be possible to seed cells within the bulk of device. Devices were also fabricated from the biopolymer gelatin (Fig. 11.7) (Paguirigan and Beebe 2006). Cross-linking of the gelatin to render it suitable for device fabrication was accomplished using the enzyme transglutaminase, rather than by chemical methods. In this way, the entire device was composed of natural components and was directly amenable for integration of live cells.

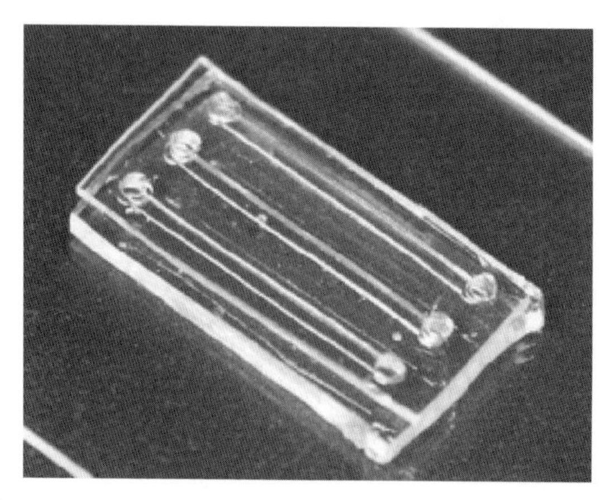

Fig. 11.7: A device consisting of an enzyme-cross-linked gelatin. Reproduced by permission of The Royal Society of Chemistry (Paguirigan and Beebe 2006).

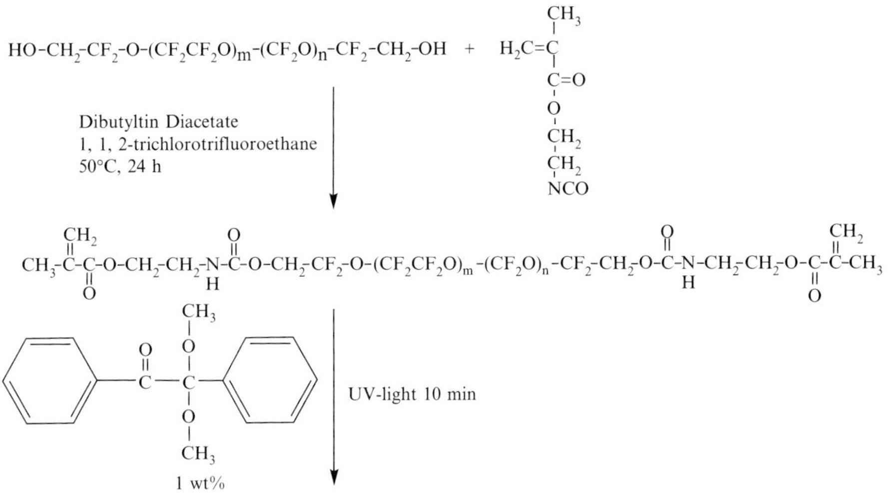

Fig. 11.8: Synthesis of a photocurable perfluoropolyether for preparing solvent-resistant microfluidic devices. Reproduced with permission from *J. Am. Chem. Soc.* Copyright 2004 American Chemical Society (Rolland *et al.* 2004).

A microfluidic device was prepared from a photocurable perfluoropolyether (PFPE) (Fig. 11.8) (Rolland *et al.* 2004). A commercially available PFPE diol was added to isocyanatoethyl methacrylate to form the dimethacrylate species. This was then cured by exposure to ultraviolet (UV) light in the presence of a photo-initiator, 2,2-dimethoxy-2-phenylacetophenone. The motivation behind this research was to prepare materials that do not swell in organic solvents. Upon testing, negligible swelling was observed when methylene chloride, acetonitrile, and methanol were perfused through the channels. Potential applications of the device include organic synthesis within the channels.

11.6.2 *Dynamic coatings (including oxidation)*

A simple method to alter the properties of a microfluidic device is to coat the surface. This can be accomplished after the device is fabricated by flow of the desired molecule through the channels. The surfaces can be pre-oxidized to help adhesion, and oxidation of PDMS surfaces is discussed below (Section 11.6.3). A wide variety of molecules and macromolecules have been adsorbed on the inside of channels including polymers, polysaccharides, and surfactants, and these dynamic surface coatings have been extensively reviewed (Belder and Ludwig 2003; Dolnik 2004; Makamba *et al.* 2003). A few unconventional approaches have also been taken. For example, citrate-stabilized gold nanoparticles were adsorbed onto glass surfaces coated with poly(diallyldimethylammonium chloride) (Pumera *et al.* 2001). Electrophoresis resolutions and plate numbers doubled with the nanoparticles present.

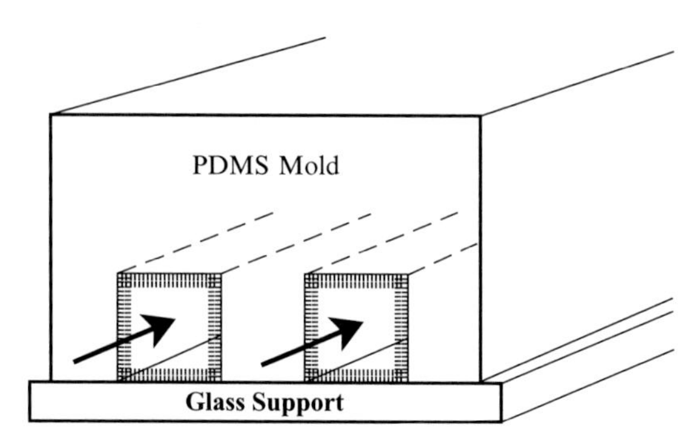

Fig. 11.9: Uniform bilayer formation following injection of small unilamellar vesicles into channels. Reproduced with permission from *Anal. Chem.* Copyright 2001 American Chemical Society (Yang *et al.* 2001).

The device was reproducible and stabile during storage for more than a month. Supported phospholipid bilayers on glass and PDMS microchannels were prepared using the vesicle fusion method (Fig. 11.9) (Mao *et al.* 2002; Yang *et al.* 2001). Uniform bilayers formed spontaneously inside the channels by injecting solutions containing small unilamellar vesicles (SUVs) into oxidized PDMS channels. Importantly, the SUVs were injected less than four minutes after oxidation of the PDMS to ensure that the resulting supported bilayer was uniform. The setup was flexible, and functionalized lipids were incorporated at low concentrations in order to provide reactive handles or ligands for bioconjugation.

11.6.3 *Covalent modifications: oxidation and small molecules*

Permanent alteration of surfaces is more difficult to achieve than simple adsorption, but often leads to more stable structures. The overall procedure and choice of small molecule for covalent modification depends primarily on the device substrate. A reactive surface is required. Some surfaces are inert and must be treated prior to incubation with a small molecule. For example, PDMS is often oxidized by subjection to oxygen plasma, ultraviolet light, and corona discharge (Makamba *et al.* 2003; Duffy *et al.* 1998; Thorslund and Nikolajeff 2007; Wang *et al.* 2003). More recently, a solution approach that employed aqueous hydrogen chloride and hydrogen peroxide ($H_2O/HCl/H_2O_2$) was reported (Sui *et al.* 2006). Oxidation of PDMS forms a hydrophilic layer that allows PDMS to be sealed to other materials and renders channels amenable to uptake of aqueous solutions. Importantly, reactive Si-OH or Si-O^- functionalities are formed. This treatment is transient and in many cases must be exploited quickly.

Small molecules are then incubated with the treated channels. Common species are trialkoxysilanes, monochlorosilanes, dichlorosilanes, and trichlorosilanes (Fig. 11.10) and the stabilities of films from these different coating compounds have been compared (Munro *et al.* 2001). These same compounds are useful for glass or Si/SiO_2 surfaces, where the surface is typically pre-activated with a basic

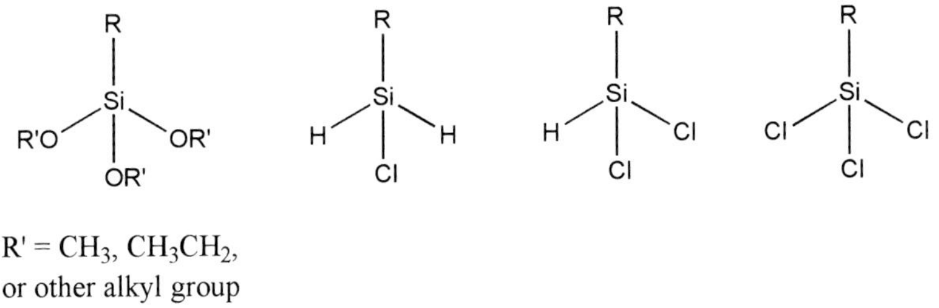

R
Si
R'O OR'
OR'

R
Si
H H
Cl

R
Si
H Cl
Cl

R
Si
Cl Cl
Cl

R' = CH_3, CH_3CH_2,
or other alkyl group

FIG. 11.10: Common reactive functionality to tether molecules to oxidized PDMS and glass surfaces. Left to right: trialkoxysilane, monochlorosilane, dichlorosilane, and trichlorosilane.

sodium hydroxide solution or hydrogen peroxide and sulfuric acid ('piranha') in order to maximize the silanol groups (Papra *et al.* 2001; Xiong and Regnier 2001).

Fortunately, many such silanes are available for purchase, and the versatility of these materials allows one to greatly vary the functionality on the surface. 3-Aminopropyl trimethoxysilane is a common reagent for introduction of amine groups for further reaction with biomolecules. The amines can be utilized directly for conjugation to carboxylic acids or can be further modified. Introduction of a bis-aldehyde species produced Schiff base-linked aldehydes on the amine surfaces (Fig. 11.11) (Xiong and Regnier 2001). The aldehydes were free to react with amino groups of biomolecules to form additional Schiff bases; the resulting imines were reduced to form a stable layer. In another example, incubation with thiophosgene produced isothiocyanate groups for direct reaction with amines (Fig 11.12) (Sui *et al.* 2006).

For devices made of PMMA, the ester groups are reactive and can be modified directly. Exposure of PMMA to 3-aminopropane that had been pretreated

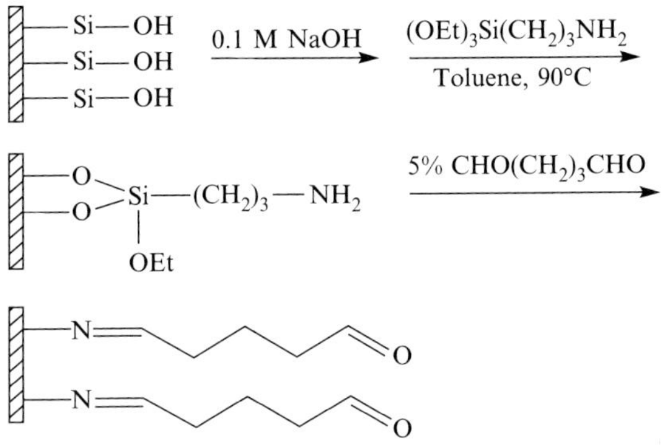

Si—OH
Si—OH
Si—OH
0.1 M NaOH
$(OEt)_3Si(CH_2)_3NH_2$
Toluene, 90°C

O
O
Si—$(CH_2)_3$—NH_2
OEt
5% $CHO(CH_2)_3CHO$

N
O
N
O

FIG 11.11: Modification with glass channels with 3-aminopropyl triethoxysilane, followed by glutaric dialdehyde. Reprinted from the *Journal of Chromatography A* Copyright (2001) with permission from Elsevier (Xiong and Regnier 2001).

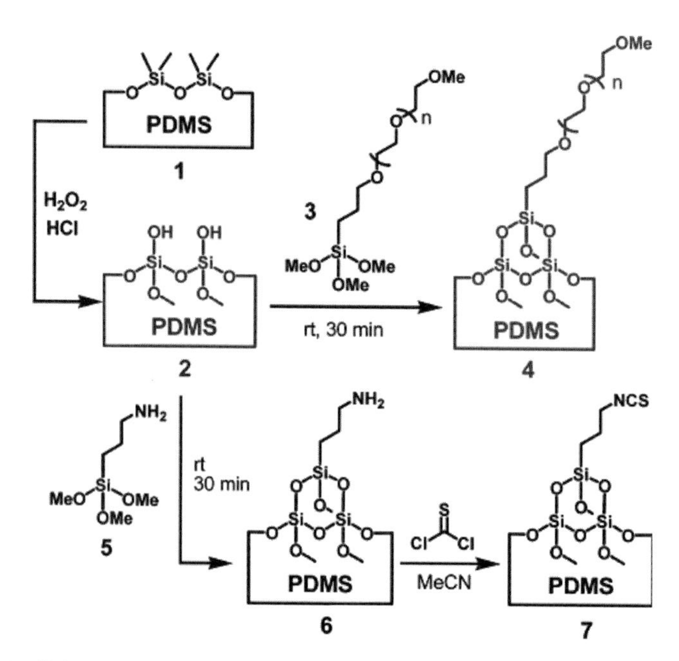

Fig 11.12: Solution treatment of PDMS channels and incubation with either 2-[methoxy(polyethylenxyl)propyl] trimethoxysilane (3) to form the PEG modified surface or 3-aminopropyl trimethoxysilane (5) followed by thiophosgene to form the amine-reactive isothiocyanate species. Reproduced with permission from *Anal. Chem.* Copyright 2006 American Chemical Society (Sui *et al.* 2006).

with butyl lithium to deprotonate the amine formed the propylamide species (Fig. 11.13) (Henry *et al.* 2000; Soper *et al.* 2002). Likewise, incubation with 1,3-diaminopropane formed channels with a high surface coverage of amine functionality. Alternatively, PMMA was exposed to UV light to generate acid functionality; amines were coupling utilizing *N*-(3-dimethyl-aminopropyl)-*N*-ethylcarbodiimide (EDC) to facilitate the amide formation (Llopis *et al.* 2007). PMMA was also exposed to base, such as sodium hydroxide, to generate the acid functionality (Bai *et al.* 2006).

Other reported strategies include chemical vapor deposition (CVD) onto the surface of a device to form a coating containing a reactive functional group. CVD of [2.2]paracyclophane pentafluorophenol ester (PPF) produced a film of poly (p-xylylene carboxylic acid pentafluorophenolester-*co*-*p*-xylylene) (PPX-PPF) on PDMS devices (Fig. 11.14) (Lahann *et al.* 2001a, 2003). The deposited film contained active ester moieties for further conjugation with amines.

11.6.4 *Covalent modification: polymers and hydrogels*

Similar to small molecule modification, covalent attachment of polymers to microfluidic device channels is common. This is typically accomplished in two ways:

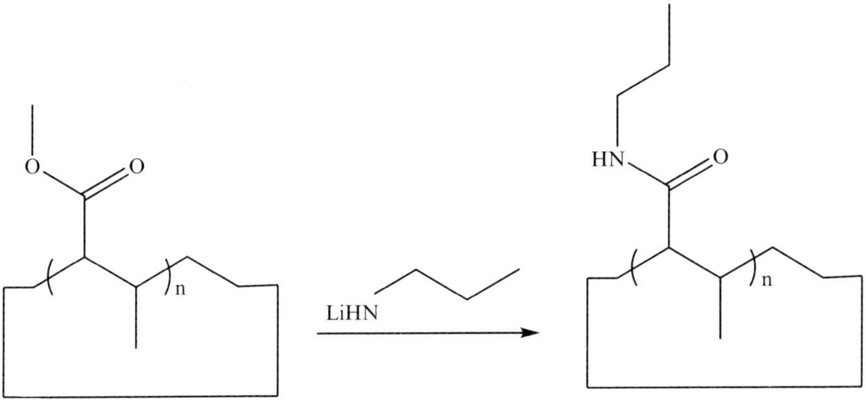

Fig. 11.13: Amidation of PMMA surfaces.

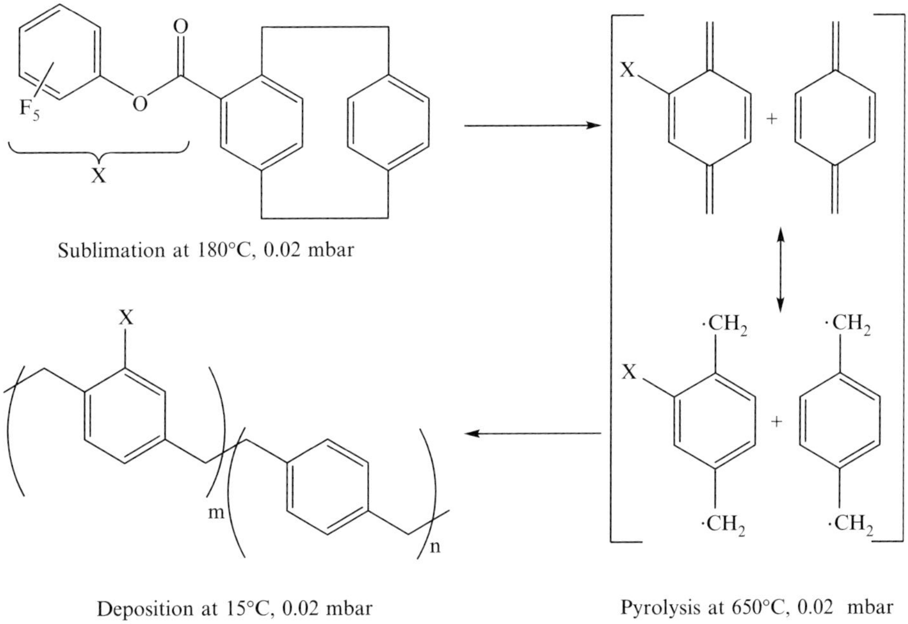

Fig. 11.14: Deposition of activated ester via CVD polymerization of PPF. Reproduced with permission from *Anal. Chem.* Copyright 2003 American Chemical Society (Lahann *et al.* 2003).

grafting to and grafting from the surface. The former is easier to carry out because the polymers are often commercially available, or if the polymers are made, they can be fully characterized prior to conjugation to the surface. The advantage of polymerizing from a surface is that high grafting densities or surface coverage can be achieved.

Grafting a pre-formed polymer is carried out using a polymer modified at the end or side chain with a specific group. Similar to small molecules, polymers with reactive silane end groups are a common reagent. For biomedical applications PEG silanes are used to reduce biofouling (Fig. 11.12) (Sui *et al.* 2006; Papra *et al.*

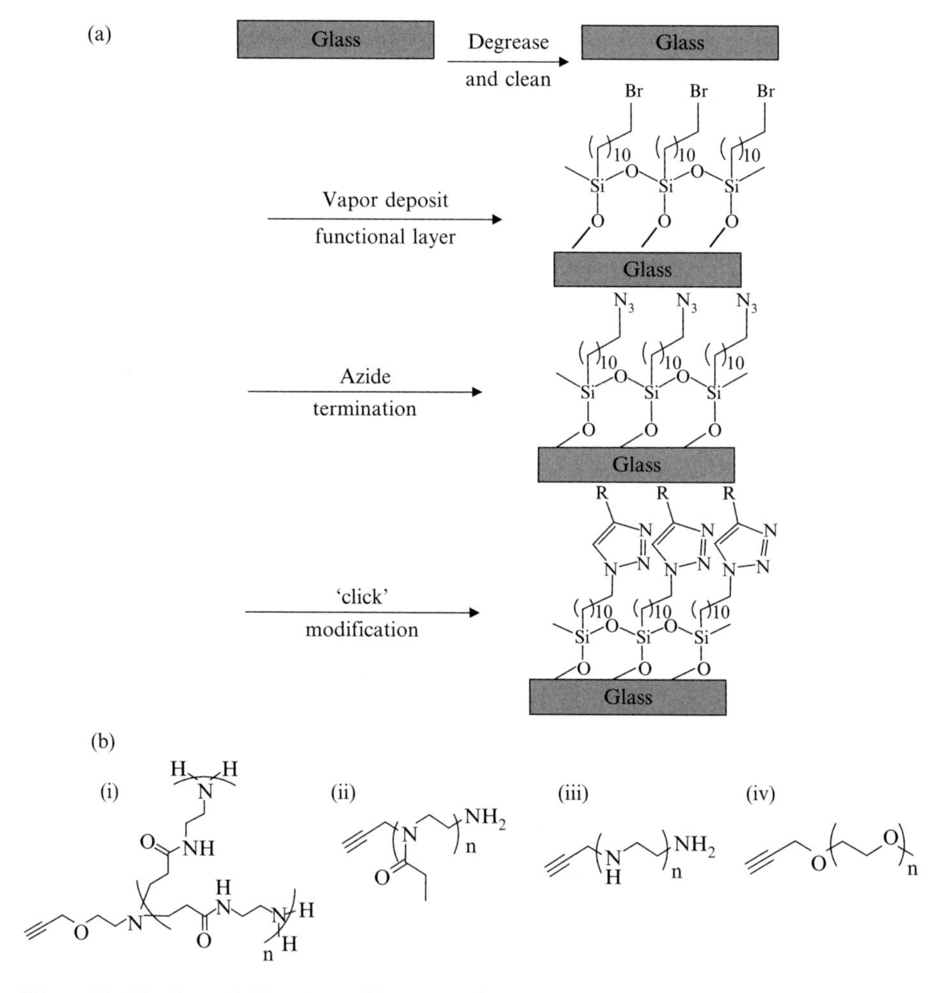

FIG. 11.15: Immobilization of linear polymers and dendrimers on microfluidic channels using 'click' chemistry. Reproduced with permission from *Anal. Chem.* Copyright 2007 American Chemical Society (Prakash *et al.* 2007).

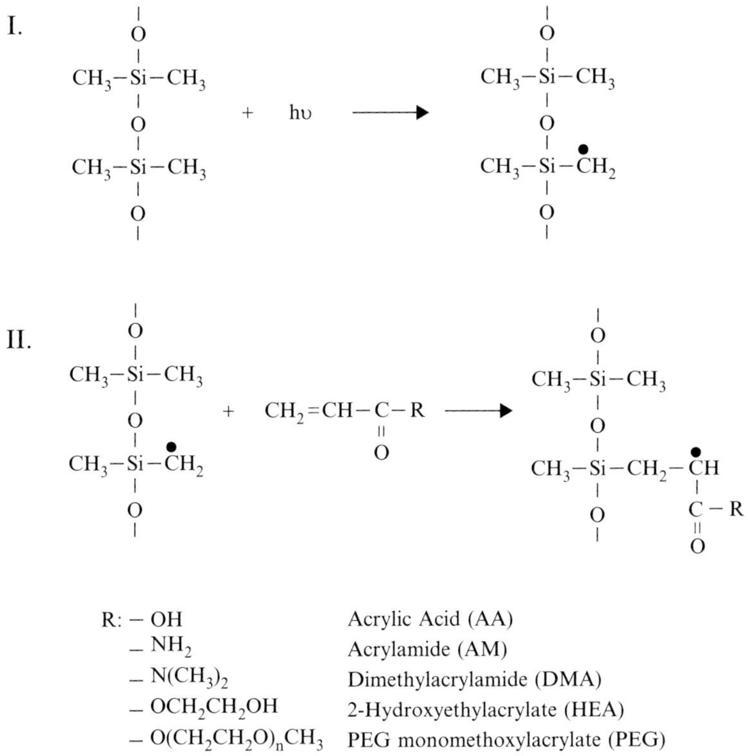

I. $-O-Si(CH_3)_2-O-Si(CH_3)_2-O-$ + hυ → $-O-Si(CH_3)_2-O-Si(CH_3)(\dot{C}H_2)-O-$

II. $-O-Si(CH_3)_2-O-Si(CH_3)(\dot{C}H_2)-O-$ + $CH_2{=}CH-C(=O)-R$ → $-O-Si(CH_3)_2-O-Si(CH_3)(CH_2-\dot{C}H-C(=O)-R)-O-$

R:	
$-OH$	Acrylic Acid (AA)
$-NH_2$	Acrylamide (AM)
$-N(CH_3)_2$	Dimethylacrylamide (DMA)
$-OCH_2CH_2OH$	2-Hydroxyethylacrylate (HEA)
$-O(CH_2CH_2O)_nCH_3$	PEG monomethoxylacrylate (PEG)

Fig. 11.16: Surface modification by in situ grafting to process. Reproduced with permission from *Anal. Chem.* Copyright 2002 American Chemical Society (Hu *et al.* 2002).

2001; Munro *et al.* 2001; Jon *et al.* 2003; Hellmich *et al.* 2005). A recent report took advantage of the copper-catalyzed Huisgen cycloaddition, which is called 'click' chemistry because of the high yields and robust nature of the reaction (Prakash *et al.* 2007). Sodium azide was introduced to a surface that had been pre-treated by vapor deposition of 11-bromoundecyl trichlorosilane (Fig. 11.15). The resulting azide formed via S_N2 nucleophilic substitution was then available for reaction with alkyne end-functionalized polymers. In this way, the surface was modified with different generation poly(amido amine) (PAMAM) dendrimers and linear polymers poly(2-ethyl-2-oxazoline) (PEOX), poly(ethylene imine) (PEI), and PEG. Some residual azide was observed with the dendrimer systems.

Forming the polymer layer by in situ polymerization is carried out by a variety of methods. The simplest strategy is to synthesize the polymer and graft to the surface simultaneously. For example, PDMS microfluidics were modified in one step with poly(acrylic acid) (polyAA), poly(acrylamide) (polyAM), poly(dimethylacrylamide) (polyDMA), poly(2-hydroxylethyl acrylate) (polyHEA), and PEG

(Fig. 11.16) (Hu *et al.* 2002). Exposure to UV light polymerized the monomer and generated radicals on the device surface for covalent conjugation. In a similar fashion, plasma polymerization formed poly(AA) coatings on PDMS (Barbier *et al.* 2006). A resin-gas injection technique was effective to introduce a layer of poly(2-hydroxyethyl methacrylate) (poly(HEMA)) onto PMMA devices (Lai *et al.* 2004).

An alternative way to generate a polymer coating is to introduce the monomer or initiator onto the surface prior to polymerization (Llopis *et al.* 2007; Ebara *et al.* 2007; Stachowiak *et al.* 2007; Idota *et al.* 2005). The latter approach has been exploited to grow polymers by controlled/'living' polymerization methods (CRPs) such as atom transfer radical polymerization (ATRP) (Kamigaito *et al.* 2001; Matyjaszewski and Xia 2001). The advantage of these methods over conventional radical polymerizations is that the resulting polymer has a narrow molecular weight distribution and the molecular weight is determined by the initial conditions. This allows for control over the thickness of the polymer brush layer and provides uniform coatings. An alkyl halide is typically employed as the initiator in ATRP and a reversible redox reaction catalyzed by a metal halide mediates the polymerization. PDMS was modified with a chloromethylbenzene initiator prior to ATRP of acrylamide to form hydrophilic devices (Fig. 11.17) (Xiao *et al.* 2002). Electrophoretic separations of proteins using this modified device were demonstrated (Xiao *et al.* 2004). ATRP of PEGMA from PMMA surfaces modified with a 2-bromoisobutyryl bromide initiator was also demonstrated (Fig. 11.18) (Liu *et al.* 2004; Sun *et al.* 2008). Grafted films of PEGMA/MMMA copolymers

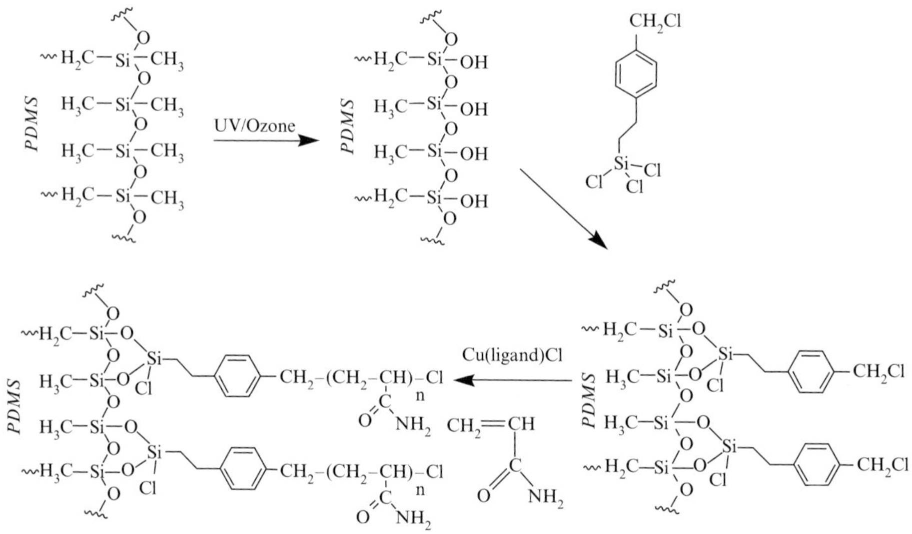

Fig 11.17: ATRP of acrylamide from an initiator modified PDMS surface. Reproduced with permission from *Langmuir* Copyright 2002 American Chemical Society (Xiao *et al.* 2002).

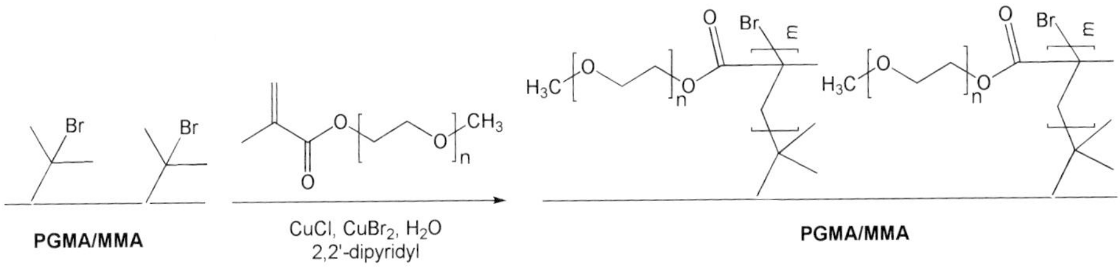

Fig. 11.18: ATRP of PEGMA from a modified PMMA surface. Reproduced with alteration with permission from *Anal. Chem.* Copyright 2008 American Chemical Society (Sun *et al.* 2008).

exhibited protein resistance, but also appeared to form hydrophobic domains that reduced their protein-repelling character (Stadler *et al.* 2008). CRPs using a photolabile, dithiocarbamate initiator bound to the surface were also reported (Hutchison *et al.* 2004; Sebra *et al.* 2006).

11.7 Immobilization and capture of proteins and peptides

Microfluidic devices that are modified inside the channels with proteins, peptides, and antibodies are utilized in numerous applications including medical diagnostics, immunoassays, proteomics, and ligand discovery (Freire and Wheeler 2006; Lion *et al.* 2003; Tokeshi *et al.* 2003). These chips have certain advantages over traditional systems, for example microarrays, in that the entire sample processing and analysis can be automated on a single device (Situma *et al.* 2006). With the chemical strategies described above, immobilization of proteins and peptides is straightforward.

Gradients of proteins were prepared by a simple physical adsorption process (Jiang *et al.* 2005). The device was oxidized by air plasma prior to formation of a gradient of avidin with bovine serum albumin as a blocking agent. Avidin, neutravidin, and streptavidin have four binding sites and bind with exceptionally high affinity ($K_a = 10^{15}$ M^{-1}) to the ligand biotin; thus these proteins are often employed as adaptors between surfaces and biomolecules of interest (Weber *et al.* 1989). Indeed, the device was used to immobilize biotinylated DNA and dextran gradients. Proteins laminin and fibronectin from the extracellular matrix (ECM) were also directly adsorbed. These are useful to study effects of gradients on cell behavior. In another example, biotinylated goat anti-mouse immunoglobulin G (IgG) was adsorbed, this time onto native PDMS (Linder *et al.* 2001, 2002). Neutravidin was added as a linker for immobilization of biotinylated goat anti-human IgG. The background was passivated with dextran, and the immobilized probe bound to human IgG with a signal to noise ratio of greater than 200:1.

Dynamic coatings of lipid bilayers are easily modified to support protein immobilization by incorporation of a small percentage of a modified lipid. These dynamic coatings have several advantages including the ability of the two-dimensional fluid to reorganize and to accommodate membrane proteins. Bilayers prepared from dinitrophenyl (DNP) capped lipids supported binding of anti-DNP antibodies (Yang *et al.* 2001), whereas biotinylated bilayers bound streptavidin (Mao *et al.* 2002). The latter was exploited to immobilize biotinylated alkaline phosphatase (Fig. 11.19) that turned over a fluorescent substrate approximately a factor of six lower than the solution enzyme. Lipid coatings were utilized to make a simple glucose sensor by placing glucose oxidase and horseradish peroxidase in series (Mao *et al.* 2002) and for detection of cholera toxin by incorporation of the GM1 (monosialotetrahexosylganglioside) receptor into the bilayer (Taylor *et al.* 2007).

Covalent coatings are also amenable to biomolecule conjugation. Amines, aldehydes, isothiocyanate, and other reactive functionality introduced onto the channel surfaces as described above are readily modified with proteins and peptides. For example, it was determined that amidation of a PMMA surface with PEI provided a substrate with ten times more active antibodies on the surface than with simple small molecule diamines. This was attributed to the amplification of the functional groups and greater retention of bioactivity because of the polymeric spacer (Bai *et al.* 2006). In another example, antibodies or peptides were grafted to isothiocyanate-modified PDMS microchannels (Sui *et al.* 2006). The cell adhesive peptide RGD was immobilized on the surface. This peptide is derived from the ECM protein fibronectin and binds to a class of cell surface receptors called integrins. Colon cancer cell adhesion was greatly enhanced on the inside of RGD-modified channels compared to PDMS, PDMS-amine, and PDMS-PEG surfaces (Fig. 11.20). Other groups have investigated cell properties of different integrin ligands and fluid shear stress on selective cell capture (Plouffe *et al.* 2007). This is important for cell enrichment and tissue engineering, in addition to fundamental studies on the effect of shear on receptor ligand interactions. Using spatially addressable solid-phase peptide synthesis, peptide arrays were synthesized and used to screen for novel ligands of murine B lymphoma cells (Mandal *et al.* 2007). This combination of diversity oriented synthesis and microfluidics is likely to be powerful for disease marker identification for drug delivery applications.

'Smart' polymer modification of microfluidic devices has also been carried out for reversible protein conjugation. Poly(*N*-isopropylacrylamide) (polyNIPAAm) is within this class of polymers and responds to changes in temperature (Alarcon *et al.* 2005). Above the lower critical solution temperature (LCST) of approximately 32°C, the polymer undergoes a hydrophobic collapse and phase separates from water; below this temperature the polymer is hydrophilic and soluble. The transition is reversible and results in dramatic alterations of the surface contact angle of water from 30°C below the LCST to close to 90°C above the LCST (Huber *et al.* 2003). This temperature-induced hydrophobicity change was exploited for the capture and release of proteins within a microfluidic device (Huber *et al.* 2003). The film was generated by exploiting chain transfer between solution initiators and surface-bound thiols to produce a high grafting density of polyNIPAAm. The adsorption and release of proteins myoglobin, BSA, hemoglobin, and cytochrome C onto individual heater lines within the device was demonstrated. Others have

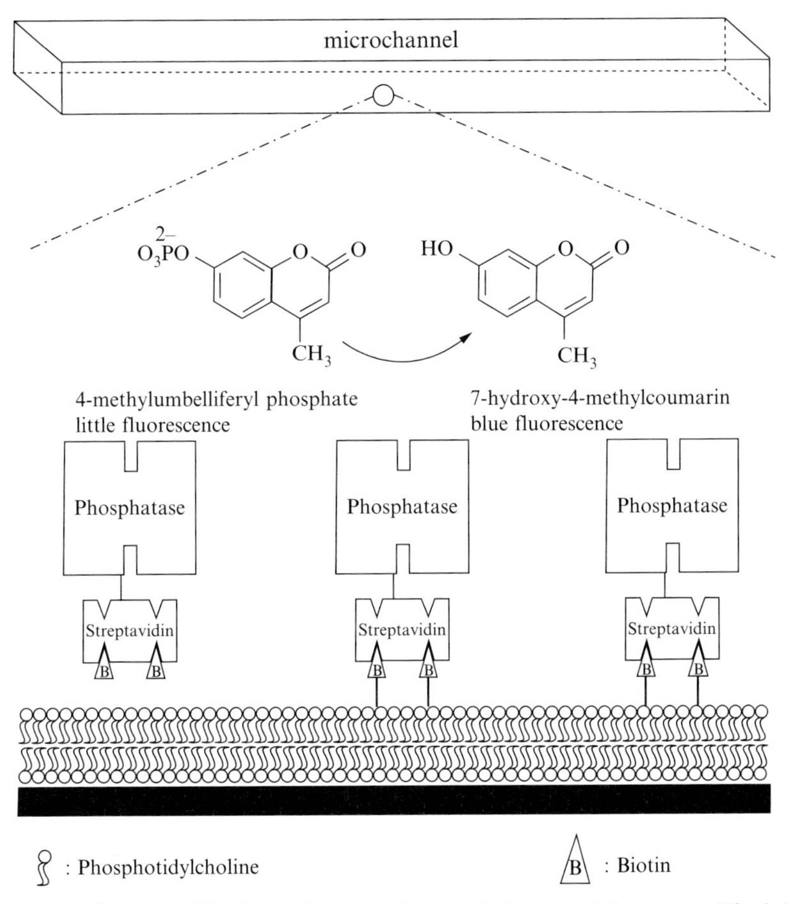

Fig: 11.19: Streptavidin in used as an adaptor between a bilayer modified device channel ad an active enzyme. Reproduced with permission from *Anal. Chem.* Copyright 2002 American Chemical Society (Mao *et al.* 2002).

shown that streptavidin coated polyNIPAAm beads bound to heated poly(ethylene terephthalate) (PET) devices (Malmstadt *et al.* 2003). In another configuration, the same group formed a polyNIPAAm-coated device by photo-grafting the polymer; polyNIPAAm beads reversibly adhered to the 'smart' polymer surfaces (Ebara *et al.* 2006). These strategies are useful for enrichment of proteins from samples and for forming highly selective bioassays that are reusable.

11.8 Integrating live cells with devices

The size, transparency, and straightforward modification of microfluidic devices permit integration with live cells (Sia and Whitesides 2003; Whitesides 2006). This is an exciting avenue of research, and devices can be applied in cell sorting, single cell manipulation, proteomics, tissue engineering, cell based screening assays, and for cell biology investigations (Whitesides 2006; Price and Culbertson 2007; Weibel and Whitesides 2006). Some of this research is highlighted below.

Microfluidic devices are excellent tools to systematically study cell behavior both in two dimensional and three dimensional formats. Gradients of chemotactic factors were easily generated and migration and differentiation of cells investigated (Chung *et al.* 2005; Jeon *et al.* 2002). Other important cell behaviors, such as cell dependent adhesion response to shear, were readily observed (Lu *et al.* 2004). Three-dimensional cell culture systems allowed for co-culture of multiple cell types in order to reproduce some activity of whole tissues including liver (Gottwald *et al.* 2007). Arrays of micropillars within channels provided optimized cell–matrix and cell–cell contacts, and were used to entrap primary hepatocytes and mesenchymal stem cells (Toh *et al.* 2007). Compared to two dimensional culture systems, the cells in these three dimensional environments produced equal or greater concentrations of markers of hepatocyte function and differentiation, respectively. Cells were seeded within hydrogel scaffolds inside microfluidic channels in order to produce a three dimensional microenvironment (Kim *et al.* 2007). These devices are useful for both fundamental studies and applied tissue engineering.

Microfluidic devices are an ideal platform for high throughput cell-based screening because small molecules are readily introduced in a dose and time dependent manner to cells adhered inside the channels. An in vitro assay to study the effect of the disintegrin echistatin was developed by modifying PPX-PPF-coated devices with antibodies that bind to integrins (Lahann *et al.* 2003). The minimum dose required to disrupt endothelial cell adhesion was easily ascertained by counting adherent cells after perfusion with the drug. The dose dependent activity on inhibition of adhesion was determined. Cancer drug screening on live cells was performed in a biomimetic multiplexed tissue culture system (Lee *et al.* 2007). It was demonstrated that in solid tumor-like cultures of HeLa cancer cells, paclitaxel had no measureable toxicity effect until day three. This was different to what was observed in two dimensional culture formats where cytotoxicity was observed at day one. The authors suggested that the microfluidic environment was able to mimic key properties of tumors, and that the device might be useful as an in vitro model of solid tumors.

The systems described above all contained assembles of cells. Yet, with microfluidic devices, bioassays can be translated to single cells because single cells are readily isolated and perfused with reagents (Roman *et al.* 2007). For example, a live-dead assay was performed (Fig. 11.21), and intracellular calcium ion concentrations were measured on a Jurkat T-cell (Wheeler *et al.* 2003). The extracellular potentials from a cardiac myocyte was also investigated (Werdich *et al.* 2004).

Not only can single cells be studied, but reagents can be delivered to subcellular micro domains with precision (Whitesides 2006; Weibel and Whitesides 2006). This is incredibly useful to understand cell sub-structure and biology. Mitochondria in different areas of a single endothelial cell were labeled with different colored fluorescent dyes and intermixing of the dyes was studied over

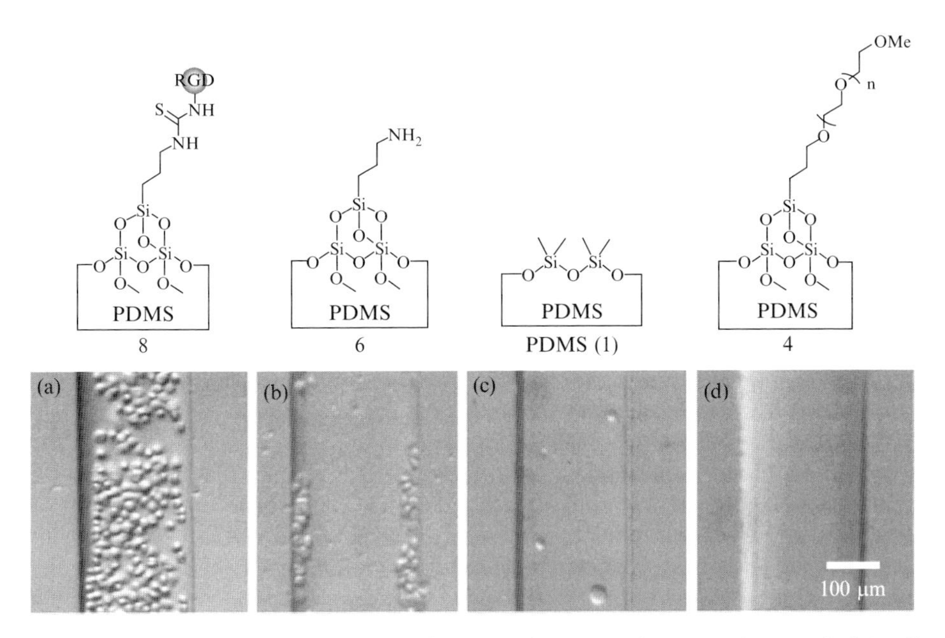

Fig: 11.20: A427 cells binding on a) RGD, b) amine, c) unmodified, and d) PEG modified PDMS channels. Reproduced with permission from *Anal. Chem.* Copyright 2006 American Chemical Society (Sui *et al.* 2006).

several hours (Takayama *et al.* 2001). The microfluidic system was dubbed PARTCELL (partial treatment of cells using laminar flows) (Fig. 11.22). Movement of mitochondria, after disruption of actin filaments and attachment and respreading of a cell partially treated with trypsin to disrupt integrin binding to the surface, were also investigated (Takayama *et al.* 2003). Microfluidic devices were designed to isolate the somal (cell body) of a neuron from the neuritic outgrowths (Taylor *et al.* 2003). In this way different dyes were selectively delivered to neurite extensions. Subcellular environment manipulation around a single cardiac myocytes was also demonstrated (Klauke *et al.* 2007). The effects of localized delivery of reagents to cause membrane permeabilization, to initiate monophasic action potentials, and to generate intracellular ions were investigated. Opposite to subcellular delivery, microfluidic devices have also been shown to be useful in controlling the environment at different parts of embryos (Lucchetta *et al.* 2005) and to trap and study whole organisms such as *C. elegans* (Chronis *et al.* 2007).

11.9 Conclusions

In biosensors and microfluidics, the choice of materials often centers on the ease or feasibility of fabricating the device, and secondarily on optimizing the biomolecule–surface interactions. In this chapter, we hope to have shown that by controlling surface chemistry, it is possible not only to minimize problems, but to enable new

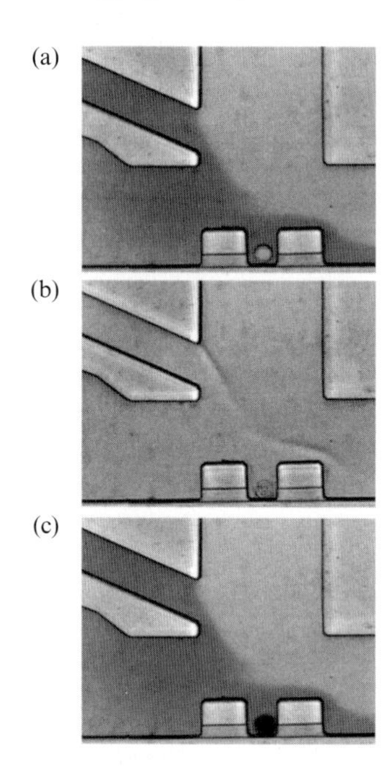

Fig. 11.21: An isolated Jurkat T-cell is first exposed to (a) trypan blue, then (b) methanol, and finally (c) trypan blue. Trypan blue is a dead cell stain and the cell is colored only after exposure to methanol. Reproduced with permission from *Anal. Chem.* Copyright 2003 American Chemical Society (Wheeler *et al.* 2003).

features, functions, and applications. An understanding of protein–surface interactions and surface wettability provides the foundation for designing surfaces that minimize non-specific adsorption of biomolecules and cells. A wide range of techniques is now available for characterizing surfaces before and after chemical modifications, and for characterizing the rate and extent of biomolecular adsorption.

In recent years, diverse chemical approaches have been developed to enable the immobilization of biomolecules and cells on almost any device surface, to pattern surfaces to achieve temporal and spatial control of adsorption, and to selectively adsorb different species in prescribed locations. Together, these methods represent a toolbox for designing new sensors and devices for diverse biomedical and bioanalytical applications.

Acknowledgments

HDM appreciates the National Science Foundation through SINAM (DMI-0327077) for funding. RLG acknowledges support from the National Institutes of Health (RR20070). The authors thank Gregory Grover (UCLA) for obtaining the copyright permissions.

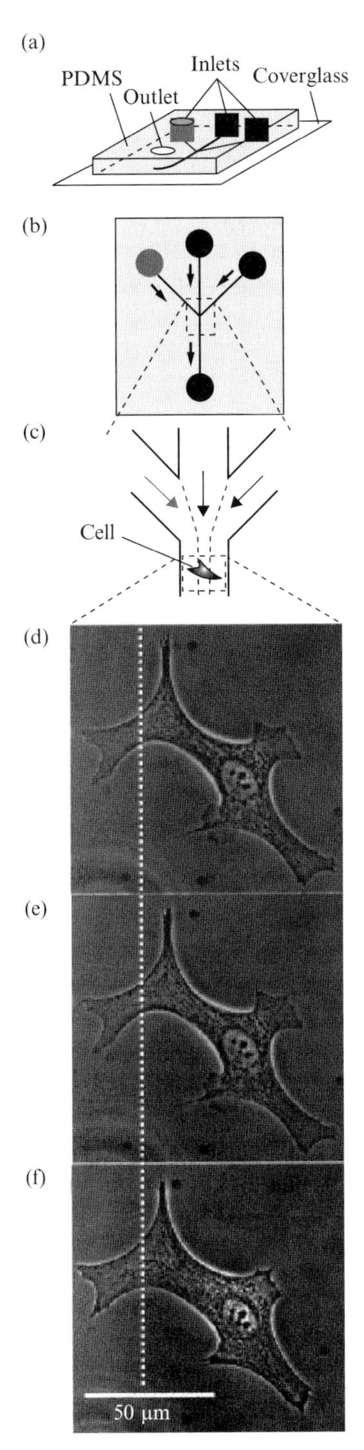

Fig. 11.22: The mitochondria in one part of a single cell is stained with a fluorescent dye using PARTCELL, and the movement is observed over 11 minutes. Reprinted from *Chemistry and Biology*, Takayama, S. *et al.* 2003. Copyright (2003) with permission from Elsevier (Takayama *et al.* 2003).

References

Aangi, R., and Engberts, J. B. F. N., 2005. 'Physisorption of hydroxide ions from aqueous solution to a hydrophobic surface.' *Journal of the American Chemical Society*, **127**, pp. 2272–2276.

Alarcon, C. D. H., Pennadam, S., and Alexander, C., 2005. 'Stimuli responsive polymers for biomedical applications.' *Chemical Society Reviews*, **34**, pp. 276–285.

Alentiev, A. Y., Shantarovich, V. P., Merkel, T. C., Bondar, V. I., Freeman, B. D., and Yampolskii, Y. P., 2002. 'Gas and vapor sorption, permeation and diffusion in glassy amorphous Teflon Af 1600.' *Macromolecules*, **35**, pp. 9513–9522.

Amiji, M., and Park, K., 1992. 'Prevention of Protein adsorption and platelet adhesion on surfaces by PEO/POP/PEO triblock copolymers.' *Biomaterials*, **13**, pp. 682–692.

Bai, Y. L., Koh, C. G., Boreman, M., Juang, Y. J., Tang, I. C., Lee, L. J. *et al.*, 2006. 'Surface modification for enhancing antibody binding on polymer-based microfluidic device for enzyme-linked immunosorbent assay.' *Langmuir*, **22**, pp. 9458–9467.

Barbier, V., Tatoulian, M., Li, H., Arefi-Khonsari, F., Ajdari, A., and Tabeling, P., 2006. 'Stable modification of PDMS surface properties by plasma polymerization: Application to the formation of double emulsions in microfluidic systems.' *Langmuir*, **22**, pp. 5230–5232.

Belder, D., and Ludwig, M., 2003. 'Surface modification in microchip electrophoresis.' *Electrophoresis*, **24**, pp. 3595–3606.

Belengrinou, S., Mannelli, I., Lisboa, P., Bretagnol, F., Valsesia, A., Ceccone, G. *et al.*, 2008. 'Ph-dependent immobilization of proteins on surfaces functionalized by plasma-enhanced chemical vapor deposition of poly(acrylic acid)- and poly(ethylene oxide)-like films.' *Langmuir*, **24**, pp. 7251–7261.

Benhabbour, S. R., Liu, L., Sheardown, H., and Adronov, A., 2008. 'Protein resistance of surfaces prepared by chemisorption of monothiolated poly(ethylene glycol) to gold and dendronization with aliphatic polyester dendrons: Effect of hydrophilic dendrons.' *Macromolecules*, **41**, pp. 2567–2576.

Bera, M., Rivaton, A., Gandon, C., and Gardette, J. L., 2000. 'Photooxidation of poly(para-xylylene).' *European Polymer Journal*, **36**, pp. 1753–1764.

Bico, J., Marzolin, C., and Quére, D., 1999. 'Pearl drops.' *Europhysics Letters*, **47**, pp. 220–226.

Brash, J. L., and Horbett, T. A., eds. 1987. American Chemical Society symposium series, *Proteins at Interfaces. Physicochemical and Biochemical Studies.* Washington DC, p. 706.

Cabodi, M., Choi, N. W., Gleghorn, J. P., Lee, C. S. D., Bonassar, L. J., and Stroock, A D., 2005. 'A microfluidic biomaterial.' *Journal of the American Chemical Society*, **127**, pp. 13788–13789.

Cassie, A. B. D., and Baxter, S., 1944. 'Wettability of porous surfaces.' *Transactions of the Faraday Society*, **40**, pp. 546–541.

Chang, T. Y., Yadav, V., De Leo, S., Mohedas, A., Rajalingam, B., Chen, C. L. *et al.*, 2007. 'Cell and protein compatibility of parylene-C surfaces.' *Langmuir*, 23, pp. 11718–11725.

Chapman, R. G., Ostuni, E., Takayama, S., Holmlin, R. E., Yan, L., and Whitesides, G M., 2000. 'Surveying for surfaces that resist the adsorption of proteins.' *Journal of the American Chemical Society*, **122**, pp. 8303–8304.

Chen, W., Fadeev, A. Y., Hsieh, M. C., Oner, D., Youngblood, J. P., and McCarthy, T J., 1999. 'Ultrahydrophobic and ultralyophobic surfaces: Some comments and examples.' *Langmuir*, **15**, pp. 3395–3399.

Chronis, N., Zimmer, M., and Bargmann, C. I., 2007. 'Microfluidics for in vivo imaging of neuronal and behavioral activity in *Caenorhabditis elegans*.' *Nature Methods*, **4**, pp. 727–731.

Chung, B. G., Flanagan, L. A., Rhee, S. W., Schwartz, P. H., Lee, A. P., Monuki, E. S. *et al.*, 2005. 'Human neural stem cell growth and differentiation in a gradient-generating microfluidic device.' *Lab on a Chip*, **5**, pp. 401–406.

Cottin-Bizonne, C., Barrat, J. L., Bocquet, L., and Charlaix, E., 2003. 'Low-friction flows of liquid at nanopatterned interfaces.' *Nature Materials*, **2**, pp. 237–240.

Davies, J., Maynes, D., Webb, B. W., and Woolford, B., 2006. 'Laminar flow in a microchannel with superhydrophobic walls exhibiting transverse ribs.' *Physics of Fluids*, **18**, 087110/087111–087110/087111.

Dill, K. A., and Brombert, S., 2003. *Molecular Driving Forces: Statistical Thermodynamics in Chemistry & Biology*. Garland Science, New York, p. 666.

Dolnik, V., 2004. 'Wall coating for capillary electrophoresis on microchips.' *Electrophoresis*, **25**, pp. 3589–3601.

Drelich, J., Laskowski, J. S., and Mittal, K. L., ed. 2000. *Apparent and Microscopic Contact Angles*. VSP., Zeist, The Netherlands, p. 522.

Duffy, D. C., McDonald, J. C., Schueller, O. J. A., and Whitesides, G. M., 1998. 'Rapid prototyping of microfluidic systems in poly(dimethylsiloxane).' *Analytical Chemistry*, **70**, pp. 4974–4984.

Dyer, M. A., Ainslie, K. M., and Pishko, M. V., 2007. 'Protein adhesion on silicon-supported hyperbranched poly(ethylene glycol) and poly(allylamine) thin films.' *Langmuir*, **23**, pp. 7018–7023.

Ebara, M., Hoffman, J. M., Hoffman, A. S., and Stayton, P. S., 2006. 'Switchable surface traps for injectable bead-based chromatography in PDMS microfluidic channels.' *Lab on a Chip*, **6**, pp. 843–848.

Ebara, M., Hoffman, J. M., Stayton, P. S., and Hoffman, A. S., 2007. 'Surface modification of microfluidic channels by UV-mediated graft polymerization of non-fouling and 'smart' polymers.' *Radiation Physics and Chemistry*, **76**, pp. 1409–1413.

Fortin, J. G., and Lu, T. M., 2001. 'Ultraviolet radiation induced degradation of poly-para-xylylene (parylene) thin films.' *Thin Solid Films*, **397**, pp. 223–228.

Freire, S. L. S., and Wheeler, A. R., 2006. 'Proteome-on-a-chip: Mirage, or on the horizon?' *Lab on a Chip*, **6**, pp. 1415–1423.

Gao, L., and McCarthy, T. J., 2008. 'Teflon is hydrophilic. Comments on definitions of hydrophobic, shear versus tensile hydrophobicity, and wettability characterization.' *Langmuir*, **24**, pp. 9183–9188.

Ge, S., Kojio, K., Takahara, A., and Kajiyama, T., 1998. 'Bovine serum albumin adsorption onto immobilized organotrichlorosilane surface: Influence of the phase separation on protein adsorption patterns.' *Journal of Biomaterials Science, Polymer Edition*, **9**, pp. 131–150.

Gölander, C. G., Lin, Y. S., Hlady, V., and Andrade, J. D., 1990. 'Wetting and plasma-protein adsorption studies using surfaces with a hydrdophobicity gradient.' *Colloids and Surfaces*, **49**, pp. 289–302.

Gottwald, E., Giselbrecht, S., Augspurger, C., Lahni, B., Dambrowsky, N., Truckenmuller, R. *et al.*, 2007. 'A chip-based platform for the in vitro generation of tissues in three-dimensional organization.' *Lab on a Chip*, **7**, pp. 777–785.

Grant, W. H., Smith, L. E., and Strombert, R. R., 1977. 'Radiotracer techniques for protein adsorption measurements.' *Journal of Biomedical Materials Research*, **11**, pp. 33–38.

Green, R. J., Frazier, R. A., Shakesheff, K. M., Davies, M. C., Roberts, C. J., and Tendler, S. J. B., 2000. 'Surface plasmon resonance analysis of dynamic biological interactions with biomaterials.' *Biomaterials*, **21**, pp. 1823–1835.

Hanein, Y., Pan, Y. V., Ratner, B. D., Denton, D. D., and Böhringer, K. F., 2001. 'Macromachining of non-fouling coatings for bio-MEMS applications.' *Sensors and Actuators B*, **81**, pp. 49–54.

Harder, P., Grunze, M., Dahint, R., Whitesides, G. M., and Laibinies, P. E., 1998. 'Molecular conformation in oligo(ethylene glycol)-terminated self-assembled monolayers on gold and silver surfaces determine their ability to resist protein adsorption.' *Journal of Physical Chemistry B*, 102, pp. 426–436.

Harris, J. M., Ed. 1992. *Poly(Ethylene Glycol) Chemistry: Biotechnical and Biomedical Applications.* Plenum Press, New York.

Hartland, S., ed. 2004. *Surface and Interfacial Tension: Measurement, Theory and Applications.* Marcel Dekker, New York, p. 619.

Haynes, C. A., and Norde, W., 1994. 'Globular proteins at solid/liquid interfaces.' *Colloids and Surfaces B: Biointerfaces*, **2**, pp. 517–566.

Hellmich, W., Regtmeier, J., Duong, T. T., Ros, R., Anselmetti, D., and Ros, A., 2005. 'Poly(oxyethylene) based surface coatings for poly(dimethylsiloxane) microchannels.' *Langmuir*, **21**, pp. 7551–7557.

Henry, A. C., Tutt, T. J., Galloway, M., Davidson, Y. Y., McWhorter, C. S., Soper, S. A., *et al.*, 2000. 'Surface modification of poly(methyl methacrylate) used in the fabrication of microanalytical devices.' *Analytical Chemistry*, **72**, pp. 5331–5337.

Herrera-Alonso, M., and McCarthy, T. J., 2004. 'Chemical surface modification of poly(p-xylylene) thin films.' *Langmuir*, **20**, pp. 9184–9189.

Herrwerth, S., Eck, W., Reinhardt, S., and Grunze, M., 2003. 'Factors that determine the protein resistance of oligoether self-assembled monolayers – internal hydrophilicity, terminal hydrophilicity, and lateral packing density.' *Journal of the American Chemical Society*, **125**, pp. 9359–9366.

Heuberger, M., Drobek, T., and Spencer, N. D., 2005. 'Interaction forces and morphology of a protein-resistant poly(ethylene glycol) layer.' *Biophysical Journal*, **88**, pp. 495–504.

Hiemenz, P. C., and Rajagopalan, R., 1997. *Principles of Colloid and Surface Chemistry.* Marcel Dekker, New York, p. 650.

Holden, M. A., and Cremer, P. S., 2005. 'Microfluidic tools for studying the specific binding, adsorption, and displacement of proteins at interfaces.' *Annual Review of Physical Chemistry*, **56**, pp. 369–387.

Horbett, T. A., and Brash, J. L., ed. 1995. *Proteins at Interfaces I: Fundamentals and Applications.* (Developed from a Symposium Sponsored by the Division of Colloid and Surface Science at the 207th National Meeting of the American Chemical Society, San Diego, California, March 13–17, 1994.) ACS., Washington, DC., p. 561.

Horbett, T. A., Ratner, B. D., Schakenraad, J. M., and Schoen, F. J., 1996. 'Some background concepts' in *Biomaterials Science: An Introduction to Materials in Medicine*, ed. B D Ratner. Academic Press, New York, p. 484.

Hu, S. W., Ren, X. Q., Bachman, M., Sims, C. E., Li, G. P., and Allbritton, N., 2002. 'Surface modification of poly(dimethylsiloxane) microfluidic devices by ultraviolet polymer grafting.' *Analytical Chemistry*, **74**, pp. 4117–4123.

Huber, D. L., Manginell, R. P., Samara, M. A., Kim, B. I., and Bunker, B. C., 2003. 'Programmed adsorption and release of proteins in a microfluidic device.' *Science*, **301**, pp. 352–354.

Hutchison, J. B., Haraldsson, K. T., Good, B. T., Sebra, R. P., Luo, N., Anseth, K. S. *et al.*, 2004. 'Robust polymer microfluidic device fabrication via contact liquid photolithographic polymerization (Clipp).' *Lab on a Chip*, **4**, pp. 658–662.

Idota, N., Kikuchi, A., Kobayashi, J., Sakai, K., and Okano, T., 2005. 'Microfluidic valves comprising nanolayered thermoresponsive polymer-grafted capillaries.' *Advanced Materials*, **17**, pp. 2723–2727.

Irene, E. A., 2008. *Surfaces, Interfaces and Thin Films for Microelectronics.* Wiley Interscience, Hoboken, NJ., p. 515.

Janocha, B., Hegemann, D., Oehr, C., Brunner, H., Rupp, F., and Geis-Gerstorfer, J., 2001. 'Adsorption of protein on plasma-polysiloxane layers of different surface energies.' *Surface and Coatings Technology*, 142–144, pp. 1051–1055.

Jeon, N. L., Baskaran, H., Dertinger, S. K. W., Whitesides, G. M., Van De Water, L., and Toner, M., 2002. 'Neutrophil chemotaxis in linear and complex gradients of interleukin-8 formed in a microfabricated device.' *Nature Biotechnology*, **20**, pp. 826–830.

Jiang, X. Y., Xu, Q. B., Dertinger, S. K. W., Stroock, A. D., Fu, T. M., and Whitesides, G. M., 2005. 'A general method for patterning gradients of biomolecules on surfaces using microfluidic networks.' *Analytical Chemistry*, **77**, pp. 2338–2347.

Jon, S., Seong, J., Khademhosseini, A., Tran, T. N. T., Laibinis, P. E., and Langer, R., 2003. 'Construction of nonbiofouling surfaces by polymeric self-assembled monolayers.' *Langmuir*, **19**, pp. 9989–9993.

Kamigaito, M., Ando, T., and Sawamoto, M., 2001. 'Metal-catalyzed living radical polymerization.' *Chemical Reviews*, **101**, pp. 3689–3745.

Kim, M. S., Yeon, J. H., and Park, J. K., 2007. 'A microfluidic platform for 3-dimensional cell culture and cell-based assays.' *Biomedical Microdevices*, **9**, pp. 25–34.

Kim, P., Jeong, H. E., Khademhosseini, A., and Suh, K. Y., 2006. 'Fabrication of non-biofouling polyethylene glycol micro- and nanochannels by ultraviolet-assisted irreversible sealing.' *Lab on a Chip*, **6**, pp. 1432–1437.

King, K. R., Wang, C. C. J., Kaazempur-Mofrad, M. R., Vacanti, J. P., and Borenstein, J. T., 2004. 'Biodegradable microfluidics.' *Advanced Materials*, **16**, 2007–2012.

Kingshott, P., and Griesser, H. J., 1999. 'Surfaces that resist bioadhesion.' *Current Opinion in Solid State and Materials Science*, **4**, pp. 403–412.

Klauke, N., Smith, G. L., and Cooper, J. M., 2007. 'Microfluidic partitioning of the extracellular space around single cardiac myocytes.' *Analytical Chemistry*, **79**, pp. 1205–1212.

Krishnan, A., Liu, Y. H., Cha, P., Woodward, R., Allara, D., and Vogler, E. A., 2005. 'An evaluation of methods contact angle measurement.' *Colloids and Surfaces B: Biointerfaces*, **43**, pp. 95–98.

Lahann, J., 2006. 'Reactive polymer coatings for biomimetic surface engineering.' *Chemical Engineering Communiations*, **193**, pp. 1457–1468.

Lahann, J., Klee, D., Bienert, T. H., Vorwerk, D., and Höcker, H., 1999. 'Improvement of haemocompatibility of metallic stents by polymer coating.' *Journal of Materials Science: Materials in Medicine*, **10**, pp. 443–448.

Lahann, J., Choi, I. S., Lee, J., Jensen, K. F., and Langer, R., 2001a. 'A new method toward microengineered surfaces based on reactive coating 13.' *Angewandte Chemie International Edition*, **40**, pp. 3166–3169.

Lahann, J., Klee, D., Pluester, W., and Hoecker, H., 2001b. 'Bioactive immobilization of R-hirudin on CVC-coated metallic implant devices.' *Biomaterials*, **22**, pp. 817–826.

Lahann, J., Balcells, M., Lu, H., Rodon, T., Jensen, K. F., and Langer, R., 2003. 'Reactive polymer coatings: A first step toward surface engineering of microfluidic devices.' *Analytical Chemistry*, **75**, pp. 2117–2122.

Lai, S. Y., Cao, X., and Lee, L. J., 2004. 'A packaging technique for polymer microfluidic platforms.' *Analytical Chemistry*, **76**, pp. 1175–1183.

Lander, L. M., Siewierski, L. M., Britain, W. J., and Vogler, E. A., 1993. 'A systematic comparison of contact angle methods.' *Langmuir*, **9**, pp. 2237–2239.

Lee, J. N., Park, C., and Whitesides, G. M., 2003. 'Solvent compatibility of poly (dimethylsiloxane)-based microfluidic devices.' *Analytical Chemistry*, **75**, pp. 6544–6554.

Lee, P. J., Gaige, T. A., Ghorashian, N., and Hung, P. J., 2007. 'Microfluidic tissue model for live cell screening.' *Biotechnology Progress*, **23**, pp. 946–951.

Lee, S., and Voeroes, J., 2005. 'An aqueous-based surface modification of poly (dimethylsiloxane) with poly(ethylene glycol) to prevent biofouling.' *Langmuir*, **21**, pp. 11957–11962.

Lenghaus, K., Dale, J. W., Henderson, J. C., Henry, D. C., Loghin, E. R., and Hickman, J. J., 2003. 'Enzymes as ultrasensitive probes for protein adsorption in flow systems.' *Langmuir*, **19**, pp. 5971–5974.

Li, L., Chen, S., Zheng, J., Ratner, B. D., and Jiang, S., 2005. 'Protein adsorption on oligo(ethylene glycol)-terminated alkanethiolate self-assembled monolayers: The molecular basis for nonfouling behavior.' *Journal of Physical Chemistry B*, **109**, pp. 2934–2941.

Linder, V., Verpoorte, E., de Rooij, N. F., Sigrist, H., and Thormann, W., 2002. 'Application of surface biopassivated disosable poly(dimethylsiloxane)/glass chips to a heterogeneous competitive human serum immunoglobulin g immunoassay with incorporated internal standard.' *Electrophoresis*, **23**, pp. 740–749.

Linder, V., Verpoorte, E., Thormann, W., de Rooij, N. F., and Sigrist, M., 2001. 'Surface biopassivation of replicated poly(dimethylsiloxane) microfluidic channels and application to heterogeneous immunoreaction with on-chip fluorescence detection.' *Analytical Chemistry*, **73**, pp. 4181–4189.

Lion, N., Rohner, T. C., Dayon, L., Arnaud, I. L., Damoc, E., Youhnovski, N. *et al.*, 2003. 'Microfluidic systems in proteomics.' *Electrophoresis*, **24**, pp. 3533–3562.

Liu, J., and Lee, M. L., 2006. 'Permanent surface modification of polymeric capillary electrophoresis microchips for protein and peptide analysis.' *Electrophoresis*, **27**, pp. 3533–3546.

Liu, J. K., Sun, X. F., and Lee, M. L., 2007. 'Adsorption-resistant acrylic copolymer for prototyping of microfluidic devices for proteins and peptides.' *Analytical Chemistry*, **79**, pp. 1926–1931.

Liu, J., Pan, T., Woolley, A. T., and Lee, M. L., 2004. 'Surface-modified poly (methyl methacrylate) capillary electrophoresis microchips for protein and peptide analysis.' *Analytical Chemistry*, **76**, pp. 6948–6955.

Llopis, S. L., Osiri, J., and Soper, S. A., 2007. 'Surface modification of poly(methyl methacrylate) microfluidic devices for high-resolution separations of single-stranded DNA.' *Electrophoresis*, **28**, pp. 984–993.

Locascio, L. E., Henry, A. C., Johnson, T. J., and Ross, D., 2003. 'Surface chemistry in polymer microfluidic systems.' In *Lab-on-a-Chip*, ed. R E Oosterbroek, R. E. and A. van den Berg. Elsevier B. V., Amsterdam, The Netherlands, p. 402.

Locascio, L. E., Hong, J. S., and Gaitan, M., 2002. 'Liposomes as signal amplification reagents for bioassays in microfluidic channels.' *Electrophoresis*, **23**, pp. 799–804.

Lu, H., Koo, L. Y., Wang, W. M., Lauffenburger, D. A., Griffith, L. G., and Jensen, K. F., 2004. 'Microfluidic shear devices for quantitative analysis of cell adhesion.' *Analytical Chemistry*, **76**, pp. 5257–5264.

Lubarsky, G. V., Browne, M. M., Mitchell, S. A., Davidson, M. R., and Bradley, R. H., 2005. 'Thevinfluence of electrostatic forces on protein adsorption.' *Colloids and Surfaces B: Biointerfaces*, **44**, pp. 56–63.

Lucchetta, E. M., Lee, J. H., Fu, L. A., Patel, N. H., and Ismagilov, R. F., 2005. 'Dynamics of drosophila embryonic patterning network perturbed in space and time using microfluidics.' *Nature*, **434**, pp. 1134–1138.

Luk, V. N., Mo, G. C., and Wheeler, A. R., 2008. 'Pluronic additives: A solution to sticky problems in digital microfluidics.' *Langmuir*, **24**, pp. 6382–6389.

Ma, M., and Hill, R. M., 2006. 'Superhydrophobic surfaces.' *Current Opinion in Colloid and Interface Science*, **11**, pp. 193–202.

Makamba, H., Kim, J. H., Lim, K., Park, N., and Hahn, J. H., 2003. 'Surface modification of poly(dimethylsiloxane) microchannels.' *Electrophoresis*, **24**, pp. 3607–3619.

Malmstadt, N., Yager, P., Hoffman, A. S., and Stayton, P. S., 2003. 'A smart microfluidic affinity chromatography matrix composed of poly(*N*-isopropylacrylamide)-coated beads.' *Analytical Chemistry*, **75**, pp. 2943–2949.

Malmsten, M., 1998. 'Formation of adsorbed protein layers.' *Journal of Colloid and Interface Science*, **207**, pp. 186–199.

Mandal, S., Rouillard, J. M., Srivannavit, O., and Gulari, E., 2007. 'Cytophobic surface modification of microfluidic arrays for in situ parallel peptide synthesis and cell adhesion assays.' *Biotechnology Progress*, **23**, pp. 972–978.

Mao, H. B., Yang, T. L., and Cremer, P. S., 2002. 'Design and characterization of immobilized enzymes in microfluidic systems.' *Analytical Chemistry*, **74**, pp. 379–385.

Marsh, L. H., Coke, M., Dettmar, P. W., Ewen, R. J., Havler, M., Nevell, T. G. *et al.*, 2002. 'Adsorbed poly(ethyleneoxide)-poly(propyleneoxide) copolymers on synthetic surfaces: Spectroscopy and microscopy of polymer structures and effects on adhesion of skin-borne bacteria.' *Journal of Biomedical Materials Research Part B: Applied Biomaterials*, **61**, pp. 641–652.

Masel, R. I., 1996. *Principles of Adsorption and Reaction on Solid Surfaces*. Wiley Interscience, New York, p. 804.

Matsuno, H., Nagasaka, Y., Kurita, K., and Serizawa, T., 2007. 'Protein adsorption is dependent on substrate polymer polymorphs.' *Chemistry Letters*, **36**, pp. 1238–1239.

Matyjaszewski, K., and Xia, J. H., 2001. 'Atom transfer radical polymerization.' *Chemical Reviews*, **101**, pp. 2921–2990.

McClellan, S. J., and Franses, E. I., 2005. 'Adsorption of bovine serum albumin at solid/aqueous interfaces.' *Colloids and Surfaces, A: Physicochemical and Engineering Aspects*, **260**, pp. 265–275.

McGurk, S. L., Green, R. J., Sanders, G. H. W., Davies, M. C., Roberts, C. J., Tendler, S. J. B. *et al.*, 1999. 'Molecular interactions of biomolecules with surface-engineered interfaces using atomic force microscopy and surface plasmon resonance.' *Langmuir*, **15**, pp. 5136–5140.

Michel, R., Pasche, S., Textor, M., and Castner, D. G., 2005. 'Influence of PEG architecture on protein adsorption and conformation.' *Langmuir*, **21**, pp. 12327–12332.

Miller, R., Fainerman, V. B., Makievski, A. V., Lser, M., Michel, M., and Aksenenko, E. V., 2004. 'Determination of protein adsorption by comparative drop and bubble profile analysis tensiometry.' *Colloids and Surfaces B: Biointerfaces*, **36**, pp. 123–136.

Mukhopadhyay, R., 2005. 'When microfluidic devices go bad.' *Analytical Chemistry*, **77**, 429A–432A.

Munro, N. J., Huhmer, A. F. R., and Landers, J. P., 2001. 'Robust polymeric microchannel coatings for microchip-based analysis of neat PCR products.' *Analytical Chemistry*, **73**, pp. 1784–1794.

Norde, W., 1995. 'Adsorption of proteins at solid-liquid interfaces.' *Cells and Materials*, **5**, pp. 97–112.

Nowlin, T. E., and Smith, D. F., 1980. 'Surface characterization of plasma-treated poly-P-xylylene films.' *Journal of Applied Polymer Science*, **25**, pp. 1619–1632.

Oner, D., and McCarthy, T. J., 2000. 'Ultrahydrophobic surfaces. Effects of topography length scales on wettability.' *Langmuir*, **16**, pp. 7777–7782.

Paguirigan, A., and Beebe, D. J., 2006. 'Gelatin based microfluidic devices for cell culture.' *Lab on a Chip*, **6**, pp. 407–413.

Papra, A., Bernard, A., Juncker, D., Larsen, N. B., Michel, B., and Delamarche, E., 2001. 'Microfluidic networks made of poly(dimethylsiloxane), Si, and Au coated with polyethylene glycol for patterning proteins onto surfaces.' *Langmuir*, **17**, pp. 4090–4095.

Pasche, S., Vörös, J., Griesser, H. J., Spencer, N. D., and Textor, M., 2005. 'Effects of ionic strength and surface charge on protein adsorption at pegylated surfaces.' *Journal of Physical Chemistry B*, **109**, pp. 17545–17552.

Pavey, K. D., and Olliff, C. J., 1999. 'SPR analysis of the total reduction of protein adsorption to surfaces coated with mixtures of long- and short-chain polyethylene oxide block copolymers.' *Biomaterials*, **20**, pp. 885–890.

Peterson, S. L., McDonald, A., Courley, P. L., and Sasaki, D. Y., 2005. 'Poly (dimethylsiloxane) thin films as biocompatible coatings for microfluidic devices: Cell culture and flow studies with glial cells.' *Journal of Biomedical Materials Research, Part A*, **72**, pp. 10–18.

Plouffe, B. D., Njoka, D. N., Harris, J., Liao, J. H., Horick, N. K., Radisic, M. *et al.*, 2007. 'Peptide-mediated selective adhesion of smooth muscle and endothelial cells in microfluidic shear flow.' *Langmuir*, **23**, pp. 5050–5055.

Popat, K. C., and Desai, T. A., 2004. 'Poly(ethylene glycol) interfaces: An approach for enhanced performance of microfluidic systems.' *Biosensors and Bioelectronics*, **19**, pp. 1037–1044.

Prakash, A. R., Amrein, M., and Kaler, K. V. I. S., 2008. 'Characteristics and impact of *taq* enzyme adsorption on surfaces in microfluidic devices.' *Microfluidics and Nanofluidics*, **4**, p. 295.

Prakash, S., Long, T. M., Selby, J. C., Moore, J. S., and Shannon, M. A., 2007. 'Click'' modification of silica surfaces and glass microfluidic channels.' *Analytical Chemistry*, **79**, pp. 1661–1667.

Price, A. K., and Culbertson, C. T., 2007. 'Strategies for culturing, sorting, trapping, and lysing cells and separating their contents on chips.' *Analytical Chemistry*, **79**, pp. 2615–2621.

Pruden, K. G., Sinclair, K., and Beaudoin, S., 2003. 'Characterization of parylene-N and parylene-C photooxidation.' *Journal of Polymer Science Part a-Polymer Chemistry*, **41**, pp. 1486–1496.

Pumera, M., Wang, J., Grushka, E., and Polsky, R., 2001. 'Gold nanoparticle-enhanced microchip capillary electrophoresis.' *Analytical Chemistry*, **73**, pp. 5625–5628.

Rolland, J. P., Van Dam, R. M., Schorzman, D. A., Quake, S. R., and DeSimone, J. M., 2004. 'Solvent-resistant photocurable "liquid teflon" for microfluidic device fabrication.' *Journal of the American Chemical Society*, **126**, pp. 2322–2323.

Roman, G. T., Chen, Y. L., Viberg, P., Culbertson, A. H., and Culbertson, C. T., 2007. 'Single-cell manipulation and analysis using microfluidic devices.' *Analytical and Bioanalytical Chemistry*, **387**, pp. 9–12.

Sapsford, K. E., and Ligler, F. S., 2004. 'Real-time analysis of protein adsorption to a variety fo thin films.' *Biosensors and Bioelectronics*, **19**, pp. 1045–1055.

Scheiner, S., 1997. *Hydrogen Bonding: A Theoretical Perspective.* Oxford University Press, New York, p. 400.

Schneider, B. H., Dickenson, E. L., Vach, M. D., Hoijer, J. V., and Howard, L. V., 2000. 'Highly sensitive optical chip immunoassays in human serum.' *Biosensors and Bioelectronics*, **15**, pp. 13–22.

Sebra, R. P., Anseth, K. S., and Bowman, C. N., 2006. 'Integrated surface modification of fully polymeric microfluidic devices using living radical photopolymerization chemistry.' *Journal of Polymer Science Part A – Polymer Chemistry*, **44**, pp. 1404–1413.

Senkevich, J. J., Yang, G. R., and Lu, T. M., 2003. 'The facile surface modification of poly(*P*-xylylene) ultrathin films.' *Colloids and Surfaces A: Physicochem Eng Aspects*, **216**, pp. 167–173.

Sheller, N. B., Petrash, S., Foster, M. D., and Tsukruk, V. V., 1998. 'Atomic force microscopy and x-ray reflectivity studies of albumin adsorbed onto self-assembled monolayers of hexadecyltrichlorosilane.' *Langmuir*, **14**, pp. 4535–4544.

Sia, S. K., and Whitesides, G. M., 2003. 'Microfluidic devices fabricated in poly (dimethylsiloxane) for biological studies.' *Electrophoresis*, **24**, pp. 3563–3576.

Situma, C., Hashimoto, M., and Soper, S. A., 2006. 'Merging microfluidics with microarray-based bioassays.' *Biomolecular Engineering*, **23**, pp. 213–231.

Smith, E. A., and Corn, R. M., 2003. 'Surface plasmon resonance imaging as a tool to monitor biomolecular interactions in ana rray based format.' *Applied Spectroscopy*, **57**, pp. 320A–332A.

Soper, S. A., Henry, A. C., Vaidya, B., Galloway, M., Wabuyele, M., and McCarley, R. L., 2002. 'Surface modification of polymer-based microfluidic devices.' *Analytica Chimica Acta*, **470**, pp. 87–99.

Southall, N. T., and Dill, K. A., 2000. 'The mechanism of hydrophobic solvation depends on solvent radius.' *Journal of Physical Chemistry B*, **204**, pp. 1326–1331.

Stachowiak, T. B., Mair, D. A., Holden, T. G., Lee, L. J., Svec, F., and Frechet, J. M. J., 2007. 'Hydrophilic surface modification of cyclic olefin copolymer microfluidic chips using sequential photografting.' *Journal of Separation Science*, **30**, pp. 1088–1093.

Stadler, V., Kirmse, R., Beyer, M., Breitling, F., Ludwig, T., and Bischoff, F. R., 2008. 'Pegma/MMA copolymer graftings: Generation, protein resistance, and a hydrophobic domain.' *Langmuir*, **24**, pp. 8151–8157.

Sui, G. D., Wang, J. Y., Lee, C. C., Lu, W. X., Lee, S. P., Leyton, J. V. *et al.*, 2006. 'Solution-phase surface modification in intact poly(dimethylsiloxane) microfluidic channels.' *Analytical Chemistry*, **78**, pp. 5543–5551.

Sun, X., Liu, J., and Lee, M. L., 2008. 'Surface modification of glycidyl-containing poly(methyl methacrylate) microchips using surface-initiated atom-transfer radical polymerization.' *Analytical Chemistry*, **80**, pp. 856–863.

Takayama, S., Ostuni, E., LeDuc, P., Naruse, K., Ingber, D. E., and Whitesides, G. M., 2001. 'Laminar flows – subcellular positioning of small molecules.' *Nature*, **411**, pp. 1016–1016.

Takayama, S., Ostuni, E., LeDuc, P., Naruse, K., Ingber, D. E., and Whitesides, G. M., 2003. 'Selective chemical treatment of cellular microdomains using multiple laminar streams.' *Chemistry & Biology*, **10**, pp. 123–130.

Taylor, A. M., Rhee, S. W., Tu, C. H., Cribbs, D. H., Cotman, C. W., and Jeon, N. L., 2003. 'Microfluidic multicompartment device for neuroscience research.' *Langmuir*, **19**, pp. 1551–1556.

Taylor, J. D., Phillips, K. S., and Cheng, Q., 2007. 'Microfluidic fabrication of addressable tethered lipid bilayer arrays and optimization using SPR with silane-derivatized nanoglassy substrates.' *Lab on a Chip*, **7**, pp. 927–930.

Thorslund, S., and Nikolajeff, F., 2007. 'Instant oxidation of closed microchannels.' *Journal of Micromechanics and Microengineering*, **17**, N16-N21.

Tie, Y., Calonder, C., and Van Tassel, P. R., 2003. 'Protein adsorption: Kinetics and history dependence.' *Journal of Colloid and Interface Science*, **268**, pp. 1–11.

Toh, Y. C., Zhang, C., Zhang, J., Khong, Y. M., Chang, S., Samper, V. D. *et al.*, 2007. 'A novel 3D mammalian cell perfusion-culture system in microfluidic channels.' *Lab on a Chip*, **7**, pp. 302–309.

Tokeshi, M., Kikutani, Y., Hibara, A., Sato, K., Hisamoto, H., and Kitamori, T., 2003. 'Chemical processing on microchips for analysis, synthesis, and bioassay.' *Electrophoresis*, **24**, pp. 3583–3594.

Unsworth, L. D., Sheardown, H., and Brash, J. L., 2008. 'Protein-resistant poly (ethylene oxide)-grafted surfaces: Chain density-dependent multiple mechanisms of action.' *Langmuir*, 24, pp. 1924–1929.

Vanderah, D. J., La, H., Naff, J., Silin, V., and Rubinson, K. A., 2004. 'Control of protein adsorption: Molecular level structural and spatial variables.' *Journal of the American Chemical Society*, **126**, pp. 13639–13641.

Vilkner, T., Janasek, D., and Manz, A., 2004. 'Micro total analysis systems. Recent developments.' *Analytical Chemistry*, **76**, pp. 3373–3385.

Vogler, E. A., 1999. 'Water and the acute biological response to surfaces.' *Journal of Biomaterials Science, Polymer Edition*, **10**, pp. 1015–1045.

Wagner, M. S., McArthur, S. L., Shen, M., Horbette, T. A., and Castner, D. G., 2002. 'Limits of detection for time of flight secondary ion mass spectrometry (ToF-SIMS) and X-ray photoelectron spectroscopy (XPS): Detection of low amounts of adsorbed protein.' *Journal of Biomaterials Science, Polymer Edition*, **13**, pp. 407–428.

Wang, B., Abdulali-Kanji, Z., Dodwell, E., Horton, J. H., and Oleschuk, R. D., 2003. 'Surface characterization using chemical force microscopy and the flow performance of modified polydimethylsiloxane for microfluidic device applications.' *Electrophoresis*, **24**, pp. 1442–1450.

Weber, P. C., Ohlendorf, D. H., Wendoloski, J. J., and Salemme, F. R., 1989. 'Structural origins of high-affinity biotin binding to streptavidin.' *Science*, **243**, pp. 85–88.

Weibel, D. B., and Whitesides, G. M., 2006. 'Applications of microfluidics in chemical biology.' *Current Opinion in Chemical Biology*, **10**, pp. 584–591.

Wenzel, R. N., 1949. 'Surface roughness and contact angle.' *Journal of Physical and Colloid Chemistry*, **53**, pp. 1466–1467.

Werdich, A. A., Lima, E. A., Ivanov, B., Ges, I., Anderson, M. E., Wikswo, J. P. *et al.*, 2004. 'A microfluidic device to confine a single cardiac myocyte in a subnanoliter volume on planar microelectrodes for extracellular potential recordings.' *Lab on a Chip*, **4**, pp. 357–362.

Wheeler, A. R., Throndset, W. R., Whelan, R. J., Leach, A. M., Zare, R. N., Liao, Y. H. *et al.*, 2003. 'Microfluidic device for single-cell analysis.' *Analytical Chemistry*, **75**, pp. 3581–3586.

Whitesides, G. M., 2006. 'The origins and the future of microfluidics.' *Nature*, **442**, pp. 368–373.

Xiao, D. Q., Van Le, T., and Wirth, M. J., 2004. 'Surface modification of the channels of poly(dimethylsiloxane) microfluidic chips with polyacrylamide for fast electrophoretic separations of proteins.' *Analytical Chemistry*, **76**, pp. 2055–2061.

Xiao, D., Zhang, H., and Wirth, M., 2002. 'Chemical modification of the surface of poly(dimethylsiloxane) by atom-transfer radical polymerization of acrylamide.' *Langmuir*, **18**, pp. 9971–9976.

Xiong, L., and Regnier, F. E., 2001. 'Channel-specific coatings on microfabricated chips.' *Journal of Chromatography A*, **924**, pp. 165–176.

Yang, T. L., Jung, S. Y., Mao, H. B., and Cremer, P. S., 2001. 'Fabrication of phospholipid bilayer-coated microchannels for on-chip immunoassays.' *Analytical Chemistry*, **73**, pp. 165–169.

Yang, Z., Galloway, J. A., and Yu, H., 1999. 'Protein interactions with poly (ethylene glycol) self-assembled monolayers on glass substrates: Diffusion and adsorption.' *Langmuir*, **15**, pp. 8405–8411.

Yoon, J. Y., and Garrell, R. L., 2008. 'Biomolecular adsorption in microfluidics.' In *Encyclopedia of Microfluidics and Nanofluidics*, ed. D. Li. New York, Springer, pp. 68–76.

Yoon, J. Y., and Garrell, R. L., 2003. 'Preventing biomolecular adsorption in electrowetting-based microchips.' *Analytical Chemistry*, **75**, pp. 5097–5102.

Zhao, H., Ismail, K., and Weber, S. G., 2004. 'Poly(2,2-bis(trifluoromethyl)-4,5-difluoro-1,3-dioxide-co-tetrafluoroethylene) (Teflon AF)?' *Journal of the American Chemical Society*, **126**, pp. 13184–13185.

INDEX